Organofluorine Chemistry

Organofluorine Chemistry

Synthesis, Modeling, and Applications

Edited by

Kálmán J. Szabó and Nicklas Selander

Editors

Prof. Kálmán J. Szabó
Stockholm University
Arrhenius Laboratory
Svante Arrhenius väg 16C
106 91 Stockholm
Sweden

Dr. Nicklas Selander
Stockholm University
Arrhenius Laboratory
Svante Arrhenius väg 16C
106 91 Stockholm
Sweden

Cover credits:
Shutterstock ID 1389609920 / Ad Gr /
Crystallized liquid crystal under
polarized light microscope forming
a rainbow texture. Abstract squares
filled with rainbow colors.

All books published by **Wiley-VCH** are carefully
produced. Nevertheless, authors, editors, and
publisher do not warrant the information contained
in these books, including this book, to be free of
errors. Readers are advised to keep in mind that
statements, data, illustrations, procedural details or
other items may inadvertently be inaccurate.

Library of Congress Card No.:
applied for

British Library Cataloguing-in-Publication Data
A catalogue record for this book is available from the
British Library.

**Bibliographic information published by the Deutsche
Nationalbibliothek**
The Deutsche Nationalbibliothek lists this publication
in the Deutsche Nationalbibliografie; detailed
bibliographic data are available on the Internet at
<http://dnb.d-nb.de>.

© 2021 WILEY-VCH GmbH, Boschstr. 12, 69469
Weinheim, Germany

Print ISBN: 978-3-527-34711-7
ePDF ISBN: 978-3-527-82513-4
ePub ISBN: 978-3-527-82514-1
oBook ISBN: 978-3-527-82515-8

Cover Design SCHULZ Grafik-Design, Fußgönheim,
Germany
Typesetting SPi Global, Chennai, India
Printing and Binding CPI Group (UK) Ltd, Croydon CR0 4YY

Printed on acid-free paper

10 9 8 7 6 5 4 3 2 1

Contents

Preface

Fluorine is the 13th most abundant element in the Earth's crust (the most abundant halogen), yet only a couple of biologically produced natural organofluorine compounds are known. Fluorine has the highest electronegativity of the elements – generating a number of specific problems in the synthesis (and biosynthesis) of organofluorine compounds. The majority of fluorine is present in its reduced form, mainly as the fluoride ion (F^-). This leads to at least two problems when creating C—F bonds: the very high heat of hydration of the fluoride ion (F^-), which makes it a very poor nucleophile; and the very high oxidation potential of fluorine which makes synthesis and handling of (highly reactive) electrophilic fluorination reagents usually difficult. The upside of the lack of efficient biosynthetic pathways to C—F bonds is that organofluorines have very useful pharmacological and pharmacokinetic properties, such as high metabolic stability against enzymatic oxidation. The C—F bonds usually increase the lipophilicity and modify the acid-base properties of pharmacophores. In addition, the C—F bond is the strongest single bond that a carbon atom can form with another element. These beneficial properties substantially increase the bioavailability of small molecules, which allows the use of fluorinated drugs and agrochemicals in lower doses than their non-fluorinated counterparts, ensuring optimal bioactivity with a lower environmental impact for these types of chemicals.

As mentioned above, all useful organofluorine compounds are human made. Therefore, there is an enormous pressure on synthetic organic chemists to produce a wide variety of new fluorinated compounds. This book is intended to cover the most important state-of-the-art synthetic methods and addresses current challenges within the field. At the same time, it gives in-depth analysis of the underlying chemical processes controlling the formation of new bonds between carbon and fluorine or fluorinated functional groups. A further important aspect is the application of organofluorine compounds in various areas of medicinal and agrochemistry.

This book has a very strong emphasis on the classification, application, and properties of various modern fluorinating reagents. The most important requirement for these reagents is the possibility to use them under mild conditions for selective transformations. Another important aspect is the safe and operationally simple handling of the fluorine transfer reagents. Accordingly, the first chapter by Jianbo Hu and coworkers analyzes the basic aspects of the fluorine effects in organic transformation with special emphasis on the introduction of difluoro- and trifluoromethyl groups. In Chapter 2 Sodeoka and coworker review the trifluoromethylation and fluoroalkylation reactions using selective, inexpensive reagents. Functional groups with OCF_3, SCF_3 and $SeCF_3$ and related structures are very important in organofluorines targeted in medicinal- and agrochemical applications. Introduction of these functional groups is reviewed in Chapter 3 by Billard and coworkers. A broad review about the development, synthesis and application of SCF_3 transfer reagents is given by Shen and coworkers in Chapter 4. The substrate scope of the fluorination and fluoroalkylation reactions is also discussed in detail. In Chapter 5 Liu and coworkers summarize the fluorination- and trifluoromethylation-based difunctionalization reactions. Chapter 6 by Szabó is focused on different substrate classes including alkenes, cyclopropanes, and diazo compounds in fluorination, trifluoromethylation, and trifluoromethylthiolation reactions. Modern methodology, such as photoredox catalysis, is an important tool even in fluorination reactions. This topic is summarized in Chapter 7 by Akita and coworkers. Control of stereochemistry to obtain enantiopure organofluorines is one of the most important requirements in medicinal applications. Toste and coworker summarized the most important development in asymmetric fluorination in Chapter 8. Chiral organofluorine compounds have some extraordinary properties, such as the ability for self-disproportionation of enantiomers (SDE), which can be exploited for purification. Klika and coworkers described this phenomenon in Chapter 9. Mechanistic and modeling studies are presented in many of the above chapters. In Chapter 10 by Xue and coworkers, DFT modeling studies on the special mechanistic features for the introduction of fluorine and fluoroalkyl groups are summarized with a particular focus on the catalytic reactions and application of the new fluorinating reagents. As mentioned above, application of organofluorines is particularly important in chemical industry. A large part of the commercially applied agrochemicals introduced in the last decade are fluorinated compounds. In Chapter 11 by Jeschke, the modern trends in design of fluorine-containing compounds are described analyzing the most important classes of herbicides, fungicides, and insecticides. Fluorine-18 labeling of positron emission tomography (PET) tracers is one of the most challenging areas in organofluorine chemistry. Application of new methodology and reagents is reviewed in Chapter 12 by Liang and coworkers.

This book is a result of a global collaboration by excellent research groups from Asia, Europe, and North America indicating worldwide efforts of the organofluorine community to forward this field, which is very important in synthetic and mechanistic research as well as in industrial applications. We would like to appreciate all authors of this book for their enthusiasm and for sharing their knowledge in organofluorine chemistry with the readers and their help in the realization of this book. In addition, the editors are indebted to the Knut och Alice Wallenbergs Foundation for granting the project "Organofluorines: anthropogenic small-molecules for life sciences" (Dnr: 2018.0066).

Stockholm, April 2020

Kálmán J. Szabó
Stockholm University, Arrhenius Laboratory
Department of Organic Chemistry
Stockholm, Sweden

1

The Development of New Reagents and Reactions for Synthetic Organofluorine Chemistry by Understanding the Unique Fluorine Effects

Qiqiang Xie and Jinbo Hu

CAS Key Laboratory of Organofluorine Chemistry, Shanghai Institute of Organic Chemistry, Chinese Academy of Sciences, 345 Ling-Ling Road, Shanghai 200032, China

1.1 Introduction

For a long period of time, fluorine chemistry research was deemed to be very dangerous, in part owing to the notorious fluorination reagents such as F_2 and HF, which are highly corrosive and reactive. In this context, synthetic organofluorine chemistry was only pursued by a handful of researchers for several decades after the first synthesis of F_2 by Moissan in 1886. However, great changes had taken place over the past decades. With the successful application of chlorofluorocarbons (although banned later owing to their ozone-depleting character by Montreal Protocol in 1987) as refrigerants, the development of high-performance fluorinated materials such as Teflon and the development of fluorinated pharmaceuticals and agrochemicals, organofluorine chemistry has become increasingly important in satisfying the huge demand for fluorinated molecules with diverse functions [1]. To date, organofluorine chemistry has become an indispensable branch of organic chemistry and enriched many aspects of related areas such as medicinal chemistry and materials science, among others. Perhaps the most prominent application of organofluorine compounds that has close relation with our everyday life is fluorinated drugs. For example, among the 59 drugs (39 of them are small molecule drugs) approved by FDA in 2018, 18 contain at least one fluorine atom; among the 48 drugs approved by FDA in 2019, 13 contain at least one fluorine atom [2]. Some representative fluorinated drugs approved in 2018 and 2019 are shown in Figure 1.1.

In spite of the importance of fluorinated molecules in modern society, naturally occurring organofluorine compounds are rare [3]. Basically, all the commercially supplied organofluorine compounds are manmade. Thus, the development of new reagents and reactions to introduce fluorine atoms or fluorine-containing moieties into organic molecules is one of the central goals in modern synthetic organofluorine chemistry.

Organofluorine Chemistry: Synthesis, Modeling, and Applications,
First Edition. Edited by Kálmán J. Szabó and Nicklas Selander.
© 2021 WILEY-VCH GmbH. Published 2021 by WILEY-VCH GmbH.

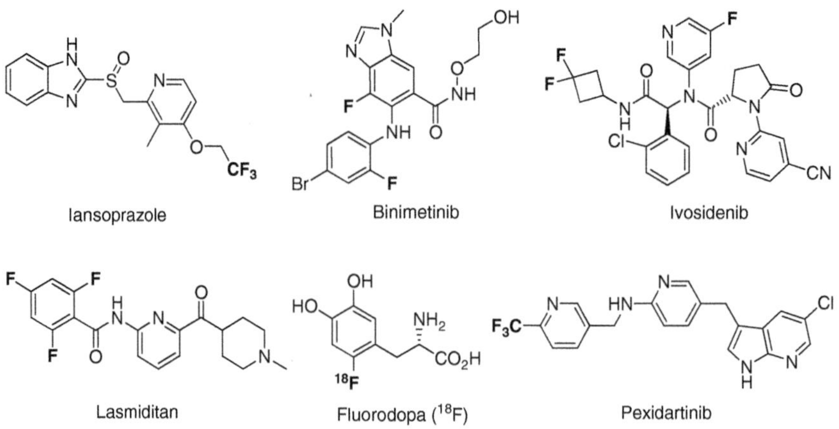

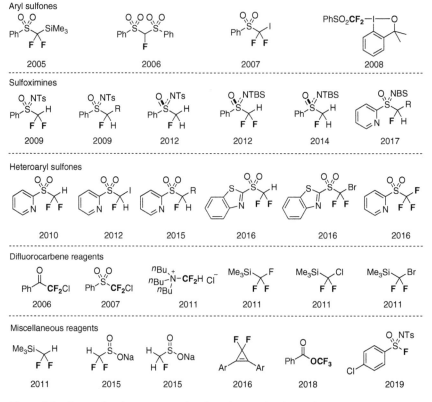

Figure 1.1 Fluorinated drugs newly approved by FDA in 2018 and 2019.

Figure 1.2 Organofluorine reagents developed or co-developed by us.

Over the past 15 years, our group has been interested in reaction mechanism-inspired synthetic organofluorine chemistry [4]. We have developed or co-developed a variety of reagents, including fluoroalkyl aryl sulfones [5], fluoroalkyl sulfoximines [6], fluoroalkyl heteroaryl sulfones [7], difluorocarbene reagents [8], fluorinated sulfinate salts [9], fluorinated esters [10], CpFluors [11], and SulfoxFluor [12], among others (for a review, see Ref. [4d]), for fluoroalkylation, fluoroolefination, and fluorination reactions (Figure 1.2). Our research program was triggered by two questions: (i) What are the unique features of organofluorine reactions (compared with regular organic reactions)? (ii) Is there any relationship among fluoroalkylation, fluoroolefination, and fluorination? In this review, we intend to answer these two questions by illustrating representative reagents and reactions developed (or co-developed) by us. In particular, understanding of the unique fluorine effects in organic reactions is helpful in addressing these two questions [4g].

1.2 The Unique Fluorine Effects in Organic Reactions

It is now gradually accepted that organofluorine reactions are usually distinct from regular organic reactions, and in many cases, fluorine substitution in an organic molecule imparts unique reactivity to the latter. As a result, direct application of the knowledge and experience acquired from regular organic reactions to organofluorine reactions often leads to failure or unexpected results. In this section, we provide some selected examples to highlight the unique features of organofluorine reactions. For more examples and discussions, one may refer to our recent tutorial review [4e].

1.2.1 Fluorine-Enabled Stability of "CuCF$_3$" in Water, and the Unusual Water-Promoted Trifluoromethylation

Organocopper reagents are widely used in organic synthesis [13]; however, these reagents are typically water sensitive. In 2012, we disclosed a water-promoted trifluoromethylation of α-diazo esters to access α-trifluoromethyl esters, representing the first example of fluoroalkylation of a non-fluorinated carbene precursor (Scheme 1.1) [14]. We found that "CuCF$_3$" (prepared from CuI/CsF/TMSCF$_3$) in n-methyl-2-pyrrolidone (NMP) is stable even in the presence of 66 equiv of water at room temperature within five hours and only about 5% of "CuCF$_3$" decomposed. Taking advantage of the unusual stability of "CuCF$_3$" in water (fluorine effect), we could therefore use water (CuI is also applicable) as the iodide scavenger to enhance the activity of "CuCF$_3$" prepared from CuI/CsF/TMSCF$_3$ significantly by changing "CuCF$_3$" in the form of [Cu(CF$_3$)I]$^-$ to "ligandless" [Cu(CF$_3$)], which is more reactive toward α-diazo esters. From an organometallic chemistry point of view, water promotes the ligand exchange in "CuCF$_3$" by eliminating iodide, making the ligation of α-diazo esters to "CuCF$_3$" more favorable. The stability of "CuCF$_3$" toward water and the instability of alkylcopper intermediate **1** toward water ensure the success of this reaction.

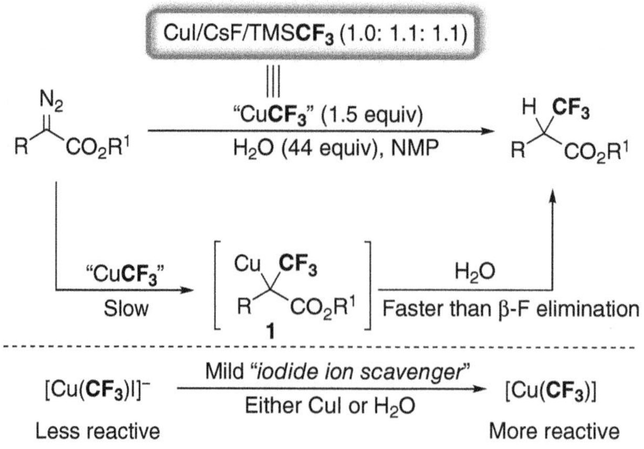

Scheme 1.1 Water-promoted trifluoromethylation.

1.2.2 Fluorine Enables β-Fluoride Elimination of Organocopper Species

In the abovementioned water-promoted trifluoromethylation, a stoichiometric amount of copper is required and *gem*-difluoroolefin was found to be formed under strictly anhydrous conditions. Inspired by this fact, we developed a copper-catalyzed *gem*-difluoroolefination of diazo compounds, concisely [15] (Scheme 1.2a). In this catalytic reaction, diaryl diazomethanes were used instead of α-diazo esters for two reasons: (i) diaryl diazomethanes possess higher reactivity than α-diazo esters and can react with unactivated "CuCF₃" directly (without the need for eliminating ligated iodide in "CuCF₃"), making the addition of excess water or CuI as iodide scavenger no longer necessary, and (ii) diaryl-substituted alkylcopper intermediate

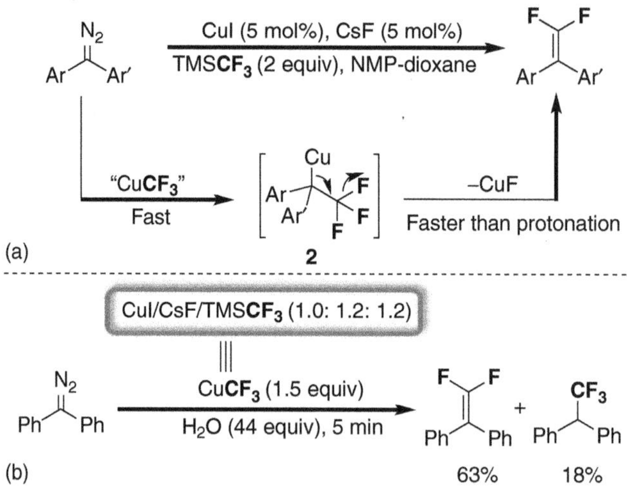

Scheme 1.2 Copper-catalyzed *gem*-difluoroolefination of diazo compounds.

2 readily undergoes β-fluoride elimination (even in the presence of excess amount of water) (Scheme 1.2b), which is critical to regenerate the copper catalyst and close the catalytic cycle.

1.2.3 The "Negative Fluorine Effect" Facilitates the α-Elimination of Fluorocarbanions to Generate Difluorocarbene Species

Based on the results described in Sections 1.2.1 and 1.2.2, we were able to achieve *gem*-difluoroolefination of diazo compounds under transition metal-free conditions via direct cross-coupling between a non-fluorinated carbene precursor from a diazo compound and difluorocarbene from $TMSCF_3$ or $TMSCF_2Br$ (Scheme 1.3a) [16]. This reaction proceeds through nucleophilic addition of diazo compounds to difluorocarbene, followed by β-N_2 elimination from intermediate **3**, to form *gem*-difluoroolefins. The most significant feature of this protocol is the broad substrate scope (compared with our previous copper-catalyzed method of diazo compounds, which is only efficient for diaryl diazomethanes). α-Diazo acetates, diaryl diazomethanes, as well as diazirines are all suitable substrates. Moreover, by simply tuning the molar ratio of diazirines and $TMSCF_2Br$, either *gem*-difluoroolefins or tetrafluoropropanes can be obtained selectively (Scheme 1.3b). It should be noted that the *in situ* generation of difluorocarbene occurs via an α-elimination of fluoride ion from "CF_3^-" (in the case of $TMSCF_3$), or via an α-elimination of bromide ion from "CF_2Br^-" (in the case of $TMSCF_2Br$), which can be explained by the "negative fluorine effect (NFE)," that is, fluorine substitution on a carbanionic center will often have a negative (unfavorable) effect on the carbanion's thermal stability and its nucleophilic reactions with many electrophiles.

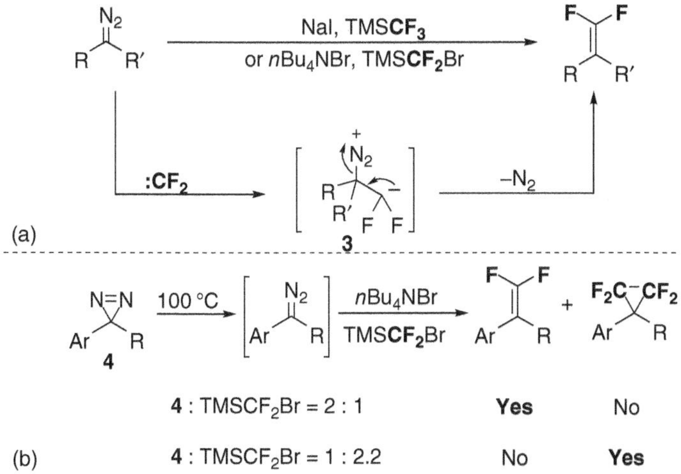

Scheme 1.3 Transition metal-free *gem*-difluoroolefination of diazo compounds.

Alkylsilane reagents are useful in cross-coupling reactions. In the presence of a Lewis base, alkylsilane can act as "R^-" donor to undergo alkylation under transition metal catalysis [17]. In this context, fluoroalkylsilanes often serve as nucleophilic fluoroalkylation agents. For instance, (trifluoromethyl)trimethylsilane (TMSCF$_3$) has been recognized as an efficient "CF$_3^-$" donor since 1989 [18] and has been studied extensively ever since [19]. However, because of the fluorine substitution, TMSCF$_3$ is not merely a "CF$_3^-$" equivalent. In 2011, our group, in collaboration with the Prakash group, found that TMSCF$_3$ is also an efficient difluorocarbene precursor [8d]. By employing nonmetallic fluoride TBAT (tetrabutylammonium triphenyldifluorosilicate) as an initiator at low temperature (−50 °C to r.t.) or NaI as a promoter at higher temperature (65 or 110 °C), efficient synthesis of *gem*-difluorocyclopropa(e)nes can be achieved with TMSCF$_3$ (Scheme 1.4).

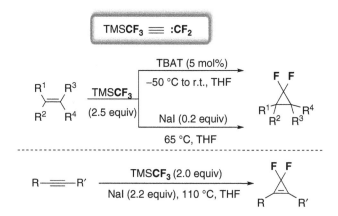

Scheme 1.4 TMSCF$_3$ as a difluorocarbene precursor for *gem*-difluorocyclopropa(e)nation.

Difluorocarbene species derived from TMSCF$_3$ can dimerize to form tetrafluoroethylene (TFE) in an efficient and mild manner, providing a convenient and safe protocol for the generation and handling of TFE in academic laboratories (Scheme 1.5a) [20]. Previously, TFE was normally inaccessible for common laboratories due to its suspected carcinogenic and explosive nature. The application of on-site prepared TFE was demonstrated by the pentafluoroethylation and aryloxytetrafluoroethylation of aryl iodides and 1,1,2,2-tetrafluoroethylation of phenols, alcohols, thiophenols, and heterocyclic amines (Scheme 1.5b–d).

More recently, we also accomplished fluorocarbon chain elongation process with TMSCF$_3$ as a difluorocarbene source. Selective one-carbon elongation from trifluoromethyl to pentafluoroethyl [21], and two-carbon elongation from pentafluorophenyl to perfluorophenylethyl [22] were efficiently realized under copper-mediated conditions, providing a non-TFE pathway (Scheme 1.6). The key issue of this process is harnessing "CuCF$_3$" as a copper difluorocarbene ("Cu=CF$_2$") equivalent.

TMSCF$_2$Br was developed by us as a privileged difluorocarbene reagent, which has been used to difluoromethylate a plethora of O-, S-, N-, P-, and C-nucleophiles as well as to cyclopropa(e)nate alkenes or alkynes [8e, f, 23] (Scheme 1.7). Because of the much better leaving ability of Br$^-$ than F$^-$, the life time of BrCF$_2^-$ is short enough

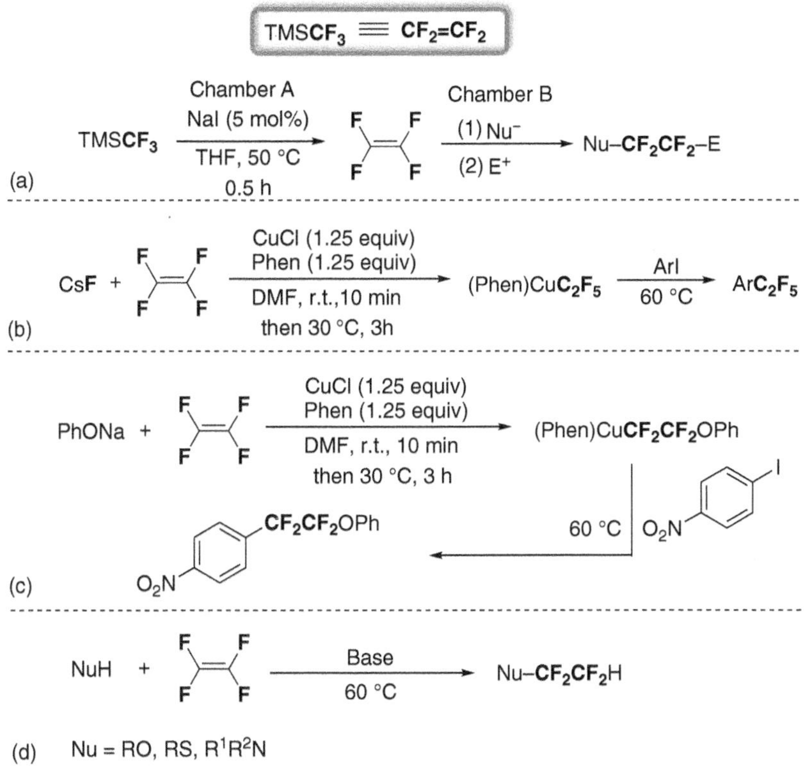

$$\boxed{\text{TMSCF}_3 \equiv \text{CF}_2=\text{CF}_2}$$

(a)

Chamber A: TMSCF$_3$ $\xrightarrow[\text{THF, 50 °C} \atop \text{0.5 h}]{\text{NaI (5 mol\%)}}$ [CF$_2$=CF$_2$] Chamber B: $\xrightarrow[\text{(2) E}^+]{\text{(1) Nu}^-}$ Nu–CF$_2$CF$_2$–E

(b)

CsF + CF$_2$=CF$_2$ $\xrightarrow[\text{DMF, r.t.,10 min} \atop \text{then 30 °C, 3h}]{\text{CuCl (1.25 equiv)} \atop \text{Phen (1.25 equiv)}}$ (Phen)CuC$_2$F$_5$ $\xrightarrow[\text{60 °C}]{\text{ArI}}$ ArC$_2$F$_5$

(c)

PhONa + CF$_2$=CF$_2$ $\xrightarrow[\text{DMF, r.t., 10 min} \atop \text{then 30 °C, 3 h}]{\text{CuCl (1.25 equiv)} \atop \text{Phen (1.25 equiv)}}$ (Phen)CuCF$_2$CF$_2$OPh

O$_2$N–C$_6$H$_4$–CF$_2$CF$_2$OPh $\xleftarrow[\text{60 °C}]{\text{O}_2\text{N–C}_6\text{H}_4\text{–I}}$

(d) NuH + CF$_2$=CF$_2$ $\xrightarrow[\text{60 °C}]{\text{Base}}$ Nu–CF$_2$CF$_2$H

Nu = RO, RS, R^{1}R^{2}N

Scheme 1.5 TMSCF$_3$ as tetrafluoroethylene precursor.

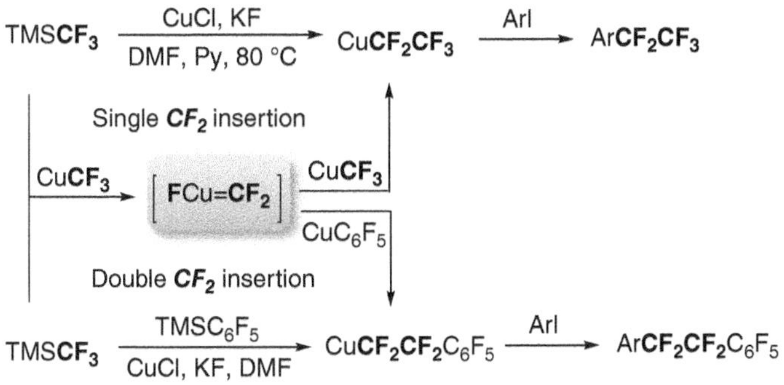

TMSCF$_3$ $\xrightarrow[\text{DMF, Py, 80 °C}]{\text{CuCl, KF}}$ CuCF$_2$CF$_3$ $\xrightarrow{\text{ArI}}$ ArCF$_2$CF$_3$

Single **CF$_2$** insertion

CuCF$_3$ → [FCu=CF$_2$] $\xrightarrow{\text{CuCF}_3}$ / $\xrightarrow{\text{CuC}_6\text{F}_5}$

Double **CF$_2$** insertion

TMSCF$_3$ $\xrightarrow[\text{CuCl, KF, DMF}]{\text{TMSC}_6\text{F}_5}$ CuCF$_2$CF$_2$C$_6$F$_5$ $\xrightarrow{\text{ArI}}$ ArCF$_2$CF$_2$C$_6$F$_5$

Scheme 1.6 TMSCF$_3$ for fluorocarbon elongation.

that it readily eliminates Br$^-$ to produce difluorocarbene even in the presence of large amount of water or acid. TMSCF$_2$Br can generate difluorocarbene under a wide range of conditions, ranging from strongly basic to weakly acidic, from aqueous to anhydrous, and from low temperatures to high temperatures [8e, f, 23, 24]. This unique feature renders TMSCF$_2$Br particularly versatile and makes its orthogonal reactions with ambident substrates possible [23] (Scheme 1.8).

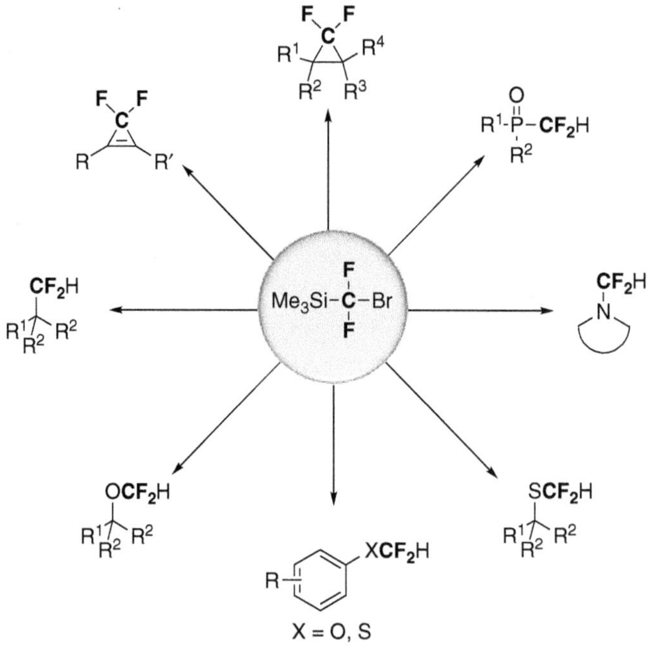

Scheme 1.7 The application of TMSCF$_2$Br as a difluorocarbene reagent.

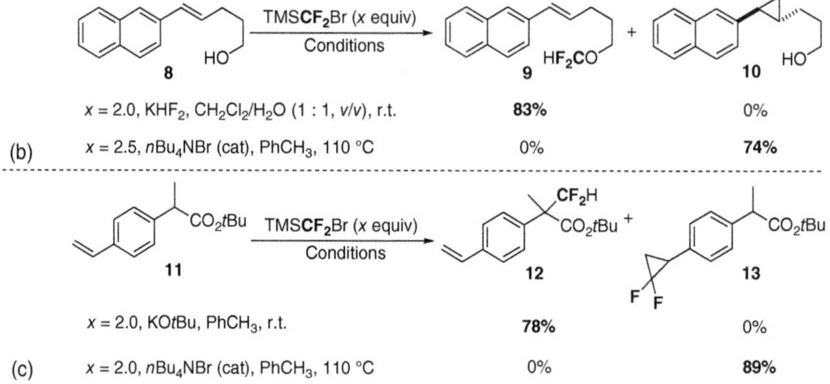

(a)

		6	7
x = 2.5, KHF$_2$, CH$_2$Cl$_2$/H$_2$O (1 : 1, v/v), r.t.		80%	0%
x = 2.5, KOH (20% aq.), CH$_2$Cl$_2$, r.t.		0%	71%

(b)

		9	10
x = 2.0, KHF$_2$, CH$_2$Cl$_2$/H$_2$O (1 : 1, v/v), r.t.		83%	0%
x = 2.5, nBu$_4$NBr (cat), PhCH$_3$, 110 °C		0%	74%

(c)

		12	13
x = 2.0, KOtBu, PhCH$_3$, r.t.		78%	0%
x = 2.0, nBu$_4$NBr (cat), PhCH$_3$, 110 °C		0%	89%

Scheme 1.8 Orthogonal reactivity of TMSCF$_2$Br with ambident substrates.

Let us take alkene-ester substrate **11** (see Scheme 1.8c) as an example to showcase the unusual orthogonal reactivity of TMSCF$_2$Br with different functional groups: in the presence of KO*t*Bu, the C—H bond α to the ester group could undergo deprotonation to generate the corresponding nucleophilic carbanion, which is highly reactive toward difluorocarbene, while the alkene group shows low reactivity toward difluorocarbene at room temperature because elevated temperature is required to conquer the substantial activation barrier for most alkenes (except the most electron-rich ones); when using *n*Bu$_4$NBr to activate TMSCF$_2$Br at elevated temperature (in this case, 110 °C), the alkene could efficiently react with the difluorocarbene, while the C—H bond α to the ester group is unreactive toward difluorocarbene in the absence of *n*Bu$_4$NBr at high temperature.

1.2.4 Tackling the β-Fluoride Elimination of Trifluoromethoxide Anion via a Fluoride Ion-Mediated Process

The trifluoromethoxy (CF$_3$O) group is increasingly important in drug development. Although methanol is definitely stable at room temperature, its perfluorinated analog, CF$_3$OH, is unstable at room temperature and decomposes (eliminating HF to form COF$_2$) even at −20 °C [25]. Therefore, trifluoromethoxylation with CF$_3$OH is impractical and the development of new trifluoromethoxylation reagents is highly desired. In 2018, we developed trifluoromethyl benzoate (TFBz) as a new type of shelf-stable trifluoromethoxylation reagent [10]. TFBz can be readily prepared from triphosgene, KF, and PhCOBr, with COF$_2$ being an *in situ*-generated key intermediate (Scheme 1.9a). Notably, all the fluorine atoms in TFBz come from the cheap fluoride source KF. A variety of other perfluoroalkoxylation reagents can be obtained in a similar manner (Scheme 1.9b). The versatility of TFBz as a trifluoromethoxylation reagent was demonstrated by trifluoromethoxylation-halogenation of arynes, nucleophilic substitution of alkyl(pseudo)halides, cross-coupling with aryl stannanes, and asymmetric difunctionalization of alkenes (Scheme 1.9c).

1.3 The Relationships Among Fluoroalkylation, Fluoroolefination, and Fluorination

Although numerous fluoroalkylation, fluoroolefination, and fluorination methods have been elegantly established, the relationships among these three reactions have been ignored. As part of our longstanding interest in studying new fluoroalkylation, fluoroolefination, and fluorination reagents and reactions by probing the unique fluorine effects, we realized that there are close relationships among fluoroalkylation, fluoroolefination, and fluorination in many cases (Scheme 1.10). In this section, we intend to discuss these relationships by providing some examples.

1.3.1 From Fluoroalkylation to Fluoroolefination

A typical olefination reaction is started with a nucleophilic alkylation step (Scheme 1.11a). It is also true for fluoroolefination reactions. PhSO$_2$CF$_2$H is a

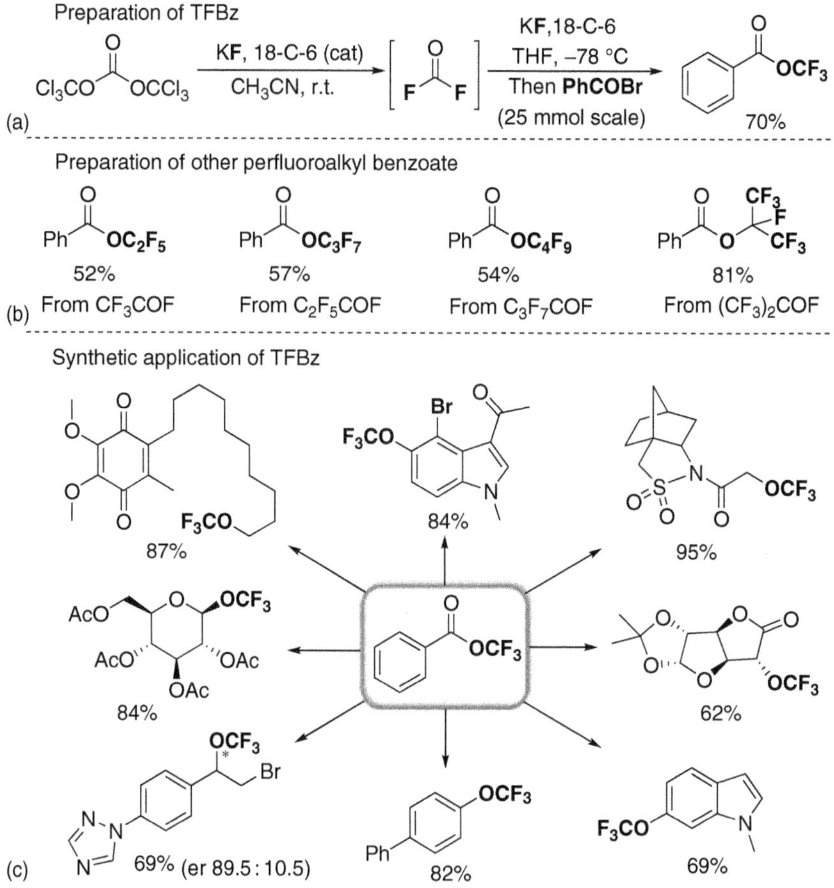

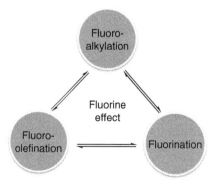

Preparation of TFBz

(a)

Preparation of other perfluoroalkyl benzoate

Ph–OC₂F₅	Ph–OC₃F₇	Ph–OC₄F₉	Ph–CF₃/F/CF₃
52%	57%	54%	81%
From CF₃COF	From C₂F₅COF	From C₃F₇COF	From (CF₃)₂COF

(b)

Synthetic application of TFBz

87% 84% 95%

84% 62%

(c) 69% (er 89.5 : 10.5) 82% 69%

Scheme 1.9 Synthesis and application of TFBz as a new trifluoromethoxylation reagent.

Fluoro-alkylation

Fluorine effect

Fluoro-olefination Fluorination

Scheme 1.10 The relationships among fluoroalkylation, fluoroolefination, and fluorination.

powerful difluoromethylation reagent [26] and was studied extensively by us and others [5b, 27]. Inspired by the Julia–Kocienski reaction that uses heteroaryl sulfones for the synthesis of olefins [28], it is natural to envision whether a fluorinated version can be achieved for fluoroolefination. Indeed, by changing phenyl to 2-pyridyl, we developed difluoromethyl 2-pyridyl sulfone (2-PySO$_2$CF$_2$H) as a novel and efficient *gem*-difluoroolefination reagent [7a] (Scheme 1.11b).

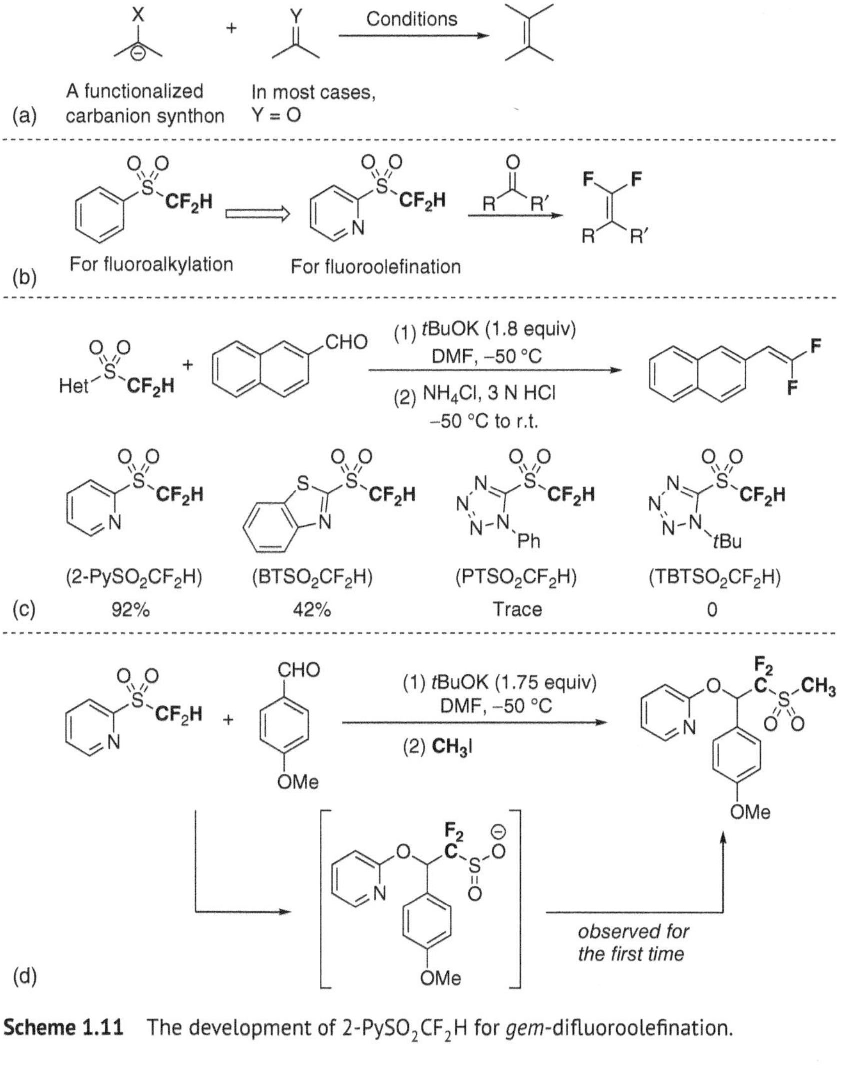

Scheme 1.11 The development of 2-PySO$_2$CF$_2$H for *gem*-difluoroolefination.

Notably, there is a unique fluorine effect in this olefination reaction. In non-fluorinated Julia–Kocienski olefination reactions, 2-pyridyl sulfones generally give lower yields of products than other heteroaryl sulfones, such as 1,3-benzothiazol-2-yl (BT), 1-phenyl-1*H*-tetrazol-5-yl (PT), and 1-*tert*-butyl-1*H*-tetrazol-5-yl (TBT) sulfones [28]. However, the Julia–Kocienski type difluoroolefination reaction shows unusual reactivity: 2-PySO$_2$CF$_2$H shows the highest reactivity, but PTSO$_2$CF$_2$H and TBTSO$_2$CF$_2$H possess almost no reactivity (Scheme 1.11c). This sharp contrast between 2-PySO$_2$CF$_2$H and the other two reagents may be attributed to the much higher stability and better nucleophilicity of 2-PySO$_2$CF$_2^-$ than other HetSO$_2$CF$_2^-$ anions. Another important feature of this reaction is that the sulfinate salt intermediate (Scheme 1.11d), which has never been observed in regular Julia–Kocienski reactions, was detected and captured by us for the first time, highlighting that organofluorine research enables intriguing insights into regular organic reactions.

Although an efficient *gem*-difluoroolefination has been realized with 2-PySO$_2$CF$_2$H, when it comes to monofluoroolefination, the issue of how to control the stereoselectivity arose.

In 2015, "a magic reaction" was discovered by us by the reaction of 2-PySO$_2$CHFR and aldehydes to prepare monofluoroalkenes [7c]. In this reaction, both *Z*- and *E*-isomers can be obtained and easily separated in an efficient and stereoselective manner (Scheme 1.12a). The key issue of this unique reaction is the significant stability difference between the two diastereoisomeric sulfinate salt intermediates (Scheme 1.12b), which enables spontaneous resolution and phase labeling of the two diastereoisomeric sulfinate salts, thus allowing separation of *Z*- and *E*-monofluoroalkenes by liquid–liquid extraction.

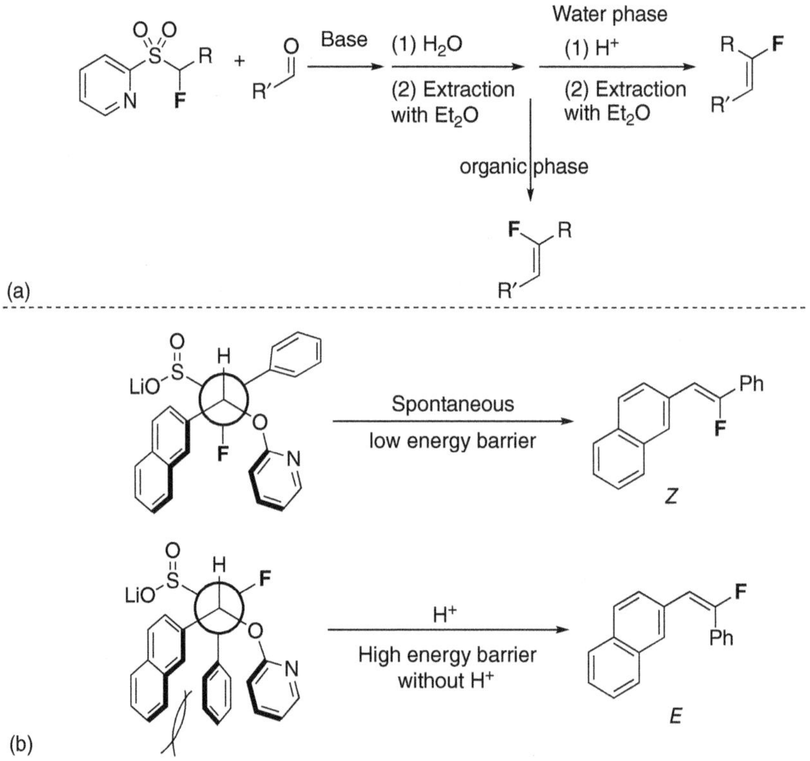

(a)

(b)

Scheme 1.12 Spontaneous resolution and phase separation to deliver *Z*- and *E*-monofluoroalkenes.

The synthesis of terminal monofluoroalkenes was regarded as a formidable challenge because of the minimal energy difference between the two stereoisomers. Encouraged by the excellent stereocontrol in the fluoroalkylation of carbonyl compounds with chiral sulfoximine reagents [6c–e] and the efficient fluoroolefination with 2-pyridyl sulfone reagents [7a] (Scheme 1.13a), we developed a novel heteroaryl sulfoximine reagent, S-monofluoromethyl-S-(2-pyridyl)sulfoximine, to access di- and trisubstituted terminal monofluoroalkenes with high stereoselectivity, concisely [6f] (Scheme 1.13b). The reaction proceeds through a highly diastereoselective addition of S-monofluoromethyl-S-(2-pyridyl)sulfoximine to carbonyls, followed by Smiles rearrangement and anti-1,2-elimination (Scheme 1.13c). The 2-pyridyl group plays a critical role in promoting the olefination, whereas the sulfoximidoyl group is pivotal for controlling the stereoselectivity.

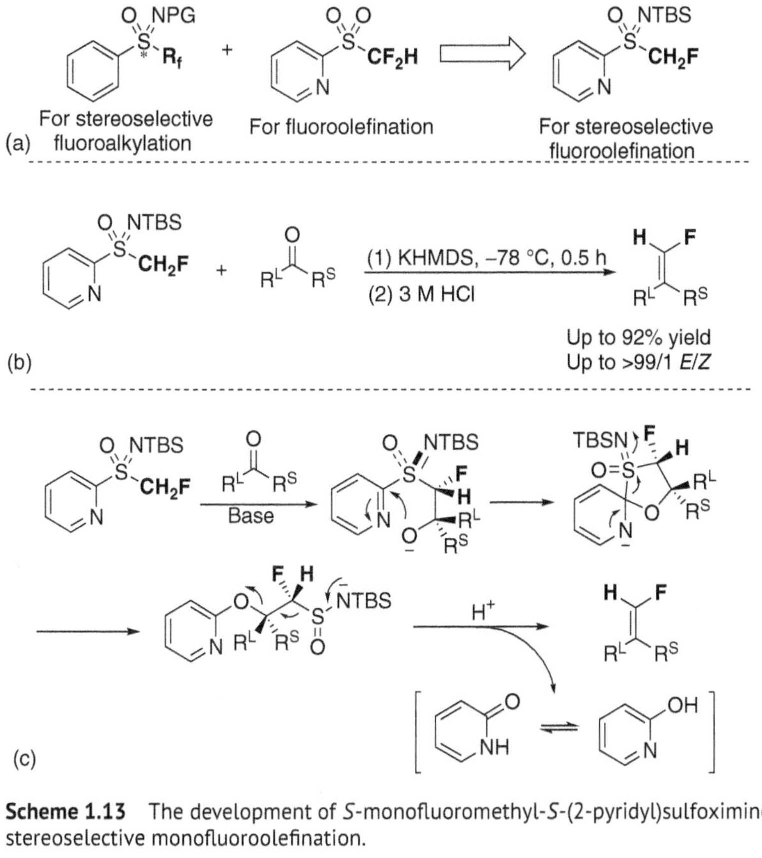

Scheme 1.13 The development of S-monofluoromethyl-S-(2-pyridyl)sulfoximine for stereoselective monofluoroolefination.

1.3.2 From Fluoroolefination to Fluoroalkylation

As mentioned in Section 1.3.1, 2-PySO$_2$CF$_2$H was developed as a new *gem*-difluoroolefination reagent and the sulfinate salt intermediate was found to be relatively stable and can be observed (see Scheme 1.11). Based on this fact, we realized a formal nucleophilic iodo- and bromodifluoromethylation of carbonyl compounds with 2-PySO$_2$CF$_2$H via *in situ* halogenation of the sulfinate salt intermediates delivered from Smiles rearrangement (Scheme 1.14). By simply using "X$^+$" reagents to quench the reaction instead of "H$^+$," we could, therefore, tune the pathway from fluoroolefination to fluoroalkylation.

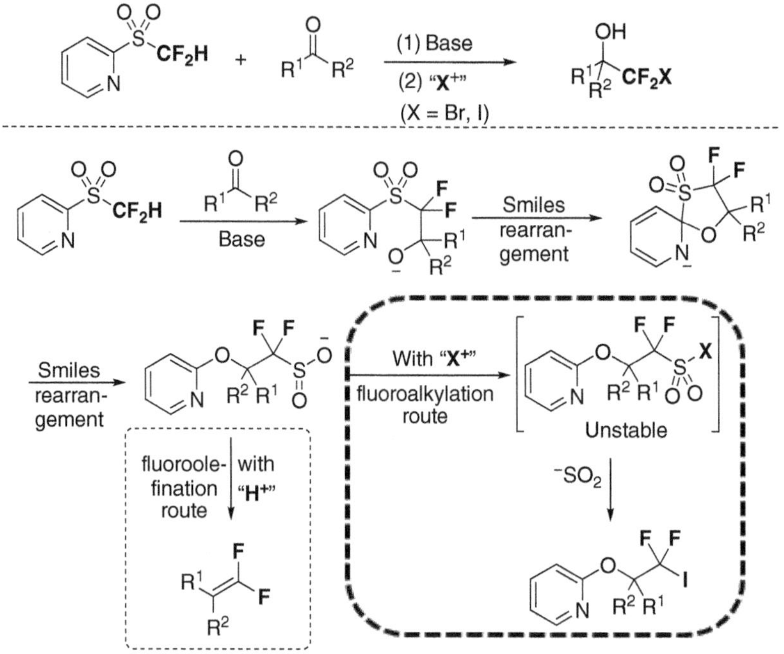

Scheme 1.14 Halodifluoromethylation with 2-PySO$_2$CF$_2$H.

Remarkably, a unique fluorine effect was observed in this reaction [29]. Unlike the reaction between 2-PySO$_2$CF$_2$H and carbonyls (after a subsequent halogenation) giving formal nucleophilic halodifluoromethylated products, a similar reaction using non-fluorinated 2-PySO$_2$CH$_3$ results in a totally different kind of product, a *E*-alkene (Scheme 1.15). This product may be formed via intramolecular cyclization of the *in situ*-generated sulfonyl iodide intermediate produced from iodination of the sulfinate salt intermediate, followed by an elimination process. The change of reaction pathway using 2-PySO$_2$CH$_3$ is probably due to the different stability between the non-fluorinated sulfonyl iodide intermediate and the fluorinated ones.

As we can see from the mechanism shown in Scheme 1.14, Smiles rearrangement is one of the key steps in Julia–Kocienski *gem*-difluoroolefination reaction. In order to realize fluoroalkylation with 2-PySO$_2$CF$_2$H, inhibiting the Smiles rearrangement is a viable strategy. Indeed, by lowering the temperature to −78 or −98 °C and changing the solvent from DMF (*N,N*-dimethylformamide) to THF (tetrahydrofuran), the Smiles rearrangement can be inhibited and the nucleophilic addition products can be obtained in good yields [30] (Scheme 1.16). The obtained addition products can

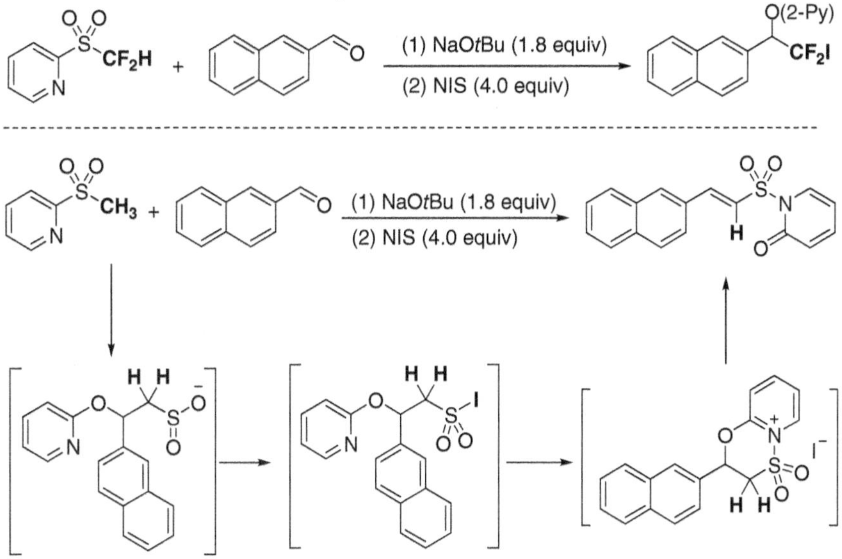

Scheme 1.15 Different reactivity of 2-PySO$_2$CF$_2$H and 2-PySO$_2$CH$_3$.

readily undergo depyridination to give sulfinate salts, which can be transformed to iododifluoromethylated products by treating with I$_2$.

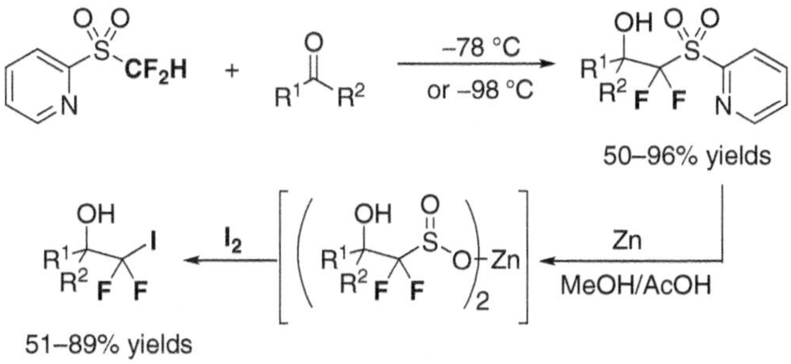

Scheme 1.16 Direct nucleophilic addition of 2-PySO$_2$CF$_2$H to carbonyls for iododifluoromethylation via non-Smiles rearrangement pathway.

The abovementioned fluoroalkylation strategies with 2-PySO$_2$CF$_2$H are largely dependent on the nucleophilic 2 - PySO$_2$CF$_2$$^-$ anion. However, difluoromethylation with 2-PySO$_2$CF$_2$H via direct CF$_2$H transfer is also possible. In 2018, we reported the first iron-catalyzed difluoromethylation of arylzincs with 2-PySO$_2$CF$_2$H, providing a facile method to structurally diverse difluoromethylated arenes at low tempera-ture [31] (Scheme 1.17a). Mechanistic studies revealed that a difluoromethyl radical is involved in the reaction, and the direct transfer of CF$_2$H as a whole was sup-ported by deuterium labeling experiment. A proposed catalytic cycle is shown in Scheme 1.17b.

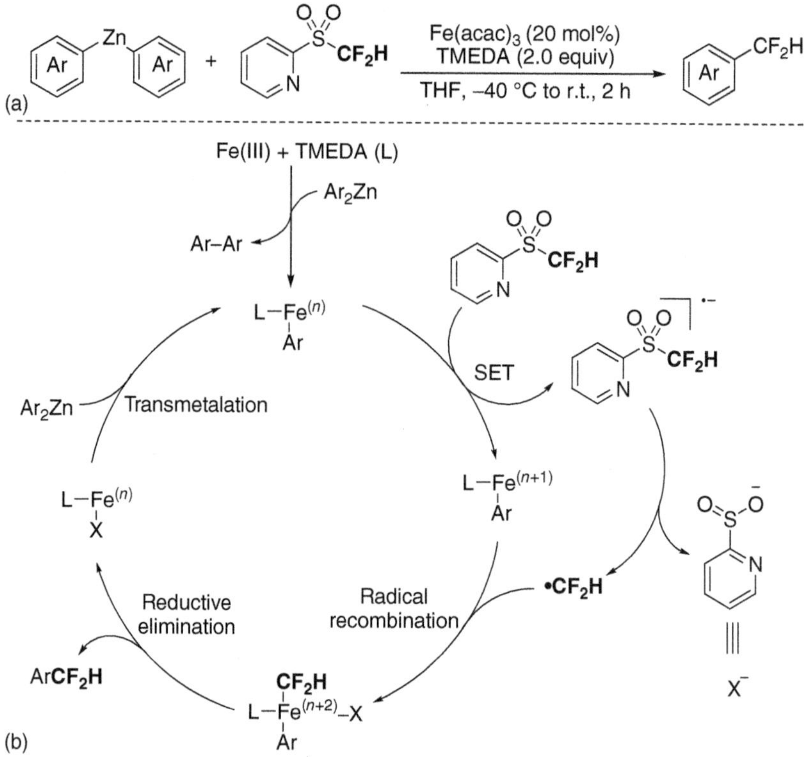

Scheme 1.17 Iron-catalyzed difluoromethylation of arylzincs with 2-PySO$_2$CF$_2$H.

Except for 2-PySO$_2$CF$_2$H, many other fluorinated heteroaryl sulfones have been extensively investigated. As we can see from Scheme 1.11c, although BTSO$_2$CF$_2$H was found to be an inefficient fluoroolefination reagent, its use for radical fluoroalkylation is a great success. In 2015, we developed a novel method for the preparation of sodium fluoroalkanesulfinates by the reduction of the corresponding benzo[*d*]thiazol-2-yl sulfones (BTSO$_2$R$_f$) [9]. This method enables a highly efficient and rapid large-scale synthesis of sodium di- and monofluoromethanesulfinates (Scheme 1.18a). Synthetic application of these sulfinates in radical fluoroalkylation is exemplified by the silver-catalyzed cascade fluoroalkylation/aryl migration/SO$_2$ extrusion of conjugated N-arylsulfonylated amides (Scheme 1.18b).

The versatility of heteroaryl sulfones as fluoroalkyl radical precursors may be best illustrated by the readily tunable reactivity of heteroaryl sulfones that can generate a wide range of fluoroalkyl radicals such as monofluoromethyl, difluoromethyl, 1,1-difluoroethyl, phenyldifluoromethyl, benzoyldifluoromethyl, and trifluoromethyl radicals via visible light photoredox catalysis under mild conditions [7d]. The synthetic application of heteroaryl sulfones in radical fluoroalkylation of isocyanides was demonstrated to highlight the versatility of heteroaryl sulfones as fluoroalkyl radical precursors (Scheme 1.19a). The most prominent feature using heteroaryl sulfones as fluoroalkyl radical precursors is that they are

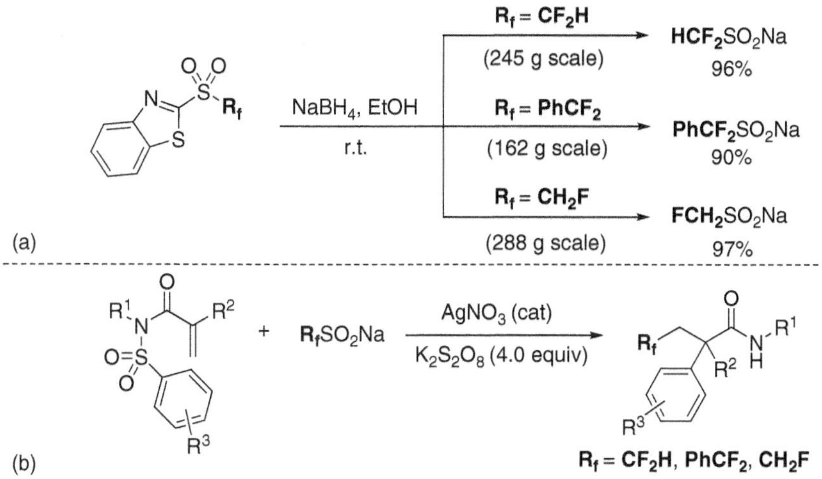

(a)

(b)

Scheme 1.18 The preparation and application of sodium fluoroalkanesulfinates.

reactivity-tunable. By slightly changing the heteroaryl rings, the redox potential of fluorinated heteroaryl sulfones can be varied, ensuring the efficient generation of various fluoroalkyl radicals (Scheme 1.19b).

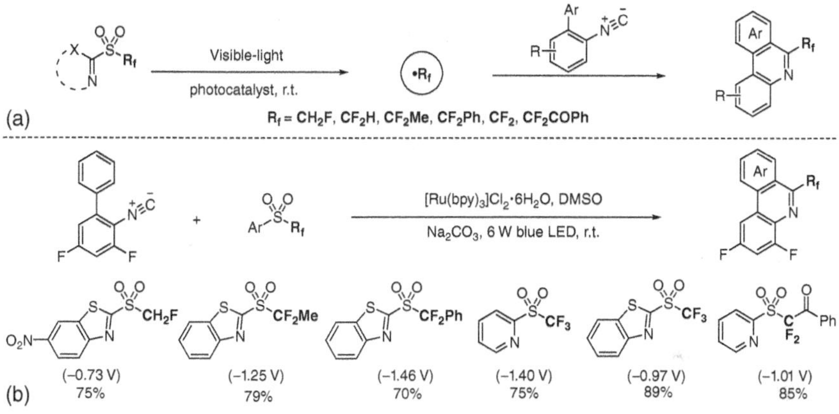

(a)

(b)

Scheme 1.19 Heteroaryl sulfones as fluoroalkyl radical precursors.

The combination of *gem*-difluoroolefination and fluorination can also be used as a good synthetic strategy for fluoroalkylation. In 2015, we reported an AgF-mediated fluorination of *gem*-difluoroolefination, followed by subsequent cross-coupling with a non-fluorinated olefin, to access α-CF$_3$ alkenes and β-CF$_3$ ketones [32] (Scheme 1.20a). Mechanistic studies revealed that α-CF$_3$-substituted benzyl radicals are involved, concluded by 2,2,6,6-tetramethylpiperidine 1-oxyl (TMEPO) trapping and radical clock experiments. A proposed mechanism is shown in Scheme 1.20b: addition of AgF to *gem*-difluoroolefination gives rise to α-CF$_3$-benzylsilver intermediate, which can undergo C—Ag bond homolysis to give α-CF$_3$-substituted benzyl radical. Addition of this radical to non-fluorinated alkene generates a new radical, which is oxidized to a carbocation followed by deprotonation to afford α-CF$_3$ alkene.

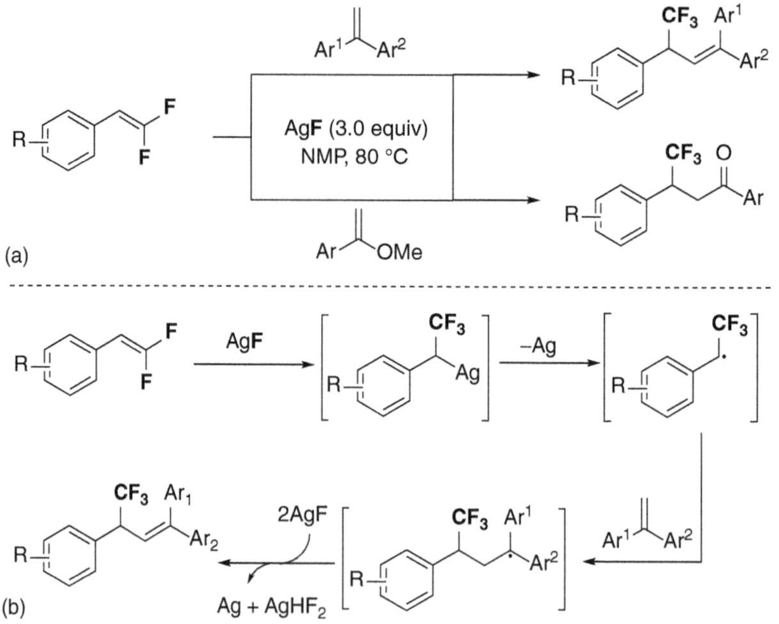

Scheme 1.20 AgF-mediated cross-coupling of *gem*-difluoroolefins and non-fluorinated olefins.

1.3.3 From Fluoroalkylation to Fluorination

3,3-Difluorocyclopropenes are readily available by the reaction of alkynes and difluorocarbene reagents, such as TMSCF$_3$, TMSCF$_2$Cl, and TMSCF$_2$Br [8d–f]. Based on the fact that 3,3-difluorocyclopropenes can be hydrolyzed to cyclopropenones in wet atmosphere and Lambert's elegant work in deoxychlorination of alcohols using 3,3-dichlorocyclopropenes [33], we successfully realized the use of safe and readily available difluorocarbene reagents for fluorination by means of 3,3-difluorocyclopropenes [11] (Scheme 1.21). To ensure the high efficiency of deoxyfluorination of aliphatic alcohols with 3,3-difluorocyclopropenes, the reaction should be carried out in a non-glass vessel because the *in situ*-generated HF will be consumed by the glassware (mainly SiO$_2$), thereby retarding the desired reaction.

The electronic nature of CpFluors is critical. For monoalcohols, utilizing electron-rich aryl substituent on CpFluors to stabilize the cyclopropenium cation intermediate is a key issue (Scheme 1.21, path a). However, for 1,2- and 1,3-diols, the reaction proceeds through cyclopropenone acetal intermediates, and is thus less dependent on the electronic nature of CpFluors (Scheme 1.21, path b). The most intriguing feature of CpFluors is that they are more sensitive to the electronic nature of alcohols than many other deoxyfluorination reagents; hence, selective

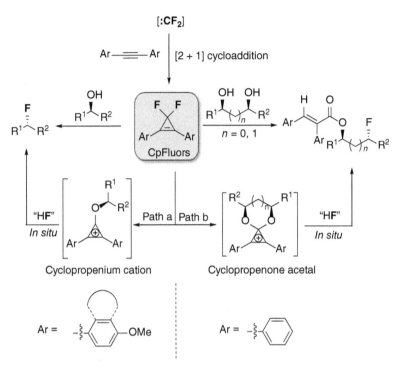

Scheme 1.21 Deoxyfluorination of alcohols with 3,3-difluorocyclopropenes.

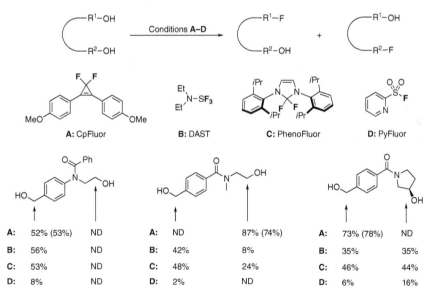

A:	52% (53%)	ND	**A:**	ND	87% (74%)	**A:**	73% (78%)	ND
B:	56%	ND	**B:**	42%	8%	**B:**	35%	35%
C:	53%	ND	**C:**	48%	24%	**C:**	46%	44%
D:	8%	ND	**D:**	2%	ND	**D:**	6%	16%

Scheme 1.22 Selective deoxyfluorination of longer diols.

fluorination of electron-rich OH groups of longer diols (non-1,2- and 1,3-diols) can be realized (Scheme 1.22).

Based on our work on fluoroalkylation chemistry using fluorinated sulfones and sulfoximines, and also inspired by Doyle's work [34], we developed *N*-tosyl-4-chlorobenzenesulfonimidoyl fluoride (SulfoxFluor) as a new bench-stable and highly reactive deoxyfluorination reagent [12] (Scheme 1.23). The prominent features of SulfoxFluor include a rapid fluorination rate, fluorine economy, selective monofluorination at the least steric hindered site of diols, and high fluorination/elimination selectivity.

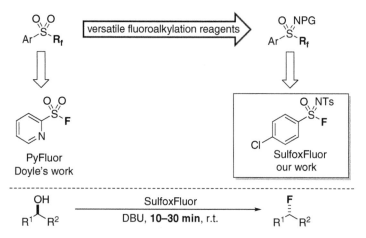

Scheme 1.23 Development of SulfoxFluor for deoxyfluorination.

1.4 Conclusions

Our efforts in the development of novel reagents for fluoroalkylation, fluoroolefination, and fluorination by probing the unique fluorine effects have been summarized. During our research work, we realized that (i) there are often unique fluorine effects in organic reactions, (ii) tackling the unique fluorine effect and unveiling the relationships among fluoroalkylation, fluoroolefination, and fluorination enable us to develop various reagents for synthetic organofluorine chemistry, and (iii) organofluorine reactions are not only practically useful but also provide fundamentally intriguing insights into generally organic reactions.

References

1 (a) Uneyama, K. (2006). *Organofluorine Chemistry*. Oxford: Blackwell. (b) Ojima, I. (2009). *Fluorine in Medicinal Chemistry and Chemical Biology*. Chichester, UK: Wiley-Blackwell. (c) Gouverneur, V. and Müller, K. (2011). *Fluorine in Pharmaceutical and Medicinal Chemistry: From Biophysical Aspects to Clinical*

Applications. London: Imperial College Press. (d) Kirsch, P. (2013). *Modern Fluoroorganic Chemistry: Synthesis. Reactivity, Applications*, 2e. Weinheim: Wiley-VCH.

2 The newly approved drugs can be found by searching in the following website: https://www.fda.gov/drugs/development-approval-process-drugs/new-drugs-fda-cders-new-molecular-entities-and-new-therapeutic-biological-products.

3 O'Hagan, D. and Deng, H. (2015). *Chem. Rev.* 115: 634–649.

4 (a) Hu, J. (2009). *J. Fluorine Chem.* 130: 1130–1139. (b) Zhang, W., Ni, C., and Hu, J. (2012). Selective fluoroalkylation of organic compounds by tackling the "negative fluorine effect". In: *Fluorous Chemistry* (ed. I.T. Horváth), 25–44. Berlin: Springer. (c) Ni, C. and Hu, J. (2011). *Synlett.*: 770–782. (d) Hu, J., Zhang, W., and Wang, F. (2009). *Chem. Commun.*: 7465–7478. (e) Shen, X. and Hu, J. (2014). *Eur. J. Org. Chem.* 2014: 4437–4451. (f) Ni, C., Hu, M., and Hu, J. (2015). *Chem. Rev.* 115: 765–825. (g) Ni, C. and Hu, J. (2016). *Chem. Soc. Rev.* 45: 5441–5454. (h) Zeng, Y. and Hu, J. (2016). *Synthesis* 48: 2137–2150.

5 (a) Ni, C. and Hu, J. (2005). *Tetrahedron Lett.* 46: 8273–8277. (b) Ni, C., Li, Y., and Hu, J. (2006). *J. Org. Chem.* 71: 6829–6833. (c) Li, Y., Liu, J., Zhang, L. et al. (2007). *J. Org. Chem.* 72: 5824. (d) Zhang, W., Zhu, J., and Hu, J. (2008). *Tetrahedron Lett.* 49: 5006–5008.

6 (a) Zhang, W., Huang, W., and Hu, J. (2009). *Angew. Chem. Int. Ed.* 48: 9858–9861. (b) Zhang, W., Wang, F., and Hu, J. (2009). *Org. Lett.* 11: 2109–2112. (c) Shen, X., Zhang, W., Ni, C. et al. (2012). *J. Am. Chem. Soc.* 134: 16999–17002. (d) Shen, X., Zhang, W., Zhang, L. et al. (2012). *Angew. Chem. Int. Ed.* 51: 6966–6970. (e) Shen, X., Miao, W., Ni, C., and Hu, J. (2014). *Angew. Chem. Int. Ed.* 53: 775–779. (f) Liu, Q., Shen, X., Ni, C., and Hu, J. (2017). *Angew. Chem. Int. Ed.* 56: 619–623.

7 (a) Zhao, Y., Huang, W., Zhu, L., and Hu, J. (2010). *Org. Lett.* 122: 1444–1447. (b) Zhao, Y., Gao, B., Ni, C., and Hu, J. (2012). *Org. Lett.* 14: 6080–6083. (c) Zhao, Y., Jiang, F., and Hu, J. (2015). *J. Am. Chem. Soc.* 137: 5199–5203. (d) Rong, J., Deng, L., Tan, P. et al. (2016). *Angew. Chem. Int. Ed.* 55: 2743–2747.

8 (a) Zhang, L., Zheng, J., and Hu, J. (2006). *J. Org. Chem.* 71: 9845–9848. (b) Zheng, J., Li, Y., Zhang, L. et al. (2007). *Chem. Commun.*: 5149–5151. (c) Wang, F., Huang, W., and Hu, J. (2011). *Chin. J. Chem.* 29: 2717–2721. (d) Wang, F., Luo, T., Hu, J. et al. (2011). *Angew. Chem. Int. Ed.* 50: 7153–7157. (e) Wang, F., Zhang, W., Zhu, J. et al. (2011). *Chem. Commun.* 47: 2411–2413. (f) Li, L., Wang, F., Ni, C., and Hu, J. (2013). *Angew. Chem. Int. Ed.* 52: 12390–12394.

9 He, Z., Tan, P., Ni, C., and Hu, J. (2015). *Org. Lett.* 17: 1838–1841.

10 Zhou, M., Ni, C., Zeng, Y., and Hu, J. (2018). *J. Am. Chem. Soc.* 140: 6801–6805.

11 Li, L., Ni, C., Wang, F., and Hu, J. (2016). *Nat. Commun.* 7: 13320–13330.

12 Guo, J., Kuang, C., Rong, J. et al. (2019). *Chem. Eur. J.* 25: 7259–7264.

13 Lipshutz, B.H. and Sengupta, S. (1992). Organocopper reagents: substitution, conjugate addition, carbo/metallocupration, and other reactions. In: *Organic Reactions*, vol. 41 (ed. L.A. Paquette), 135–631. Wiley.

14 Hu, M., Ni, C., and Hu, J. (2012). *J. Am. Chem. Soc.* 134: 15257–15260.

15 Hu, M., He, Z., Gao, B. et al. (2013). *J. Am. Chem. Soc.* 135: 17302–17305.

16 Hu, M., Ni, C., Li, L. et al. (2015). *J. Am. Chem. Soc.* 137: 14496–14501.

17 Nakao, Y., Takeda, M., Matsumoto, T., and Hiyama, T. (2010). *Angew. Chem. Int. Ed.* 49: 4447.

18 Prakash, G.K.S., Krishnamuri, R., and Olah, G.A. (1989). *J. Am. Chem. Soc.* 111: 393–395.

19 (a) Liu, X., Xu, C., Wang, M., and Liu, Q. (2015). *Chem. Rev.* 115: 683–730. (b) Singh, R.P. and Shreeve, J.M. (2000). *Tetrahedron* 56: 7613–7632.

20 Li, L., Ni, C., Xie, Q. et al. (2017). *Angew. Chem. Int. Ed.* 56: 9971–9975.

21 Xie, Q., Li, L., Zhu, Z. et al. (2018). *Angew. Chem. Int. Ed.* 57: 13211–13215.

22 Xie, Q., Zhu, Z., Li, L. et al. (2020). *Chem. Sci.* 11: 276–280.

23 (a) Xie, Q., Ni, C., Zhang, R. et al. (2017). *Angew. Chem. Int. Ed.* 56: 3206–3210. (b) Xie, Q., Zhu, Z., Li, L. et al. (2019). *Angew. Chem. Int. Ed.* 58: 6405–6410.

24 Dilman, A.D. and Levin, V.V. (2018). *Acc. Chem. Res.* 51: 1272–1280.

25 Seppelt, K. (1977). *Angew. Chem. Int. Ed.* 16: 322–323.

26 (a) Prakash, G.K.S., Hu, J., Wang, Y., and Olah, G.A. (2004). *Angew. Chem. Int. Ed.* 43: 5203–5206. (b) Prakash, G.K.S., Hu, J., Wang, Y., and Olah, G.A. (2004). *Org. Lett.* 6: 4315–4317. (c) Prakash, G.K.S., Hu, J., Mathew, T., and Olah, G.A. (2003). *Angew. Chem. Int. Ed.* 42: 5216–5219. (d) Stahly, G.P. (1989). *J. Fluorine Chem.* 43: 53–66. (e) Hine, J. and Porter, J.J. (1960). *J. Am. Chem. Soc.* 82: 6178–6181.

27 (a) Li, Y. and Hu, J. (2005). *Angew. Chem. Int. Ed.* 44: 5882–5886. (b) Liu, J., Li, Y., and Hu, J. (2007). *J. Org. Chem.* 72: 3119–3121. (c) Ni, C., Liu, J., Zhang, L., and Hu, J. (2007). *Angew. Chem. Int. Ed.* 46: 786–789.

28 Aïssa, C. (2009). *Eur. J. Org. Chem.*: 1831–1844.

29 Zhao, Y., Gao, B., and Hu, J. (2012). *J. Am. Chem. Soc.* 134: 5790–5793.

30 Miao, W., Ni, C., Zhao, Y., and Hu, J. (2016). *Org. Lett.* 18: 2766–2769.

31 Miao, W., Zhao, Y., Ni, C. et al. (2018). *J. Am. Chem. Soc.* 140: 880–883.

32 Gao, B., Zhao, Y., and Hu, J. (2015). *Angew. Chem. Int. Ed.* 54: 638–642.

33 (a) Kelly, B.D. and Lambert, T.H. (2009). *J. Am. Chem. Soc.* 131: 13930–13931. (b) Hardee, D.J., Kovalchuke, L., and Lambert, T.H. (2010). *J. Am. Chem. Soc.* 132: 5002–5003. (c) Vanos, C.M. and Lambert, T.H. (2011). *Angew. Chem. Int. Ed.* 50: 12222–12226.

34 (a) Nielsen, M.K., Ugaz, C.R., Li, W., and Doyle, A.G. (2015). *J. Am. Chem. Soc.* 137: 9571–9574. (b) Nielsen, M.K., Ahneman, D.T., Riera, O., and Doyle, A.G. (2018). *J. Am. Chem. Soc.* 140: 5004–5008.

2

Perfluoroalkylation Using Perfluorocarboxylic Acids and Anhydrides

Shintaro Kawamura[1,2] and Mikiko Sodeoka[1,2]

[1] Catalysis and Integrated Research Group, RIKEN Center for Sustainable Resource Science, 2-1 Hirosawa, Wako, Saitama, 351-0198 Japan
[2] Synthetic Organic Chemistry Laboratory, RIKEN Cluster for Pioneering Research, 2-1 Hirosawa, Wako, Saitama, 351-0198 Japan

2.1 Introduction

In recent years, increasing numbers of perfluoroalkyl group-containing medicines, agrochemicals, and functional materials have been reported [1], in large part due to the development of sophisticated perfluoroalkylating reagents, such as the Ruppert–Prakash, Togni, Umemoto, and Langlois reagents, as reviewed elsewhere in this book. Although synthetic routes to many perfluoroalkylated molecules can now be designed based on reported perfluoroalkylations using these reagents, it remains preferable to employ readily available, inexpensive, multipurpose perfluoroalkyl compounds that can be conveniently stored in the laboratory. In particular, perfluoroalkyl group-containing building blocks that can be prepared by means of scalable perfluoroalkylation reactions are needed. In this context, we focus here on perfluoroalkylation reactions utilizing perfluorocarboxylic acids and anhydrides as user-friendly perfluoroalkyl sources. Although several excellent secondary reagents, including the corresponding esters prepared from perfluorocarboxylic acids and anhydrides, are known [2], we will review only methods directly using the carboxylic acids and anhydrides themselves.

2.2 Perfluoroalkylation with Perfluorocarboxylic Acids

The history of perfluoroalkylation reactions of organic compounds using perfluorocarboxylic acids began in the 1970s, and since then various methods for the generation of perfluoroalkyl radicals or perfluoroalkyl metals as reactive species for perfluoroalkylations have been developed. In this section, we describe perfluoroalkylation reactions with perfluorocarboxylic acids, classified according to the following reaction modes: electrochemical reactions (2.1), reactions using XeF_2 (2.2), reactions using copper and silver salts (2.3), photochemical reactions (2.4), and other methods (2.5).

Organofluorine Chemistry: Synthesis, Modeling, and Applications,
First Edition. Edited by Kálmán J. Szabó and Nicklas Selander.
© 2021 WILEY-VCH GmbH. Published 2021 by WILEY-VCH GmbH.

2.2.1 Electrochemical Reactions

Electrolysis of perfluorocarboxylic acids and their metal salts can generate a perfluoroalkyl radical by anodic reaction, which was inspired by Kolbe electrosynthesis (Scheme 2.1) [3, 4]. The perfluoroalkyl radical can be electrophilically added to alkenes, alkynes, and aromatic compounds, and some examples are described below.

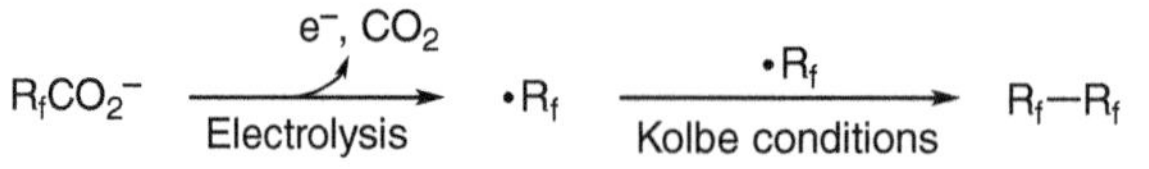

$$R_fCO_2^- \xrightarrow[\text{Electrolysis}]{e^-, CO_2} \cdot R_f \xrightarrow[\text{Kolbe conditions}]{\cdot R_f} R_f-R_f$$

Scheme 2.1 Electrochemical generation of perfluoroalkyl radical.

2.2.1.1 Reactions of Alkenes and Alkynes

In 1973, Renaud and Sullivan found that electrochemical oxidation of a mixture of sodium trifluoroacetate and propionate gave a mixture of 3,3,3-trifluoropropene, 1,2-bis(trifluoromethyl)ethane, 1,1,2-tris(trifluoromethyl)ethane, and 1,2,4-tris(trifluoromethyl)butane (Scheme 2.2) [5]. These products were considered to be produced by addition of the CF_3 radical to ethylene formed *in situ* from propionic acid under electrolysis conditions; this was confirmed by the reaction of ethylene gas with trifluoroacetate [6].

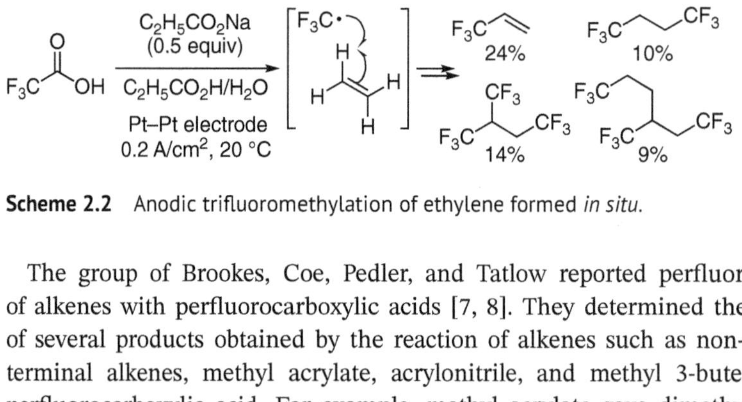

Scheme 2.2 Anodic trifluoromethylation of ethylene formed *in situ*.

The group of Brookes, Coe, Pedler, and Tatlow reported perfluoroalkylation of alkenes with perfluorocarboxylic acids [7, 8]. They determined the structures of several products obtained by the reaction of alkenes such as non-substituted terminal alkenes, methyl acrylate, acrylonitrile, and methyl 3-butenoate with perfluorocarboxylic acid. For example, methyl acrylate gave dimethyl succinate derivatives as major products in most cases via homocoupling of alkyl radicals generated by the reaction of alkene and perfluoroalkyl radical (Scheme 2.3). Notably, β-perfluoroalkylated methyl acrylate was also isolated as a major product when pentafluorobutanoic acid was used.

On the other hand, cyclopentene did not give the trifluoromethylated product, but instead trifluoroacetoxylation occurred. Furthermore, they also examined the reactivity of 1-hexyne, but obtained a complex mixture containing 1,1,1-trifluoro-2-heptene ($E/Z = 1 : 4$) and a vicinal double trifluoromethylated product, 1,1,1-trifluoro-3-(trifluoromethyl)-2-heptene. In 1975, Renaud and Champagne independently reported a similar reaction of trifluoroacetic acid with alkenes possessing an ester group, affording the corresponding trifluoromethylated meso-dimers and vicinal bis-trifluoromethylated products [9]. They also

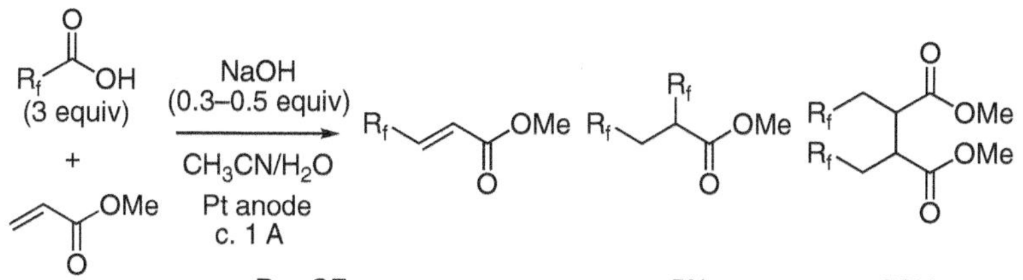

$R_f = CF_3$	—	5%	50%
$R_f = C_2F_5$	55%	15%	35%
$R_f = C_3F_7$	—	14%	36%
$R_f = C_7F_{15}$	—	Trace	34%

Scheme 2.3 Perfluoroalkylation of methyl acrylate.

synthesized ethyl 3,3,3-trifluoropropionate by decarboxylative electrochemical oxidation of malonic acid monoethyl ester in the presence of trifluoroacetic acid. Dmowski et al. reported the dimer-forming electrochemical trifluoromethylation of acrylonitrile and crotonitrile in 1997 [10].

Renaud extensively studied the electrochemical trifluoromethylation of alkenes, using mono- and disubstituted alkenes and heterocyclic alkenes [11]. In contrast to acyclic alkenes such as methyl vinyl ketone, vinyl acetate, diethyl fumarate, and diethyl maleate, which gave the dimer as the major product, heterocyclic alkenes such as *N*-ethylmaleimide and 2,5-dihydrothiophene 1,1-dioxide efficiently afforded the vicinal bis-trifluoromethylated products (Scheme 2.4a). They also succeeded in demonstrating an intramolecular carbo-trifluoromethylation via radical cyclization of bis-alkenes bearing ester groups (Scheme 2.4b) [12]. Grinberg's group performed electrolysis of sodium trifluoroacetate in the presence of the dimethyl ester of maleic acid, obtaining the corresponding bis-trifluoromethylated product and the trifluoromethylated dimer [13]. They also examined the influence of the current density on the product yields and selectivity, finding that an increase of the current density reduced the yields of the trifluoromethylated products without significantly changing the product ratio and facilitated the formation of hexafluoroethane (C_2F_6).

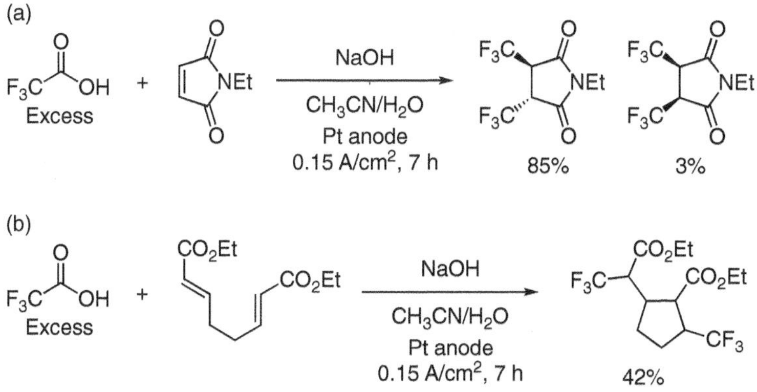

Scheme 2.4 (a) Vicinal bis-trifluoromethylation and (b) carbo-trifluoromethylation.

From the viewpoint of synthesizing new trifluoromethylated molecules, bis-trifluoromethylated products could be a good building block; for example, Muller reported the derivatization of a bis-trifluoromethylated product, 2-(trifluoromethyl)-4,4,4-trifluorobutyric acid, obtained by the anodic trifluoromethylation of acrylic acid (Scheme 2.5) [14]. The trifluoromethyl group at the α-position of the carboxylic acid was hydrolyzed selectively to a carboxyl group under basic conditions, and the obtained malonic acid was transformed into a barbital analog in several steps.

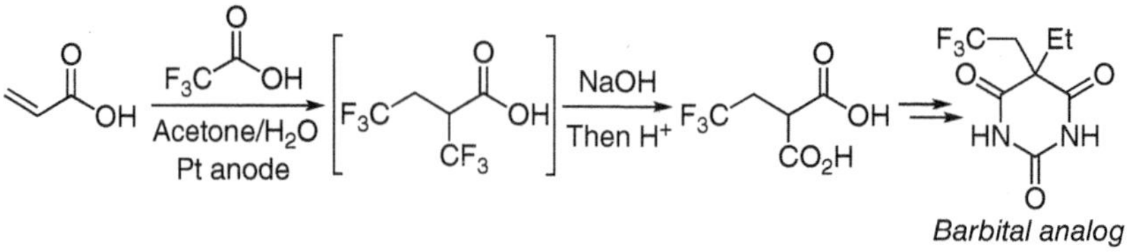

Scheme 2.5 Synthesis of a barbital analog using a bis-trifluoromethylated product.

Uneyama et al. demonstrated the synthetic utility of a bis-trifluoromethylated amide prepared by anodic trifluoromethylation of acrylamide with trifluoroacetic acid, successfully generating cyanoester, amino acid, and β-lactam analogs (Scheme 2.6) [15, 16]. In the above early reports, the products of electrochemical trifluoromethylation of alkenes using trifluoroacetic acid were usually limited to bis-trifluoromethylated products or trifluoromethylated dimeric products. Muller overcame this limitation, achieving the synthesis of mono-trifluoromethylated products by means of anodic trifluoromethylation (Scheme 2.7) [17]. When isopropenyl acetate was employed in electrochemical trifluoromethylation with trifluoroacetic acid in the presence of sodium hydroxide, 4,4,4-trifluoro-2-butanone was efficiently obtained [17a]. In addition, 12,12,12-trifluorododecanoic acid [17b] and 4,4,4-trifluorobutanal [17c] were synthesized from undecylenic acid and allyl alcohol, respectively.

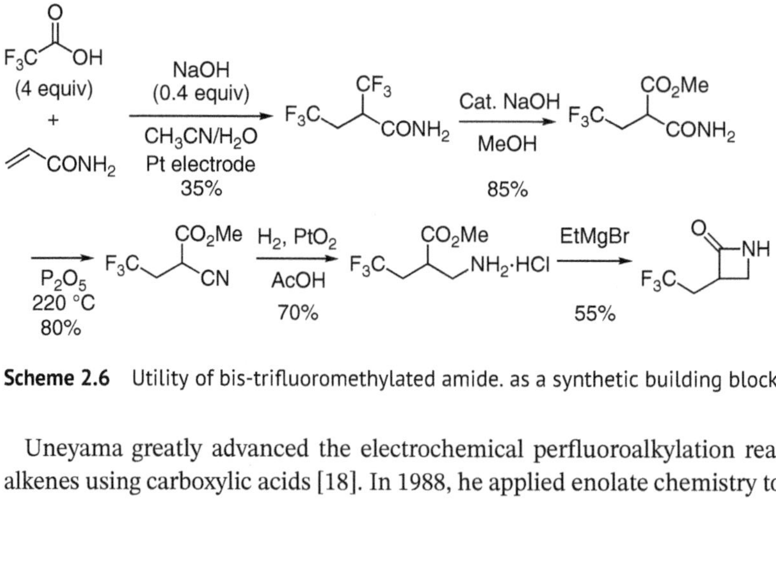

Scheme 2.6 Utility of bis-trifluoromethylated amide. as a synthetic building block

Uneyama greatly advanced the electrochemical perfluoroalkylation reaction of alkenes using carboxylic acids [18]. In 1988, he applied enolate chemistry to anodic

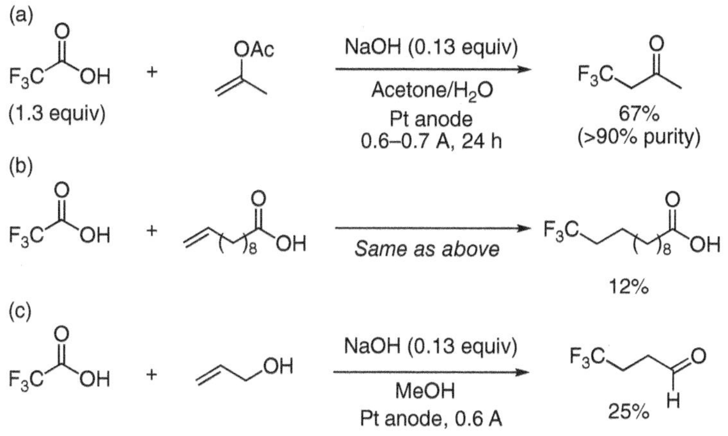

(a)

(b)

(c)

Scheme 2.7 Mono-trifluoromethylations of alkenes.

electrophilic trifluoromethylation; electrochemically generated trifluoromethyl radical was added to an enol formed *in situ* from β-ketoester (Scheme 2.8a) [19]. While the reaction at 60 °C gave α-trifluoromethylated β-ketoester in 31% yield as the sole product, interestingly, the α-trifluoromethylated ester was generated via elimination of acetic acid at −40 °C. In addition, the use of enol acetate instead of the ketoester substrate was found to give trifluoromethylated β-ketoester exclusively in better yield (Scheme 2.8b), which suggests that the acetate group is a better leaving group to facilitate the C—O bond cleavage. Uneyama also reported pioneering

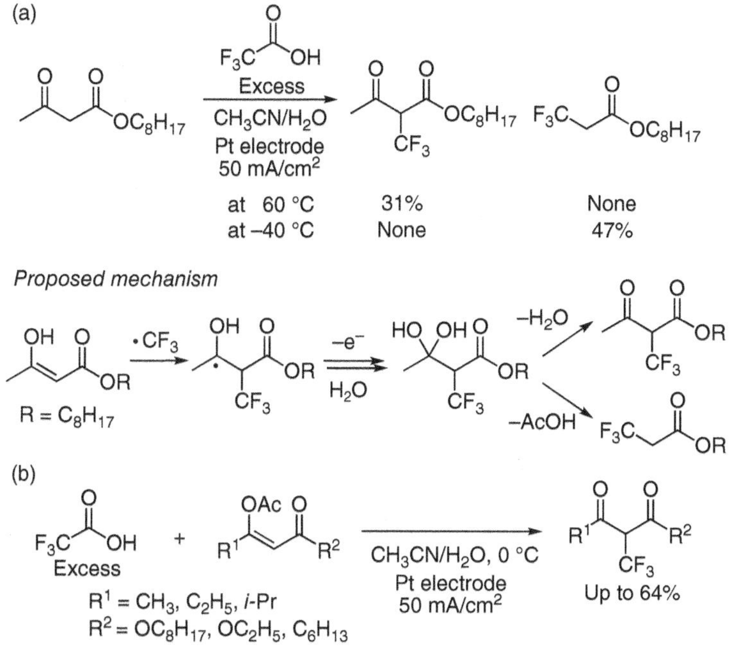

(a)

Proposed mechanism

(b)

Scheme 2.8 Application of enolate chemistry; reactions of (a) β-ketoester (b) enol acetate.

work on bifunctionalization-type perfluoroalkylation reactions [20]. In 1988, amino-trifluoromethylation of methyl methacrylate with trifluoroacetic acid under basic conditions was developed (Scheme 2.9a) [20a], in which trifluoromethyl and acetamide groups were installed simultaneously in an acetonitrile–H_2O cosolvent system. Notably, the dimeric product was not obtained under these conditions.

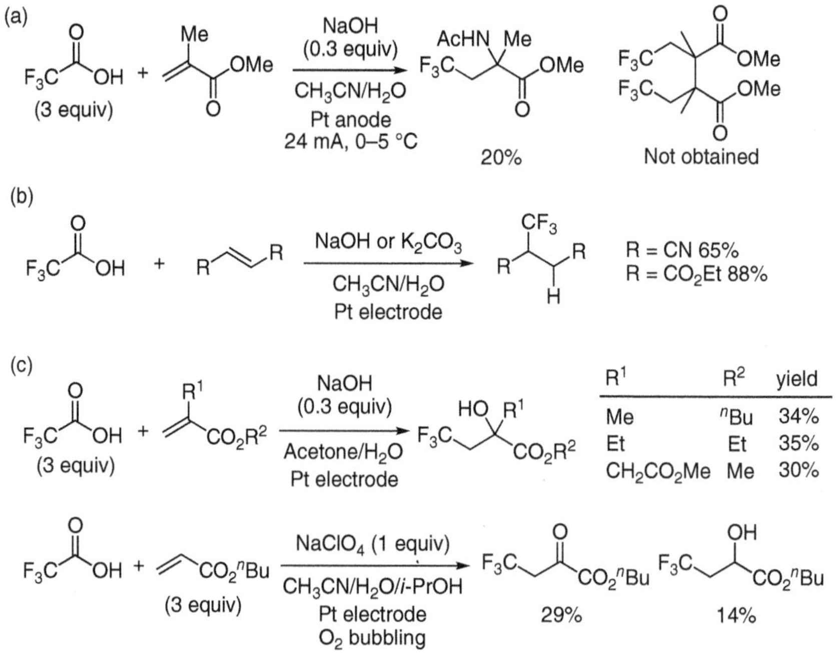

Scheme 2.9 Bifunctionalization-type trifluoromethylations.

He then reported hydro-trifluoromethylation of fumaronitrile [20b] and dialkyl fumarates [20c] (Scheme 2.9b). The hydrogen atom was proposed to come from the water cosolvent via protonation of an anionic intermediate. Furthermore oxy-trifluoromethylations affording alcohol [20d] and ketone [20e] products were developed (Scheme 2.9c). In the reaction, water and oxygen were utilized as oxygen sources for the bifunctionalization-type trifluoromethylation. These conditions of electrochemical trifluoromethylation could be applied to perfluoroalkylations of electron-deficient alkenes with perfluoroalkanoic acids (R_fCO_2H: $R_f = CF_3$, C_3F_7, C_7F_{15}, CHF_2, and CH_2F) [20d].

In 1996, Dmowski and Biernacki found that electrochemical trifluoromethylation of 2,5-dihydrothiophene 1,1-dioxide in H_2O provided the allylic trifluoromethylation product in 44% yield as the major product, in contrast to the reaction in a CH_3CN/H_2O cosolvent system, which afforded the bis-trifluoromethylation product (Scheme 2.10) [21].

Although many electrochemical perfluoroalkylations of alkenes have been reported, reactions of alkynes are extremely rare. Dmowski and Biernacki reported

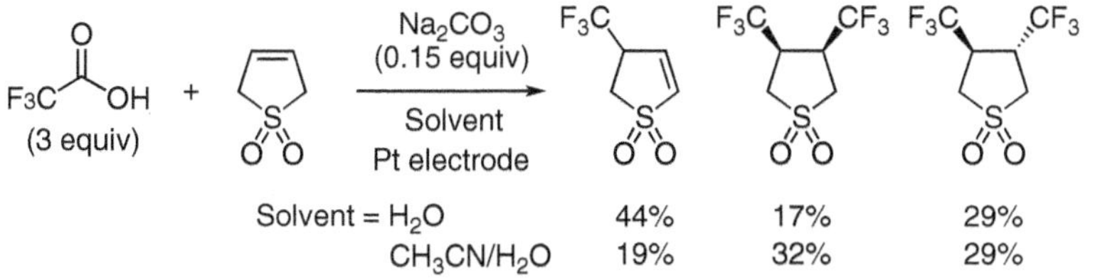

Scheme 2.10 Solvent-controlled allylic trifluoromethylations.

the reaction of dimethyl acetylenedicarboxylate, affording an isomeric mixture of bis-trifluoromethylated alkenes together with the tris-trifluoromethylated product and polymers (Scheme 2.11) [21].

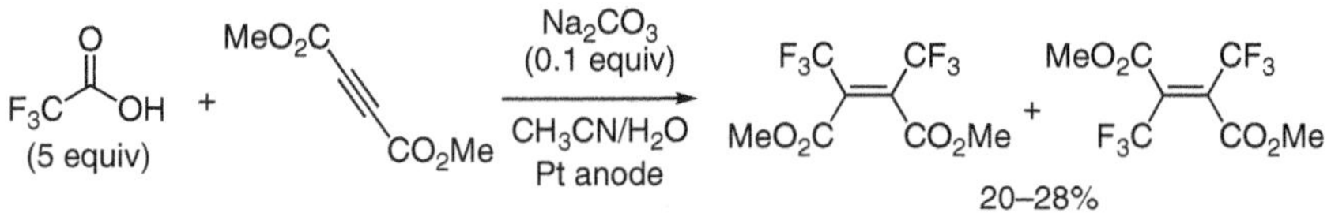

Scheme 2.11 Reaction with alkyne.

Several types of apparatus for electrochemical trifluoromethylation have been developed. In 2009, Kaurova's group applied glassy carbon as the anode material, instead of platinum, for the trifluoromethylation of ethylene, affording 1,1,1,6,6,6-hexafluorohexane (Scheme 2.12) [22]. Under these conditions, the glassy carbon anode showed higher efficiency for the trifluoromethylation than a platinum anode, which was considered to be due to reduced absorption on the carbon anode. Grinberg used a Pt-10% Ir anode for electrochemical trifluoromethylation and found that the rate of electrolysis of trifluoroacetate was four times faster than with the platinum electrode [23].

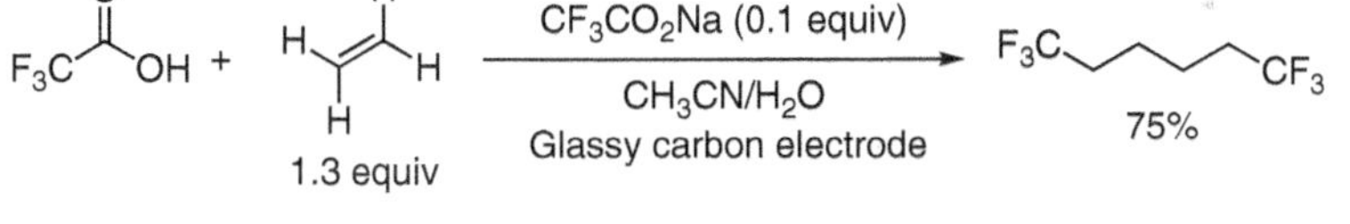

Scheme 2.12 Reaction using glassy carbon electrode.

In 2014, Wirth and coworkers designed an electrochemical microflow reactor for trifluoromethylation (Scheme 2.13) [24]. The reactor gave the trifluoromethyl and difluoromethyl group-containing dimeric products from carboxylic acids and alkenes within 69 seconds, although the batch reaction required 16 hours to obtain a comparable result. In addition, very rapid amino-fluoroalkylation of methyl methacrylate and bis-fluoroalkylation of acrylamides were performed with this system.

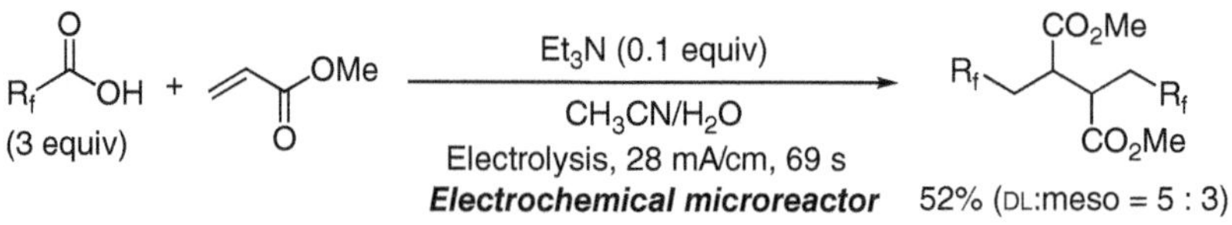

Scheme 2.13 Rapid alkene perfluoroalkylation with an electrochemical microflow reactor.

2.2.1.2 Reaction of Aromatic Compounds

In contrast to perfluoroalkylations of alkenes, the reaction of aromatic compounds with perfluorocarboxylic acids via electrolysis is less well studied. This is due to the occurrence of undesired acetoxylation via radical cation formation of the aromatic substrates in the electrooxidation, and thus Kolbe-type electrolysis of the carboxylic acid fails in the usual cases (Scheme 2.14a). Exceptionally, in 1978, Grinberg et al. demonstrated the trifluoromethylation of monosubstituted benzenes possessing a trifluoromethyl or cyano group with trifluoroacetate in aqueous acetonitrile on a Pt electrode (Scheme 2.14b) [25a]. Acetonitrile, used as the solvent, was found to suppress oxidation of the substrate as well as the acetoxylation [25b]. Trevin and coworkers investigated the trifluoromethylation of several aromatic compounds possessing electron-withdrawing groups with trifluoroacetic acid by means of preparative electrolysis using Pt electrodes in pure organic solvent [26]. Under the nonaqueous conditions, pyridine as a base forming the trifluoroacetate salt promoted the anodic trifluoromethylation.

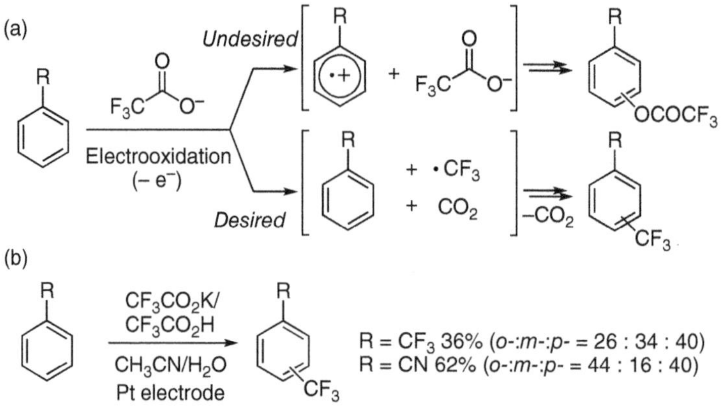

Scheme 2.14 Aromatic trifluoromethylation: (a) problem in reaction development; (b) trifluoromethylation of electron-deficient aromatic compounds.

2.2.2 Reactions Using XeF$_2$

Some early examples of perfluoroalkylations with perfluorocarboxylic acids used XeF$_2$ as an activator to generate perfluoroalkyl radicals. Eisenberg and DesMarteau reported that xenon fluoride trifluoroacetate and xenon bis(trifluoroacetate), which detonate when thermally or mechanically shocked, were prepared by the reaction of trifluoroacetic acid and XeF$_2$. These compounds were found to give hexafluoroethane via decomposition on standing at 23 °C [27]. Zupan and coworker reported that in the course of studies on alkene fluorination using XeF$_2$ in the presence of trifluoroacetic acid, styrene was transformed into trifluoromethylated products together with fluorination products (Scheme 2.15) [28].

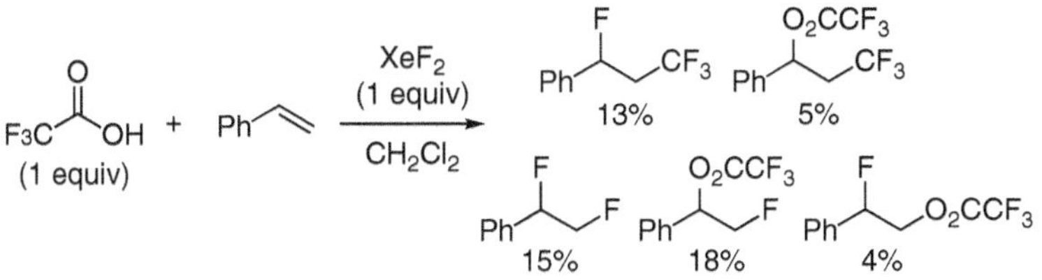

Scheme 2.15 Trifluoromethylation of alkene. in the presence of XeF$_2$

In addition, diphenylacetylene gave 1-(trifluoromethyl)-2-(trifluoroacetoxy)-1,2-diphenylethylene and 1-fluoro-2-(trifluoromethyl)-1,2-diphenylethylene, although the yields were low. It was suggested that the reaction proceeds via trifluoromethyl radical generation from xenon trifluoroacetate species. In 1988, Matsuo developed an efficient perfluoroalkylation of electron-deficient and heterocyclic aromatic compounds with perfluorocarboxylic acids promoted by XeF$_2$ (Scheme 2.16) [29]. More than 10 perfluoroalkylated products, including trifluoromethyl derivatives, were synthesized in up to 72% yield. Popkov applied Matsuo's conditions to the synthesis of trifluoromethylated furan and thiophene derivatives in order to evaluate their activity against phytopathogenic fungi *in vitro* [30].

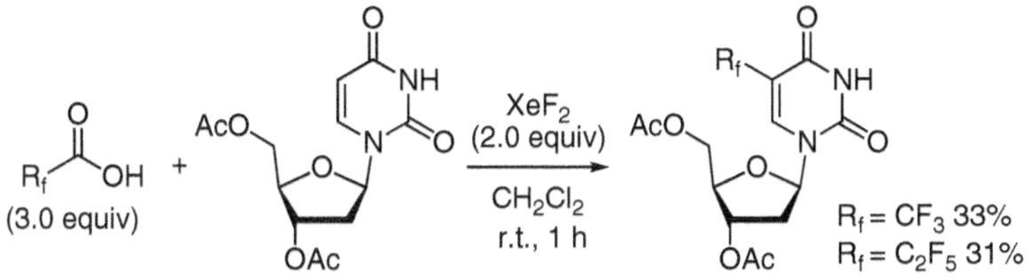

Scheme 2.16 Perfluoroalkylation of heterocyclic aromatic compounds.

2.2.3 Reactions Using Copper and Silver Salts

Perfluoroalkylation with the aid of transition metals is more practical, not only because of the ready availability of the chemicals but also because no special equipment is needed. In particular, copper and silver salts have been frequently used for the reactions, but these two transition metals tend to induce different types of reactivity: ionic and radical types, respectively. Some examples are shown below.

2.2.3.1 Using Copper Salts

Cross-coupling-type perfluoroalkylations of pre-functionalized aromatic compounds can selectively afford target molecules bearing a perfluoroalkyl group at a specific position of the aromatic ring, in contrast to the electrochemical perfluoroalkylations of aromatic compounds mediated by perfluoroalkyl radicals. In 1981, Kondo and coworkers achieved a convenient copper iodide-mediated trifluoromethylation of aromatic halides with sodium trifluoroacetate, obtaining the desired trifluoromethylarenes from aryl iodide and bromide in up to 88% yield (Scheme 2.17a) [31, 32]. Suzuki independently reported a similar reaction using the combination of sodium trifluoroacetate and copper iodide [33]. Chambers further investigated the scope of the reaction under Kondo's conditions; perfluoroalkylations using sodium pentafluoropropionate and heptafluorobutyrate were found to be viable, and pentafluoroethylation of β-bromostyrene and 1-iodopentane, as well as various aryl halides, was demonstrated (Scheme 2.17b) [34]. A nucleophilic intermediate $[CF_3CuI]^-$ was proposed as the reactive species [35]. Kondo's conditions were also applied by Miller and coworkers [36] and Ammann and coworkers [37], who synthesized trifluoromethylated aromatic compounds for evaluation of their biological activity.

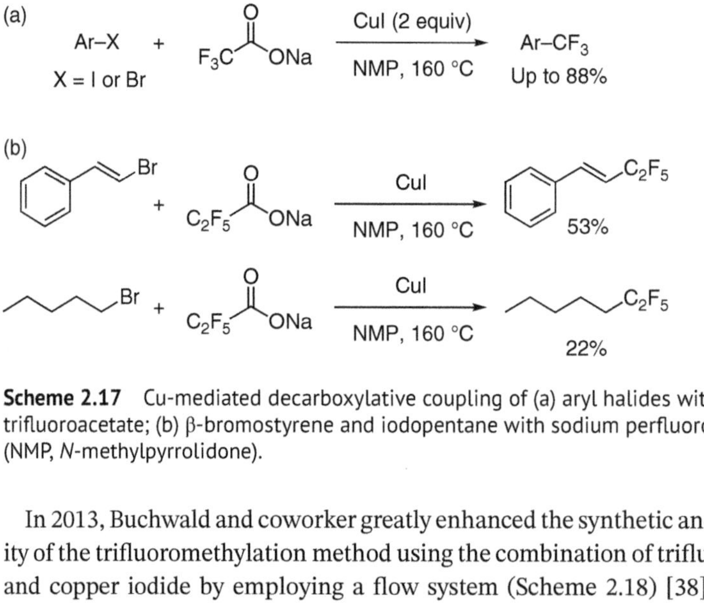

Scheme 2.17 Cu-mediated decarboxylative coupling of (a) aryl halides with trifluoroacetate; (b) β-bromostyrene and iodopentane with sodium perfluorocarboxylates (NMP, *N*-methylpyrrolidone).

In 2013, Buchwald and coworker greatly enhanced the synthetic and practical utility of the trifluoromethylation method using the combination of trifluoroacetate salt and copper iodide by employing a flow system (Scheme 2.18) [38]. These conditions showed excellent efficiency and substrate scope, and a diverse array of trifluoromethylated arenes and heteroarenes could be obtained in up to 96% yield with only a 16 minutes residence time (t_r).

There have been a few attempts to control the reactivity of copper perfluoro carboxylates by using ligands. Vicic and coworkers prepared *N*-heterocyclic carbene (NHC)–copper-trifluoroacetate and -chlorodifluoroacetate complexes and examined their reactivity for trifluoromethylation of aryl halides (Scheme 2.19) [39]. (SI*i*Pr)Cu(O$_2$CCF$_3$) greatly enhanced the yield of trifluoromethylation compared to "ligandless" conditions using CuI and CF$_3$CO$_2$Na in aryl halide solvent (neat conditions). In addition, (SI*i*Pr)Cu(O$_2$CCF$_2$Cl) was found to be available as a precursor for trifluoromethylation in the presence of CsF.

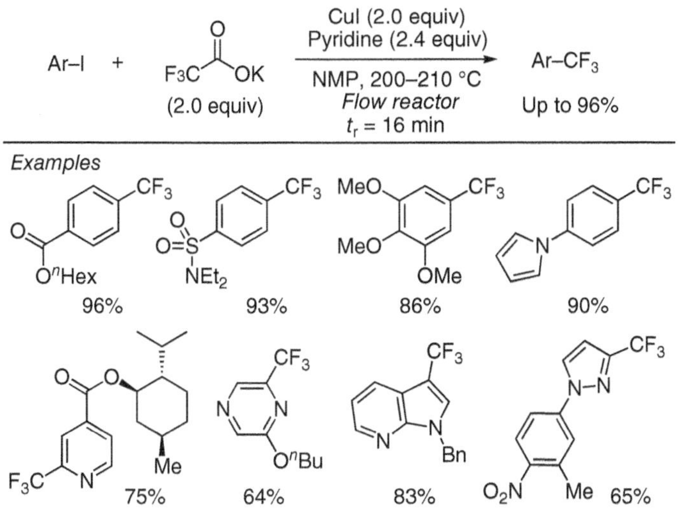

Scheme 2.18 Application of a flow system to coupling-type trifluoromethylation.

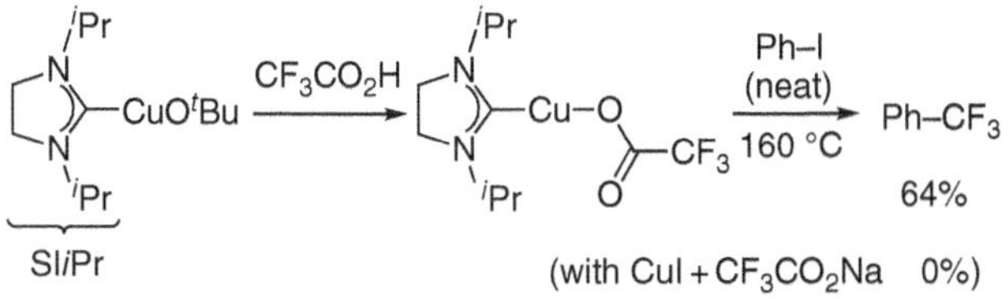

Scheme 2.19 N-heterocyclic carbene-ligand enhanced trifluoromethylation with copper trifluoroacetate.

Weng and coworkers developed efficient trifluoromethylations with copper-trifluoroacetate and chlorodifluoroacetate complexes possessing diamine ligands: 1,10-phenanthroline (phen) and 2,2′-bipyridyl (bpy) (Scheme 2.20) [40].

The reaction of aryl iodide or heteroaryl bromide with well-defined [(phen)Cu(O_2CCF$_3$)] in the presence of NaF as an additive afforded the desired trifluoromethylated products (Scheme 2.20a) [40a]. Because the "ligandless" conditions previously reported required a higher reaction temperature (160 °C) for the decarboxylation, the ligand was considered to facilitate the reaction. Mechanistic studies indicated that radical intermediates are unlikely to be involved in the reaction – instead a reactive species with nucleophilic character was proposed. In addition, copper-perfluoro carboxylate complexes [(phen)$_2$Cu](O_2CR$_f$) (R$_f$ = C$_2$F$_5$, C$_3$F$_7$, C$_4$F$_9$, C$_5$F$_{11}$), as well as the combination of [(phen)Cu(O_2CCF$_3$)] and Na$_2$CO$_3$, were found to show sufficient reactivity in decarboxylative perfluoroalkylations of E-β-bromostyrenes, affording the desired trifluoromethylated styrenes in up to 98% yield in a stereospecific manner [40b]. Weng's group reported trifluoromethylation of (hetero)aryl iodide and bromides with a copper chlorodifluoroacetate complex (Scheme 2.20b) [40c]; in the presence of NaOH and CsF, the reaction of aromatic halides and

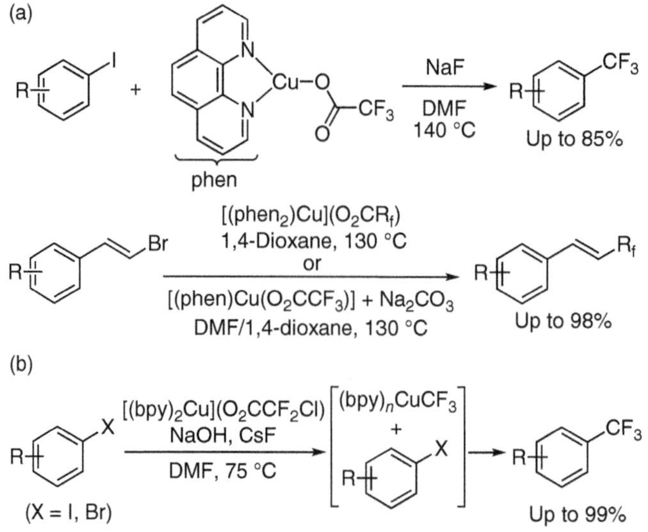

(a)

(b)

Scheme 2.20 Trifluoromethylations using diamine–copper complexes: reaction of (a) copper-trifluoroacetate, (b) copper-difluorochloroacetate (DMF, *N,N*,-dimethylformamide).

[(bpy)$_2$Cu](O$_2$CCF$_2$Cl) gave trifluoromethylated products in excellent yield, via difluorocarbene and subsequent formation of CuCF$_3$ species.

Chen and Liu employed Cu(O$_2$CCF$_2$SO$_2$F)$_2$, which was decomposed in *N,N*,-dimethylformamide (DMF) at room temperature to generate CF$_2$ carbene and fluoride, in the trifluoromethylation of aryl iodides and aryl bromide. The reaction required a stoichiometric amount of additional copper metal in order to form a reactive Cu^{I}CF$_3$ intermediate (Scheme 2.21) [41].

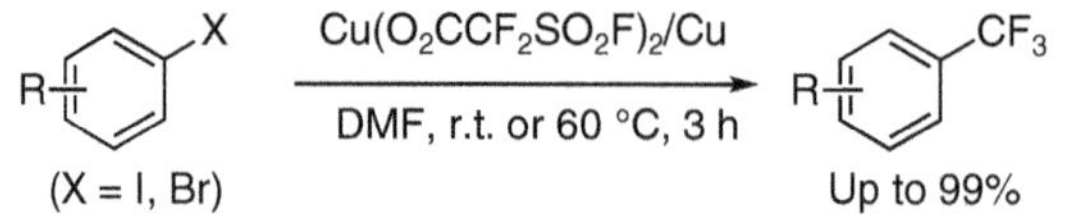

Scheme 2.21 Trifluoromethylation with Cu(O$_2$CCF$_2$SO$_2$F)$_2$.

Although various trifluoromethylations with combinations of copper salts and trifluoroacetate salts had been developed, a stoichiometric amount of copper salts was necessary in almost all cases. To overcome this limitation, Li, Duan, and coworkers developed a copper-catalyzed trifluoromethylation of aryl iodides with sodium trifluoroacetate using Ag$_2$O as a promoter (Scheme 2.22) [42]. The use of copper catalyst alone gave the trifluoromethylation product in low yield, but the addition of Ag$_2$O dramatically improved the yield. Notably, the use of Ag$_2$O alone did not give any product at all.

Beside the cross-coupling-type trifluoromethylations, the combination of trifluoroacetate and copper salts could be used as an efficient nucleophilic reagent for addition reaction to carbonyl compounds, as reported by Chang and Cai (Scheme 2.23)

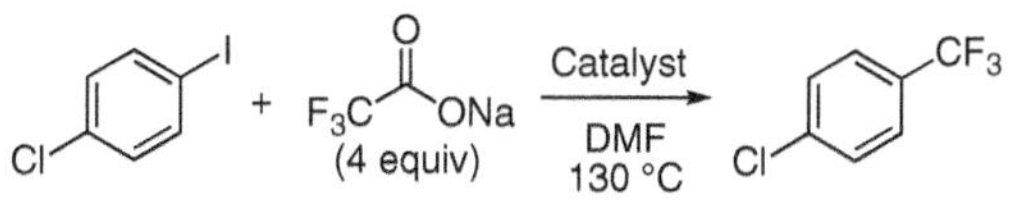

	Catalyst	Yield
	CuI (20 mol%)	17%
	CuI (20mol%)/Ag$_2$O (10 mol%)	54%
	Ag$_2$O	n.r.
	Cu (20 mol%)/Ag$_2$O (10 mol%)	64%
	Cu (30 mol%)/Ag$_2$O (30 mol%)	90%

Scheme 2.22 Cu-catalyzed trifluoromethylation of aryl iodides. using Ag$_2$O as a promoter.

[43]. Not only aldehydes but also ketones, acetyl chloride, and phthalic anhydride were available for this reaction.

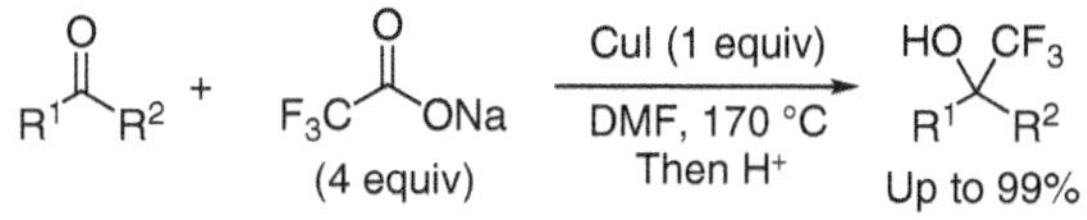

Scheme 2.23 Nucleophilic addition of trifluoromethyl group to carbonyl compounds.

2.2.3.2 Using Silver Salts

Interestingly, an early example of perfluoroalkylations using silver salts was the synthesis of multi-trifluoromethylated fullerene [44]. In 2001, Boltalina's group prepared mixtures of multi-trifluoromethylated fullerenes by solid-state trifluoromethylation with silver trifluoroacetate at 300 °C [45]. The silver salt could add up to 22 trifluoromethyl groups to C60, although other metal trifluoroacetates such as copper, palladium, and chromium salts could add only less than 8 trifluoromethyl groups. Several multi-trifluoromethylated fullerenes, including isomers, were independently characterized by means of nuclear magnetic resonance (NMR) and mass spectroscopies by Boltalina and coworkers [46] and Taylor and coworkers [47]. In 2007, Goryunkov and coworkers successfully determined the X-ray crystal structures of some of them and discussed the observed isomeric distribution in mixtures of C60(CF$_3$)$_n$ compounds up to $n = 6$ [48].

In 2015, Zhang and coworkers reported an electrophilic trifluoromethylation of aromatic compounds with trifluoroacetic acid by using a silver catalyst (Scheme 2.24) [49]. The Ag$_2$CO$_3$ catalyst was considered to facilitate the generation of CF$_3$ radical from trifluoroacetic acid via decarboxylation, and then this radical mediates aromatic trifluoromethylation. The resulting Ag(I) species is reoxidized by K$_2$S$_2$O$_8$, used as an additive.

Zhang's conditions have been applied to several types of fluoroalkylations using fluorine-containing carboxylic acids. Nielsen and coworkers performed a decarboxylative difluoromethylation of *N*-heteroaromatic compounds with difluoroacetic acid (Scheme 2.25a) [50]. Wan, Hao, and coworkers reported an aryldifluoromethylation of isocyanides with potassium difluoroarylacetate, affording phenanthridines bearing an arylated difluoromethene motif (Scheme 2.25b) [51, 52]. Wan, Hao, and coworkers [53] and Deng and coworkers [54] independently reported an oxindole synthesis by the reaction of *N*-arylacrylamides with potassium difluoroarylacetate or its acid form in the presence of persulfate salts (Scheme 2.25c).

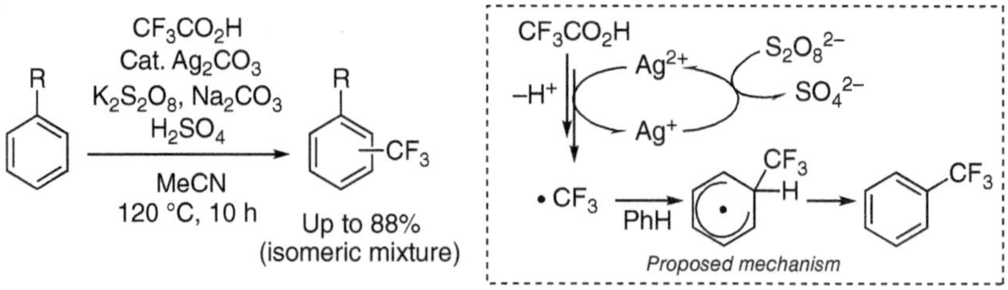

Scheme 2.24 Silver-catalyzed electrophilic trifluoromethylation.

Hashmi and coworker employed ethynyl benziodoxolone, as the coupling partner, in a decarboxylative aryldifluoromethylation with difluoroarylacetic acids (Scheme 2.25d) [55].

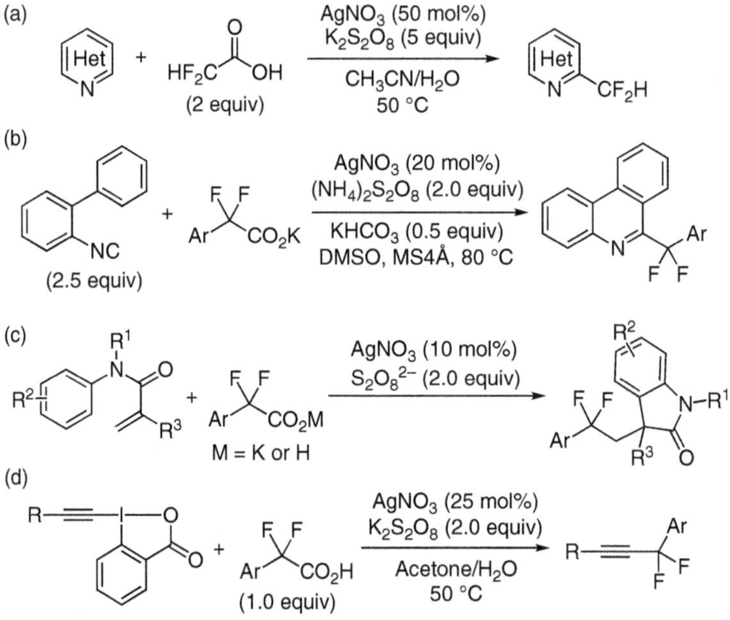

Scheme 2.25 Silver-catalyzed fluoroalkylations with various fluorinated carboxylic acids. DMSO, dimethylsulfoxide; MS4Å, molecular sieves 4Å.

2.2.4 Photochemical Reactions

In 1993, Mallouk and Lai developed a photochemical approach for trifluoromethylation with silver trifluoroacetate, in which TiO_2 was used as a photocatalyst for trifluoromethylation of aromatic compounds (Scheme 2.26a) [56]. The TiO_2 catalyst promotes photolysis of trifluoroacetate to generate trifluoromethyl radical. In 2017, Su, Li, and coworkers developed a new catalytic system using Rh-modified TiO_2 nanoparticles for photochemical trifluoromethylation with trifluoroacetic acid in the presence of $Na_2S_2O_8$ as an additive (Scheme 2.26b) [57].

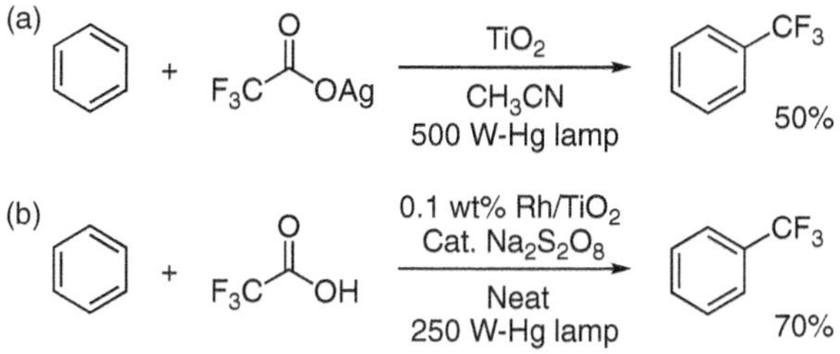

(a)

$$TiO_2$$

$$CH_3CN$$
500 W-Hg lamp

CF₃ ... 50%

(b)

0.1 wt% Rh/TiO₂
Cat. Na₂S₂O₈

Neat
250 W-Hg lamp

CF₃ ... 70%

Scheme 2.26 Photochemical trifluoromethylation using TiO_2.

Not only simple arenes but also *N*-containing heteroarenes were available, affording the desired products in modest to good yields. Very recently, Hosseini-Sarvari and Bazyar achieved a photocatalytic trifluoromethylation using sodium trifluoroacetate under blue LED irradiation in the presence of Au-modified ZnO catalyst (Au@ZnO core–shell nanoparticles) (Scheme 2.27) [58]. Au@ZnO could catalyze not only aromatic trifluoromethylation but also coupling-type trifluoromethylations of aryl halides as well as boronic acids under appropriate conditions.

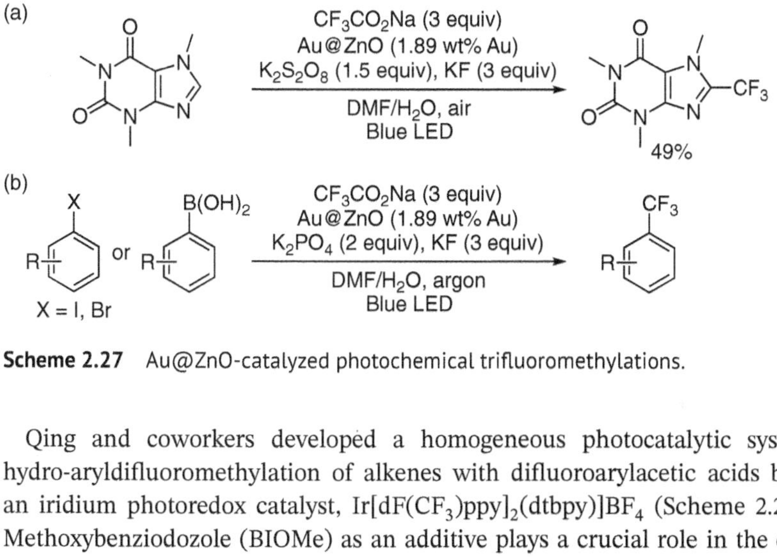

(a)

CF₃CO₂Na (3 equiv)
Au@ZnO (1.89 wt% Au)
K₂S₂O₈ (1.5 equiv), KF (3 equiv)

DMF/H₂O, air
Blue LED

49%

(b)

X = I, Br

CF₃CO₂Na (3 equiv)
Au@ZnO (1.89 wt% Au)
K₂PO₄ (2 equiv), KF (3 equiv)

DMF/H₂O, argon
Blue LED

Scheme 2.27 Au@ZnO-catalyzed photochemical trifluoromethylations.

Qing and coworkers developed a homogeneous photocatalytic system for hydro-aryldifluoromethylation of alkenes with difluoroarylacetic acids by using an iridium photoredox catalyst, Ir[dF(CF₃)ppy]₂(dtbpy)]BF₄ (Scheme 2.28) [59]. Methoxybenziodozole (BIOMe) as an additive plays a crucial role in the catalytic cycle; it accelerates photolysis of the carboxylic acid by forming a hypervalent iodine intermediate possessing carboxylate as a ligand, and promotes turnover of the photoredox catalytic cycle by oxidizing excited-state Ir^{III*} to Ir^{IV} species. The CF₃ radical generated by photolysis reacts with alkene and the resulting alkyl radical affords the hydro-aryldifluoromethylated product via hydrogen abstraction from *N*-methylpyrrolidone (NMP).

Zhu and coworkers also developed a photocatalytic fluoroalkylation of alkenes bearing benzaldehyde or propenal functionalities on the side chain with fluorinated carboxylic acids by using the combination of Ir-catalyst and the oxidant PhI(OAc)₂

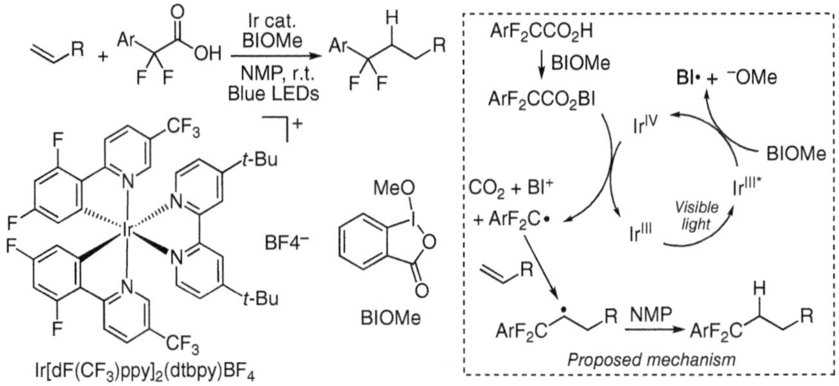

Scheme 2.28 Ir-photocatalyzed aryldifluoromethylation.

(Scheme 2.29) [60]. Cyclic ketones bearing fluoroalkyl groups, such as CF_2Ar, CF_2H, and CF_2Me groups, were obtained up to 90% yield.

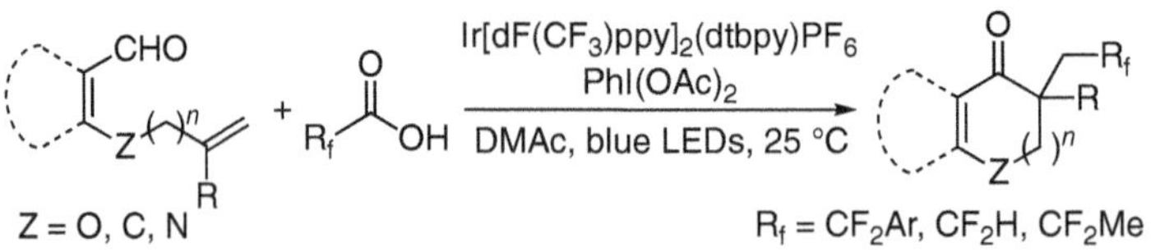

Scheme 2.29 Carbo and heterocyclic ketone synthesis by Ir-photocatalyzed carbo-fluoroalkylation of alkenyl aldehydes.

Recently, Gouverneur and coworkers developed a photocatalyst-free hydro-difluoromethylation [61a] as well as -chlorofluoromethylation [61b] of simple alkenes under blue LED irradiation (Scheme 2.30); the reactions were accomplished by the use of a combination of fluorine-containing carboxylic acid and iodobenzene diacetate.

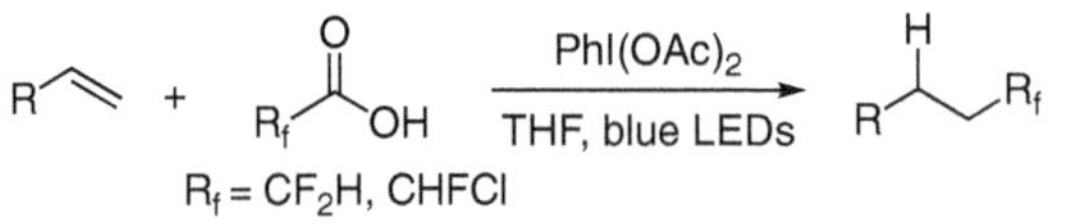

Scheme 2.30 Metal-free photochemical hydro-fluoroalkylations.

2.2.5 Other Methods

2.2.5.1 Hydro-Trifluoromethylation of Fullerene

A few exceptional reactions that do not fall into the categories described above (2.1–2.4) have been reported recently and are introduced herein. In fullerene trifluoromethylation reactions, silver trifluoroacetate was reported to show high reactivity for installing multiple trifluoromethyl groups (see Section 2.2.3) [45–48]. Very recently, Garyunkov and coworkers achieved hydro-trifluoromethylation of C60 (Scheme 2.31) [62]. In this work, C60 reacted readily with potassium or cesium trifluoroacetate in o-dichlorobenzene/benzonitrile at c.180 °C in the presence of a crown ether, affording a [C60–CF$_3$]M (M = K or Cs) intermediate. Treatment of the intermediate with acid gave the hydro-trifluoromethylation product.

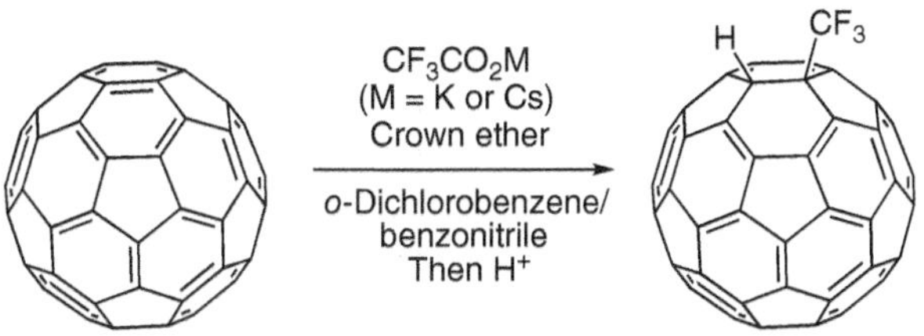

Scheme 2.31 Hydro-trifluoromethylation of fullerene.

2.2.5.2 Metal-Free Aryldifluoromethylation Using $S_2O_8^{2-}$

Persulfate ($S_2O_8^{2-}$) salts were reported to promote metal-free aryldifluoromethylations using aryldifluoroacetic acids (Scheme 2.32); in 2016, Wu and coworkers reported an aryldifluoromethylation of ethynyl benziodoloxolone with aryldifluoroacetic acids by using potassium persulfate in acetonitrile/H$_2$O cosolvent (Scheme 2.32a) [63]; in contrast to Hashmi's conditions using silver catalyst (Scheme 2.25d) [56], Wu's conditions did not require any transition metal catalyst. Zhang and coworkers developed an aryldifluoromethylation of quinoxaline-2(1H)-ones with aryldifluoroacetic acid in dimethylsulfoxide (DMSO) (Scheme 2.32b) [64]. In both studies, persulfate salts were considered to oxidize the carboxylic acids and to generate the corresponding reactive difluorobenzyl radical via oxidative decarboxylation.

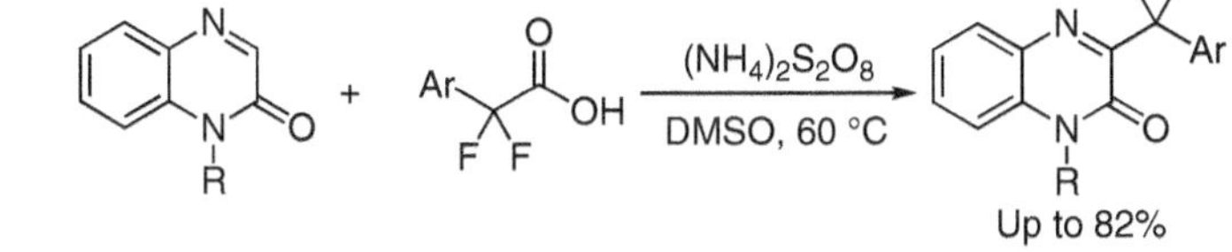

(a)

(b)

Scheme 2.32 Metal-free aryldifluoromethylation using $S_2O_8^{2-}$ as an oxidant.

2.3 Perfluoroalkylation with Perfluorocarboxylic Anhydride

Perfluorocarboxylic anhydrides are readily available and widely used perfluoroalkyl sources for organic syntheses, like the carboxylic acids. However, it was initially not easy to use the anhydrides directly as perfluoroalkyl sources, although a few secondary reagents prepared from them were reported [2]. Recently, several methods that are synthetically highly useful have been developed.

2.3.1 Reactions Using Perfluorocarboxylic Anhydride/Urea·H_2O_2

Bräse and coworkers reported a radical perfluoroalkylation of aromatic compounds by using a combination of perfluorocarboxylic anhydrides and urea-hydrogen peroxide; a mixture of 10 equiv of urea-hydrogen peroxide (urea·H_2O_2) and 20 equiv of the carboxylic anhydrides generated diacyl peroxide *in situ*, and this reacted with arene substrates (Scheme 2.33) [65]. In that work, 15 perfluoroalkylated compounds were synthesized in up to 50% yield.

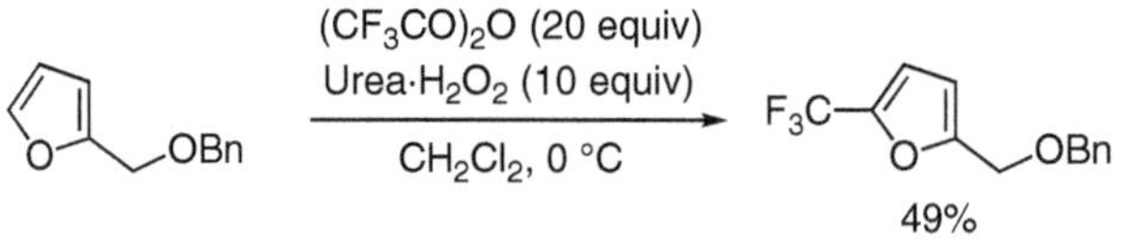

Scheme 2.33 Aromatic perfluoroalkylation using perfluorocarboxylic anhydride/urea·H_2O_2.

The group of Sodeoka and Kawamura independently developed perfluoroalkylations using diacyl peroxide generated *in situ* from anhydrides/urea·H_2O_2, and described various transformations, in particular bifunctionalization-type perfluoroalkylations of alkenes (Schemes 2.34 and 2.35) [66, 67]. Although the reactive perfluoroalkyl radical could be generated via thermal fragmentation of the diacyl

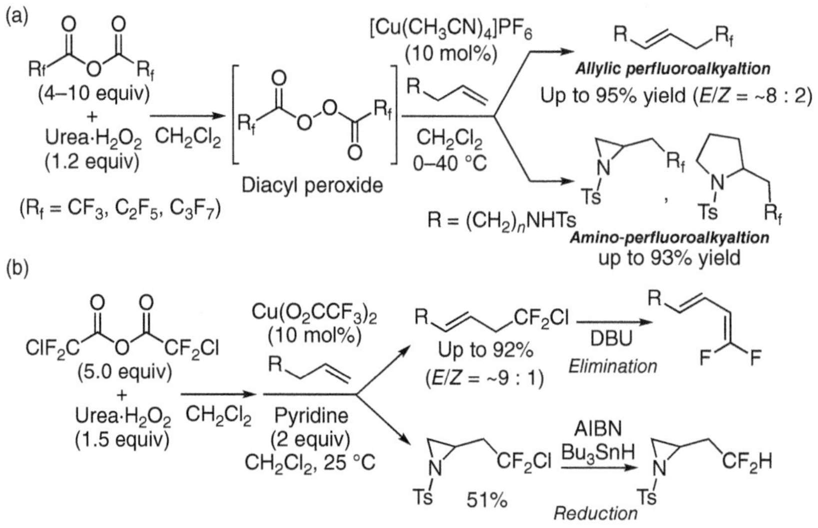

Scheme 2.34 Cu-catalyzed perfluoroalkylations of alkenes. DBU, 1,8-diazabicyclo[5.4.0]undec-7-ene; AIBN, 2,2′-azobis(isobutyronitrile).

peroxide, heating of a mixture of alkene and *in situ*-generated diacyl peroxide gave a complex mixture. They hypothesized that reactivity control of the alkyl radical formed by the reaction of alkene with perfluoroalkyl radical would be the key to a successful reaction [68a–c]. First, catalytic control using copper catalyst was demonstrated; when a catalytic amount of $[Cu(CH_3CN)_4]PF_6$ was added to a mixture of substrate and the peroxide, not only the selectivity but also the conversion was dramatically improved, affording allylic perfluoroalkylation products in up to 95% yield (Scheme 2.34a) [68a].

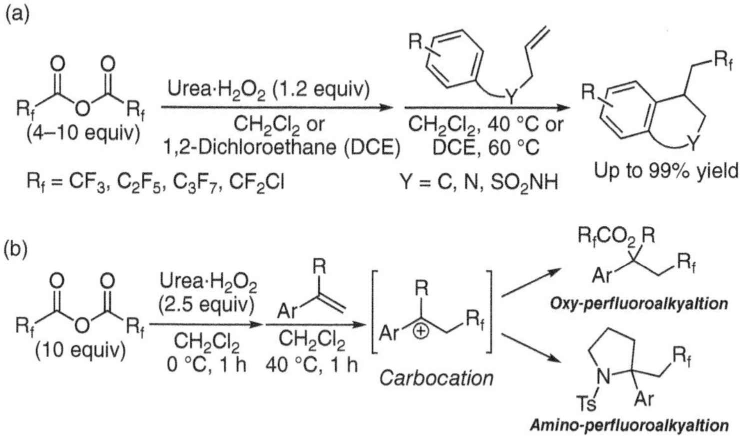

(a)

(b)

Scheme 2.35 Metal-free alkene perfluoroalkylations: (a) carbo-perfluoroalkylation of aromatic alkenes; (b) bifunctionalization-type perfluoroalkylations of styrenes.

In addition, alkenes bearing a pendant sulfonamide group efficiently gave a wide variety of intramolecular amino-perfluoroalkylation products: perfluoroalkyl group-containing aziridines and pyrrolidines [68b]. In particular, the aziridine product proved to be a good building block; it was derivatized to various amines, including indole alkaloid analogs. Furthermore, their group developed allylic and amino-chlorodifluoromethylations of alkenes, in which the use of $Cu(O_2CCF_3)_2$ as the catalyst with pyridine additive was found to improve the yield (Scheme 2.34b) [68c]. The chlorodifluoromethyl group of the products was transformed into difluorodiene, difluoromethyl-, or trimethylsilyldifluoromethyl groups in order to confirm the utility of these products as synthetic building blocks.

They also performed metal-free perfluoroalkylations by using perfluorocarboxylic anhydride/urea·H_2O_2, focusing on the structure of the substrates (Scheme 2.35). When an alkene bearing an aromatic ring at an appropriate position of the carbon side chain was reacted with *in situ*-generated diacyl peroxide, intramolecular carbo-perfluoroalkylation via radical cyclization occurred (Scheme 2.35a) [68a–c].

This method provides simple access to benzo-fused carbo- and heterocyclic products in excellent yields. Notably, switching from amino- to carbo-perfluoroalkylation of the same substrate, e.g. *N*-tosyl allylamine, by the removal of copper catalyst remarkably increased the diversity of available perfluoroalkylated molecules. Furthermore, styrene derivatives were found to undergo bifunctionalization-type perfluoroalkylation, oxy- and amino-perfluoroalkylations, via a carbocation, which is a rare intermediate under metal-free conditions (Scheme 2.35b) [68d]. Mechanistic studies suggested that the unique redox properties of the peroxide reagent, styrene, and the radical cation formed accounted for the extraordinary carbocation formation under metal-free conditions.

2.3.2 Photocatalytic Reactions Using Perfluorocarboxylic Anhydride/Pyridine *N*-oxide

Stephenson applied photochemistry to perfluoroalkylation using perfluorocarboxylic anhydrides (Scheme 2.36) [68], in which the perfluorocarboxylic anhydride-pyridine *N*-oxide adduct is generated *in situ* as a reactive intermediate [69]. The reaction was applicable to a diverse array of substrates, such as vinyl, aryl, and heteroaryl compounds, and could be run on a kilogram scale by the use of a flow system. Their conditions are also available for chlorodifluoromethylation of aromatic compounds with chlorodifluoroacetic anhydride. In addition, the reaction of alkyne was found to give *gem*-difluoroenones via oxy-chlorodifluoromethylation and subsequent elimination of chloride (Scheme 2.37).

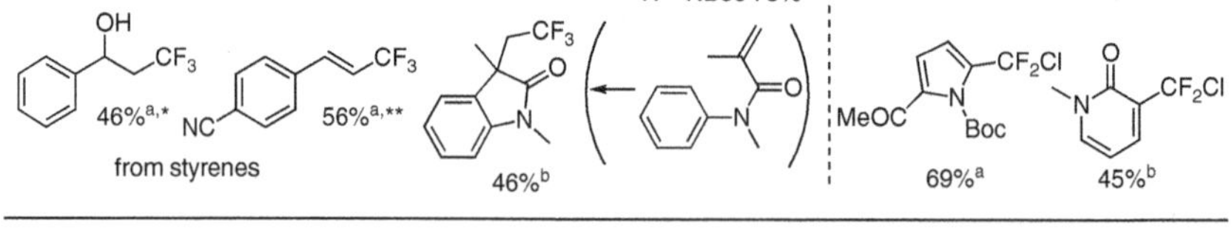

Scheme 2.36 Photocatalytic reaction using anhydrides/pyridine *N*-oxides. a. With pyridine *N*-oxide, b. With 4-Ph-pyrdine *N*-oxide, c. Stirred with MeOH on reaction completion, d. Stirred with DBU on reaction completion.

Schaub and coworkers very recently reported a photocatalytic α-trifluoromethylation of aromatic ketones by using trifluoroacetic anhydride/pyridine *N*-oxide (Scheme 2.38) [70].

The reaction proceeds via oxy-trifluoromethylation, which installs a CF_3 group and trifluoroacetate on *in situ*-formed vinyl trifluoroacetate; the resulting trifluoroacetyl-protected acetal was transformed upon workup to the trifluoromethylated ketone product.

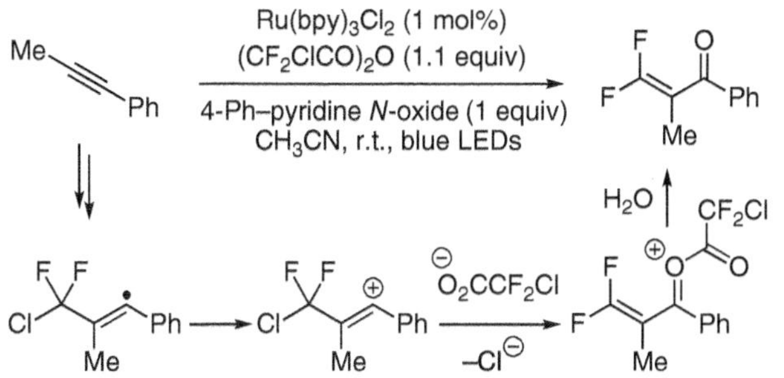

Scheme 2.37 *gem*-Difluoroenone synthesis by chlorodifluoromethylation of alkynes.

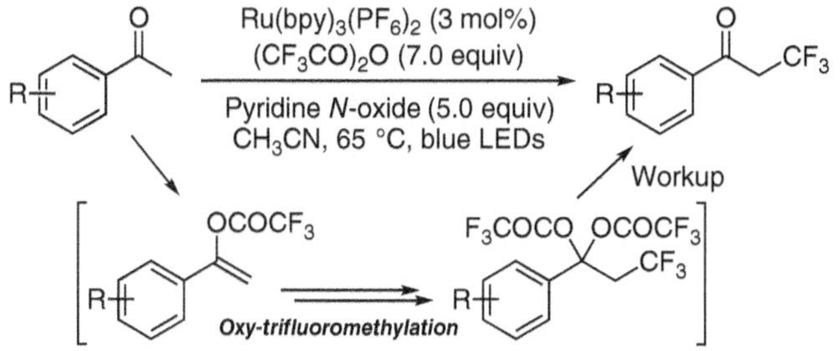

Scheme 2.38 Photocatalytic α-trifluoromethylation of aromatic ketones.

2.4 Summary and Prospects

We have reviewed developments in perfluoroalkylation reactions with perfluorocarboxylic acids and anhydrides as perfluoroalkylating reagents. Early work tended to focus on methodologies for the generation of reactive species, such as perfluoroalkyl radicals and perfluoroalkyl metal species, and their reactivities. More recent reports have dealt with precise control of the reactivity of reactive intermediates and efficient production of perfluoroalkylated molecules containing important skeletons as candidate pharmaceuticals and functional materials. Based on the ready availability of the perfluoroalkyl sources and the high synthetic utility of recently reported reactions, we consider that perfluorocarboxylic acids and anhydrides will become the first choice of perfluoroalkylating reagents for practical organic syntheses in the near future.

References

1 Recent reviews: (a) Han, J., Fustero, S., Soloshonok, V.A. et al. (2019). *Chem. Eur. J.* 25: 11797. (b) Pan, Y. (2019). *ACS Med. Chem. Lett.* 10: 1016.
2 Selected reviews: (a) Sawada, H. (1996). *Chem. Rev.* 96: 1779. (b) Zard, S. (2016). *Z. Org. Biomol. Chem.* 14: 6891. Selected examples: (c) Schareina, T., Wu, X.-F., Beller, M. et al. (2012). *Top. Catal.* 55: 426. (d) Sakamoto, R., Kashiwagi, H.,

and Maruoka, K. (2017). *Org. Lett.* 19: 5126. (e) Yang, B., Yu, D., and Qing, F.-L. (2018). *ACS Catal.* 8: 2839.

3 Selected reviews: (a) Vijh, A.K. and Conway, B.E. (1967). *Chem. Rev.* 67: 623. (b) Svadkovskaya, G.E. and Voitkevich, S.A. (1960). *Russ. Chem. Rev.* 29: 161. (c) Banks, R.E. and Tatlow, J.C. (1986). *J. Fluorine Chem.* 33: 71. For a selected book: (d) Barlow, M.G. and Taylor, D.R. Per- and poly-fluorinated olefins, dienes, heterocumulenes and acetylenes. In: *Fluorocarbon and Related Chemistry*, vol. 2, 1974 (eds. R.E. Banks and M.G. Barlow), 37–123. London, UK: Chemical Society.

4 Renaud, R.N. and Sullivan, D.E. (1972). Renaud reported cross-coupling of alkyl radicals generated by co-electrolysis of potassium trifluoroacetate in deuterated carboxylic acid as a solvent, obtaining trifluoroethane-1,1,1-d_3 and pentafluoropropane-1,1,1-d_3. *Can. J. Chem.* 50: 3084.

5 Renaud, R.N. and Sullivan, D.E. (1973). *Can. J. Chem.* 51: 772.

6 Grinberg, V.A. and Vassiliev, Y.B. (1992). Grinberg and Vassiliev examined in detail the mechanism of electrochemical perfluoroalkylation by means of kinetic and EPR studies, focusing on the electrochemical behavior and reactivity of the reactants and radical intermediates. *J. Electroanal. Chem.* 325: 167.

7 Brookes, C.J., Coe, P.L., Tatlow, J.C. et al. (1974). *J. Chem. Soc., Chem. Commun.* 3: 323.

8 Brookes, C.J., Pedler, A.E., Tatlow, J.C. et al. (1978). *J. Chem. Soc., Perkin I* 3: 202.

9 Renaud, R.N. and Champagne, P.J. (1975). *Can. J. Chem.* 53: 529.

10 Dmowski, W., Biernacki, A., Kozlowski, T. et al. (1997). *Tetrahedron* 53: 4437.

11 Renaud, R.N., Champagne, P.J., and Savard, M. (1979). *Can. J. Chem.* 57: 2617.

12 Renaud, R.N., Stephens, C.J., and Bérubé, D. (1982). *Can. J. Chem.* 60: 1687.

13 Vassiliev, Y.B., Bagotzky, V.S., Grinberg, V.A. et al. (1982). *Electrochim. Acta* 27: 919.

14 Muller, N. (1986). In this work, when acetone was used as co-solvent of the reaction, a unique bifunctionalization-type trifluoromethylation installing a methyl group derived from acetone was found to occur. *J. Org. Chem.* 51: 263.

15 Uneyama, K., Morimoto, O., and Nanbu, H. (1989). *Tetrahedron Lett.* 30: 109.

16 Uneyama, K., Makio, S., and Nanbu, H. (1989). Uneyama demonstrated derivatization of dimethyl 2,3-bis(2,2,2-trifluoroethyl)succinate prepared by electrochemical oxidation of trifluoroacetic acid and methyl acrylate under basic conditions. *J. Org. Chem.* 54: 872.

17 (a) Muller, N. (1983). *J. Org. Chem.* 48: 1370. (b) Muller, N. (1984). *J. Org. Chem.* 49: 2826. (c) Muller, N. (1984). *J. Org. Chem.* 49: 4559.

18 (a) Uneyama, K. (1991). *Tetrahedron* 47: 555. (b) Uneyama, K. (2000). *J. Fluorine Chem.* 105: 209.

19 Uneyama, K. and Ueda, K. (1988). *Chem. Lett.* 17: 853.

20 (a) Uneyama, K. and Nanbu, H. (1988). *J. Org. Chem.* 53: 4598. (b) Uneyama, K. and Watanabe, S. (1990). *J. Org. Chem.* 55: 3909. (c) Dan-oh, Y. and

Uneyama, K. (1995). *Bull. Chem. Soc. Jpn.* 68: 2993. (d) Uneyama, K., Watanabe, S., Tokunaga, Y. et al. (1992). *Bull. Chem. Soc. Jpn.* 65: 1976. (e) Sato, Y., Watanabe, S., and Uneyama, K. (1993). *Bull. Chem. Soc. Jpn.* 66: 1840.

21 Dmowski, W. and Biernacki, A. (1996). *J. Fluorine Chem.* 78: 193.

22 Krasil'nikov, A.A., Kaurova, G.I., Matalin, V.A. et al. (2009). *Russ. J. Appl. Chem.* 82: 2127.

23 (a) Andreev, V.N., Grinberg, V.A., Dedov, A.G. et al. (2013). *Russ. J. Electrochem.* 49: 996. See also, (b) Maiorova, N.A., Kagramanov, N.D., Grinberg, V.A. et al. (2013). *Russ. J. Electrochem.* 49: 181.

24 (a) Arai, K., Watts, K., and Wirth, T. (2014). *ChemistryOpen* 3: 23. See also, (b) Elsherbini, M. and Wirth, T. (2019). *Acc. Chem. Res.* 52: 3287.

25 (a) Grinberg, V.A., Polishchuk, V.R., German, L.S. et al. (1978). *Bull. Acad. Sci. USSR, Div. Chem. Sci.* 27: 580. (b) Grinberg, V.A., Lundgren, S.A., Sterlin, S.R. et al. (1997). *Russ. Chem. Bull.* 46: 1131.

26 Depecker, C., Trevin, S., Devynck, J. et al. (1999). *New J. Chem.* 23: 739.

27 Eisenberg, M. and DesMarteau, D.D. (1970). *Inorg. Nucl. Chem. Lett.* 6: 29.

28 Gregorčič, A. and Zupan, M. (1979). *J. Org. Chem.* 44: 4120.

29 Tanabe, Y., Matsuo, N., and Ohno, N. (1988). *J. Org. Chem.* 53: 4582.

30 Popkov, S.V. and Kuzenkov, A.V. (2005). *Russ. Chem. Bull.* 54: 1672.

31 Matsui, K., Tobita, E., Kondo, K. et al. (1981). *Chem. Lett.* 10: 1719.

32 Lin, R.W. and Davidson, R.I. (1989). Improved conditions using potassium trifluoroacetate were reported. Trifluoromethylation process. US Patent 4808748, filed 12 December 1985 and issued 28 February 1989.

33 Suzuki, H., Yoshida, Y., and Osuka, A. (1982). *Chem. Lett.* 11: 135.

34 Carr, G.E., Chambers, R.D., Holmes, T.F. et al. (1988). *J. Chem. Soc., Perkin Trans. 1* 4: 921.

35 Rijs, N.J. and O'Hair, R.A. (2012). O'Hair discussed the decomposition pathways of metal trifluoroacetates. *J. Dalton Trans.* 41: 3395.

36 Markovich, K.M., Tantishaiyakul, V., Miller, D.D. et al. (1992). *J. Med. Chem.* 35: 466.

37 Dong, L.C., Crowe, M., Ammann, J.R. et al. (2004). *Tetrahedron Lett.* 45: 2731.

38 Chen, M. and Buchwald, S.L. (2013). *Angew. Chem. Int. Ed.* 52: 11628.

39 McReynolds, K.A., Lewis, R.S., Vicic, D.A. et al. (2010). *J. Fluorine Chem.* 131: 1108.

40 (a) Lin, X., Hou, C., Weng, Z. et al. (2016). *Chem. Eur. J.* 22: 2075. (b) Wu, C., Huang, Y., and Weng, Z. (2016). *Asian J. Org. Chem.* 5: 1406. (c) Lin, X., Han, X., Weng, Z. et al. (2016). *RSC Adv.* 6: 75465.

41 Zhao, G., Wu, H., Liu, C. et al. (2016). *RSC Adv.* 6: 50250.

42 Li, Y., Chen, T., Duan, C. et al. (2011). *Synlett* 12: 1713.

43 (a) Chang, Y. and Cai, C. (2005). *Tetrahedron Lett.* 46: 3161. (b) Chang, Y. and Cai, C. (2005). *J. Fluorine Chem.* 126: 937.

44 (a) Boltalina, O.V., Popov, A.A., and Stauss, S.H. (2015). *Chem. Rev.* 115: 1051. See also, (b) Pimenova, A.S., Kozlov, A.A., Sidorov, L.N. et al. (2007). *Dalton Trans.*: 5322.

45 Uzkikh, I.S., Dorozhkin, E.I., Boltalina, O.V. et al. (2001). *Dokl. Akad. Nauk* 379: 344.

46 (a) Goryunkov, A.A., Kuvychko, I.V., Boltalina, O.V. et al. (2003). *J. Fluorine Chem.* 124: 61. (b) Dorozhkin, E.I., Strauss, S.H., Boltalina, O.V. et al. (2006). *Chem. Eur. J.* 12: 3876.

47 (a) Darwish, A.D., Avent, A.G., Taylor, R. et al. (2003). *Chem. Commun.* 3: 1374. (b) Darwish, A.D., Abdul-Sada, A.K., Taylor, R. et al. (2003). *Org. Biomol. Chem.* 1: 3102.

48 Dorozhkin, E.I., Goryunkov, A.A., Ioffe, I.N. et al. (2007). *Eur. J. Org. Chem.* 2007: 5082.

49 Shi, G., Shao, C., Zhang, Y. et al. (2015). *Org. Lett.* 17: 38.

50 Tung, T.T., Christensen, S.B., and Nielsen, J. (2017). *Chem. Eur. J.* 23: 18125.

51 Wan, W., Ma, G., Hao, J. et al. (2016). *Chem. Commun.* 52: 1598.

52 Wang, Y., Hao, J., Wan, W. et al. (2019). Wan and Han also reported a decarboxylative homo-coupling from potassium difluoroarylacetate by using the combination of AgNO$_3$, (NH$_4$)$_2$S$_2$O$_8$, and KHCO$_3$, affording tetrafluoroethylene-bridging aromatic products. *Synth. Commun.* 49: 2961.

53 Wan, W., Li, J., Hao, J. et al. (2017). *Org. Biomol. Chem.* 15: 5308.

54 Li, Y.-L., Wang, J.-B., and Deng, J. (2017). *Eur. J. Org. Chem.* 2017: 6052.

55 Chen, F. and Hashmi, A.S.K. (2016). *Org. Lett.* 18: 2880.

56 Lai, C. and Mallouk, T.E. (1993). *J. Chem. Soc., Chem. Commun.*: 1359.

57 Lin, J., Su, W., Li, Y. et al. (2017). *Nat. Commun.* 8: 14353.

58 Bazyar, Z. and Hosseini-Sarvari, M. (2019). *Org. Process Res. Dev.* 23: 2345.

59 Yang, B., Xu, X.H., and Qing, F.L. (2016). *Org. Lett.* 18: 5956.

60 Zhou, Y., Xiong, Z., Zhu, G. et al. (2019). *Org. Chem. Front.* 6: 1022.

61 (a) Meyer, C.F., Hell, S.M., Gouverneur, V. et al. (2019). *Angew. Chem. Int. Ed.* 58: 8829. (b) Meyer, C.F., Hell, S.M., Gouverneur, V. et al. (2019). *Tetrahedron* 75: 130679.

62 Bogdanov, V.P., Dmitrieva, V.A., Goryunkov, A.A. et al. (2019). *J. Fluorine Chem.* 226: 109344.

63 Yang, F., Wu, Y., Wu, Y. et al. (2016). *Adv. Synth. Catal.* 358: 1699.

64 Hong, G., Yuan, J., Zhang, X. et al. (2019). *Org. Chem. Front.* 6: 1173.

65 Zhong, S., Hafner, A., Bräse, S. et al. (2015). *RSC Adv.* 5: 6255.

66 Kawamura, S. and Sodeoka, M. (2019). *Bull. Chem. Soc. Jpn.* 92: 1245.

67 (a) Kawamura, S. and Sodeoka, M. (2016). *Angew. Chem. Int. Ed.* 55: 8740. (b) Kawamura, S., Dosei, K., Sodeoka, M. et al. (2017). *J. Org. Chem.* 82: 12539. (c) Kawamura, S., Henderson, C.J., Sodeoka, M. et al. (2018). *Chem. Commun.* 54: 11276. (d) Valverde, E., Kawamura, S., Sodeoka, M. et al. (2018). *Chem. Sci.* 9: 7115.

68 Staveness, D., Bosque, I., and Stephenson, C.R.J. (2016). *Acc. Chem. Res.* 49: 2295.

69 (a) Beatty, J.W., Douglas, J.J., Stephenson, C.R.J. et al. (2015). *Nat. Commun.* 6: 7919. (b) Beatty, J.W., Douglas, J.J., Stephenson, C.R.J. et al. (2016). *Chem* 1: 456.

(c) McAtee, R.C., Beatty, J.W., Stephenson, C.R.J. et al. (2018). *Org. Lett.* 20: 3491. See also, (d) Sun, A., McClain, E.J., Stephenson, C.R.J. et al. (2018). *Org. Lett.* 20: 3487.

70 Das, S., Hashmi, A.S.K., and Schaub, T. (2019). *Adv. Synth. Catal.* 361: 720.

3

Chemistry of OCF₃, SCF₃, and SeCF₃ Functional Groups

Fabien Toulgoat[1,2], François Liger[3] and Thierry Billard[1,3]

[1]CNRS-University of Lyon, Institute of Chemistry and Biochemistry (ICBMS), 43 Bd du 11 novembre 1918, 69622 Lyon, France
[2]CPE – Lyon, 43 Bd du 11 novembre 1918, 69616 Villeurbanne, France
[3]CERMEP – In vivo imaging, Groupement Hospitalier Est, 59 Bd Pinel, 69677 Bron, France

3.1 Introduction

Fluorinated compounds have during the last decades received considerable attention due to their specific properties, opening the way to new applications from materials to life sciences. This fascinating interest has contributed to the development of new emerging groups with new characteristics. Merging the trifluoromethyl group with chalcogens fulfills this objective since this association led to new substituents with high lipophilicity.

However, despite such an interest the synthesis of molecules bearing these groups remains challenging. In this chapter, the diverse methods and reagents used to perform CF_3X chemistry will be described. A classification by reagents has been selected with, at the top of each part, a report of reactivity types.

3.2 CF₃O Chemistry

With a Hansch–Leo lipophilicity parameter $\pi_R = 1.04$ [1], specific electronic and conformational properties [2], and high stability [3], the CF_3O group has fascinated fluorine chemists since several years. However, despite this high interest the synthesis of CF_3O molecules remains one of the highest challenges in fluorine chemistry [4]. Three strategies can be envisaged to achieve CF_3O molecules and in the following parts, the more efficient and more recent approaches will be presented (more specific methods could be found in recent reviews [4]).

3.2.1 De Novo Construction

3.2.1.1 Trifluorination of Alcohol Derivatives

Since the first synthesis of trifluoromethoxy-arene compounds in 1955 through a halogen exchange between trichloromethoxy aromatic derivatives and anhydrous

Organofluorine Chemistry: Synthesis, Modeling, and Applications,
First Edition. Edited by Kálmán J. Szabó and Nicklas Selander.
© 2021 WILEY-VCH GmbH. Published 2021 by WILEY-VCH GmbH.

HF or SbF₃/SbCl₅, several methods using phenol derivatives and various fluorinating reagents have been described. However, from a practical point of view, only a few methods are useful. In the aromatic series, fluorination of trichloromethoxy-arenes or -heteroarenes and (Het)aryl dithiocarbonates appeared to be the favored approaches (Scheme 3.1) [5]. In the aliphatic series, only dithiocarbonate fluorination gave rise to the expected products [5e–h, 6]. It is noteworthy that this method is clearly more efficient with primary alcohol derivatives (Scheme 3.1).

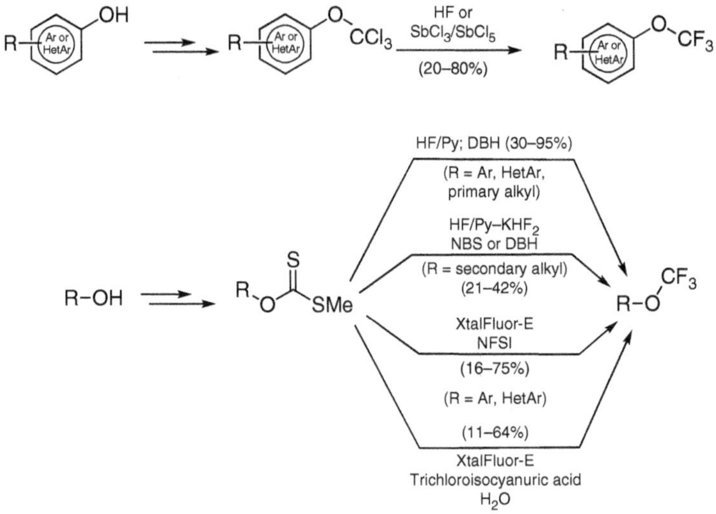

Scheme 3.1 Construction of OCF₃ group.

3.2.1.2 Fluorination of Difluorinated Compounds

Difluoroether derivatives are often easier to obtain because of the availability of several difluorinated reagents that can react with alcohols.

Thus, fluorination of bromodifluoroaryl ethers [7] and fluorodecarboxylation of difluoroaryloxy acids [8] have been performed to obtain the corresponding trifluoromethoxy arenes (Scheme 3.2).

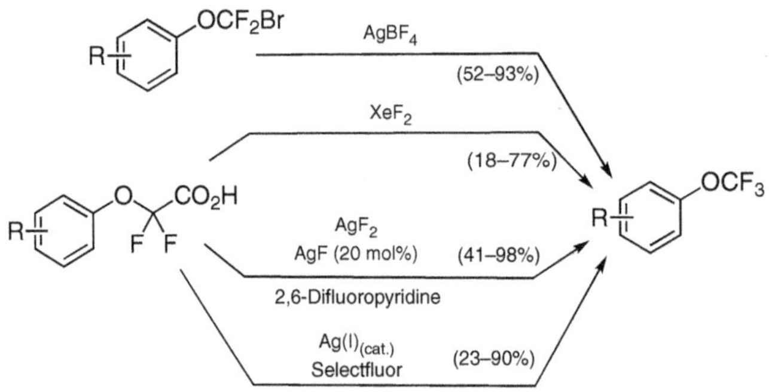

Scheme 3.2 Fluorination of difluorinated aryl ethers.

3.2.2 Indirect Methods

The second strategy, which has been considered to synthesize CF$_3$O molecules, is the trifluoromethylation of alcohols or hydroxylated derivatives.

3.2.2.1 *O*-(Trifluoromethyl)dibenzofuranium Salts

Reactivity	Nucleophilic	Radical	Electrophilic
CF$_3$O-(dibenzofuran)	☒	☑	☑

To realize the electrophilic O-trifluoromethylation of alcohols, Umemoto designed an *O*-(trifluoromethyl)dibenzofuranium salt. However, this reagent is not stable and must be prepared *in situ* at very low temperature (Scheme 3.3) [9].

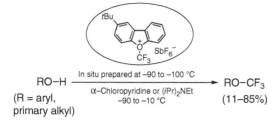

Scheme 3.3 Electrophilic trifluoromethylation of alcohols with Umemoto reagents.

3.2.2.2 Hypervalent Iodine Trifluoromethylation Reagents

Reactivity	Nucleophilic	Radical	Electrophilic
CF$_3$–iodine(III)	☒	☑	☑

Another well-known family of reagents used to perform electrophilic trifluoromethylation has been developed by Togni, based on hypervalent iodine-(III) derivatives. In specific conditions, these reagents are able to trifluoromethylate aliphatic alcohols to lead to the corresponding CF$_3$O compounds (Scheme 3.4) [10]. In aromatic series, only α-hydroxy *N*-heteroaromatics have been successfully trifluoromethylated (Scheme 3.4) [11].

These reagents have been also used to trifluoromethylate sulfonic acids [12], phosphates [13], and *N*-aryl hydroxylamines (Scheme 3.4) [14].

3.2.2.3 CF$_3$SiMe$_3$

Reactivity	Nucleophilic	Radical	Electrophilic
CF$_3$SiMe$_3$	☒	☒	☑

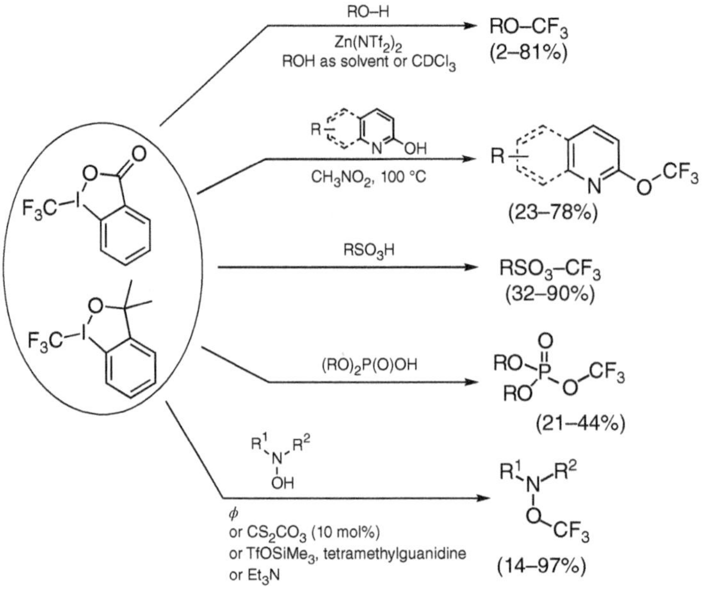

Scheme 3.4 Electrophilic trifluoromethylation of OH with Togni reagents.

Although initially developed as a nucleophilic trifluoromethylating reagent, this reagent has been successfully used under oxidative conditions to trifluoromethylate alcohols. This reaction requires a silver salt and various oxidants, fluoride anion sources, and some other additives to circumvent undesirable side reactions. Despite this complex reactive mixture, interesting results are generally observed even with some elaborated molecules (Scheme 3.5) [15].

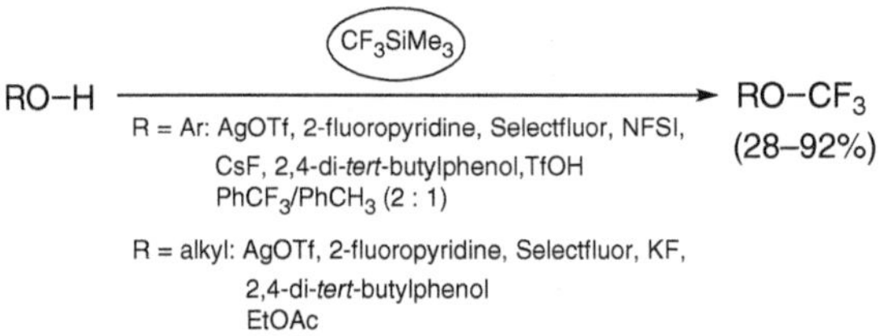

Scheme 3.5 Oxidative trifluoromethylation of alcohols with CF₃SiMe₃.

3.2.3 Direct Trifluoromethoxylation

Among the different methods to prepare trifluoromethyl ethers, direct trifluoromethoxylation (i.e. formation of the C—OCF₃ bond) appeared to be a promising alternative to trifluoromethylation of alcohols or derivatives, or de novo synthesis. Therefore, a series of reagents have been developed to perform direct trifluoromethoxylation reactions through different pathways.

3.2.3.1 Difluorophosgene and Derivatives

Reactivity	Nucleophilic	Radical	Electrophilic
$O-CF_2$	☑	☒	☒

Historically, difluorophosgene represents one of the first reagents used to perform trifluoromethoxylation, although it is a gaseous and toxic compound. Actually, an equilibrium (Scheme 3.6) exists between difluorophosgene and fluoride anion on one hand and trifluoromethoxide anion on the other [16]. Taking advantage of such equilibrium, a few trifluoromethoxide salts have been prepared [17]. In addition, a few examples of nucleophilic substitution involving these salts preformed *in situ* [18], or isolated (see Section 3.2.3.4), have been reported. Similarly, a procedure involving triphosgene and potassium fluoride was reported for the preparation of *ex situ* difluorophosgene [19] and its use as precursor of trifluoromethylbenzoate (TFBz), a new trifluoromethoxylating reagent (see Section 3.2.3.6) [19b].

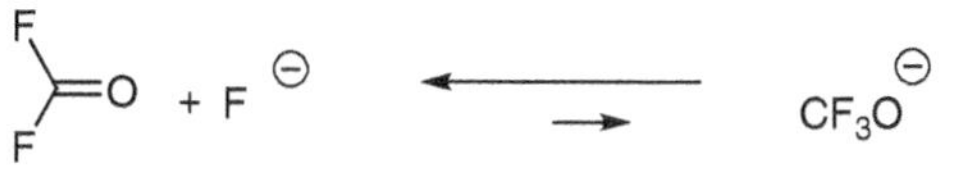

Scheme 3.6 Equilibrium between difluorophosgene/fluoride and trifluoromethoxide anion.

3.2.3.2 Trifluoromethyl Hypofluorite and Derivatives

Reactivity	Nucleophilic	Radical	Electrophilic
CF_3OX	☑	☑	☒

Trifluoromethyl hypofluorite (CF_3OF) can also be considered among the earliest reagents available to perform direct trifluoromethoxylation. However, its toxicity and potential explosiveness considerably restrict its use [4c, 20]. Nevertheless, it has been used to introduce the trifluoromethoxy moiety into not only alkenes and arenes but also carbohydrates or aziridines [4c, 20b]. From a mechanistic point of view, radical pathways, as well as concerted electrophilic fluorination–trifluoromethoxylation can occur. In addition to trifluoromethyl hypofluorite, analogs such as trifluoromethyl hypochlorite (CF_3OCl) or bis(trifluoromethyl)peroxide (CF_3OOCF_3) have seldom been used [4c].

3.2.3.3 Trifluoromethyl Triflate (TFMT)

Reactivity	Nucleophilic	Radical	Electrophilic
$CF_3SO_2OCF_3$	☑	☒	☒

Trifluoromethyl triflate (TFMT) is a commercially available reagent that can be considered as a source of trifluoromethoxide anion [21]. Indeed, its reaction with various nucleophiles [22], especially with fluoride salts, furnished trifluoromethoxide salts that can be used as trifluoromethoxylating reagents. Moreover, this anion could be involved in trifluoromethoxylation reaction without prior isolation (Scheme 3.7).

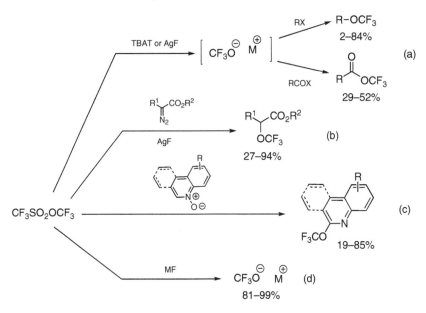

Scheme 3.7 TFMT as nucleophilic trifluoromethoxylating reagent.

Indeed, activation of TFMT with silver fluoride or tetrabutylammonium triphenyldifluorosilicate (TBAT), followed by reaction of the resulting trifluoromethoxide salts with alkyl halides or benzoyl halides [23], allows preparing the corresponding trifluoromethyl ethers (Scheme 3.7a). These reactions appeared to be limited to primary or activated secondary alkyl iodides or bromides and extension to alkyl chlorides appeared difficult as only benzyl chlorides were converted in modest yield [23]. Nevertheless, this approach was applied in the preparation of α-trifluoromethoxy 1-(2-furyl)ethanone, up to 50 g scale [24]. Similarly, cesium fluoride was used to activate TFMT during the preparation of α-trifluoromethoxyacetophenone [25]. Benzoyl bromide and allyl chloroformate were also converted into their corresponding trifluoromethoxy derivatives [23]. Trifluoromethoxylation of α-diazoesters [26] was also performed using TFMT as trifluoromethoxide anion source through a nucleophilic process (Scheme 3.7b). It should be noted that trifluoromethoxylation of α-diazo vinyl esters occurred on the γ position with C—C double bond migration. Interestingly, preformation of the trifluoromethoxide anion was not necessary as better yields were achieved when TFMT was added to a solution containing α-diazoesters and silver fluoride. In addition to fluoride anion, silver is also required as almost no product was observed without any silver salts.

An alternative strategy was developed using heterocyclic *N*-oxides as activators as well as electrophiles. In such a process, these substrates are able to activate TFMT, leading to more electrophilic heterocycles (*N*-triflate cations) as intermediates, which react further with the trifluoromethoxide generated during the activation step [27]. Quinolines and phenanthridines were thus trifluoromethoxylated with moderate to good yields (Scheme 3.7c).

In addition, TFMT could be considered as a precursor of trifluoromethoxide salts (Scheme 3.7d) [16, 28], which could themselves be used as trifluoromethoxylating reagents.

3.2.3.4 Trifluoromethoxide Salts Derived from TFMT or Difluorophosgene

Reactivity	Nucleophilic	Radical	Electrophilic
CF₃OM	☑	☒	☒

As mentioned earlier, TFMT and difluorophosgene were used to prepare trifluoromethoxide salts. However, one of the main difficulties in handling trifluoromethoxide salts is their low stability [16] due to the equilibrium between the trifluoromethoxide anion and gaseous difluorophosgene and fluoride anion (Scheme 3.6). Thus, only some salts could be isolated in solid state, for example, cesium or tris(dimethylamino)-sulfonium trifluoromethoxide (TAS·OCF₃) [28a, b, d]. Nevertheless, these "isolated" salts remain highly sensitive. Concerning silver trifluoromethoxide, one of the most popular salt claimed as trifluoromethoxylating reagent, it should be noted that it was isolated in solution in acetonitrile, stored at low temperature (−20 °C) in darkness in the procedures reported [16, 28c, e]. Actually, such salts are usually prepared just before their use as trifluoromethoxylating reagent. Finally, it should be noticed that except for the trifluoromethoxylations of some alkyl halides [17c, 29], methods described below are based on use of trifluoromethoxide salts prepared from TFMT.

Trifluoromethoxide salts were found to be able to convert alkyl bromides, iodides, or triflates into the corresponding trifluoromethyl ethers in moderate to good yields [17c, 28a, 29]. Alternatively, benzyl and alkyl alcohols have been used as starting material in dehydroxytrifluoromethoxylation (Scheme 3.8) [30]. In such reaction, the developed system based on triphenylphosphine, 1,2-diiodoethane in DMF activates the alcohols, to perform rapid trifluoromethoxylation within 15 minutes.

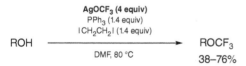

Scheme 3.8 Dehydroxytrifluoromethoxylation of alcohols.

While nucleophilic aromatic substitution failed (only fluoroaryls were observed), two examples of trifluoromethoxylation of arynes were reported (Scheme 3.9a) [28a]. However, a mixture of fluoro and trifluoromethoxy products or a mixture of regioisomers was obtained. A more convenient access to aryl trifluoromethyl ethers was achieved thanks to a silver-mediated trifluoromethoxylation of aryl stannanes or arylboronic acids using tris(dimethylamino)-sulfonium trifluoromethoxide

(TAS·OCF$_3$) as trifluoromethoxylating reagent (Scheme 3.9b,c) [28b]. While aryl stannanes can be converted into the desired ethers in one step, sequential steps are required when aryl boronic acids are used as starting material. These latter ones have to be transformed into their corresponding aryl silver complexes prior to the trifluoromethoxylation step. The reaction proved to be efficient for a broad range of substituted arenes with the exception of basic substituents such as amines or pyridines. In addition, aryl trifluoromethyl ethers can be synthesized via Sandmeyer-type reactions involving aryl and heteroaryl diazonium salts and AgOCF$_3$ in excess as trifluoromethoxylating reagent (Scheme 3.9d) [31]. Best yields were obtained for substrates bearing an electron-withdrawing group at the meta position. An S$_{RN}$1 mechanism was proposed for this additive-free trifluoromethoxylation. Of note, a combination of silver fluoride and TFMT could be used in such transformation.

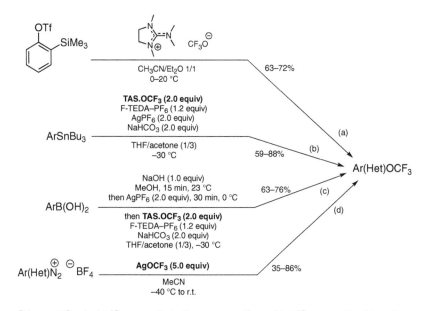

Scheme 3.9 Aryl trifluoromethyl ethers preparation with trifluoromethoxide salts.

To circumvent the low-stability issues of the trifluoromethoxide anion, Liu and coworkers developed high-valent palladium-catalyzed reactions. Indeed, with such high-valent intermediates, the reductive elimination is more favored than β-fluoride elimination. Such strategy has allowed developing intramolecular amino trifluoromethoxylation of unactivated alkenes to prepare trifluoromethoxylated piperidines via reductive elimination of a Pd^{IV}OCF$_3$ intermediate (Scheme 3.10a) [28c]. Initially developed as achiral synthesis, an enantioselective version was reported using the more stable CsOCF$_3$ instead of AgOCF$_3$ as trifluoromethoxylating reagent [32]. This replacement was crucial in that context as the ee values were 9% and 79% with AgOCF$_3$ and CsOCF$_3$, respectively. These ee values were finally increased thanks to the design of a specific Pybox ligand bearing a sterically bulky ortho substituent. The intermolecular di-trifluoromethoxylation of unactivated alkenes is another reaction developed thanks to Pd(IV) intermediates

Figure 3.1 Well-defined silver trifluoromethoxide complexes.

With Ar = 2,6-*i*Pr-C₆H₃

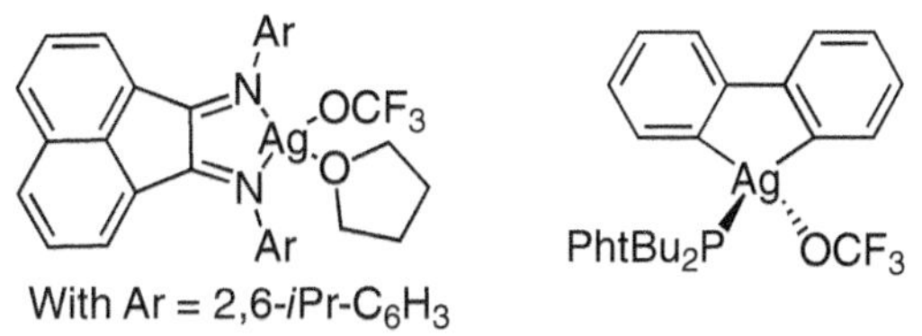

(Scheme 3.10b) [28e]. However, in addition to the di-trifluoromethoxylated product, a Wacker-type product was observed as side product in that reaction.

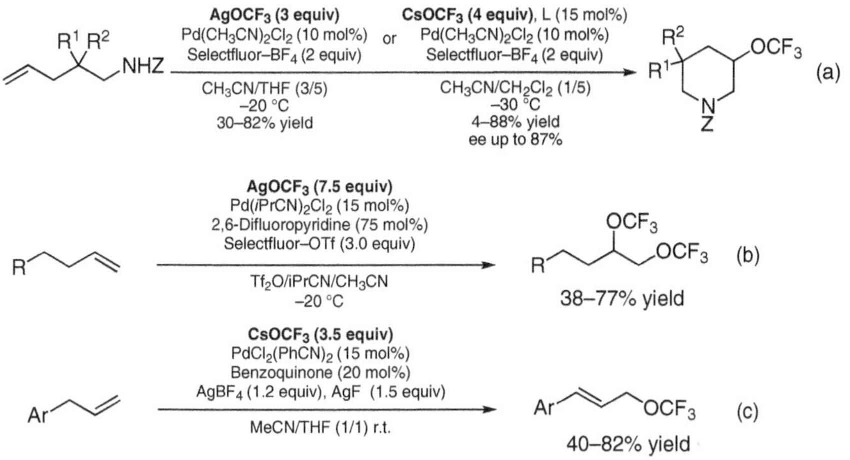

Scheme 3.10 Palladium-catalyzed trifluoromethoxylations of alkenes.

In a different strategy, AgOCF₃ was generated slowly from the more stable and less soluble CsOCF₃ to perform a catalytic allylic C–H trifluoromethoxylation of alkenes through a Pd⁰/Pdᴵᴵ catalytic cycle at room temperature (Scheme 3.10c) [28d]. Silver fluoride was used here as additive to the combination of AgBF₄ and CsOCF₃ to slow down decomposition of the trifluoromethoxide anion, and addition of 2,4,6-trimethylbenzoic acid accelerates the key allylic C–H activation step. Thus, a series of (*E*)-allylic trifluoromethyl ethers were prepared in moderate to good yields.

In addition to these salts, well-defined silver trifluoromethoxide complexes were prepared starting from TFMT as the trifluoromethoxide anion source (Figure 3.1) [33]. These isolated complexes were able to perform trifluoromethoxylation of benzyl bromide derivatives [33a] and secondary alkyl nosylates [33b] without any additives. Interestingly, such reaction was found to be stereoselective, given access to enantioenriched trifluoromethyl ethers [33b].

3.2.3.5 Trifluoromethyl Arylsulfonates (TFMSs)

Reactivity	Nucleophilic	Radical	Electrophilic
ArSO₂OCF₃	☑	☑	☒

More recently, a new class of trifluoromethoxylating reagents has been developed by P. Tang and coworkers: trifluoromethyl arylsulfonates (TFMSs) [12b]. Actually, TFMSs present a structural analogy with TFMT, which allows the liberation of the trifluoromethoxide anion thanks to S—O bond cleavage. In addition, the trifluoromethyl group of TFMT is replaced by a heavier and less electron-withdrawing aryl group in TFMS, which can be a way to decrease their volatility and to modulate their reactivity, and thus to facilitate their handling. Their synthesis (Scheme 3.11) [12b] is based on the key electrophilic trifluoromethylation of sulfonic acids (see Section 3.2.2.2) with hypervalent iodine reagents (namely Togni's reagents) [12].

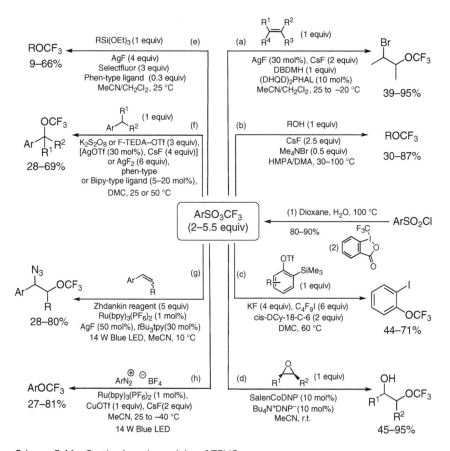

Scheme 3.11 Synthesis and reactivity of TFMS.

TFMS can act as a nucleophilic trifluoromethoxylating reagent to perform a series of reactions. The reaction initially developed (Scheme 3.11a) was an asymmetric silver-catalyzed bromo trifluoromethoxylation of alkenes [12b]. In such process, an electrophilic bromine source activates the alkenes to form a reactive bromonium intermediate that could trap the trifluoromethoxide anion released in parallel by the reaction of TFMS with cesium fluoride. Best results in terms of regioselectivity were achieved with trisubstituted alkenes or benzylic alkenes. In addition, dimeric cinchona alkaloid (DHQD)₂PHAL was used as chiral ligand to obtain enantioenriched vicinal bromo trifluoromethoxylated compounds. Following this first report, several reactions still based on TFMS activation by fluoride anion have been developed to introduce the OCF₃ moiety. In that context, dehydroxytrifluoromethoxylation of alcohols (Scheme 3.11b) furnished the corresponding trifluoromethyl ethers through the intermediate formation of fluoroformates [34]. Indeed, the latter ones are the products of the reaction of alcohols with difluorophosgene generated by degradation of part of the trifluoromethoxide anion, itself released by TFMS. Then, the trifluoromethoxide anion reacts with fluoroformates, a good leaving group, to form the desired trifluoromethyl ethers. Primary benzylic, propargylic, allylic, as well as some secondary alcohols were thus transformed in moderate to good yields. Inspired by previous reports about aryne chemistry [19b, 35], a simultaneous activation of TFMS and 2-(trimethylsilane)phenyl trifluoromethane sulfonate by potassium fluoride in the presence of crown ethers, and an iodide donor to trap the intermediate aryl anion, led to the formation of *o*-iodoaryl trifluoromethyl ethers (Scheme 3.11c) [36]. Nevertheless, the fluoride requirement could be considered as a limitation. For example, ring opening of epoxides could occur in the presence of fluoride anions. Therefore, TFMS activation by the axial anions ArO⁻ of cobalt complexes SalenCoOAr was developed in order to perform trifluoromethoxylation of epoxides (Scheme 3.11d) [37]. With the optimized conditions in hand, *meso*-epoxides as well as racemic epoxides were converted into vicinal trifluoromethoxyhydrins with good yields and regioselectivity. Moreover, one example of stereoselective trifluoromethoxylation of epoxide was reported as well as one example of an enantioselective transformation. Nevertheless, it could be noticed that a limitation appeared with styrene epoxide in terms of regioselectivity.

In addition to these nucleophilic trifluoromethoxylations, radical processes can be involved in reactions based on TFMS as the trifluoromethoxylating reagent. Such mechanism was proposed to describe the silver-mediated oxidative trifluoromethoxylation of alkylsilanes by TFMS (Scheme 3.11e) [38]. To perform such reactions, Selectfluor was chosen as oxidant while silver fluoride was the mediator as well as fluoride source. Although quite good yields were obtained with primary alkylsilanes, yields remained low with secondary ones. TFMS was also involved in a silver-promoted oxidative benzylic C–H trifluoromethoxylation (Scheme 3.11f) [39]. The TFMS reagent was proposed to react with an *in situ* formed Ag^{II}–F_2 complex, which is thereafter converted into a new F–Ag^{II}–OCF₃ complex. This complex is then involved in a reaction with a benzylic radical to form the desired trifluoromethyl ethers. After vicinal bromo-trifluoromethoxy compounds [12b], vicinal trifluoromethoxyhydrins [37], Tang and coworkers [40] succeeded in preparing vicinal azido-trifluoromethoxy compounds through a combination of silver catalysis and ruthenium photoredox catalysis (Scheme 3.11g). As in the previous methods, silver fluoride activates TFMS to generate a trifluoromethoxide silver complex. In parallel, a ruthenium complex is excited by visible light irradiation to form an active species, which is involved in a catalytic cycle where an azide radical is formed intermediately and a benzyl carbocation is liberated as a product of this first cycle. Finally, this carbocation reacts with the trifluoromethoxide silver complex to furnish the desired vicinal azido-trifluoromethoxy compounds. A dual activation was also involved to perform the trifluoromethoxylation of diazonium salts with TFMS as the trifluoromethoxylating reagent (Scheme 3.11h) [41]. Here, cesium fluoride activates TFMS while a catalyzed photoredox reaction generates an aryl radical that combined during a copper-mediated trifluoromethoxylation.

3.2.3.6 Trifluoromethylbenzoate (TFBz)

Reactivity	Nucleophilic	Radical	Electrophilic
TFBz	☑	☒	☒

By analogy with the S—O bond of the sulfonic ester function of TFMT or TFMS, the C—O bond of the ester function of TFBz can be cleaved to generate the trifluoromethoxide anion. In that context, preparation of TFBz has been developed via a specific procedure involving *ex situ* formation of difluorophosgene from triphosgene, followed by, in the presence of potassium fluoride and crown ethers, formation of a trifluoromethoxide salt, which is finally trapped by benzoyl bromide to furnish the desired TFBz (Scheme 3.12a) [19b]. By analogy to trifluoromethylation and trifluoromethylthiolation of arynes [35], the efficiency of TFBz as trifluoromethoxylating reagent has been proved thanks to the synthesis of a series of *ortho*-halogenated aryl trifluoromethyl ethers (Scheme 3.12b), involving aryne intermediates [19b]. Moreover, some trifluoromethoxylation methods developed with TFMT or trifluoromethoxide salts were applied to TFBz. Thus, trifluoromethoxylations of an alkyl iodide, a benzyl bromide, α-bromoacetyl derivatives, or an arylstannane were described. Bromo trifluoromethoxylation of an alkene as well as trifluoromethoxylation–protonation of an aryne were also performed [19b]. TFBz has also been proved to be a possible precursor of CsOCF$_3$ or AgOCF$_3$ salts (Scheme 3.12c–f).

3.2.3.7 2,4-Dinitro(trifluoromethoxy)benzene (DNTFB)

Reactivity	Nucleophilic	Radical	Electrophilic
DNTFB	☑	☒	☒

2,4-Dinitro(trifluoromethoxy)benzene (DNTFB) is an alternative commercial reagent to introduce OCF$_3$ group. DNTFB is cheaper and easier to handle than TFMT, which is volatile (boiling point [b.p.] = 21 °C) and highly sensitive to any nucleophiles [42]. Similarly to esters and sulfonic esters such as TFMT, TFMS, or TBz, DNTFB is activated by fluoride to generate trifluoromethoxide salts but in contrast to previous reagents, an S$_N$Ar mechanism was implied here. Unfortunately, the scope of DNTFB as trifluoromethoxylating reagent seems limited as trifluoromethoxylation of only benzylic or allylic bromide occurs in 45–70% yields (Scheme 3.13) [42]. Only small amounts (<10%) of trifluoromethoxylated

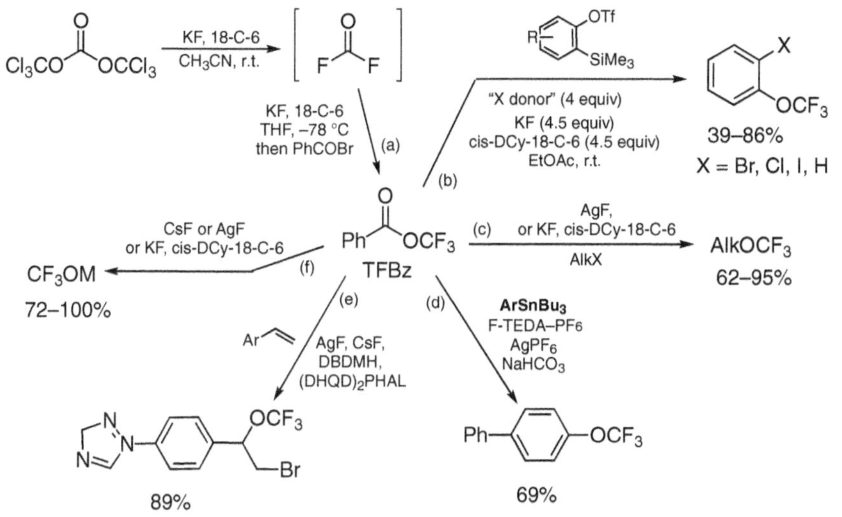

Scheme 3.12 Preparation and reactivity of trifluoromethylbenzoate (TFBz).

products were observed with an unactivated alkyl iodide or an α-iodocarbonyl compound.

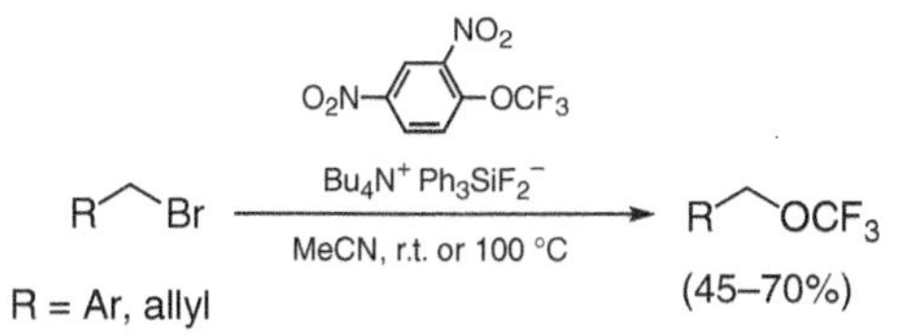

Scheme 3.13 Trifluoromethoxylation based on DNTFB use.

3.2.3.8 (Triphenylphosphonio)difluoroacetate (PDFA)

Reactivity	Nucleophilic	Radical	Electrophilic
PDFA	☑	☒	☒

(Triphenylphosphonio)difluoroacetate (PDFA), a reagent prepared from potassium bromodifluoroacetate and triphenylphosphine (Scheme 3.14) [43], was found to release difluorocarbene through a decarboxylation process. Recently, this reagent was applied in a trifluoromethoxylation procedure (Scheme 3.14) involving oxidation of difluorocarbene by diphenylsulfoxide to form intermediately difluorophosgene [44]. In presence of silver fluoride, difluorophosgene is converted into silver trifluoromethoxide, which could act as a trifluoromethoxylating reagent. Therefore, such methodology allowed conversion of various primary as well secondary benzyl bromides or iodides, primary allylic bromides, and primary alkyl iodides into their corresponding trifluoromethyl ethers in moderate to good yields.

Moreover, the scope was extended to the preparation of ^{18}O-labeled trifluoromethyl ethers starting from Ph$_2$S=^{18}O as oxidant.

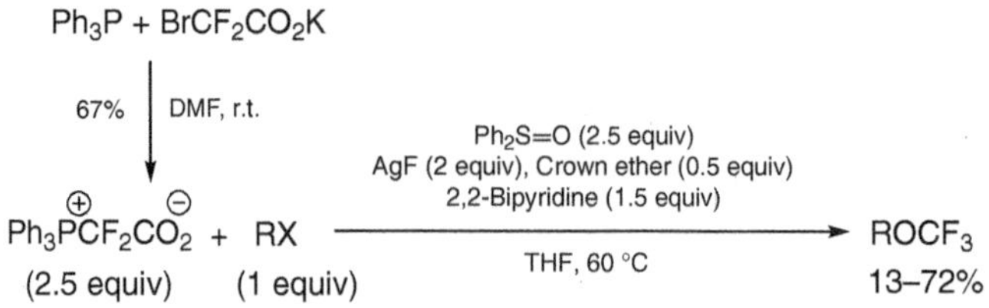

Ph$_3$P + BrCF$_2$CO$_2$K

67% | DMF, r.t.

Ph$_2$S=O (2.5 equiv)
AgF (2 equiv), Crown ether (0.5 equiv)
2,2-Bipyridine (1.5 equiv)

Ph$_3$PCF$_2$CO$_2$ + RX $\xrightarrow{\text{THF, 60 °C}}$ ROCF$_3$
(2.5 equiv) (1 equiv) 13–72%

Scheme 3.14 PDFA preparation and its use as trifluoromethoxylating reagent.

3.2.3.9 N-Trifluoromethoxylated Reagents

Reactivity	Nucleophilic	Radical	Electrophilic
O-CF$_3$-hydroxylamine	☑ (intermolecular)	☒	☒
N-OCF$_3$-benzimidazole	☒	☑	☒
N-OCF$_3$-benzimidazolium	☒	☑	☒
N-OCF$_3$-pyridinium	☒	☑	☑

Interestingly, N-trifluoromethoxylated arenes, prepared from N-hydroxylamine (see Section 3.2.2.2), are thermally sensitive as an intramolecular OCF$_3$ migration occurs upon heating, leading to the formation of aryl trifluoromethoxylated ethers [14b]. Moreover, a one-pot procedure involving trifluoromethylation of hydroxylamine followed by an intramolecular trifluoromethoxylation step was developed (Scheme 3.15). Extension of this strategy to trifluoromethoxylation of pyridines and pyrimidines derivatives was successfully achieved [45]. In contrast to aryl-anilines, no base and room temperature were required with pyridine or pyrimidine derivatives. Mechanistic studies indicated that the trifluoromethylation step is a radical process wherein a heterolytic cleavage of the N—OCF$_3$ bond, followed by a rapid recombination of short-lived ion pair (nitrenium and trifluoromethoxide ions), is involved in the intramolecular step [46]. Such mechanism opened the route to photocatalytic trifluoromethylation of hydroxylamines using iodotrifluoromethane instead of Togni reagents, followed by intramolecular migration of the trifluoromethoxy moiety as earlier. For that purpose, hydroxylamines and iodotrifluoromethane in the presence of potassium carbonate and a ruthenium photoredox catalyst were irradiated by visible light to furnish the desired aryl trifluoromethyl ethers (Scheme 3.15). One of the critical parameters of such reactions appeared to be the temperature, which has to be decreased during the trifluoromethylation step [47].

Based on the weak N—OCF$_3$ bond (BDE (bond dissociation energy) = 53.1 kcal/mol), development of trifluoromethoxylating reagents able to undergo a homolytic cleavage of this bond received attention [4g]. In order to perform radical trifluoromethoxylation, the challenge was to get an O-centered radical after cleavage. Therefore, N-trifluoromethoxy-benzimidazole [48], benzotriazole [49], and

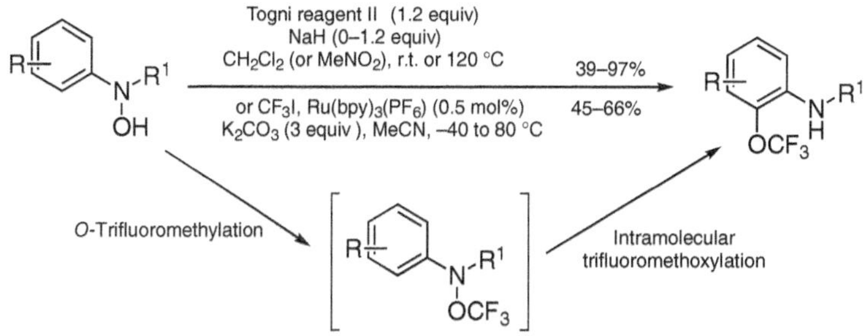

Scheme 3.15 One-pot O-trifluoromethylation–intramolecular trifluoromethoxylation reactions.

pyridinium [50] compounds were designed, synthesized (Scheme 3.16a–c), and then applied in photoredox arene trifluoromethoxylations with success (Scheme 3.16d). Indeed, a series of aryl trifluoromethyl ethers have been synthesized with a broad functional group tolerance. However, one drawback of these methods remains the lack of regioselectivity of such reactions. Concerning mechanistic details, the photoexcitation of N-trifluoromethoxy benzimidazole derivative was required to perform the reaction, which could be considered as a drawback in comparison to N-trifluoromethoxy benzotriazole or pyridinium derivatives, which are involved in a Ru-catalyzed single electron transfer (SET) step to generate the trifluoromethoxy radical, which is thus catalytically and selectively generated.

An alternative synthesis of N-trifluoromethoxy pyridinium salts was reported in a patent, using 2-trifluoromethoxy-2'-diazonium salts and the corresponding pyridine N-oxides. In addition to these salt syntheses, the electrophilic trifluoromethoxylation of electron-rich arenes (Scheme 3.17) using N-trifluoromethoxy pyridinium salts as reagent was claimed [51]. However, modest yields were reported in the examples of this patent (trifluoromethoxylation of 2-naphtyl methyl ether, naphtol, 1-bromo-4-methoxynaphtalene, and 1,4-dimethoxybenzene).

3.3 CF₃S Chemistry

The CF₃S group possesses the highest lipophilicity parameter, with a Hansch–Leo parameter $\pi_R = 1.44$ [1]. This has certainly contributed to the renaissance and the high development of the CF₃S chemistry over the last 10 years [4g, 52]. Consequently, a lot of efficient methods have been developed and are now available in the toolbox of organic chemists. In this part, we present a schematic view of the most used (and easily available and easy to handle) reagents. For a detailed review and other reagents, see Chapter 4 by Shen and coworkers in this book.

The synthesis of CF₃S molecules can be mainly divided into two main approaches.

3.3.1 Indirect Methods

The first one is commonly called "indirect strategy," consisting of using a trifluoromethylating reagent and sulfur derivatives [52b, c, g]. In this topic, the

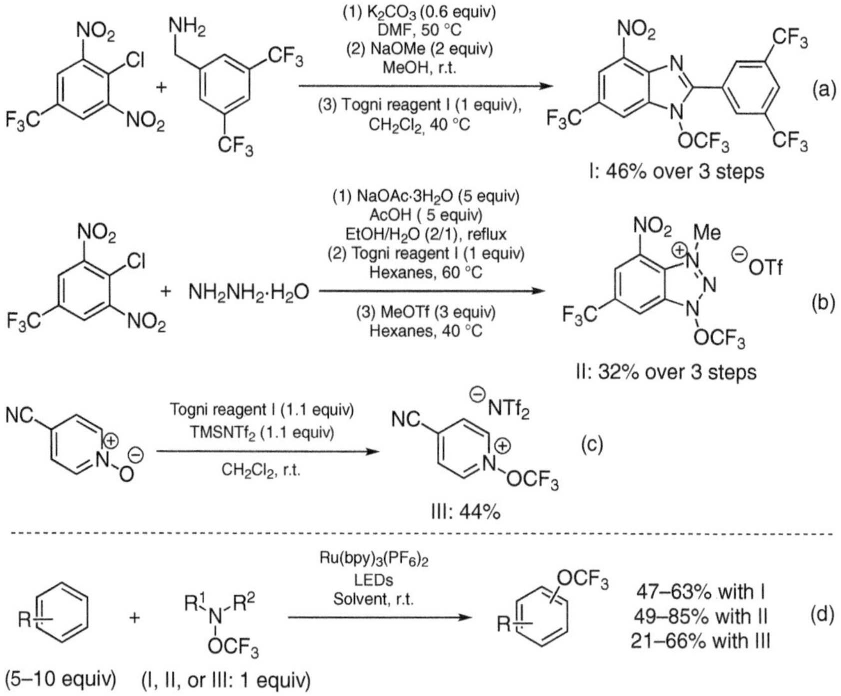

Scheme 3.16 Photoredox arenes trifluoromethoxylations.

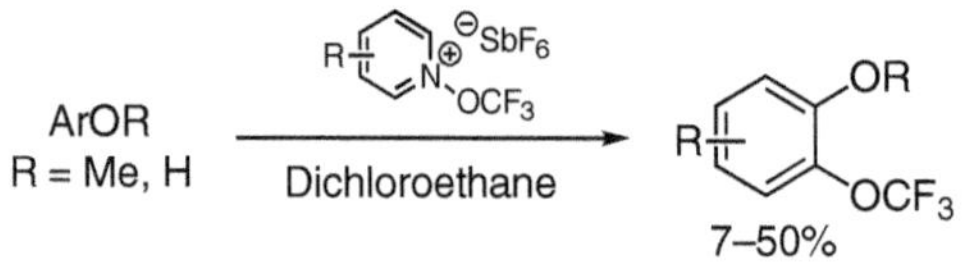

Scheme 3.17 Electrophilic trifluoromethoxylation of electron-rich arenes.

trifluoromethylation of thiocyanates (easy to synthesize) with Ruppert–Prakash reagent (CF₃SiMe₃) remains the more efficient way [53]. Based on this method, a one-pot method (thiocyanation then trifluoromethylation) has also been developed [54] (Scheme 3.18).

3.3.2 Direct Trifluoromethylthiolation

The second strategy is the direct introduction of CF₃S group onto organic substrates. This direct disconnection of CF₃S— bond is probably more intuitive from a retrosynthetic point of view and more in accordance with late-stage trifluoromethylthiolation strategies. That has certainly contributed to the huge development of various very efficient reagents to perform such reactions [52]. Herein, only the most used (in term of literature references) have been graphically summarized. For a detailed overview, see Chapter 4 by Shen and coworkers in this book.

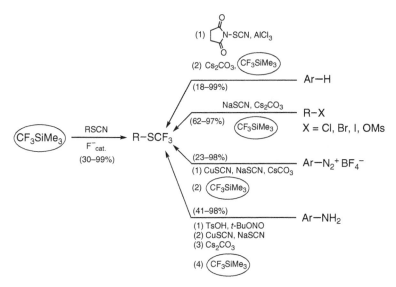

Scheme 3.18 Indirect trifluoromethylthiolation.

3.3.2.1 CF₃SAg, CF₃SCu, CF₃SNR₄

Reactivity	Nucleophilic	Radical	Electrophilic
CF₃SM	☑	☑	☑

Initially, nucleophilic trifluoromethylthiolation was developed. These reagents have gained popularity in recent years, mainly due to their easy access (CF₃SiMe₃ + S₈ + M⁺F⁻) and their recent reactivity under oxidative conditions to perform radical reactions. The major reactions developed with these reagents are summarized in Schemes 3.19 and 3.20 [35b, 38, 55].

3.3.2.2 Trifluoromethanesulfenamides

Reactivity	Nucleophilic	Radical	Electrophilic
CF₃SN(R¹)R²	☑	☒	☑

Trifluoromethanesulfenamides are easily obtained from CF₃SiMe₃, (diethylamino)sulfur trifluoride (DAST), and the corresponding amines. Several derivatives of this family have been synthesized (R¹ = H or Me, R² = Ph/R¹ = H or Me, R² = SO₂Tol/R¹ = H or Me, R² = SO₂Ph–*p*NO₂). Initially developed as electrophilic reagents, an umpolung of the reactivity allowing nucleophilic trifluoromethylthiolation has also been disclosed (Scheme 3.21) [56].

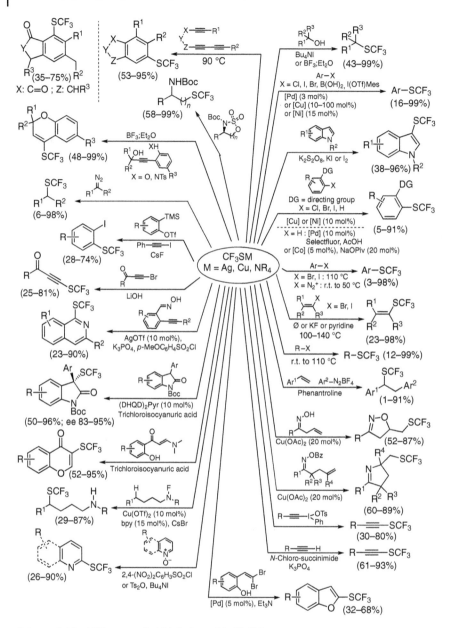

Scheme 3.19 Trifluoromethylthiolation with CF₃SM.

3.3.2.3 *N*-Trifluoromethylthiophthalimide

Reactivity	Nucleophilic	Radical	Electrophilic
CF₃SNphthalimide	☒	☑	☑

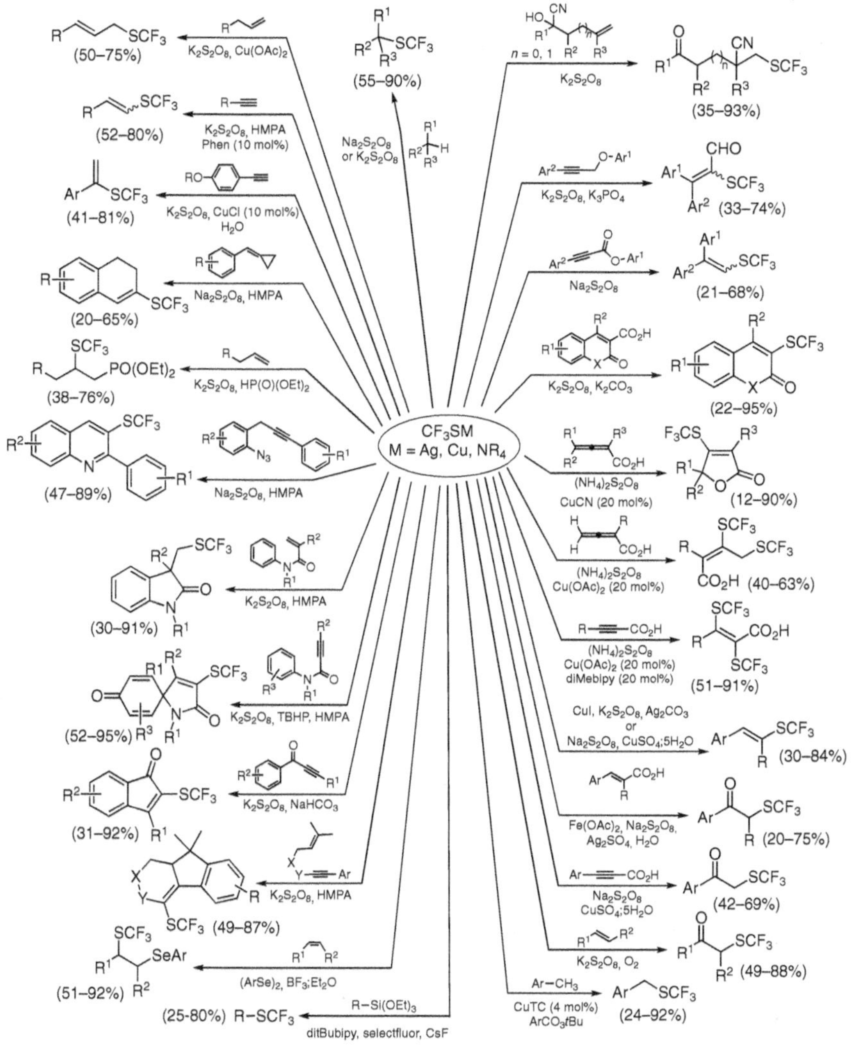

Scheme 3.20 Trifluoromethylthiolation with CF₃SM under oxidative conditions.

This reagent is obtained from CF₃SAg. Also developed initially as an electrophilic reagent, its reactivity has been extended to radical process (Scheme 3.22) [56z, 57].

3.3.2.4 *N*-Trifluoromethylthiosaccharin

Reactivity	Nucleophilic	Radical	Electrophilic
CF₃SNsaccharin	☒	☑	☑

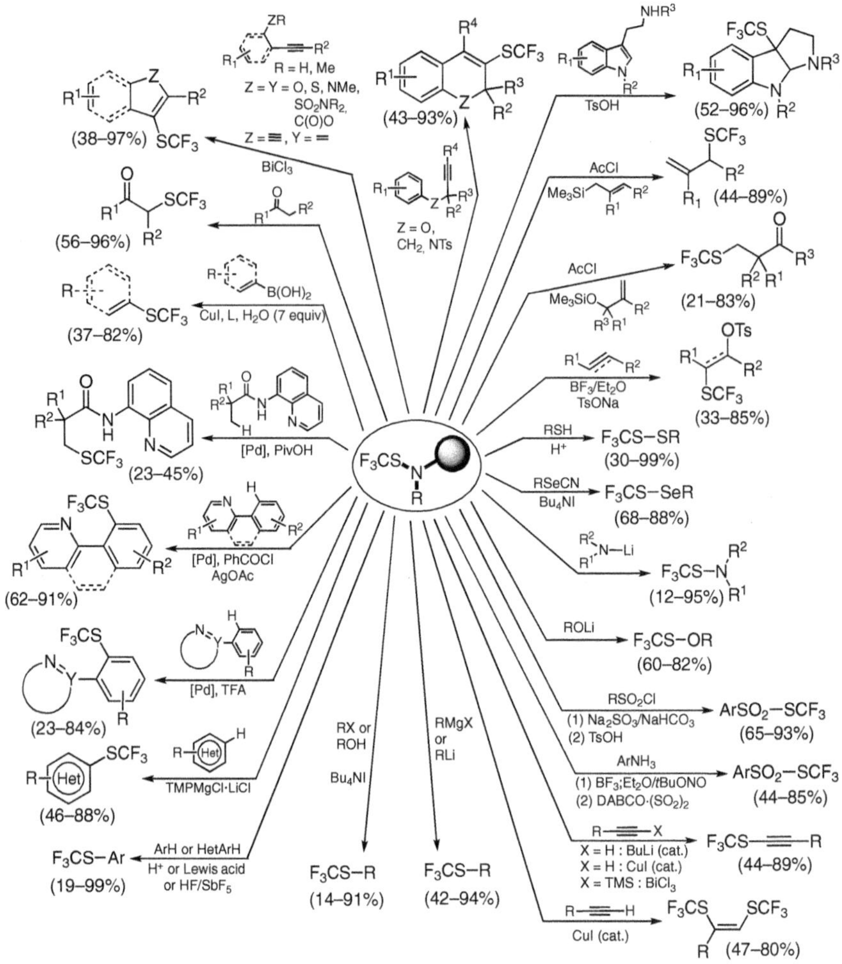

Scheme 3.21 Trifluoromethanesulfenamides reactivity.

Obtained from CF₃SAg, this reagent has been designed as a more electrophilic version of trifluoromethanesulfenamides (Section 3.3.2.2) [58]. Radical reactions have also been performed with this compound (Scheme 3.23) [59].

3.3.2.5 N-Trifluoromethylthiobis(phenylsulfonyl)amide

Reactivity	Nucleophilic	Radical	Electrophilic
CF₃SN(SO₂Ph)₂	☒	☑	☑

Developed as a more electrophilic reagent than the previous ones, its preparation also involves CF₃SAg. Electrophilicity, and also radical reactivity, has been disclosed (Scheme 3.24) [60].

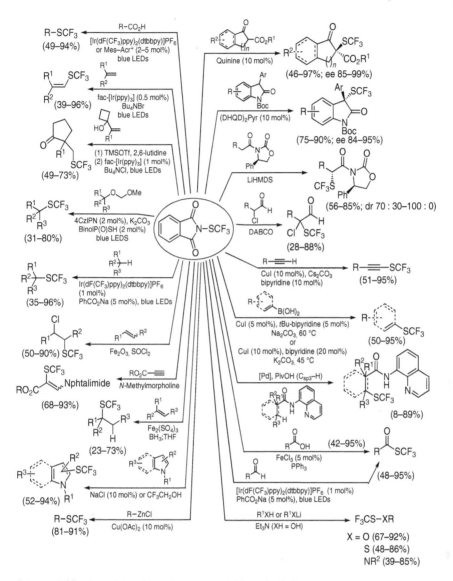

Scheme 3.22 Reactivity of *N*-trifluoromethylthiophthalimide.

3.4 CF₃Se Chemistry

3.4.1 Introduction

Because of the success story of CF_3O and CF_3S groups, the following chalcogen has logically drawn attention. Selenium is a crucial trace element in human physiology [61] and various applications in life sciences involved the use of selenolated products [62]. Compared to OCF_3 and SCF_3, $SeCF_3$ possesses slight differences in terms of electronic and lipophilicity properties (Hansch–Leo lipophilicity parameter $\pi_R = 1.29$) [1, 63]. Modulation of these parameters is fundamental in the design of

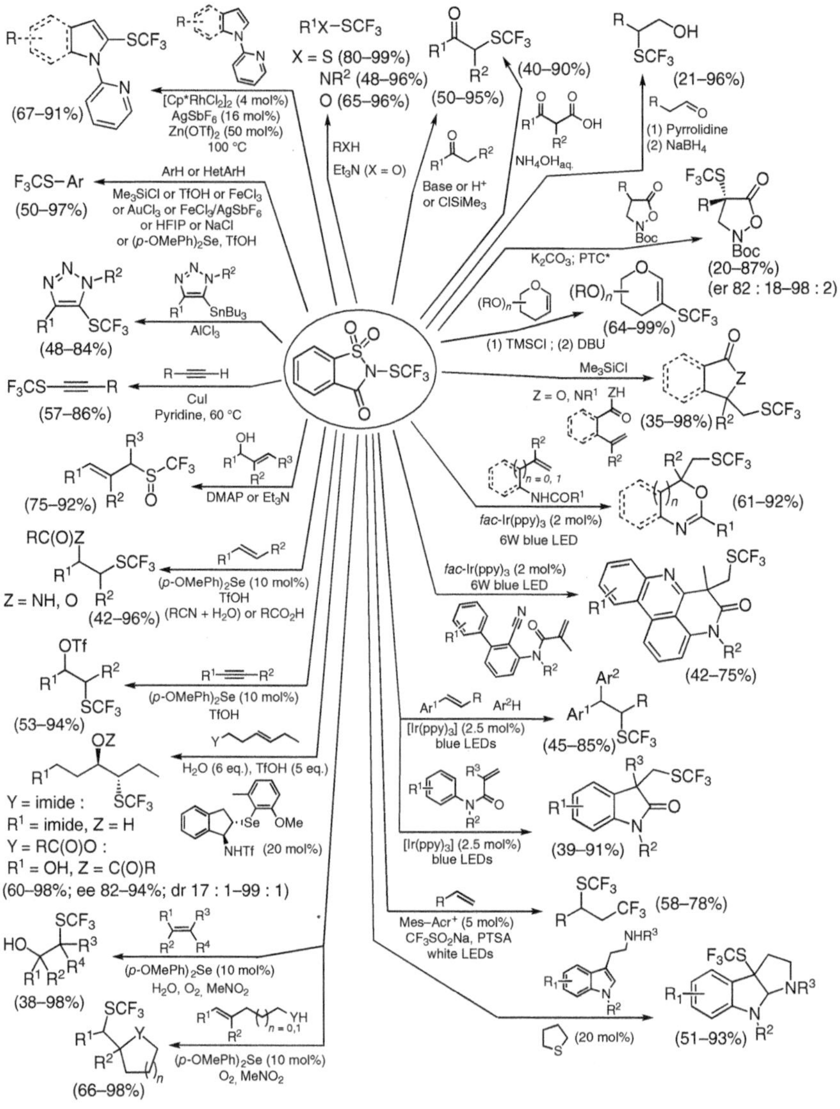

Scheme 3.23 Reactivity of *N*-trifluoromethylthiosaccharin.

agrochemicals and pharmaceutical compounds [64]. Thus, the chemistry of CF_3Se has recently gained attention [65].

3.4.2 Indirect Synthesis of CF₃Se Moiety

The first strategies developed to synthesize trifluoromethylselenolated molecules were based on an indirect approach: formation of the CF_3—Se bond. Trifluoromethylation of selenolated substrates was performed using various trifluoromethylating reagents.

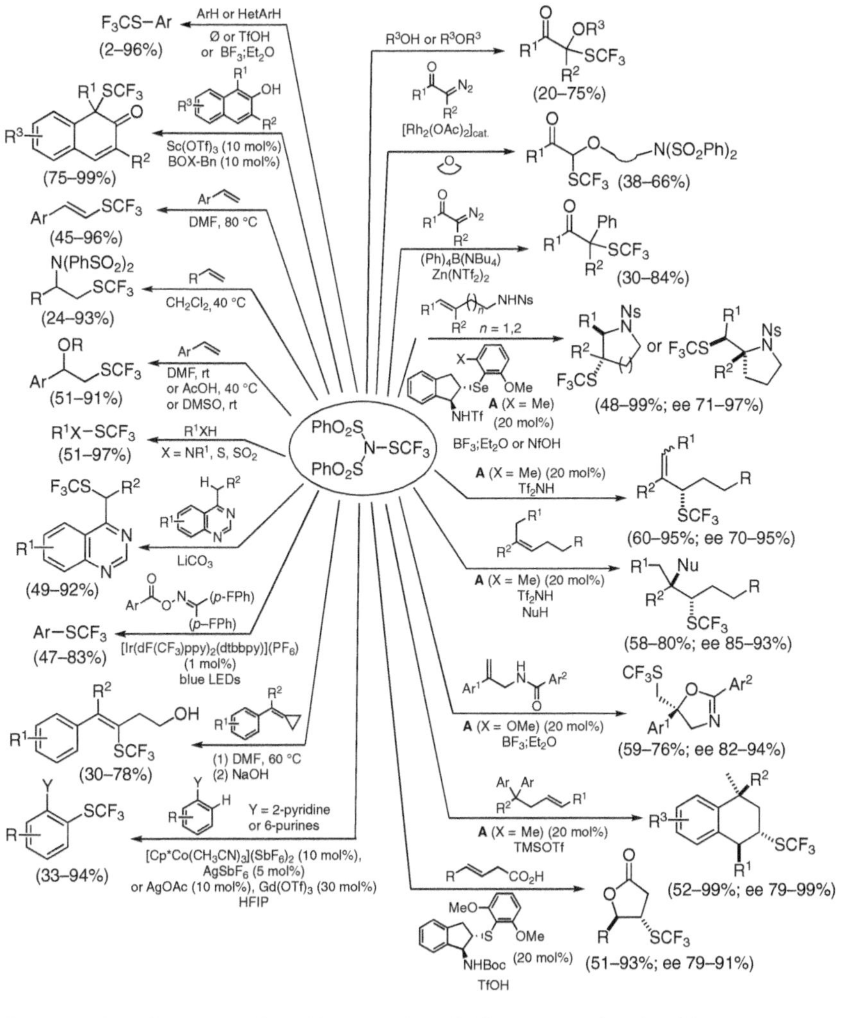

Scheme 3.24 Reactivity of *N*-trifluoromethylthiobis(phenylsulfonyl)amide.

3.4.2.1 Ruppert–Prakash Reagent (CF$_3$SiMe$_3$)

Reactivity	Nucleophilic	Radical	Electrophilic
CF$_3$SiMe$_3$	☑	☒	☒

The Ruppert–Prakash reagent (trimethyl(trifluoromethyl)silane) CF$_3$SiMe$_3$ has been one of the most used trifluoromethylating reagents for the formation of the CF$_3$—Se bond. Diselenides could be converted to the corresponding trifluoromethylselenoethers with CF$_3$SiMe$_3$, in the presence of a stoichiometric amount of fluoride anion (Scheme 3.25a) [66]. The method has been improved by replacing diselenides with selenocyanates (Scheme 3.25b) [53a, 67]. Owing to the easier synthesis of selenocyanates and the need for only catalytic amount of fluoride anion, the scope of the preparation of trifluoromethylselenolated substrates has been greatly extended, for example to some bioactive compounds [53h, m, 62a]. Two decades later, a one-pot process by *in situ* formation of selenocyanates from diazonium salts has been adapted (Scheme 3.25c) [68]. An example of one-pot process starting from *p*-nitro-aniline has also been reported (Scheme 3.25d).

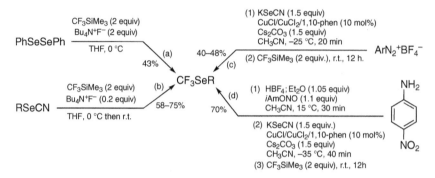

Scheme 3.25 Trifluoromethylation of selenylated compounds with CF_3SiMe_3.

3.4.2.2 Fluoroform (HCF₃)

Reactivity	Nucleophilic	Radical	Electrophilic
HCF₃	☑	☒	☒

Fluoroform (HCF₃) chemistry has been considered in the formation of trifluoromethylselenoethers. In basic conditions, HCF_3 can be deprotonated into the trifluoromethyl CF_3^- anion, which reacts with diselenides (Scheme 3.26a) [69]. This strategy has been used to prepare the radiolabeled compound $[^{18}F]CF_3SePh$ [70]. Fluoroform can also be deprotonated in the presence of copper salts to form CF_3Cu. This complex has been involved in reactions with diselenides (Scheme 3.26b) or selenocyanates (Scheme 3.26c) to afford the corresponding trifluoromethylselenoethers [71].

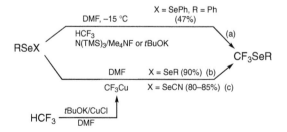

Scheme 3.26 Trifluoromethylation of selenylated compounds with HCF₃.

3.4.2.3 Other Reagents Involved in CF₃⁻ Anion Generation

Reactivity	Nucleophilic	Radical	Electrophilic
CF₃⁻ donors	☑	☒	☒

Other reagents able to release CF_3^- anion have been specifically used. Fluoral hemiaminals can react with diselenides to provide the corresponding products (Scheme 3.27a) [72]. An example of trifluoromethylation of diphenyl diselenide with trifluoromethylphosphonate, activated by *t*BuOK and leading to trifluoromethylselenylbenzene, was described (Scheme 3.27b) [73]. PhSeCF₃ was also prepared from diphenyl diselenide with a solution of [borazine–CF₃]⁻ anion (Scheme 3.27d) [74]. Tetrakis(dimethylamino)ethylene (TDAE) can reduce trifluoromethyl iodide into CF_3^- anion, which can react with diselenides (Scheme 3.27c) [75]. In this case, the released RSe⁻ (from nucleophilic attack of CF_3^-) can react with excess of CF_3I.

3.4.2.4 Sodium Trifluoromethylsulfinate (CF₃SO₂Na)

Reactivity	Nucleophilic	Radical	Electrophilic
CF₃SO₂Na	☒	☑	☒

Sodium trifluoromethanesulfinate (or triflinate), also known as the Langlois reagent, is an efficient reagent to generate trifluoromethyl radical $CF_3^·$. In oxidative conditions, sodium triflinate has been used to trifluoromethylate phenylselenol (Scheme 3.28a) [76].

Trifluoromethylselenosulfonates have been synthesized from sodium triflinate and diselenides or selenyl chlorides (Scheme 3.28) [77]. Under UV irradiation, these compounds generate the $CF_3SO_2^·$ radical, which rapidly extrudes SO₂ to release the $CF_3^·$ radical. The latter is trapped in the presence of diselenides to give the corresponding trifluoromethylselenoethers (Scheme 3.28b) [78]. This transformation could be realized without UV irradiation, but in the presence of 2 equiv of diselenide (Scheme 3.28c) [67]. A one-pot process from sodium triflinate and 3 equiv of diphenyl diselenide has been described (Scheme 3.28d) [67].

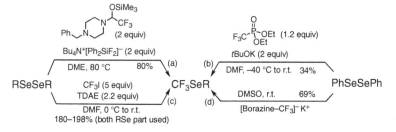

Scheme 3.27 Miscellaneous trifluoromethylation of selenylated compounds.

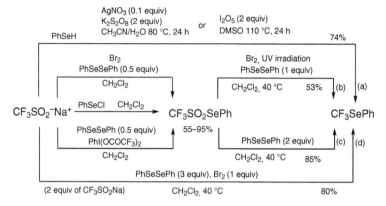

Scheme 3.28 Radical trifluoromethylation with sodium triflinate.

An alternative generation of the CF₃· radical from CF₃I and Rongalite (sodium hydroxymethanesulfonate HOCH₂SO₂Na) has also been used to perform the synthesis of trifluoromethylselenylethers from diselenide compounds [79].

3.4.3 Direct Introduction of the CF₃Se Moiety

Direct introduction of the CF₃Se moiety could be envisaged with nucleophilic substitutions, electrophilic substitutions, or by radical pathway using different reagents. Although metallic salts have been prepared [80], only a few of them have been well documented as trifluoromethylselenolating reagents.

3.4.3.1 Trifluoromethyl Selenocopper DMF Complex

Reactivity	Nucleophilic	Radical	Electrophilic
CuSeCF₃;DMF	☑	☒	☒

Yagupolskii's group reported the first synthesis of a trifluoromethylselenocopper complex in 1985 [55h]. Bis(trifluoromethyl)diselenide reacts quantitatively with copper powder in DMF to provide the CF₃SeCu;DMF complex. This reagent was used in a one-pot two-step reaction for nucleophilic trifluoromethylselenolation of iodopyridine and aryl iodides (Scheme 3.29). Trifluorometylselenolation of propargylic bromide was also reported but with a low yield (14%) [55j].

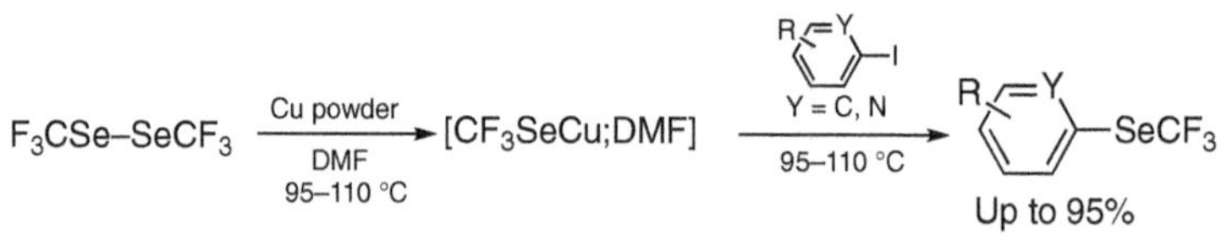

$$F_3CSe-SeCF_3 \xrightarrow[\substack{DMF \\ 95-110\ °C}]{Cu\ powder} [CF_3SeCu;DMF] \xrightarrow[95-110\ °C]{}$$

R⟨Y⟩–SeCF₃ Y = C, N Up to 95%

Scheme 3.29 Nucleophilic trifluoromethylselenolation of aryl iodides.

3.4.3.2 Trifluoromethyl Selenocopper Bipyridine Complex: [bpyCuSeCF₃]₂

Reactivity	Nucleophilic	Radical	Electrophilic
[bpyCuSeCF₃]₂	☑	☒	☒

New copper(I) trifluoromethylselenolate complexes recently emerged thanks to the convergence in copper(I) trifluoromethylthiolation development [55z] and copper(I)-catalyzed trifluoromethylselenolation [81]. Starting from the Ruppert–Prakash reagent CF₃SiMe₃, elemental selenium, copper iodide, and a ligand in the presence of a fluoride source, four new copper(I)trifluoromethylselenolate complexes have been prepared [82]. Except for the electron-rich 4,4′-di-*tert*-butylbipyridine ligand, dimeric species were isolated with the use of bipyripine and phenantroline types of ligands. All resulting complexes were air stable for months. Thanks to its reactivity toward iodotoluene, the copper trifluoromethylselenolate bipyridine complex [bpyCuSeCF₃]₂ was selected for further studies (Scheme 3.30).

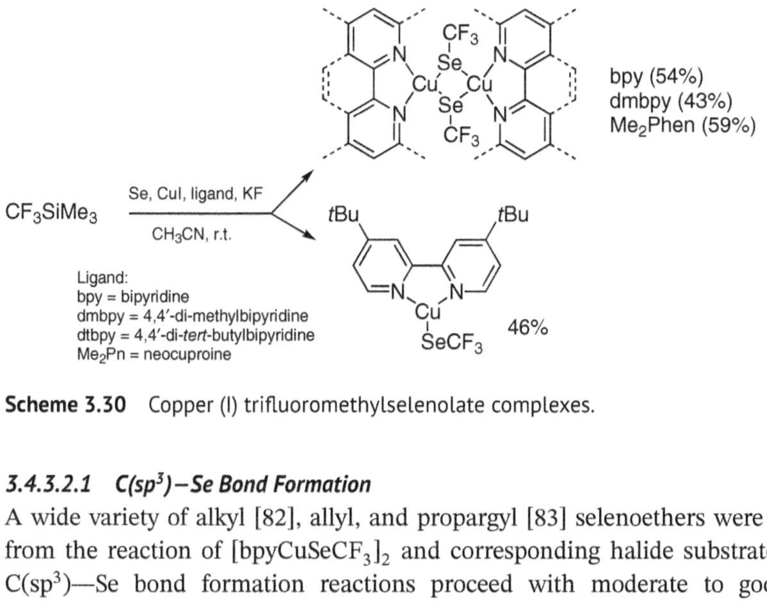

Scheme 3.30 Copper (I) trifluoromethylselenolate complexes.

3.4.3.2.1 C(sp³)–Se Bond Formation
A wide variety of alkyl [82], allyl, and propargyl [83] selenoethers were prepared from the reaction of [bpyCuSeCF₃]₂ and corresponding halide substrates. These C(sp³)—Se bond formation reactions proceed with moderate to good yields (Scheme 3.31). An excess of electrophile (1.6–1.8 equiv) was used in each case. The reaction with alkyl halides occurs at higher temperature (110 °C) than with other substrates (70 °C). α-Haloketones [84] and α-diazoesters [85] were also converted to the corresponding selenoethers in lower temperature conditions (40–45 °C).

3.4.3.2.2 C(sp²)–Se Bond Formation
The C(sp²)—Se bond formation mediated with [bpyCuSeCF₃]₂ for the preparation of aryl [82] and heteroaryl selenoethers [86] is effective on a large scope of substrates (Scheme 3.32). Aromatics halides bearing nitrogen, oxygen, sulfur heteroatoms,

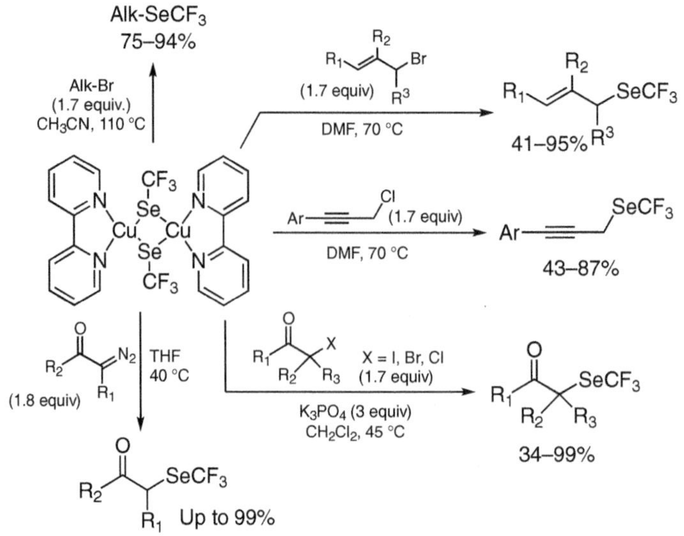

Scheme 3.31 C(sp³)–Se bond formation using [bpyCuSeCF₃]₂ reagent.

and/or sensitive functional groups were submitted to this method. Reactions were conducted in CH_3CN at 110 °C or dioxane at 80 °C for 16 hours. Yields were satisfactory and strongly dependent on the substrate. For example, 4-bromopyridines react in only 14% yield, whereas 2-bromopyridines react in more than 75% yield. The method was successfully applied to vinyl halides [87] with retention of the C—C double bond stereochemistry and to halo-pyrones [55cm]. α-Bromo-α-β-unsaturated carbonyl compounds were converted to the corresponding selenoethers but a mixture of Z/E isomers was observed [55ap]. Of note, 2 equiv of CsF were needed to accomplish this transformation. In addition, [bpyCuSeCF₃]₂ complex proved to be efficient for the conversion of β-bromo-α-β-unsaturated ketones with good to excellent yields [55av]. In addition, a palladium tandem synthesis of CF₃Se-substituted benzo-fused heterocycles from dibromovinyl phenols, thiophenols, and anilines was recently described [55df]. The overall scope of C(sp²)—Se bonds is complete with the synthesis of trifluoromethylselenoesters obtained through iron-catalyzed trifluoromethylselenolation of acid chlorides [3].

3.4.3.2.3 C(sp)–Se Bond Formation

Finally, oxidative trifluoromethylselenolation of terminal alkynes allows the formation of C(sp)—Se bonds (Scheme 3.32). Treatment of [bpyCuSeCF₃]₂ with terminal alkynes in the presence of an oxidant, DessMartin periodinane, and a base led to alkynyl trifluoromethyl selenides [88].

3.4.3.3 Tetramethylammonium Trifluoromethylselenolate [(NMe₄)(SeCF₃)]

Reactivity	Nucleophilic	Radical	Electrophilic
[(NMe₄)(SeCF₃)]	☑	☑	☑

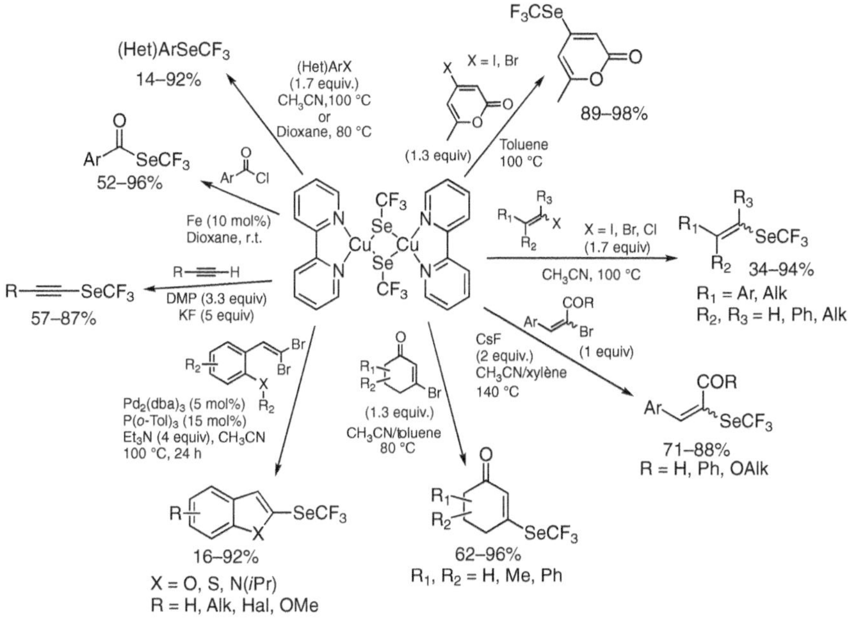

F₃CSe

(Het)ArSeCF₃
14–92%

(Het)ArX
(1.7 equiv.)
CH₃CN,100 °C
or
Dioxane, 80 °C

X = I, Br

89–98%

Ar
SeCF₃
52–96%

Ar Cl

(1.3 equiv)

Toluene
100 °C

CF₃
Se
Cu Cu
Se
CF₃

Fe (10 mol%)
Dioxane, r.t.

R₃
R₁ X
R₂
X = I, Br, Cl
(1.7 equiv)
CH₃CN, 100 °C

R₃
R₁ SeCF₃
R₂
34–94%
R₁ = Ar, Alk
R₂, R₃ = H, Ph, Alk

R SeCF₃
57–87%

R H

DMP (3.3 equiv)
KF (5 equiv)

COR
Ar Br

CsF
(2 equiv.)
CH₃CN/xylène
140 °C

(1 equiv)

COR
Ar SeCF₃
71–88%
R = H, Ph, OAlk

Br
R₂ Br
X

Pd₂(dba)₃ (5 mol%) R₂
P(o-Tol)₃ (15 mol%)
Et₃N (4 equiv), CH₃CN
100 °C, 24 h

O
R₁
R₂ Br
(1.3 equiv)
CH₃CN/toluene
80 °C

R SeCF₃
X
16–92%
X = O, S, N(iPr)
R = H, Alk, Hal, OMe

O
R₁
R₂ SeCF₃
62–96%
R₁, R₂ = H, Me, Ph

Scheme 3.32 C(sp²)–Se and C(sp)–Se bonds formation using [bpyCuSeCF₃]₂ reagent.

The first preparation of tetramethylammonium trifluoromethylselenolate [(NMe₄)(SeCF₃)] was achieved by reaction of difluoroselenophosgene with tetramethylammonium fluoride. Other trifluoromethylselenolate salts were prepared in an analogous manner but only used on scarce experiments [80b]. In 2003, an alternative preparation was reported by reaction of Me₃SiCF₃ with elemental selenium and tetramethylammonium fluoride at low temperature [89].

[(NMe₄)(SeCF₃)] is a useful reagent for the creation of C(sp³)—SeCF₃ bonds starting from alkylbromides, chlorides, or tosylates (Scheme 3.33a). The reaction occurs in only 30 minutes in acetonitrile with slight excess of selenolate salt without any additive. This simple procedure gave excellent yields, up to 80% (Scheme 3.33a). α-Diazo-esters or ketones were also trifluoromethylselenolated in acidic condition with TfOH at −20 °C [90] (Scheme 3.33b). A previous work on α-diazo-esters was reported by the group of Goossen: a catalytic amount of copper thiocyanate was able to promote this trifluoromethylselenolation in 15 hours at room temperature with good yields (64–95%) (Scheme 3.33b) [55bn].

Initially, conversion of aryl iodides to the corresponding ArSeCF₃ with [(NMe₄)(SeCF₃)] was reported in the presence of metal catalysts. The group of Schoenebeck described a palladium-catalyzed trifluoromethylselenolation of aryl iodides and heteroaryl iodides in toluene at 40–60 °C in 24 hours with moderate to good yields [91]. The reaction is tolerant to various functional groups and unprotected amines (Scheme 3.33c). Ni-catalyzed trifluoromethylselenolation of aryl halides with [(NMe₄)(SeCF₃)] was also developed (Scheme 3.33c) [92]. Ni(COD)₂ catalyst associated with bipyridine ligand proved to be efficient in converting aryl iodides and aryl bromides at room temperature. Double amount of

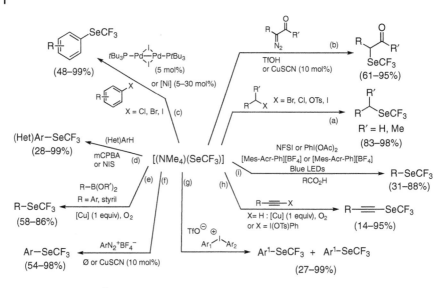

Scheme 3.33 C(sp³)−Se bond formation using [(NMe₄)(SeCF₃)].

metal catalyst (10% instead of 5%) and longer reaction times (12 hours instead of 2 hours) are needed for aryl bromides. In case of aryl chlorides, the bulkier ligand 1,1′-bis(diphenylphosphino)ferrocene (dffp) is required as well as a higher reaction temperature (50 °C) [92a]. This method gives yields up to 99% and presents a good tolerance to functional groups (Scheme 3.33c). Another approach used a dinuclear Ni(I) complex as catalyst in order to increase the chemoselectivity of the SeCF₃ functionalization with [(NMe₄)(SeCF₃)]. Aryl iodides were trifluoromethylse-lenolated with high yields; however, aryl bromides were found to be unreactive [92b].

[(NMe₄)(SeCF₃)] was also involved in a copper-catalyzed oxidative coupling (Scheme 3.33e) [93]. Aryl boronic acids or esters in the presence of Cu(OTf)₂, bipyridine as catalytic system, and O₂ as oxidant were trifluoromethylselenolated in moderate to good yields. Stoichiometric amounts of copper are needed for this transformation (Scheme 3.33e). Of note, a few examples of vinyl boronic acids were also converted to the corresponding selenoethers. Therefore, a one-pot two-step procedure involving *in situ* formation of [(NMe₄)(SeCF₃)] from Ruppert–Prakash reagent and elemental selenium was developed for this oxidative coupling reaction.

A copper-catalyzed trifluoromethylselenolation with [(NMe₄)(SeCF₃)] of aromatic diazonium salts was developed by the group of Goossen (Scheme 3.33f) [55bm]. Based on optimization experiments on trifluoromethylthiolation, reaction conditions involved the use of catalytic amount of copper thiocyanate (10%). Nevertheless, the same reaction was developed by the groups of Zhang in a metal- and additive-free procedure (Scheme 3.33f) [90]. Selenoethers were prepared from the corresponding aryldiazonium salts with 2 equiv of [(NMe₄)(SeCF₃)] in acetonitrile at −40 or −20 °C to room temperature. Yields are moderate to good and comparable to the copper-catalyzed reaction.

These conditions were also applied to convert diaryliodonium triflate into the corresponding trifluoromethyl selenoethers (Scheme 3.33g) [90]. Best results were obtained on unsymmetrical diaryliodonium salts bearing electron-deficient aryl moieties.

Tetramethylammonium trifluoromethylselenolate $[(NMe_4)(SeCF_3)]$ reacts with terminal alkynes through an oxidative coupling in the presence of oxygen and a stoichiometric amount of copper triflate and bypiridine (Scheme 3.33h) [93]. Similarly to boronic acid derivatives, a one-pot two-step selenolation of a terminal alkyne from Ruppert–Prakash reagent is reported. A metal-free approach involving the use of alkynyl(phenyl) iodonium salts has also been described [55bf]. In this case, trifluoromethylselenolation was accomplished in only 5–10 minutes at room temperature.

Although $[(NMe_4)(SeCF_3)]$ is mainly used as a nucleophilic reagent, it has been involved in trifluoromethylselenolation of electron-rich aromatic compounds under oxidative conditions (Scheme 3.33d). The authors proposed an electrophilic pathway: CF_3Se^- anion would be oxidized to form $CF_3Se^.$ radical, which is further oxidized to form the CF_3Se^+ cation [94].

Finally, an organic photoredox-catalyzed decarboxylative trifluoromethylselenolation of carboxylic acids with $[(NMe_4)(SeCF_3)]$ was recently described (Scheme 3.33i) [95]. It allowed access to $C(sp_3)$—$SeCF_3$ bond formation from primary, secondary, and tertiary aliphatic carboxylic acids. The authors suggested a radical process for the reaction, in which the reactive $CF_3Se^.$ radical might be formed.

3.4.3.4 *In Situ* Generation of CF₃Se⁻ Anion from Elemental Selenium

Reactivity	Nucleophilic	Radical	Electrophilic
CF_3^-/Se	☑	☒	☒

Generation of the trifluoromethyl anion *in situ* is an alternative to the use of trifluoromethylselenolate salt of complexes. A copper-catalyzed trifluoromethylselenolation of alkyl bromides and aryl iodides was developed from elemental selenium and Ruppert–Prakash reagent in the presence of fluoride source (Scheme 3.34a) [81]. In the presence of silver carbonate as additive and phenanthroline, these conditions led to *in situ* formation of the complex $[PhenCu(SeCF_3)]_2$. One-pot trifluoromethylselenolation of boronic acids or alkynes was also developed in copper-catalyzed oxidative coupling reactions from Ruppert–Prakash reagent and elemental selenium (see Section 3.4.2.1).

The use of a solution of trifluoromethylborazine mixed with grey elemental selenium [74] is reported for the synthesis of an example of selenoether. The CF_3Se^- anion generated *in situ* reacts with bromomethyl naphthalene to form the corresponding product (Scheme 3.34b).

An original *in situ* generation of CF_3Se^- anion was reported for the creation of C—$SeCF_3$ bonds from benzyl halides [96]. This Cu-promoted difluorocarbene-derived trifluoromethylselenolation involved the use of elemental selenium, a

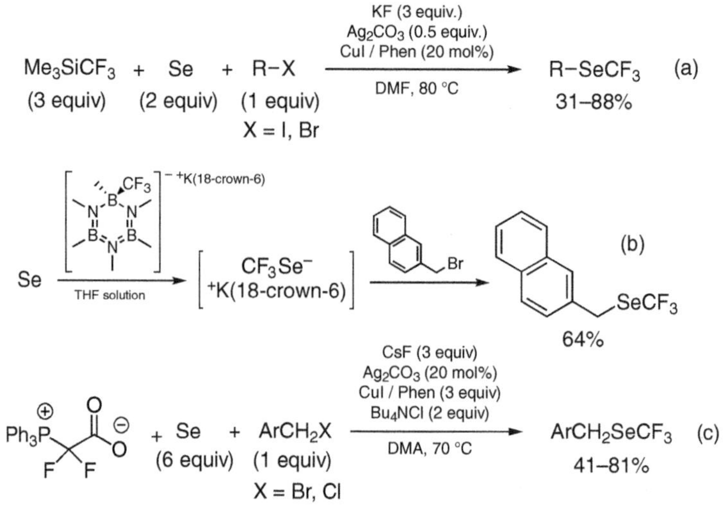

Scheme 3.34 Trifluoromethylselenolation from *in situ* generation of the CF_3Se^- anion.

fluoride source, and $Ph_3P^+CF_2CO_2^-$ for trifluoromethylselenide anion generation (Scheme 3.34c). The system was extended to alkyl, allyl, and secondary benzylic halides but aryl halides were unreactive. [CuSeCF₃] was proposed as intermediate for this transformation.

3.4.3.5 Trifluoromethylselenyl Chloride (CF_3SeCl)

Reactivity	Nucleophilic	Radical	Electrophilic
CF_3SeCl	☒	☒	☑

In contrast to the earlier reagents that were initially developed for nucleophilic trifluoromethylselenolations, trifluoromethylselenyl chloride (CF_3SeCl) was the first reagent reported to perform electrophilic trifluoromethylseleno-lation reactions. Initially, this compound was prepared via chlorination of bis(trifluoromethyl)diselenide with 90% yield (Scheme 3.35) [80a, 97]. In 2002, an alternative synthesis based on the formation of benzyltrifluoromethylselenide was reported (Scheme 3.35) [79b]. CF_3SeBn was prepared by the reaction of dibenzyld-iselenide and trifluoromethyliodide in a first step, followed by chlorination using sulfuryl chloride to give CF_3SeCl in 95% yield.

Moreover, CF_3SeBn can also be prepared in two steps, starting from potassium selenocyanate and benzyl bromide. Then, benzylselenocyanate can react with Ruppert–Prakash reagent to form benzyltrifluoromethylselenide in 82% yield (Scheme 3.35) [53a, 67, 98]. It should be noticed that a method of preparation of CF_3SeBn from $[(NMe_4)(SeCF_3)]$ has been recently presented [90].

Electrophilic trifluoromethylselenolations with CF_3SeCl were applied to various aromatic compounds including dimethylaminobenzene [99], anilines [99, 100],

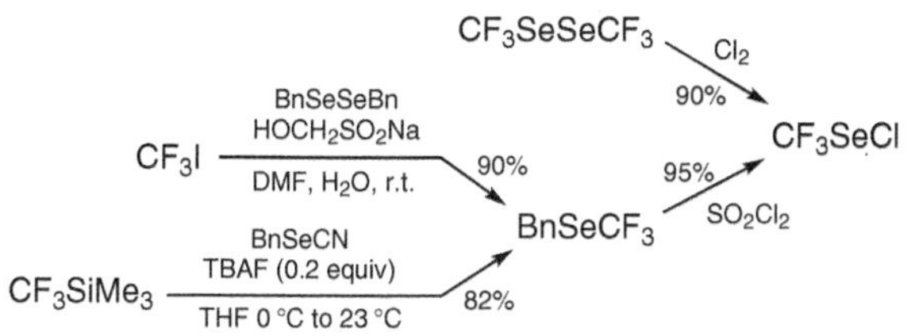

Scheme 3.35 Preparation of trifluoromethylselenyl chloride.

phenol [101], or trimethoprim (Scheme 3.36a–c) [102]. A few examples of conversion of phenyl- or tolylmagnesium bromide into the corresponding selenoethers via CF₃SeCl have been reported (Scheme 3.36d) [103]. Introduction of SeCF₃ onto malonates or orthoacetate derivatives was also described (3.36e,f) [104]. Of note, CF₃SeCl can lead to trifluoromethylselenocyanates (CF₃SeCN) and trifluoromethylselenoisocyanates (CF₃SeOCN) after reaction with, respectively, cyanide and cyanate salts [97, 105].

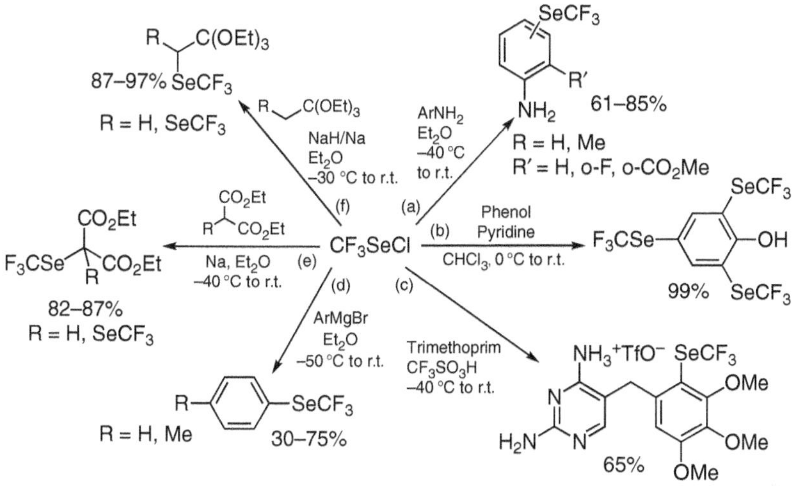

Scheme 3.36 Trifluoromethylselenolation with CF₃SeCl.

3.4.3.6 Benzyltrifluoromethylselenide (CF₃SeBn)

Reactivity	Nucleophilic	Radical	Electrophilic
CF₃SeBn	☒	☑	☑

Direct trifluoromethylselenolation with trifluoromethylselenyl chloride is limited to a few examples, maybe due to the difficulty in handling. Indeed, this electrophilic reagent is highly volatile (b.p. = 21–31 °C) [79b, 80a, 97]. Moreover, by analogy with its sulfur analogs, it could be assumed to be toxic [106]. The group of Billard reported the *in situ* generation of CF₃SeCl from CF₃SeBn, a stable and easy-to-handle liquid that can be prepared in multi-gram scale (Scheme 3.35). By using a stoichiometric amount of sulfuryl chloride in THF or dichloromethane (DCM) at room temperature, CF₃SeBn can react in one-pot two-step electrophilic trifluoromethylselenolation via "CF₃SeCl-generation" [98].

Trifluoromethylselenoethers bearing C(sp³)—Se bond were prepared from alkenes [107], ketones [108], or Grignard reagents (Scheme 3.37) [63b]. In the case of alkenes, the incorporation of a CF₃Se group was accompanied by the introduction of a chlorine atom to furnish α-chloro-β-trifluoroalkylselenoethers (Scheme 3.37a) [107]. Good yields were obtained with symmetrical alkenes, but unsymmetrical alkenes led to mixtures of regioisomers. α-Functionalization of ketones was conducted without any additives in moderate to good yields (Scheme 3.37b) [108]. Reactions with Grignard reagents have also been reported in aryl and alkyl series. Because of the formation of benzyl chloride as a side product during the *in situ* formation of CF₃SeCl, reactions required 2 equiv of Grignard reagents (Scheme 3.37c) [63b].

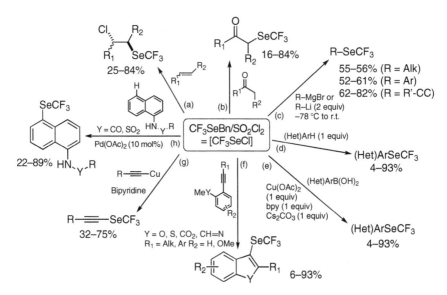

Scheme 3.37 Trifluoromethylselenolation using BnSeCF₃ as pre-reagent.

Aryl- and heteroarylselenoethers can be also prepared in one-pot two-step trifluoromethylselenolation from CF₃SeBn through SE$_{Ar}$ reactions (Scheme 3.37d) [98]. Additionally, a copper-mediated trifluoromethylselenolation of boronic acids was described using copper(II) acetate, bipyridine as ligand, and cesium carbonate as base (Scheme 3.37e) [109].

An approach for the construction of heterocycles with SeCF₃ moiety was also reported (Scheme 3.37f) [110]. This cascade reaction involved the trifluoromethylselenolation of *ortho*-functionalized alkynyl–aryl compounds followed by intramolecular ring closure. Various heterocycles were prepared such as benzo-furanes, benzothiophenes, and six-membered rings such as isochromanones, isoquinolines, or dihydroisoquinolines.

Trifluoromethylselenolations of terminal alkynes can also be done in a one-pot two-step process from CF$_3$SeBn. Lithium alkynides provided the corresponding CF$_3$Se-compounds in moderate to good yields (Scheme 3.37c) [63b]. Alkynyl copper reagents were also found very reactive toward these conditions, allowing a larger scope of application such as substrates bearing sensitive functions such as cyano or ester groups (Scheme 3.37g) [107].

Trifluoromethylselenyl chloride, generated *in situ* from CF$_3$SeBn and SO$_2$Cl$_2$, has also been involved in regioselective remote C–H trifluoromethylselenolation of 8-aminoquinolines (Scheme 3.37h) [108]. In this palladium-catalyzed reaction, the authors postulated the generation of the CF$_3$Se$^{\cdot}$ radical through a single-electron transfer process.

3.4.3.7 Trifluoromethylselenotoluenesulfonate (CF$_3$SeTs)

Reactivity	Nucleophilic	Radical	Electrophilic
CF$_3$SeTs	☑	☑	☑

3.4.3.7.1 CF$_3$SeTs as Electrophilic Reagent

Trifluoromethylselenotoluenesulfonate has been recently reported as a new electrophilic trifluoromethylselenolating reagent. Its preparation involved a one-pot two-step procedure: a first generation of CF$_3$SeCl followed by trapping with sodium toluenesulfinate (Scheme 3.38) [111].

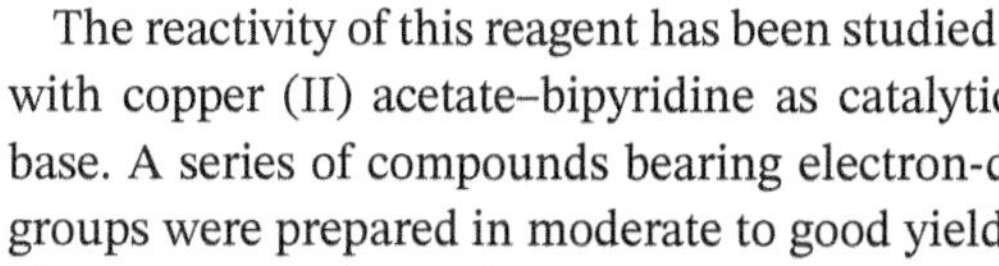

Scheme 3.38 Preparation of trifluoromethylselenotoluenesulfonate.

The reactivity of this reagent has been studied toward aryl and vinyl boronic acids with copper (II) acetate–bipyridine as catalytic system and cesium carbonate as base. A series of compounds bearing electron-donating and electron-withdrawing groups were prepared in moderate to good yields (Scheme 3.39a,b) [111]. Terminal alkynes were submitted to reaction with CF$_3$SeTs under various conditions. In THF at room temperature without any additive, difunctionalized alkenes were produced in moderate to good yields (Scheme 3.39c) [112]. In the presence of copper (II) and tetramethylethylenediamine (TMEDA) as catalytic system and cesium carbonate as base, CF$_3$SeTs leads to alkynyl trifluoromethyl selenoethers (Scheme 3.39d) [113]. Increasing the amount of copper(II) allows the bi-addition of tosyl and SeCF$_3$

groups across the triple bond. Starting from substrates bearing an oxygen-directing atom and 6 equiv of water, the corresponding vinylsulfones were obtained in good yields (Scheme 3.39e).

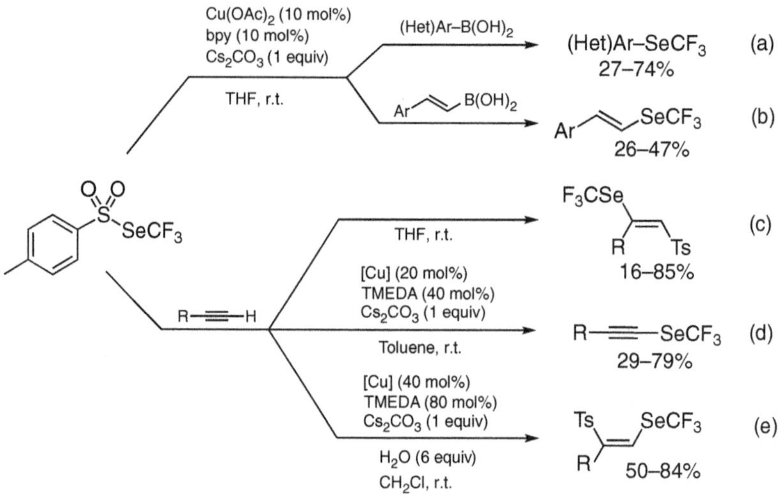

Scheme 3.39 Trifluoromethylselenolation using TsSeCF$_3$.

3.4.3.7.2 CF$_3$SeTs as Nucleophilic Reagent

Although trifluoromethylselenotoluene sulfonate CF$_3$SeTs was initially designed for electrophilic trifluoromethylselenolations [111], it has been recently involved in nucleophilic pathway. A one-pot two-step procedure was developed to convert alkyl halides into the corresponding selenoethers (Scheme 3.40) [114]. In the presence of TDAE, CF$_3$SeTs is first reduced to the trifluoromethylselenolate anion and could react without isolation with alkyl halides through nucleophilic substitutions.

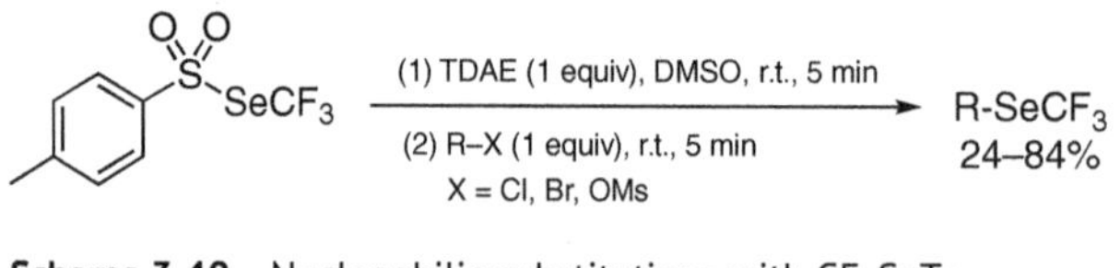

Scheme 3.40 Nucleophilic substitutions with CF$_3$SeTs.

3.4.3.7.3 CF$_3$SeTs in Radical Trifluoromethylselenolation

Under white LED (visible light) irradiation, CF$_3$SeTs could generate CF$_3$Se$^•$ radical, which rapidly dimerizes into CF$_3$SeSeCF$_3$. The generated dimer could then trap other radical species to provide a C—SeCF$_3$ bond. In the presence of Eosin Y, CF$_3$SeTs could react with diazonium salts to give (hetero)aromatic trifluoromethylselenoethers (Scheme 3.41a) [115]. Without photoredox catalyst, the radical addition of tosyl and CF$_3$Se moieties onto alkenes or alkynes has also been observed (Scheme 3.41b) [116]. Moreover, CF$_3$SeTs has recently been used in the desulfonylative functionalization of alkyl allyl sulfones leading to the formation of alkyl trifluoromethylselenoethers in a radical pathway (Scheme 3.41c) [117].

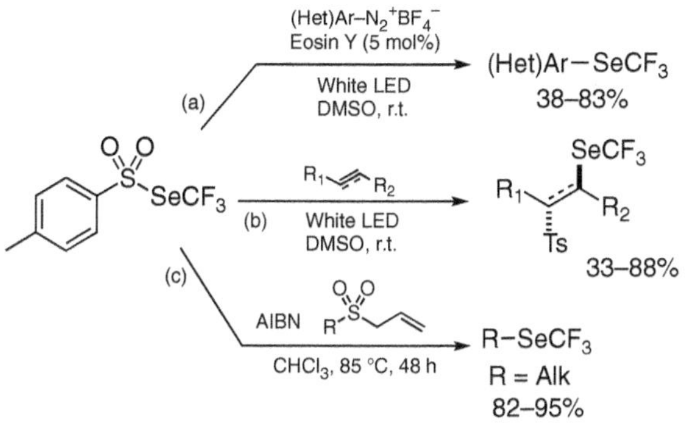

Scheme 3.41 Radical trifluoromethylselenolation with CF$_3$SeTs.

3.4.3.8 Benzylthiazolium Salt BT-SeCF$_3$

Reactivity	Nucleophilic	Radical	Electrophilic
BT-SeCF$_3$	☑	☒	☒

A recent metal-free deoxytrifluoromethylselenylation reaction using BT-SeCF$_3$, a new benzothiazolium salt, has been reported. Primary, secondary, and benzylic alcohols were converted into the corresponding trifluoromethylselenoethers under very mild condition (Scheme 3.42) [118]. BT-SeCF$_3$ is a bench-stable and easy-to-handle solid that releases the anion CF$_3$Se$^-$ after activation in the presence of an alcohol and the Hünig's base NEt(iPr)$_2$.

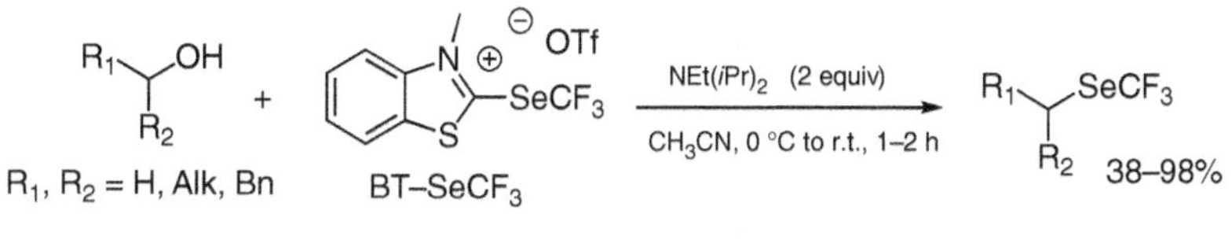

Scheme 3.42 Deoxytrifluoromethylselenylation of alcohols by using BT-SeCF$_3$.

3.5 Summary and Conclusions

Despite the huge challenge brought by the CF$_3$X chemistry, numerous elegant and efficient solutions have been brought from the beginning of this adventure (in early 1960s) to date. Noteworthy, because of the emerging interest toward fluorine chemistry in these last years, CF$_3$X chemistry has seen since around 10 years a resurgence of interest, leading to an exploding development of numerous methods and reagents to synthesize CF$_3$X-molecules. Nowadays, the organic chemistry toolbox possesses some efficient tools to introduce CF$_3$X moieties onto organic substrates. However, compared to the hydrogenated equivalent (i.e. CH$_3$X), CF$_3$X chemistry still requires developments to be able to graft these substituents "when you want, where you want"!

References

1 Leo, A., Hansch, C., and Elkins, D. (1971). *Chem. Rev.* 71: 525–616.

2 (a) Hansch, C., Leo, A., and Taft, R.W. (1991). *Chem. Rev.* 91: 165–195. (b) Leroux, F., Jeschke, P., and Schlosser, M. (2005). *Chem. Rev.* 105: 827–856.

3 Wang, J., Zhang, M., and Weng, Z. (2017). *J. Fluorine Chem.* 193: 24–32.

4 (a) Manteau, B., Pazenok, S., Vors, J.-P., and Leroux, F.R. (2010). *J. Fluorine Chem.* 131: 140–158. (b) Lin, J.-H., Ji, Y.-L., and Xiao, J.-C. (2015). *Curr. Org. Chem.* 19: 1541–1553. (c) Tlili, A., Toulgoat, F., and Billard, T. (2016). *Angew. Chem. Int. Ed.* 55: 11726–11735. (d) Besset, T., Jubault, P., Pannecoucke, X., and Poisson, T. (2016). *Org. Chem. Front.* 3: 1004–1010. (e) Basudev, S. and Hopkinson, M.N. (2018). *Angew. Chem. Int. Ed.* 57: 7942–7944. (f) Lee, K.N., Lee, J.W., and Ngai, M.-Y. (2018). *Tetrahedron* 74: 7127–7135. (g) Lee, J.W., Lee, K.N., and Ngai, M.-Y. (2019). *Angew. Chem. Int. Ed.* 58: 11171–11181.

5 (a) Yagupol'skii, L.M. (1955). *Dokl. Akad. Nauk SSSR* 105: 100–102. (b) Yarovenko, N.N. and Vasil'eva, A.S. (1958). *Zh. Obshch. Khim.* 28: 2502–2504. (c) Yagupol'skii, L.M. and Troitskaya, V.I. (1961). *Zh. Obshch. Khim.* 31: 915–924. (d) Yagupol'skii, L.M. and Orda, V.V. (1964). *Zh. Obshch. Khim.* 34: 1979–1984. (e) Kuroboshi, M., Suzuki, K., and Hiyama, T. (1992). *Tetrahedron Lett.* 33: 4173–4176. (f) Kanie, K., Tanaka, Y., Suzuki, K. et al. (2000). *Bull. Chem. Soc. Jpn.* 73: 471–484. (g) Kuroboshi, M., Kanie, K., and Hiyama, T. (2001). *Adv. Synth. Catal.* 343: 235–250. (h) Shimizu, M. and Hiyama, T. (2004). *Angew. Chem. Int. Ed.* 44: 214–231. (i) Guiadeen, D., Kothandaraman, S., Yang, L. et al. (2008). *Tetrahedron Lett.* 49: 6368–6370. (j) Manteau, B., Genix, P., Brelot, L. et al. (2010). *Eur. J. Org. Chem.*: 6043–6066. (k) Landelle, G., Schmitt, E., Panossian, A. et al. (2017). *J. Fluorine Chem.* 203: 155–165. (l) Yoritate, M., Londregan, A.T., Lian, Y., and Hartwig, J.F. (2019). *J. Org. Chem.* 84: 15767–15776.

6 (a) Kanie, K., Tanaka, Y., Shimizu, M. et al. (1997). *Chem. Commun.*: 309–310. (b) Blazejewski, J.-C., Anselmi, E., and Wakselman, C. (2001). *J. Org. Chem.* 66: 1061–1063. (c) Zriba, R., Magnier, E., and Blazejewski, J.-C. (2009). *Synlett*: 1131–1135.

7 Khotavivattana, T., Verhoog, S., Tredwell, M. et al. (2015). *Angew. Chem. Int. Ed.* 54: 9991–9995.

8 (a) Chatalova-Sazepin, C., Binayeva, M., Epifanov, M. et al. (2016). *Org. Lett.* 18: 4570–4573. (b) Zhang, Q.-W., Brusoe, A.T., Mascitti, V. et al. (2016). *Angew. Chem. Int. Ed.* 55: 9758–9762. (c) Zhou, M., Ni, C., He, Z., and Hu, J. (2016). *Org. Lett.* 18: 3754–3757. (d) Krishanmoorthy, S., Schnell, S.D., Dang, H. et al. (2017). *J. Fluorine Chem.* 203: 130–135.

9 Umemoto, T., Adachi, K., and Ishihara, S. (2007). *J. Org. Chem.* 72: 6905–6917.

10 Koller, R., Stanek, K., Stolz, D. et al. (2009). *Angew. Chem. Int. Ed.* 48: 4332–4336.

11 Liang, A., Han, S., Liu, Z. et al. (2016). *Chem. Eur. J.* 22: 5102–5106.

12 (a) Koller, R., Huchet, Q., Battaglia, P. et al. (2009). *Chem. Commun.*: 5993–5995. (b) Guo, S., Cong, F., Guo, R. et al. (2017). *Nat. Chem.* 9: 546–551.

13 Santschi, N., Geissbühler, P., and Togni, A. (2012). *J. Fluorine Chem.* 135: 83–86.

14 (a) Matoušek, V., Pietrasiak, E., Sigrist, L. et al. (2014). *Eur. J. Org. Chem.* 2014: 3087–3092. (b) Hojczyk, K.N., Feng, P., Zhan, C., and Ngai, M.-Y. (2014). *Angew. Chem. Int. Ed.* 53: 14559–14563.

15 (a) Liu, J.-B., Xu, X.-H., and Qing, F.-L. (2015). *Org. Lett.* 17: 5048–5051. (b) Liu, J.B., Chen, C., Chu, L. et al. (2015). *Angew. Chem. Int. Ed.* 54: 11839–11842.

16 Zhang, C.-P. and Vicic, D.A. (2012). *Organometallics* 31: 7812–7815.

17 (a) Redwood, M.E. and Willis, C.J. (1965). *Can. J. Chem.* 43: 1893–1898. (b) Farnham, W.B., Smart, B.E., Middleton, W.J. et al. (1985). *J. Am. Chem. Soc.* 107: 4565–4567. (c) Farnham, W.B. and Middleton, W.J. (du Pont de Nemours, E. I., and Co., USA.), EP164124A2 (1985). *Chem. Abstr.*

18 Nishida, M., Vij, A., Kirchmeier, R.L., and Shreeve, J.N.M. (1995). *Inorg. Chem.* 34: 6085–6092.

19 (a) Flosser, D.A. and Olofson, R.A. (2002). *Tetrahedron Lett.* 43: 4275–4279. (b) Zhou, M., Ni, C., Zeng, Y., and Hu, J. (2018). *J. Am. Chem. Soc.* 140: 6801–6805.

20 (a) Venturini, F., Navarrini, W., Famulari, A. et al. (2012). *J. Fluorine Chem.* 140: 43–48. (b) Francesco, V., Sansotera, M., and Navarrini, W. (2013). *J. Fluorine Chem.* 155: 2–20.

21 Billard, T. (2016). Methanesulfonic acid, 1,1,1-trifluoro-, trifluoromethyl ester. In: *Encyclopedia of Reagents for Organic Synthesis* [Online], 1–3. https://onlinelibrary.wiley.com/doi/abs/10.1002/047084289X.rn01930.

22 Taylor, S.L. and Martin, J.C. (1987). *J. Org. Chem.* 52: 4147–4156.

23 Marrec, O., Billard, T., Vors, J.-P. et al. (2010). *J. Fluorine Chem.* 131: 200–207.

24 Barbion, J., Pazenok, S., Vors, J.-P. et al. (2014). *Org. Process Res. Dev.* 18: 1037–1040.

25 Sokolenko, T.M., Davydova, Y.A., and Yagupolskii, Y.L. (2012). *J. Fluorine Chem.* 136: 20–25.

26 Zha, G.-F., Han, J.-B., Hu, X.-Q. et al. (2016). *Chem. Commun.* 52: 7458–7461.

27 Zhang, Q.-W. and Hartwig, J.F. (2018). *Chem. Commun.* 54: 10124–10127.

28 (a) Kolomeitsev, A.A., Vorobyev, M., and Gillandt, H. (2008). *Tetrahedron Lett.* 49: 449–454. (b) Huang, C., Liang, T., Harada, S. et al. (2011). *J. Am. Chem. Soc.* 133: 13308–13310. (c) Chen, C., Chen, P., and Liu, G. (2015). *J. Am. Chem. Soc.* 137: 15648–15651. (d) Qi, X., Chen, P., and Liu, G. (2017). *Angew. Chem. Int. Ed.* 56: 9517–9521. (e) Chen, C., Luo, Y., Fu, L. et al. (2018). *J. Am. Chem. Soc.* 140: 1207–1210.

29 (a) Trainor, G.L. (1985). *J. Carbohydr. Chem.* 4: 545–563. (b) Minkwitz, R. and Konikowski, D. (1996). *Z. Naturforsch., B: Chem. Sci.* 51: 147.

30 Zhang, W., Chen, J., Lin, J.-H. et al. (2018). *iScience* 5: 110–117.

31 Yang, Y.-M., Yao, J.-F., Yan, W. et al. (2019). *Org. Lett.* 21: 8003–8007.

32 Chen, C., Pflüger, P.M., Chen, P., and Liu, G. (2019). *Angew. Chem. Int. Ed.* 58: 2392–2396.

33 (a) Chen, S., Huang, Y., Fang, X. et al. (2015). *Dalton Trans.* 44: 19682–19686. (b) Chen, D., Lu, L., and Shen, Q. (2019). *Org. Chem. Front.* 6: 1801–1806.

34 Jiang, X., Deng, Z., and Tang, P. (2018). *Angew. Chem. Int. Ed.* 57: 292–295.

35 (a) Zeng, Y., Zhang, L., Zhao, Y. et al. (2013). *J. Am. Chem. Soc.* 135: 2955–2958. (b) Zeng, Y. and Hu, J. (2016). *Org. Lett.* 18: 856–859.

36 Lei, M., Miao, H., Wang, X. et al. (2019). *Tetrahedron Lett.* 60: 1389–1392.

37 Liu, J., Wei, Y., and Tang, P. (2018). *J. Am. Chem. Soc.* 140: 15194–15199.

38 Wang, F., Xu, P., Cong, F., and Tang, P. (2018). *Chem. Sci.* 9: 8836–8841.

39 Yang, H., Wang, F., Jiang, X. et al. (2018). *Angew. Chem. Int. Ed.* 57: 13266–13270.

40 Cong, F., Wei, Y., and Tang, P. (2018). *Chem. Commun.* 54: 4473–4476.

41 Yang, S., Chen, M., and Tang, P. (2019). *Angew. Chem. Int. Ed.* 58: 7840–7844.

42 Marrec, O., Billard, T., Vors, J.-P. et al. (2010). *Adv. Synth. Catal.* 352: 2831–2837.

43 Zheng, J., Cai, J., Lin, J.-H. et al. (2013). *Chem. Commun.* 49: 7513–7515.

44 Yu, J., Lin, J.-H., Yu, D. et al. (2019). *Nat. Commun.* 10: 5362.

45 Feng, P., Lee, K.N., Lee, J.W. et al. (2016). *Chem. Sci.* 7: 424–429.

46 Lee, K.N., Lei, Z., Morales-Rivera, C.A. et al. (2016). *Org. Biomol. Chem.* 14: 5599–5605.

47 Lee, J.W., Spiegowski, D.N., and Ngai, M.-Y. (2017). *Chem. Sci.* 8: 6066–6070.

48 Zheng, W., Morales-Rivera, C.A., Lee, J.W. et al. (2018). *Angew. Chem. Int. Ed.* 57: 9645–9649.

49 Zheng, W., Lee, J.W., Morales-Rivera, C.A. et al. (2018). *Angew. Chem. Int. Ed.* 57: 13795–13799.

50 Jelier, B.J., Tripet, P.F., Pietrasiak, E. et al. (2018). *Angew. Chem. Int. Ed.* 57: 13784–13789.

51 Umemoto, T., Zhou, M., and Hu, J. (Shanghai Institute of Organic Chemistry, Chinese Academy of Sciences, Peop. Rep. China.), CN105017143A (2015). *Chem. Abstr.*

52 (a) Boiko, V.N. (2010). *Beilstein J. Org. Chem.* 6: 880–921. (b) Toulgoat, F., Alazet, S., and Billard, T. (2014). *Eur. J. Org. Chem.*: 2415–2428. (c) Xu, X.-H., Matsuzaki, K., and Shibata, N. (2014). *Chem. Rev.* 115: 731–764. (d) Barata-Vallejo, S., Bonesi, S., and Postigo, A. (2016). *Org. Biomol. Chem.* 14: 7150–7182. (e) Chachignon, H. and Cahard, D. (2016). *Chin. J. Chem.* 34: 445–454. (f) Zheng, H., Huang, Y., and Weng, Z. (2016). *Tetrahedron Lett.* 57: 1397–1409. (g) Toulgoat, F. and Billard, T. (2017). Towards CF₃S group: From trifluoromethylation of sulfides to direct trifluoromethylthiolation. In: *Modern Synthesis Processes and Reactivity of Fluorinated Compounds: Progress in Fluorine Science* (eds. H. Groult, F. Leroux and A. Tressaud), 141–179. London, United Kingdom: Elsevier Science. (h) Barthelemy, A.-L., Magnier, E., and Dagousset, G. (2018). *Synthesis* 50: 4765–4776. (i) Hamzehloo, M., Hosseinian, A., Ebrahimiasl, S. et al. (2019). *J. Fluorine Chem.* 224: 52–60.

53 (a) Billard, T., Large, S., and Langlois, B.R. (1997). *Tetrahedron Lett.* 38: 65–68. (b) Bouchu, M.-N., Large, S., Steng, M. et al. (1998). *Carbohydr. Res.* 314: 37–45. (c) Granger, C.E., Félix, C.P., Parrot-Lopez, H.P., and Langlois, B.R.

(2000). *Tetrahedron Lett.* 41: 9257–9260. (d) Hess, J., Konatschnig, S., Morard, S. et al. (2014). *Inorg. Chem.* 53: 3662–3667. (e) Liang, Z., Wang, F., Chen, P., and Liu, G. (2015). *Org. Lett.* 17: 2438–2441. (f) Mitra, S., Ghosh, M., Mishra, S., and Hajra, A. (2015). *J. Org. Chem.* 80: 8275–8281. (g) Lv, Y., Pu, W., Cui, H. et al. (2016). *Synth. Commun.* 46: 1223–1229. (h) Muniraj, N., Dhineshkumar, J., and Prabhu, K.R. (2016). *ChemistrySelect* 1: 1033–1038. (i) Chen, Q., Lei, Y., Wang, Y. et al. (2017). *Org. Chem. Front.* 4: 369–372. (j) Jiang, G., Zhu, C., Li, J. et al. (2017). *Adv. Synth. Catal.* 359: 1208–1212. (k) Malik, G., Swyka, R.A., Tiwari, V.K. et al. (2017). *Chem. Sci.* 8: 8050–8060. (l) Kong, D.-L., Du, J.-X., Chu, W.-M. et al. (2018). *Molecules* 23: 2727. (m) Chao, M.N., Lorenzo-Ocampo, M.V., Szajnman, S.H. et al. (2019). *Biorg. Med. Chem.* 27: 1350–1361. (n) Chen, Y.-J., He, Y.-H., and Guan, Z. (2019). *Tetrahedron* 75: 3053–3061. (o) Dey, A. and Hajra, A. (2019). *Adv. Synth. Catal.* 361: 842–849. (p) Dyga, M., Hayrapetyan, D., Rit, R.K., and Gooßen, L.J. (2019). *Adv. Synth. Catal.* 361: 3548–3553. (q) Hoque, I.U., Chowdhury, S.R., and Maity, S. (2019). *J. Org. Chem.* 84: 3025–3035. (r) Wei, W., Liao, L., Qin, T., and Zhao, X. (2019). *Org. Lett.* 21: 7846–7850. (s) Yang, T., Song, X.-R., Li, R. et al. (2019). *Tetrahedron Lett.* 60: 1248–1253.

54 (a) Danoun, G., Bayarmagnai, B., Gruenberg, M.F., and Goossen, L.J. (2014). *Chem. Sci.* 5: 1312–1316. (b) Jouvin, K., Matheis, C., and Goossen, L.J. (2015). *Chem. Eur. J.* 21: 14324–14327.

55 (a) Emeléus, H.J. and Macduffie, D.E. (1961). *J. Chem. Soc.*: 2597. (b) Yagupolskii, L.M. and Smirnova, O.D. (1972). *Zh. Org. Khim.* 8: 1990–1991. (c) Kondratenko, N.V. and Sambur, V.P. (1975). *Ukr. Khim. Zh. (Russ. Ed.)* 41: 516–519. (d) Yagupolskii, L.M., Kondratenko, N.V., and Sambur, V.P. (1975). *Synthesis*: 721–723. (e) Remy, D.C., Rittle, K.E., Hunt, C.A., and Freedman, M.B. (1976). *J. Org. Chem.* 41: 1644–1646. (f) Hanack, M. and Kühnle, A. (1981). *Tetrahedron Lett.* 22: 3047–3048. (g) Hanack, M. and Massa, F.W. (1981). *Tetrahedron Lett.* 22: 557–558. (h) Kondratenko, N.V., Kolomeytsev, A.A., Popov, V.I., and Yagupolskii, L.M. (1985). *Synthesis*: 667–669. (i) Haas, A. and Lieb, M. (1986). *J. Heterocycl. Chem.* 23: 1079–1084. (j) Haas, A. and Krächter, H.-U. (1988). *Chem. Ber.* 121: 1833–1840. (k) Clark, J.H., Jones, C.W., Kybett, A.P. et al. (1990). *J. Fluorine Chem.* 48: 249–253. (l) Clark, J.H. and Smith, H. (1993). *J. Fluorine Chem.* 61: 223–231. (m) Adams, D.J. and Clark, J.H. (2000). *J. Org. Chem.* 65: 1456–1460. (n) Adams, D.J., Goddard, A., Clark, J.H., and Macquarrie, D.J. (2000). *Chem. Commun.*: 987–988. (o) Tyrra, W., Naumann, D., Hoge, B., and Yagupolskii, Y.L. (2003). *J. Fluorine Chem.* 119: 101–107. (p) Kremlev, M.M., Tyrra, W., Naumann, D., and Yagupolskii, Y.L. (2004). *Tetrahedron Lett.* 45: 6101–6104. (q) Kirsch, P., Lenges, M., Kühne, D., and Wanczek, K.-P. (2005). *Eur. J. Org. Chem.*: 797–802. (r) Rhode, C., Lemke, J., Lieb, M., and Metzler-Nolte, N. (2009). *Synthesis*: 2015–2018. (s) Teverovskiy, G., Surry, D.S., and Buchwald, S.L. (2011). *Angew. Chem. Int. Ed.* 50: 7312–7314. (t) Zhang, C.-P. and Vicic, D.A. (2012). *Chem. Asian J.* 7: 1756–1758. (u) Zhang, C.-P. and Vicic, D.A. (2012). *J. Am. Chem. Soc.* 134: 183–185. (v) Kong, D., Jiang, Z., Xin, S. et al. (2013). *Tetrahedron* 69:

6046–6050. (w) Rueping, M., Tolstoluzhsky, N., and Nikolaienko, P. (2013). *Chem. Eur. J.* 19: 14043–14046. (x) Tan, J., Zhang, G., Ou, Y. et al. (2013). *Chin. J. Chem.* 31: 921–926. (y) Wang, K.-P., Yun, S.Y., Mamidipalli, P., and Lee, D. (2013). *Chem. Sci.* 4: 3205–3211. (z) Weng, Z., He, W., Chen, C. et al. (2013). *Angew. Chem. Int. Ed.* 52: 1548–1552; (aa) Chen, C., Xu, X.-H., Yang, B., and Qing, F.-L. (2014). *Org. Lett.* 16: 3372–3375. (ab) Emer, E., Twilton, J., Tredwell, M. et al. (2014). *Org. Lett.* 16: 6004–6007. (ac) Hu, M., Rong, J., Miao, W. et al. (2014). *Org. Lett.* 16: 2030–2033. (ad) Huang, Y., He, X., Li, H., and Weng, Z. (2014). *Eur. J. Org. Chem.*: 7324–7328. (ae) Lefebvre, Q., Fava, E., Nikolaienko, P., and Rueping, M. (2014). *Chem. Commun.* 50: 6617–6619. (af) Lin, Q., Chen, L., Huang, Y. et al. (2014). *Org. Biomol. Chem.* 12: 5500–5508. (ag) Nikolaienko, P., Pluta, R., and Rueping, M. (2014). *Chem. Eur. J.* 20: 9867–9870. (ah) Wang, X., Zhou, Y., Ji, G. et al. (2014). *Eur. J. Org. Chem.*: 3093–3096. (ai) Xiang, H. and Yang, C. (2014). *Org. Lett.* 16: 5686–5689. (aj) Xiao, Q., Sheng, J., Ding, Q., and Wu, J. (2014). *Eur. J. Org. Chem.*: 217–221. (ak) Xu, J., Mu, X., Chen, P. et al. (2014). *Org. Lett.* 16: 3942–3945. (al) Ye, K.-Y., Zhang, X., Dai, L.-X., and You, S.-L. (2014). *J. Org. Chem.* 79: 12106–12110. (am) Yin, F. and Wang, X.-S. (2014). *Org. Lett.* 16: 1128–1131. (an) Yin, W., Wang, Z., and Huang, Y. (2014). *Adv. Synth. Catal.* 356: 2998–3006. (ao) Zhang, K., Liu, J.-B., and Qing, F.-L. (2014). *Chem. Commun.* 50: 14157–14160. (ap) Zhu, L., Wang, G., Guo, Q. et al. (2014). *Org. Lett.* 16: 5390–5393. (aq) Zhu, P., He, X., Chen, X. et al. (2014). *Tetrahedron* 70: 672–677. (ar) Zhu, S.-Q., Xu, X.-H., and Qing, F.-L. (2014). *Eur. J. Org. Chem.*: 4453–4456. (as) Zhu, X.-L., Xu, J.-H., Cheng, D.-J. et al. (2014). *Org. Lett.* 16: 2192–2195. (at) Fuentes, N., Kong, W., Fernández-Sánchez, L. et al. (2015). *J. Am. Chem. Soc.* 137: 964–973. (au) Guo, S., Zhang, X., and Tang, P. (2015). *Angew. Chem. Int. Ed.* 54: 4065–4069. (av) Hou, C., Lin, X., Huang, Y. et al. (2015). *Synthesis* 47: 969–975. (aw) Huang, Y., Ding, J., Wu, C. et al. (2015). *J. Org. Chem.* 80: 2912–2917. (ax) Jiang, M., Zhu, F., Xiang, H. et al. (2015). *Org. Biomol. Chem.* 13: 6935–6939. (ay) Liu, J.-B., Xu, X.-H., Chen, Z.-H., and Qing, F.-L. (2015). *Angew. Chem. Int. Ed.* 54: 897–900. (az) Qiu, Y.-F., Zhu, X.-Y., Li, Y.-X. et al. (2015). *Org. Lett.* 17: 3694–3697. (ba) Saravanan, P. and Anbarasan, P. (2015). *Adv. Synth. Catal.* 357: 3521–3528. (bb) Wu, H., Xiao, Z., Wu, J. et al. (2015). *Angew. Chem. Int. Ed.* 54: 4070–4074. (bc) Xie, F., Zhang, Z., Yu, X. et al. (2015). *Angew. Chem. Int. Ed.* 54: 7405–7409. (bd) Yin, G., Kalvet, I., and Schoenebeck, F. (2015). *Angew. Chem. Int. Ed.* 54: 6809–6813. (be) Zeng, Y.-F., Tan, D.-H., Chen, Y. et al. (2015). *Org. Chem. Front.* 2: 1511–1515. (bf) Fang, W.-Y., Dong, T., Han, J.-B. et al. (2016). *Org. Biomol. Chem.* 14: 11502–11509; (bg) Jin, D.-P., Gao, P., Chen, D.-Q. et al. (2016). *Org. Lett.* 18: 3486–3489. (bh) Johnson, M.W., Hannoun, K.I., Tan, Y. et al. (2016). *Chem. Sci.* 7: 4091–4100. (bi) Karmakar, R., Mamidipalli, P., Salzman, R.M. et al. (2016). *Org. Lett.* 18: 3530–3533. (bj) Luo, B., Zhang, Y., You, Y., and Weng, Z. (2016). *Org. Biomol. Chem.* 14: 8615–8622. (bk) Luo, P., Ding, Q., Ping, Y., and Hu, J. (2016). *Org. Biomol. Chem.* 14: 2924–2929. (bl) Ma, L., Cheng, X.-F., Li, Y., and Wang, X.-S. (2016). *Tetrahedron Lett.* 57: 2972–2975. (bm) Matheis, C., Bayarmagnai, B., Jouvin, K., and Goossen, L.J.

(2016). *Org. Chem. Front.* 3: 949–952. (bn) Matheis, C., Krause, T., Bragoni, V., and Goossen, L.J. (2016). *Chem. Eur. J.* 22: 12270–12273. (bo) Matheis, C., Wagner, V., and Goossen, L.J. (2016). *Chem. Eur. J.* 22: 79–82. (bp) Nguyen, T., Chiu, W., Wang, X. et al. (2016). *Org. Lett.* 18: 5492–5495. (bq) Nikolaienko, P., Yildiz, T., and Rueping, M. (2016). *Eur. J. Org. Chem.*: 1091–1094. (br) Pan, S., Huang, Y., and Qing, F.-L. (2016). *Chem. Asian J.* 11: 2854–2858. (bs) Qiu, Y.-F., Song, X.-R., Li, M. et al. (2016). *Org. Lett.* 18: 1514–1517. (bt) Song, Y.-K., Qian, P.-C., Chen, F. et al. (2016). *Tetrahedron* 72: 7589–7593. (bu) Wang, F., Zhao, L., You, J., and Wang, M.-X. (2016). *Org. Chem. Front.* 3: 880–886. (bv) Wu, W., Dai, W., Ji, X., and Cao, S. (2016). *Org. Lett.* 18: 2918–2921. (bw) Zeng, Y. and Hu, J. (2016). *Synthesis* 48: 2137–2150. (bx) Zhang, M., Chen, J., Chen, Z., and Weng, Z. (2016). *Tetrahedron* 72: 3525–3530. (by) Anselmi, E., Simon, C., Marrot, J. et al. (2017). *Eur. J. Org. Chem.*: 6319–6326. (bz) Chen, M.-T., Tang, X.-Y., and Shi, M. (2017). *Org. Chem. Front.* 4: 86–90. (ca) Gao, W., Ding, Q., Yuan, J. et al. (2017). *Chin. J. Chem.* 35: 1717–1725. (cb) Guo, C.-H., Chen, D.-Q., Chen, S., and Liu, X.-Y. (2017). *Adv. Synth. Catal.* 359: 2901–2906. (cc) Ji, M., Wu, Z., Yu, J. et al. (2017). *Adv. Synth. Catal.* 359: 1959–1962. (cd) Kovács, S., Bayarmagnai, B., and Goossen, L.J. (2017). *Adv. Synth. Catal.* 359: 250–254. (ce) Li, H., Liu, S., Huang, Y. et al. (2017). *Chem. Commun.* 53: 10136–10139. (cf) Li, M., Petersen, J.L., and Hoover, J.M. (2017). *Org. Lett.* 19: 638–641. (cg) Liu, X.-G., Li, Q., and Wang, H. (2017). *Adv. Synth. Catal.* 359: 1942–1946. (ch) Ohta, R., Kuboki, Y., Yoshikawa, Y. et al. (2017). *J. Fluorine Chem.* 201: 1–6. (ci) Pan, S., Huang, Y., Xu, X.-H., and Qing, F.-L. (2017). *Org. Lett.* 19: 4624–4627. (cj) Pan, S., Li, H., Huang, Y. et al. (2017). *Org. Lett.* 19: 3247–3250. (ck) Wu, W., Wang, B., Ji, X., and Cao, S. (2017). *Org. Chem. Front.* 4: 1299–1303. (cl) Zeng, J.-L., Chachignon, H., Ma, J.-A., and Cahard, D. (2017). *Org. Lett.* 19: 1974–1977. (cm) Zhang, Y., Yang, D.-Y., and Weng, Z. (2017). *Tetrahedron* 73: 3853–3859. (cn) Zhao, M., Zhao, X., Zheng, P., and Tian, Y. (2017). *J. Fluorine Chem.* 194: 73–79. (co) Bertoli, G., Exner, B., Evers, M.V. et al. (2018). *J. Fluorine Chem.* 210: 132–136. (cp) Cheng, Z.-F., Tao, T.-T., Feng, Y.-S. et al. (2018). *J. Org. Chem.* 83: 499–504. (cq) Guo, K., Zhang, H., Cao, S. et al. (2018). *Org. Lett.* 20: 2261–2264. (cr) Ji, M., Yu, J., and Zhu, C. (2018). *Chem. Commun.* 54: 6812–6815. (cs) Jiang, H., Zhu, R., Zhu, C. et al. (2018). *Org. Biomol. Chem.* 16: 1646–1650. (ct) Julien, M., Guillaume, F., Benjamin, D., and David, L. (2018). *Adv. Synth. Catal.* 360: 2752–2756. (cu) Liu, Y.-L., Xu, X.-H., and Qing, F.-L. (2018). *Tetrahedron* 74: 5827–5832. (cv) Xia, X., Chen, B., Zeng, X., and Xu, B. (2018). *Adv. Synth. Catal.* 360: 4429–4434. (cw) Xiao, Z., Liu, Y., Zheng, L. et al. (2018). *J. Org. Chem.* 83: 5836–5843. (cx) Xiao, Z., Liu, Y., Zheng, L. et al. (2018). *Tetrahedron* 74: 6213–6219. (cy) Zhang, B.-S., Gao, L.-Y., Zhang, Z. et al. (2018). *Chem. Commun.* 54: 1185–1188. (cz) Zhen, L., Yuan, K., Li, X.-Y. et al. (2018). *Org. Lett.* 20: 3109–3113. (da) Liu, Y.-L., Xu, X.-H., and Qing, F.-L. (2019). *Tetrahedron Lett.* 60: 953–956. (db) Modak, A., Pinter, E.N., and Cook, S.P. (2019). *J. Am. Chem. Soc.* 141: 18405–18410. (dc) Muta, R., Torigoe, T., and Kuninobu, Y. (2019). *Org. Lett.* 21: 4289–4292. (dd) Qiu, Y.-F., Niu, Y.-J., Wei, X. et al. (2019). *J. Org. Chem.* 84: 4165–4178. (de)

Saravanan, P. and Anbarasan, P. (2019). *Chem. Commun.* 55: 4639–4642. (df) Zhang, M. and Weng, Z. (2019). *Org. Lett.* 21: 5838–5842. (dg) Zhang, S.-B., Xu, X.-H., and Qing, F.-L. (2019). *J. Fluorine Chem.* 227: 109367. (dh) Zheng, C., Liu, Y., Hong, J. et al. (2019). *Synlett* 30: 1324–1328. (di) Zheng, C., Liu, Y., Hong, J. et al. (2019). *Tetrahedron Lett.* 60: 1404–1407.

56 (a) Ferry, A., Billard, T., Langlois, B.R., and Bacque, E. (2008). *J. Org. Chem.* 73: 9362–9365. (b) Ferry, A., Billard, T., Langlois, B.R., and Bacqué, E. (2009). *Angew. Chem. Int. Ed.* 48: 8551–8555. (c) Baert, F., Colomb, J., and Billard, T. (2012). *Angew. Chem. Int. Ed.* 51: 10382–10385. (d) Ferry, A., Billard, T., Bacqué, E., and Langlois, B.R. (2012). *J. Fluorine Chem.* 134: 160–163. (e) Yang, Y., Jiang, X., and Qing, F.-L. (2012). *J. Org. Chem.* 77: 7538–7547. (f) Alazet, S., Ollivier, K., and Billard, T. (2013). *Beilstein J. Org. Chem.* 9: 2354–2357. (g) Alazet, S., Zimmer, L., and Billard, T. (2013). *Angew. Chem. Int. Ed.* 52: 10814–10817. (h) Liu, J., Chu, L., and Qing, F.-L. (2013). *Org. Lett.* 15: 894–897. (i) Xiao, Q., Sheng, J., Chen, Z., and Wu, J. (2013). *Chem. Commun.* 49: 8647–8649. (j) Alazet, S., Zimmer, L., and Billard, T. (2014). *Chem. Eur. J.* 20: 8589–8593. (k) Li, Y., Li, G., and Ding, Q. (2014). *Eur. J. Org. Chem.*: 5017–5022. (l) Sheng, J., Fan, C., and Wu, J. (2014). *Chem. Commun.* 50: 5494–5496. (m) Sheng, J., Li, S., and Wu, J. (2014). *Chem. Commun.* 50: 578–580. (n) Sheng, J. and Wu, J. (2014). *Org. Biomol. Chem.* 12: 7629–7633. (o) Xiao, Q., Zhu, H., Li, G., and Chen, Z. (2014). *Adv. Synth. Catal.* 356: 3809–3815. (p) Alazet, S. and Billard, T. (2015). *Synlett* 26: 76–78. (q) Alazet, S., Ismalaj, E., Glenadel, Q. et al. (2015). *Eur. J. Org. Chem.*: 4607–4610. (r) Alazet, S., Zimmer, L., and Billard, T. (2015). *J. Fluorine Chem.* 171: 78–81. (s) Chen, D.-Q., Gao, P., Zhou, P.-X. et al. (2015). *Chem. Commun.* 51: 6637–6639. (t) Glenadel, Q., Alazet, S., and Billard, T. (2015). *J. Fluorine Chem.* 179: 89–95. (u) Glenadel, Q., Alazet, S., Tlili, A., and Billard, T. (2015). *Chem. Eur. J.* 21: 14694–14698. (v) Huang, Z., Yang, Y.-D., Tokunaga, E., and Shibata, N. (2015). *Asian J. Org. Chem.* 4: 525–527. (w) Jereb, M. and Dolenc, D. (2015). *RSC Adv.* 5: 58292–58306. (x) Jereb, M. and Gosak, K. (2015). *Org. Biomol. Chem.* 13: 3103–3115. (y) Wu, W., Zhang, X., Liang, F., and Cao, S. (2015). *Org. Biomol. Chem.* 13: 6992–6999. (z) Xiong, H.-Y., Besset, T., Cahard, D., and Pannecoucke, X. (2015). *J. Org. Chem.* 80: 4204–4212; (aa) Glenadel, Q., Alazet, S., Baert, F., and Billard, T. (2016). *Org. Process Res. Dev.* 20: 960–964. (ab) Glenadel, Q., Bordy, M., Alazet, S. et al. (2016). *Asian J. Org. Chem.* 5: 428–433. (ac) Glenadel, Q., Tlili, A., and Billard, T. (2016). *Eur. J. Org. Chem.*: 1955–1957. (ad) Ismalaj, E., Le Bars, D., and Billard, T. (2016). *Angew. Chem. Int. Ed.* 55: 4790–4793. (ae) Le, T.-N., Diter, P., Pégot, B. et al. (2016). *Org. Lett.* 18: 5102–5105. (af) Liu, T., Qiu, G., Ding, Q., and Wu, J. (2016). *Tetrahedron* 72: 1472–1476. (ag) Tlili, A., Alazet, S., Glenadel, Q., and Billard, T. (2016). *Chem. Eur. J.* 22: 10230–10234. (ah) Bonazaba Milandou, L.J.C., Carreyre, H., Alazet, S. et al. (2017). *Angew. Chem. Int. Ed.* 56: 169–172. (ai) Li, Y., Qiu, G., Wang, H., and Sheng, J. (2017). *Tetrahedron Lett.* 58: 690–693; (aj) Sheng, J., Li, Y., and Qiu, G. (2017). *Org. Chem. Front.* 4: 95–100. (ak) Arunachalam, K., Manthena, C., and Pazhamalai, A. (2018). *Eur. J. Org. Chem.*: 3276–3279.

(al) Glenadel, Q., Ayad, C., D'elia, M.-A. et al. (2018). *J. Fluorine Chem.* 210: 112–116. (am) Granados, A., Olmo, A.D., Peccati, F. et al. (2018). *J. Org. Chem.* 83: 303–313. (an) Horvat, M., Jereb, M., and Iskra, J. (2018). *Eur. J. Org. Chem.*: 3837–3843. (ao) Xi, C.-C., Chen, Z.-M., Zhang, S.-Y., and Tu, Y.-Q. (2018). *Org. Lett.* 20: 4227–4230.

57 (a) Haas, A. and Möller, G. (1996). *Chem. Ber.* 129: 1383–1388. (b) Bootwicha, T., Liu, X., Pluta, R. et al. (2013). *Angew. Chem. Int. Ed.* 52: 12856–12859. (c) Kang, K., Xu, C., and Shen, Q. (2014). *Org. Chem. Front.* 1: 294–297. (e) Pluta, R., Nikolaienko, P., and Rueping, M. (2014). *Angew. Chem. Int. Ed.* 53: 1650–1653. (f) Pluta, R. and Rueping, M. (2014). *Chem. Eur. J.* 20: 17315–17318. (g) Rueping, M., Liu, X., Bootwicha, T. et al. (2014). *Chem. Commun.* 50: 2508–2511. (h) Honeker, R., Ernst, J.B., and Glorius, F. (2015). *Chem. Eur. J.* 21: 8047–8051. (i) Liao, K., Zhou, F., Yu, J.-S. et al. (2015). *Chem. Commun.* 51: 16255–16258. (j) Shao, X., Xu, C., Lu, L., and Shen, Q. (2015). *J. Org. Chem.* 80: 3012–3021. (k) Yang, T., Lu, L., and Shen, Q. (2015). *Chem. Commun.* 51: 5479–5481. (l) Candish, L., Pitzer, L., Gómez-Suárez, A., and Glorius, F. (2016). *Chem. Eur. J.* 22: 4753–4756. (m) Honeker, R., Garza-Sanchez, R.A., Hopkinson, M.N., and Glorius, F. (2016). *Chem. Eur. J.* 22: 4395–4399. (n) Mukherjee, S., Maji, B., Tlahuext-Aca, A., and Glorius, F. (2016). *J. Am. Chem. Soc.* 138: 16200–16203. (o) Fleige, M. and Glorius, F. (2017). *Chem. Eur. J.* 23: 10773–10776. (p) Zhao, Q., Poisson, T., Pannecoucke, X. et al. (2017). *Org. Lett.* 19: 5106–5109. (q) Chachignon, H., Kondrashov, E.V., and Cahard, D. (2018). *Adv. Synth. Catal.* 360: 965–971. (r) Gelat, F., Poisson, T., Biju, A.T. et al. (2018). *Eur. J. Org. Chem.* 2018: 3693–3696. (s) Jia, Y., Qin, H., Wang, N. et al. (2018). *J. Org. Chem.* 83: 2808–2817. (t) Jin, M.Y., Li, J., Huang, R. et al. (2018). *Chem. Commun.* 54: 4581–4584. (u) Li, J., Yang, Z., Guo, R. et al. (2018). *Asian J. Org. Chem.* 7: 1784–1787. (v) Mukherjee, S., Patra, T., and Glorius, F. (2018). *ACS Catal.* 8: 5842–5846. (w) Xu, W., Ma, J., Yuan, X.-A. et al. (2018). *Angew. Chem. Int. Ed.* 57: 10357–10361. (x) Zhao, Q., Chen, M.-Y., Poisson, T. et al. (2018). *Eur. J. Org. Chem.* 2018: 6167–6175. (y) Graßl, S., Hamze, C., Koller, T.J., and Knochel, P. (2019). *Chem. Eur. J.* 25: 3752–3755. (z) Mao, R., Bera, S., Cheseaux, A., and Hu, X. (2019). *Chem. Sci.* 10: 9555–9559. (aa) Lu, S., Chen, W., and Shen, Q. (2019). *Chin. Chem. Lett.* 30: 2279–2281.

58 Zhang, J., Yang, J.-D., Zheng, H. et al. (2018). *Angew. Chem. Int. Ed.* 57: 12690–12695.

59 (a) Xu, C., Ma, B., and Shen, Q. (2014). *Angew. Chem. Int. Ed.* 53: 9316–9320. (b) Luo, J., Zhu, Z., Liu, Y., and Zhao, X. (2015). *Org. Lett.* 17: 3620–3623. (c) Maeno, M., Shibata, N., and Cahard, D. (2015). *Org. Lett.* 17: 1990–1993. (d) Wang, Q., Qi, Z., Xie, F., and Li, X. (2015). *Adv. Synth. Catal.* 357: 355–360. (e) Wang, Q., Xie, F., and Li, X. (2015). *J. Org. Chem.* 80: 8361–8366. (f) Xu, C. and Shen, Q. (2015). *Org. Lett.* 17: 4561–4563. (g) Hu, L., Wu, M., Wan, H. et al. (2016). *New J. Chem.* 40: 6550–6553. (h) Wu, J.-J., Xu, J., and Zhao, X. (2016). *Chem. Eur. J.* 22: 15265–15269. (i) Yu, Y., Xiong, D.-C., and Ye, X.-S. (2016). *Org. Biomol. Chem.* 14: 6403–6406. (j) Dagousset, G., Simon, C., Anselmi, E. et al. (2017). *Chem. Eur. J.* 23: 4282–4286. (k) Ernst, J.B., Rakers,

L., and Glorius, F. (2017). *Synthesis* 49: 260–268. (l) Fang, J., Wang, Z.-K., Wu, S.-W. et al. (2017). *Chem. Commun.* 53: 7638–7641. (m) Wei, F., Zhou, T., Ma, Y. et al. (2017). *Org. Lett.* 19: 2098–2101. (n) Zhu, Z., Luo, J., and Zhao, X. (2017). *Org. Lett.* 19: 4940–4943. (o) Abubakar, S.S., Benaglia, M., Rossi, S., and Annunziata, R. (2018). *Catal. Today* 308: 94–101. (p) Guyon, H., Chachignon, H., Tognetti, V. et al. (2018). *Eur. J. Org. Chem.*: 3756–3763. (q) Kondo, H., Maeno, M., Sasaki, K. et al. (2018). *Org. Lett.* 20: 7044–7048. (r) Nalbandian, C.J., Brown, Z.E., Alvarez, E., and Gustafson, J.L. (2018). *Org. Lett.* 20: 3211–3214. (s) Xu, J., Zhang, Y., Qin, T., and Zhao, X. (2018). *Org. Lett.* 20: 6384–6388. (t) Carthy, C.M., Tacke, M., and Zhu, X. (2019). *Eur. J. Org. Chem.*: 2729–2734. (u) Guo, R., Cai, B., Jin, M.Y. et al. (2019). *Asian J. Org. Chem.* 8: 687–690. (v) Liu, S., Zeng, X., and Xu, B. (2019). *Asian J. Org. Chem.* 8: 1372–1375. (w) Zhu, M., Li, R., You, Q. et al. (2019). *Asian J. Org. Chem.* 8: 2002–2005.

60 (a) Liu, X., An, R., Zhang, X. et al. (2016). *Angew. Chem. Int. Ed.* 55: 5846–5850. (b) Zhang, P., Li, M., Xue, X.-S. et al. (2016). *J. Org. Chem.* 81: 7486–7509. (c) Lübcke, M., Yuan, W., and Szabó, K.J. (2017). *Org. Lett.* 19: 4548–4551. (d) Luo, J., Liu, Y., and Zhao, X. (2017). *Org. Lett.* 19: 3434–3437. (e) Yoshida, M., Kawai, K., Tanaka, R. et al. (2017). *Chem. Commun.* 53: 5974–5977. (f) Chen, M., Wei, Y., and Shi, M. (2018). *Org. Chem. Front.* 5: 2030–2034. (g) Liu, X., Liang, Y., Ji, J. et al. (2018). *J. Am. Chem. Soc.* 140: 4782–4786. (h) Luo, J., Cao, Q., Cao, X., and Zhao, X. (2018). *Nat. Commun.* 9: 527. (i) Mai, B.K., Szabó, K.J., and Himo, F. (2018). *Org. Lett.* 20: 6646–6649. (j) Meanwell, M., Adluri, B.S., Yuan, Z. et al. (2018). *Chem. Sci.* 9: 5608–5613. (k) Lübcke, M., Bezhan, D., and Szabó, K.J. (2019). *Chem. Sci.* 10: 5990–5995. (l) Patra, T., Mukherjee, S., Ma, J. et al. (2019). *Angew. Chem. Int. Ed.* 58: 10514–10520. (m) Qin, T., Jiang, Q., Ji, J. et al. (2019). *Org. Biomol. Chem.* 17: 1763–1766.

61 Rayman, M.P. (2000). *Lancet* 356: 233–241.

62 (a) Block, E., Booker, S.J., Flores-Penalba, S. et al. (2016). *ChemBioChem* 17: 1738–1751. (b) Reem, M., Rebecca, N.D., and Norman, M. (2017). *Angew. Chem. Int. Ed.* 56: 15818–15827. (c) Rocha, J.B.T., Piccoli, B.C., and Oliveira, C.S. (2017). *ARKIVOC*: 457–491. (d) Anouar, Y., Lihrmann, I., Falluel-Morel, A., and Boukhzar, L. (2018). *Free Radical Biol. Med.* 127: 145–152. (e) Bertz, M., Kühn, K., Koeberle, S.C. et al. (2018). *Free Radical Biol. Med.* 127: 98–107. (f) Brigelius-Flohé, R. and Arnér, E.S.J. (2018). *Free Radical Biol. Med.* 127: 1–2. (g) Fernandes, J., Hu, X., Ryan Smith, M. et al. (2018). *Free Radical Biol. Med.* 127: 215–227. (h) Liao, C., Carlson, B.A., Paulson, R.F., and Prabhu, K.S. (2018). *Free Radical Biol. Med.* 127: 165–171. (i) Short, S.P., Pilat, J.M., and Williams, C.S. (2018). *Free Radical Biol. Med.* 127: 26–35. (j) Solovyev, N., Drobyshev, E., Björklund, G. et al. (2018). *Free Radical Biol. Med.* 127: 124–133.

63 (b) Glenadel, Q., Ismalaj, E., and Billard, T. (2017). *Eur. J. Org. Chem.*: 530–533.

64 (a) Mugesh, G., Du Mont, W.-W., and Sies, H. (2001). *Chem. Rev.* 101: 2125–2180. (b) Singh, N., Halliday, A.C., Thomas, J.M. et al. (2013). *Nat. Commun.* 4: 1332. (c) Victoria, F.N., Anversa, R., Penteado, F. et al. (2014). *Eur. J.*

Pharmacol. 742: 131–138. (d) Thangamani, S., Younis, W., and Seleem, M.N. (2015). *Sci. Rep.* 5: 11596. (e) Antoniadou, I., Kouskou, M., Arsiwala, T. et al. (2018). *Br. J. Pharmacol.* 175: 2599–2610. (f) Cheignon, C., Cordeau, E., Prache, N. et al. (2018). *J. Med. Chem.* 61: 10173–10184. (g) Gandin, V., Khalkar, P., Braude, J., and Fernandes, A.P. (2018). *Free Radical Biol. Med.* 127: 80–97.

65 Billard, T., Toulgoat, F. when fluorine meets selenium in *Emerging Fluorinated Motifs: Synthesis, Properties and Applications* (Eds.: D. Cahard, J.-A. Ma), Wiley, 2020. 691-721.

66 Billard, T. and Langlois, B.R. (1996). *Tetrahedron Lett.* 37: 6865–6868.

67 Billard, T., Langlois, B.R., and Large, S. (1998). *Phosphorus, Sulfur Silicon Relat. Elem.* 136,137&138: 521–524.

68 Nikolaienko, P. and Rueping, M. (2016). *Chem. Eur. J.* 22: 2620–2623.

69 Large, S., Roques, N., and Langlois, B.R. (2000). *J. Org. Chem.* 65: 8848–8856.

70 Carbonnel, E., Besset, T., Poisson, T. et al. (2017). *Chem. Commun.* 53: 5706–5709.

71 Potash, S. and Rozen, S. (2014). *J. Org. Chem.* 79: 11205–11208.

72 Blond, G., Billard, T., and Langlois, B.R. (2001). *Tetrahedron Lett.* 42: 2473–2475.

73 Cherkupally, P. and Beier, P. (2010). *Tetrahedron Lett.* 51: 252–255.

74 Geri, J.B., Wade Wolfe, M.M., and Szymczak, N.K. (2018). *Angew. Chem. Int. Ed.* 57: 1381–1385.

75 (a) Pooput, C., Medebielle, M., and Dolbier, W.R. (2004). *Org. Lett.* 6: 301–303. (b) Pooput, C., Dolbier, W.R., and Médebielle, M. (2006). *J. Org. Chem.* 71: 3564–3568.

76 (a) Ma, J.-J., Yi, W.-B., Lu, G.-P., and Cai, C. (2016). *Catal. Sci. Technol.* 6: 417–421. (b) Ma, J.-J., Liu, Q.-R., Lu, G.-P., and Yi, W.-B. (2017). *J. Fluorine Chem.* 193: 113–117.

77 (a) Billard, T., Langlois, B.R., Large, S. et al. (1996). *J. Org. Chem.* 61: 7545–7550. (b) Billard, T. and Langlois, B.R. (1997). *J. Fluorine Chem.* 84: 63–64.

78 Billard, T., Roques, N., and Langlois, B.R. (1999). *J. Org. Chem.* 64: 3813–3820.

79 (a) Magnier, E., Vit, E., and Wakselman, C. (2001). *Synlett*: 1260–1262. (b) Magnier, E. and Wakselman, C. (2002). *Collect. Czech. Chem. Commun.* 67: 1262–1266.

80 (a) Dale, J.W., Eméléus, H.J., and Haszeldine, R.N. (1958). *J. Chem. Soc. (Resumed)*: 2939–2945. (b) Darmadi, A., Haas, A., and Koch, B. (1980). *Z. Naturforsch., B: Chem. Sci.* 35: 526. (c) Tyrra, W.E. (2001). *J. Fluorine Chem.* 112: 149–152. (d) Naumann, D., Tyrra, W., Quadt, S. et al. (2005). *Z. Anorg. Allg. Chem.* 631: 2733–2737. (e) Kirij, N.V., Tyrra, W., Pantenburg, I. et al. (2006). *J. Organomet. Chem.* 691: 2679–2685. (f) Haas, A. (1986). *J. Fluorine Chem.* 32: 415–439. (g) Feldhoff, R., Haas, A., and Lieb, M. (1994). *J. Fluorine Chem.* 67: 245–251.

81 Chen, C., Hou, C., Wang, Y. et al. (2014). *Org. Lett.* 16: 524–527.

82 Chen, C., Ouyang, L., Lin, Q. et al. (2014). *Chem. Eur. J.* 20: 657–661.

83 Rong, M., Huang, R., You, Y., and Weng, Z. (2014). *Tetrahedron* 70: 8872–8878.

84 Yang, Y., Lin, X., Zheng, Z. et al. (2017). *J. Fluorine Chem.* 204: 1–5.

85 Chen, T., You, Y., and Weng, Z. (2018). *J. Fluorine Chem.* 216: 43–46.

86 Tian, Q. and Weng, Z. (2016). *Chin. J. Chem.* 34: 505–510.

87 Wu, C., Huang, Y., Chen, Z., and Weng, Z. (2015). *Tetrahedron Lett.* 56: 3838–3841.

88 Wang, Y., You, Y., and Weng, Z. (2015). *Org. Chem. Front.* 2: 574–577.

89 Tyrra, W., Naumann, D., and Yagupolskii, Y.L. (2003). *J. Fluorine Chem.* 123: 183–187.

90 Dong, T., He, J., Li, Z.-H., and Zhang, C.-P. (2018). *ACS Sustainable Chem. Eng.* 6: 1327–1335.

91 Aufiero, M., Sperger, T., Tsang, A.S.K., and Schoenebeck, F. (2015). *Angew. Chem. Int. Ed.* 54: 10322–10326.

92 (a) Han, J.-B., Dong, T., Vicic, D.A., and Zhang, C.-P. (2017). *Org. Lett.* 19: 3919–3922. (b) Dürr, A.B., Fisher, H.C., Kalvet, I. et al. (2017). *Angew. Chem. Int. Ed.* 56: 13431–13435.

93 Lefebvre, Q., Pluta, R., and Rueping, M. (2015). *Chem. Commun.* 51: 4394–4397.

94 Han, Q.-Y., Zhao, C.-L., Dong, T. et al. (2019). *Org. Chem. Front.* 6: 2732–2737.

95 Han, Q.-Y., Tan, K.-L., Wang, H.-N., and Zhang, C.-P. (2019). *Org. Lett.* 21: 10013–10017.

96 Chen, X.-L., Zhou, S.-H., Lin, J.-H. et al. (2019). *Chem. Commun.* 55: 1410–1413.

97 Yarovenko, N.N., Shemanina, V.N., and Gazieva, G.B. (1959). *Russ. J. Gen. Chem.* 29: 924–927.

98 Glenadel, Q., Ismalaj, E., and Billard, T. (2016). *J. Org. Chem.* 81: 8268–8275.

99 Yagupol'skii, L.M. and Voloshchuk, V.G. (1966). *Russ. J. Gen. Chem.* 36: 173–174.

100 Voloshchuk, V.G., Yagupol'skii, L.M., Syrova, G.P., and Bystrov, V.P. (1967). *Russ. J. Gen. Chem.* 37: 105–108.

101 Yagupol'skii, L.M. and Voloshchuk, V.G. (1968). *Russ. J. Gen. Chem.* 38: 2426–2429.

102 Haas, A., Lieb, M., and Schwederski, B. (1987). *Rev. Roum. Chim.* 32: 1219–1224.

103 Yagupol'skii, L.M. and Voloshchuk, V.G. (1967). *Russ. J. Gen. Chem.* 37: 1463–1465.

104 Haas, A. and Praas, H.-W. (1992). *Chem. Ber.* 125: 571–579.

105 Welcman, N. and Wulf, M. (1968). *Isr. J. Chem.* 6: 37–41.

106 Stump, E.C. (1967). *Chem. Eng. News* 45: 51, 44.

107 Ghiazza, C., Glenadel, Q., Tlili, A., and Billard, T. (2017). *Eur. J. Org. Chem.*: 3812–3814.

108 Ghiazza, C., Tlili, A., and Billard, T. (2018). *Eur. J. Org. Chem.*: 3498–3498.

109 Ghiazza, C., Tlili, A., and Billard, T. (2017). *Molecules* 22: 833/831–833/838.

110 Glenadel, Q., Ismalaj, E., and Billard, T. (2018). *Org. Lett.* 20: 56–59.

111 Glenadel, Q., Ghiazza, C., Tlili, A., and Billard, T. (2017). *Adv. Synth. Catal.* 359: 3414–3420.

112 Ghiazza, C., Tlili, A., and Billard, T. (2017). *Beilstein J. Org. Chem.* 13: 2626–2630.

113 Ghiazza, C., Debrauwer, V., Billard, T., and Tlili, A. (2018). *Chem. Eur. J.* 24: 97–100.

114 Ghiazza, C., Kataria, A., Tlili, A. et al. (2019). *Asian J. Org. Chem.* 8: 675–678.

115 Ghiazza, C., Debrauwer, V., Monnereau, C. et al. (2018). *Angew. Chem. Int. Ed.* 57: 11781–11785.

116 Ghiazza, C., Khrouz, L., Monnereau, C. et al. (2018). *Chem. Commun.* 54: 9909–9912.

117 Xia, Y. and Studer, A. (2019). *Angew. Chem. Int. Ed.* 58: 9836–9840.

118 Dix, S., Jakob, M., and Hopkinson, M.N. (2019). *Chem. Eur. J.* 25: 7635–7639.

4

Introduction of Trifluoromethylthio Group into Organic Molecules

Hangming Ge, He Liu and Qilong Shen

Key Laboratory of Organofluorine Chemistry, Shanghai Institute of Organic Chemistry, University of Chinese Academy of Sciences, Chinese Academy of Sciences, 345 Lingling Road, Shanghai 200032, China

4.1 Introduction

The excellent beneficial properties [1] of the trifluoromethylthio group have urged synthetic chemists to develop efficient methods for the incorporation of the trifluoromethylthio group into the target molecules. In the past decade, numerous trifluoromethylthiolating methods have been developed and a few reviews have already appeared [2]. Given the rapid development in this field, we think that a detailed summary of recent progress in the development of new trifluoromethylthiolating methods is necessary, which will be presented in this chapter.

4.2 Nucleophilic Trifluoromethylthiolation

4.2.1 Preparation of Nucleophilic Trifluoromethylthiolating Reagent

4.2.1.1 Preparation of $Hg(SCF_3)_2$, $AgSCF_3$, and $CuSCF_3$

Reaction of a nucleophilic trifluoromethylthiolating reagent with an electrophile is a general method for the introduction of the trifluoromethylthio group into a small molecule, which was the early focus in the field of trifluoromethylthiolation. In 1959, Muetterties and coworkers reported the development of the first nucleophilic trifluoromethylthiolating reagent $Hg(SCF_3)_2$, which was prepared by reaction of HgF_2 with CS_2 at 250 °C [3] (Scheme 4.1). Yet, the high toxicity of $Hg(SCF_3)_2$ limited its further widespread application. Consequently, two other organometallic reagents, $CuSCF_3$ and $AgSCF_3$, which are now commonly used nucleophilic trifluoromethylthiolating reagents, were developed by the treatment of $Hg(SCF_3)_2$ with copper powder or $AgNO_3$, respectively. In 1961, MacDuffie and coworker simplified the synthesis procedure of $AgSCF_3$ by the reaction of AgF with CS_2 at 140 °C, which generated $AgSCF_3$ in 70–80% yields [4] (Scheme 4.1). In addition, Yagupolskii et al. reported in 1975 that $CuSCF_3$ could be easily prepared from $AgSCF_3$ in quantitative yield when it reacted with $CuBr$ [5] (Scheme 4.1). Moreover, $CuSCF_3$ could also be accessed

Organofluorine Chemistry: Synthesis, Modeling, and Applications,
First Edition. Edited by Kálmán J. Szabó and Nicklas Selander.

from Chen's Reagent ($FO_2SCF_2CO_2Me$) when it was treated with CuI and sulfur powder [6].

$$3HgF_2 + 2CS_2 \xrightarrow[75\%]{250\ ^\circ C} 2HgS + Hg(SCF_3)_2$$

$$3AgF + CS_2 \xrightarrow[70–80\%]{140\ ^\circ C} Ag_2S + AgSCF_3$$

$$AgSCF_3 + CuBr \xrightarrow[>99\%]{140\ ^\circ C} AgBr + CuSCF_3$$

Scheme 4.1 Preparation of $Hg(SCF_3)_2$, $AgSCF_3$, and $CuSCF_3$.

4.2.1.2 Preparation of MSCF$_3$ (M = K, Cs, Me$_4$N, and S(NMe$_2$)$_3$)

A few other ionic nucleophilic trifluoromethylthiolating reagents $MSCF_3$ (M = K, Cs, Me_4N, and $S(NMe_2)_3$) have also been reported. These reagents were prepared from highly toxic starting materials such as $C(S)F_2$ or CF_3SSCF_3 via a multistep procedure, thus limiting their further applications [7]. In 2003, Tyrra et al. described an improved one-pot procedure for the preparation of these ionic nucleophilic trifluoromethylthiolating reagents from Ruppert–Prakash reagent (TMSCF$_3$, TMS = Me$_3$Si, trimethylsilyl), elemental sulfur, and the corresponding fluoride source [8] (Scheme 4.2). For example, treatment of Ruppert–Prakash reagent (Me_3SiCF_3) and elemental sulfur with Me_4NF in tetrahydrofuran (THF) at $-60\ ^\circ C$ for 30 minutes and then allowing to warm to room temperature afforded Me_4NSCF_3 in 88% yield.

$$MF + TMSCF_3 + 1/8S_8 \xrightarrow[\substack{-60\ ^\circ C\ to\ r.t. \\ 75\%}]{\substack{Diglyme \\ or\ THF}} TMSF + MSCF_3$$
$$M = K, Cs, NMe_4, etc.$$

Scheme 4.2 One-pot procedure for the preparation of $MSCF_3$ from $TMSCF_3$, S, and MF.

4.2.1.3 Preparation of Stable Trifluoromethylthiolated Copper(I) Complexes

One of the problems for the above-mentioned nucleophilic trifluoromethylthiolating reagents is their thermal stability since these compounds decompose quickly at high temperature. For example, in 1992, Munavalli's group reported the preparation of the first crystalline trifluoromethylthio copper(I) complex $(CuSCF_3)_{10}(MeCN)_8$ from CF_3SSCF_3 with copper powder in MeCN [9]. However, it was found that the solid was not stable in crystalline form for a prolonged period.

To improve the stability of the trifluoromethylthiolated copper species, in 2013, Weng et al. tried to install a bidentate nitrogen ligand on copper metal center [10]. It was found that 2,2′-bipyridine-ligated trifluoromethylthiolated copper complex was much more stable since (bpy)CuSCF$_3$ (bpy = 2,2′-bipyridine) was not sensitive to air, moisture, or light. Switching the ligand from 2,2′-bipyridine to 1,10-phenanthroline (phen) led to the formation of the dinuclear complex [(phen)CuSCF$_3$]$_2$ (Scheme 4.3).

Likewise, in 2016, Vicic and coworkers reported the preparation of air-stable phosphine-coordinated trifluoromethylthio copper(I) complexes $(Ph_3P)_2CuSCF_3$ and $(dppf)CuSCF_3$ (dppf = bis(diphenylphosphino)ferrocene) from CF_3SO_2Na using Ph_3P as the reducing reagent [11] (Scheme 4.3). These stable complexes can participate in trifluoromethylthiolation of a series of electrophilic substrates, such as aryl, alkyl, vinyl, and acyl halides [10–12].

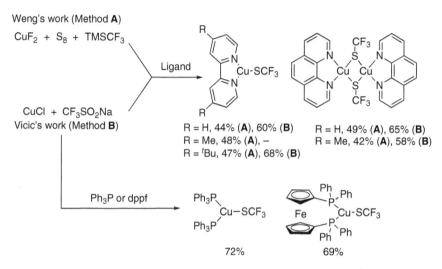

Scheme 4.3 *N,P*-ligated shelf-stable CuSCF$_3$ complexes.

4.2.2 Formation of C(sp²)-SCF₃ by Nucleophilic Trifluoromethylthiolating Reagents

4.2.2.1 Reaction of CuSCF₃ with Aryl Halides

Classic methods for the preparation of aryltrifluoromethylthioethers typically involved the chlorination of arylmethylthioether to generate trichloromethylthioether, followed by fluorine/chlorine exchange. Yet, the reaction conditions were harsh and *ortho*-substituted trifluoromethylthioethers were inaccessible. To address this problem, Yagupolskii et al. reported a general method for the preparation of aryltrifluoromethylthioethers by reactions of CuSCF$_3$ with aryl halides. In general, reactions of electron-poor aryl iodides occurred in 70–75% yields, while the same reactions with electron-rich aryl iodides occurred in much lower yields (30–55% yields) [5] (Scheme 4.4). Later on, Chen and Duan reported that *in situ*-generated CuSCF$_3$ from Chen's Reagent ($FO_2SCF_2CO_2Me$), CuI, and sulfur powder were able to react with aryl bromides when the reaction was conducted in NMP (*N*-methylpyrrolidone) or HMPA (hexamethylphosphoramide) [6] (Scheme 4.4).

In 2012, Weng et al. discovered that the air-stable nucleophilic trifluoromethylthiolating reagent [(bpy)CuSCF$_3$] was able to react with a variety of electron-rich or -poor aryl iodides in good to excellent yields, while reactions of aryl bromides were limited to electron-poor aryl bromides or activated heteroaryl bromides such as 2-bromopyridine [10] (Scheme 4.4).

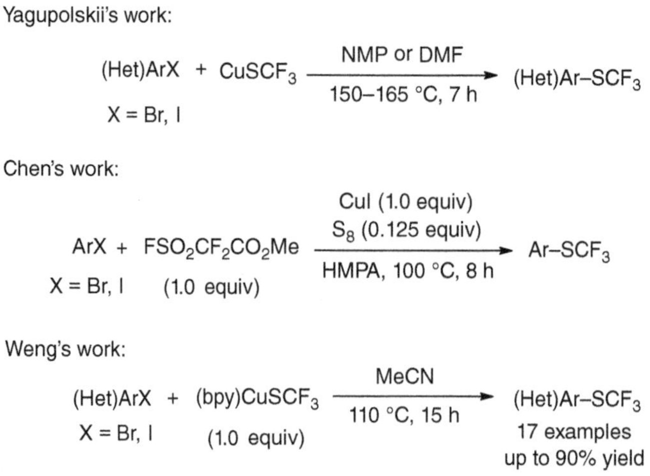

Yagupolskii's work:

$$(\text{Het})\text{ArX} + \text{CuSCF}_3 \xrightarrow[150-165\ °C,\ 7\ h]{\text{NMP or DMF}} (\text{Het})\text{Ar–SCF}_3$$

$$X = \text{Br, I}$$

Chen's work:

$$\text{ArX} + \text{FSO}_2\text{CF}_2\text{CO}_2\text{Me} \xrightarrow[\text{HMPA, 100 °C, 8 h}]{\substack{\text{CuI (1.0 equiv)} \\ \text{S}_8\ (0.125\ \text{equiv})}} \text{Ar–SCF}_3$$

$$X = \text{Br, I} \quad (1.0\ \text{equiv})$$

Weng's work:

$$(\text{Het})\text{ArX} + (\text{bpy})\text{CuSCF}_3 \xrightarrow[110\ °C,\ 15\ h]{\text{MeCN}} (\text{Het})\text{Ar–SCF}_3$$

$$X = \text{Br, I} \quad (1.0\ \text{equiv})$$

17 examples
up to 90% yield

Scheme 4.4 Reactions of $(L_n)\text{CuSCF}_3$ with aryl iodides and bromides.

4.2.2.2 Sandmeyer-Type Trifluoromethylthiolation

Sandmeyer reaction represents one of the most useful methods for the introduction of functional groups to aromatic ring. Based on the same principle, Sandmeyer-type trifluoromethylthiolation was also accomplished by using AgSCF_3 and CuSCF_3 as the nucleophilic trifluoromethylthio reagent. As early as in 1975, Kondratenko's group reported the first example of trifluoromethylthiolation of aryl diazonium salts [13]. In this case, AgSCF_3 was served as trifluoromethylthiolating reagent to deliver the corresponding trifluoromethylthioether in moderate yields. In 2000, Clark's group improved the reaction by using CuSCF_3 as the nucleophilic trifluoromethylthiolating reagent that generally occurred in higher yields [14]. Nevertheless, the reaction was limited to aryl diazonium salts with electron-withdrawing groups (Scheme 4.5). To further expand the scope of the reaction, Goossen's group developed an alternative strategy for the preparation of aryltrifluoromethylthioethers from aryl diazonium salts [15]. In this method, CuSCN instead of CuSCF_3 was initially allowed to react with aryl diazonium salts to give aryl thiocyanates, which was then reacted with TMSCF_3 using Cs_2CO_3 as the base to give the trifluoromethylthiolated arenes in good to excellent yields. This method was applicable to a variety of arenediazonium salts regardless of whether the aryl group had the electron-donating or -accepting substituted groups (Scheme 4.5).

More recently, Goossen and coworkers reported another method for the preparation of aryltrifluoromethylthioethers from aryl diazonium salts. In this method, Me_4NSCF_3, which was used as the nucleophilic trifluoromethylthiolating reagent, reacted with both electron-rich and electron-poor aryl diazonium salts in excellent yields when CuSCN was used as the catalyst [16]. Mechanistic studies suggest that Me_4NSCF_3 was served as a trifluoromethylthio radical source ($\cdot\text{SCF}_3$) in the reaction. A single-electron-transfer (SET) process takes place between arenediazonium salts and Me_4NSCF_3 to generate $\cdot\text{SCF}_3$ and an aryl radical, concurrently releasing

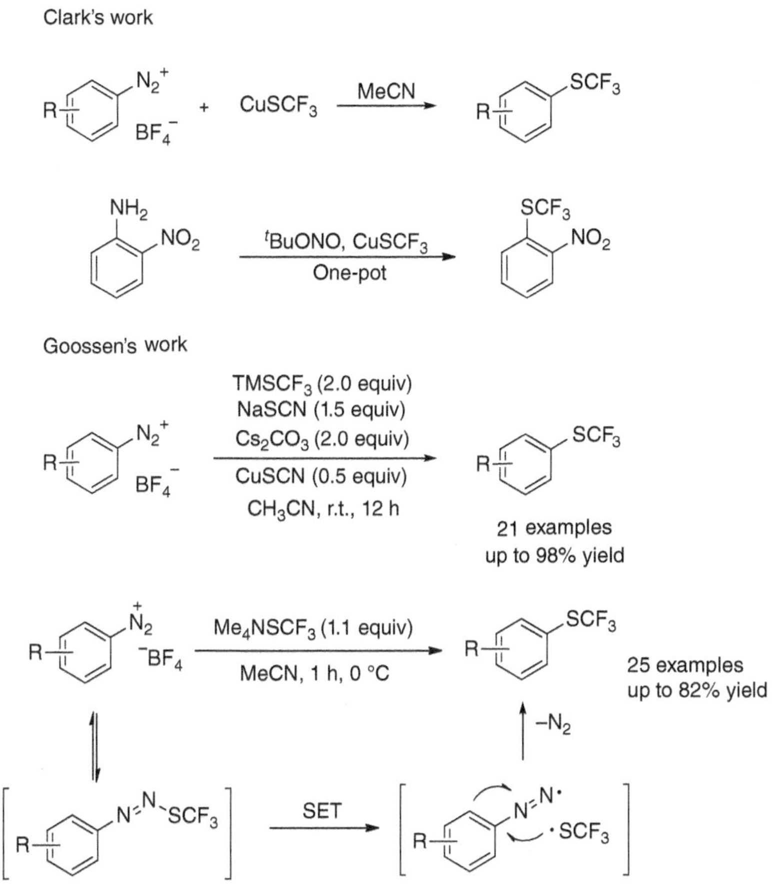

Clark's work

Goossen's work

Scheme 4.5 Sandmeyer-type trifluoromethylthiolation.

nitrogen gas. The aryl radical then reacts with trifluoromethylthio radical to form the aryltrifluoromethylthioether (Scheme 4.5).

4.2.2.3 Transition Metal-Catalyzed Trifluoromethylthiolation

One limitation of the methods for the preparation of aryltrifluoromethylthioethers from CuSCF$_3$ with aryl halides or Sandmeyer-type trifluoromethylthiolation with aryl diazonium salts is that both methods require the use of stoichiometric amount of copper salt. In addition, the substrates in both reactions were limited to activated aryl electrophiles such as aryl iodides, activated aryl bromides, or aryl diazonium salts; unactivated aryl bromides or aryl chlorides were not suitable for such a transformation. To overcome these limitations, Buchwald and coworkers made a breakthrough in 2011 by employing palladium as the catalyst, which enabled coupling a variety of aryl bromides including those with electron-donating groups, with AgSCF$_3$ in the presence of 1.3 equiv of Ph(Et)$_3$NI as an additive, for the very first time [17]. The key for the success of such a cross-coupling was the use of an electron-rich, sterically hindered alkyl phosphine Brettphos (dicyclohexyl[3,6-dimethoxy-2′,4′,6′-tri(isopropyl)[1,1′-biphenyl]]-2-yl] phosphine) as the ligand (Scheme 4.6).

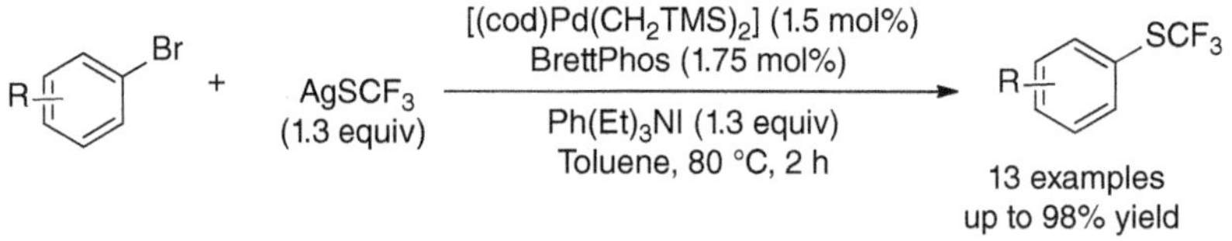

$$[(cod)Pd(CH_2TMS)_2] \ (1.5 \ mol\%)$$
$$BrettPhos \ (1.75 \ mol\%)$$

R + AgSCF₃ (1.3 equiv) $\xrightarrow{\hspace{1cm}}$ R—SCF₃

Ph(Et)₃NI (1.3 equiv)
Toluene, 80 °C, 2 h

13 examples
up to 98% yield

Scheme 4.6 Palladium-catalyzed coupling of aryl bromides with AgSCF₃.

Later on, Schoenebeck and coworkers discovered that a dinuclear palladium(I) complex was able to catalyze the same coupling reaction [18]. Not only aryl iodides but also aryl bromides reacted with Me₄NSCF₃ to give the corresponding aryltrifluoromethylthioethers in high yields. Mechanistic studies showed that the iodide-bridged palladium(I) dimer initially reacted with Me₄NSCF₃ to give trifluoromethylthio-bridged palladium(I) dimer, which then reacted with aryl halides to give the target trifluoromethylthiolated arenes [18] (Scheme 4.7). Since the Pd(I) catalyst is not air sensitive, it provides a considerable advantage over more sensitive Pd(0)-catalyzed processes for practical applications.

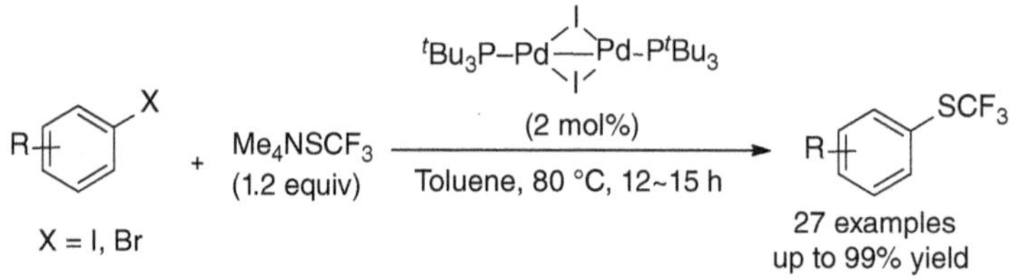

$$^tBu_3P-Pd \overset{I}{\underset{I}{\diagup\hspace{-0.3cm}\diagdown}} Pd-P^tBu_3$$

R + Me₄NSCF₃ (1.2 equiv) $\xrightarrow[\text{Toluene, 80 °C, 12~15 h}]{\text{(2 mol\%)}}$ R—SCF₃

X = I, Br

27 examples
up to 99% yield

Scheme 4.7 Dinuclear palladium-catalyzed coupling of aryl iodides and bromides with Me₄NSCF₃.

Not only palladium but also nickel was able to catalyze the reaction of aryl halides with nucleophilic trifluoromethylthiolating reagents. The first such nickel-catalyzed cross-coupling reaction was reported by Vicic and coworker in 2012 [19]. It was found that reactions of a variety of aryl iodides and bromides with Me₄NSCF₃ occurred smoothly at room temperature to give the corresponding trifluoromethylthiolated products in moderate to excellent yields when a combination of Ni(COD)₂ (COD = 1,5-cyclooctadiene) and 4,4′-dimethoxybipyridine (dmbpy) was used as the catalyst. Interestingly, reactions of aryl bromides with electron-withdrawing groups occurred in low yields and aryl chlorides did not react at all under these conditions (Scheme 4.8).

Notably, by switching the ligand from dmbpy to dppf, Schoenebeck and coworkers realized the first example of nickel-catalyzed trifluoromethylthiolation of aryl chlorides [20]. It was found that reactions of aryl chlorides with electron-withdrawing substituents occurred in high yields while the reaction of aryl chlorides with electron-rich substituents required the presence of MeCN as additive. The protocol

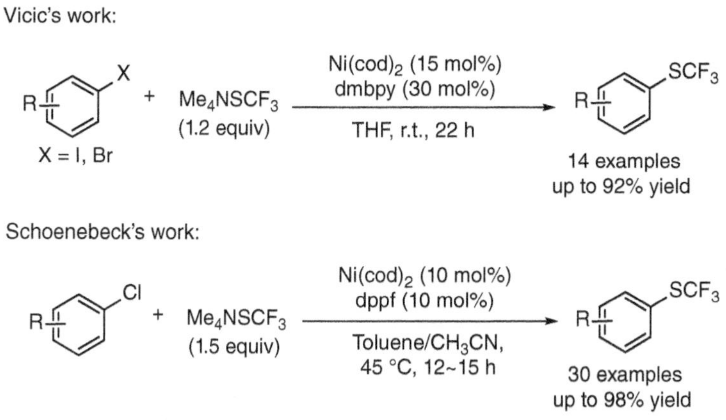

Vicic's work:

R—⟨ ⟩—X + Me₄NSCF₃ (1.2 equiv) → [Ni(cod)₂ (15 mol%), dmbpy (30 mol%), THF, r.t., 22 h] → R—⟨ ⟩—SCF₃

X = I, Br

14 examples
up to 92% yield

Schoenebeck's work:

R—⟨ ⟩—Cl + Me₄NSCF₃ (1.5 equiv) → [Ni(cod)₂ (10 mol%), dppf (10 mol%), Toluene/CH₃CN, 45 °C, 12~15 h] → R—⟨ ⟩—SCF₃

30 examples
up to 98% yield

Scheme 4.8 Nickel-catalyzed coupling of aryl halides with Me₄NSCF₃.

was then applied to the synthesis of a few trifluoromethylthiolated derivatives of drug molecules (Scheme 4.8).

In addition, the same protocol was further extended to trifluoromethylthiolate other substrates such as aryl and vinyl triflates or nonaflates with good to excellent yields [21] (Scheme 4.9).

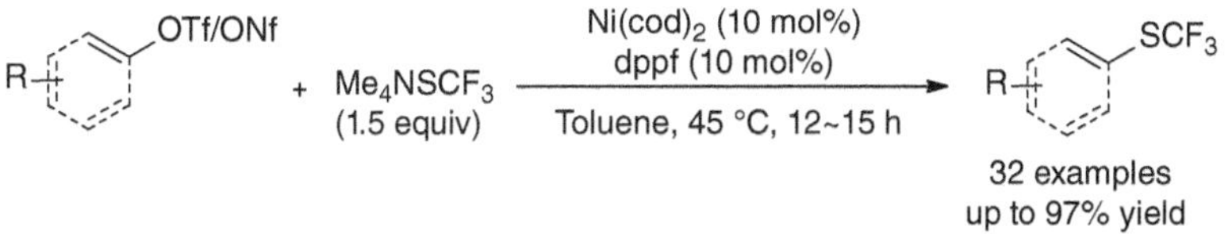

R—⟨ ⟩—OTf/ONf + Me₄NSCF₃ (1.5 equiv) → [Ni(cod)₂ (10 mol%), dppf (10 mol%), Toluene, 45 °C, 12~15 h] → R—⟨ ⟩—SCF₃

32 examples
up to 97% yield

Scheme 4.9 Nickel-catalyzed coupling of aryl, vinyl triflates, or nonaflates with Me₄NSCF₃.

Nickel-catalyzed trifluoromethylthiolation could also take place in the absence of ligand when a directing group was present at the *ortho*-position of the aryl halides. In 2016, Love and coworkers discovered that aryl halides with an *ortho*-directing group such as imine, amide, or oxazoline could react with AgSCF₃ using Ni(COD)₂ as the catalyst in the absence of any ligand [22]. The reaction occurred smoothly at room temperature. Not only (hetero)aryl bromides but also (hetero)aryl chlorides reacted to give the corresponding trifluoromethylthiolated (hetero)arenes in high yields (Scheme 4.10).

Likewise, in the presence of a directing group, copper was also able to catalyze trifluoromethylthiolative cross-coupling reaction of aryl halides with nucleophilic trifluoromethylthiolating reagents. In 2014, Liu and coworkers reported the Cu-catalyzed trifluoromethylthiolation of aryl bromides and iodides with a directing group such as amide, imine, and oxime, pyridyl, or ester [23]. The choice of the nucleophilic trifluoromethylthiolating reagents was crucial for the transformation since Me₄NSCF₃ was not as effective as AgSCF₃. A Cu(I/III) catalytic cycle was proposed for the formation of aryl—SCF₃ bond (Scheme 4.11).

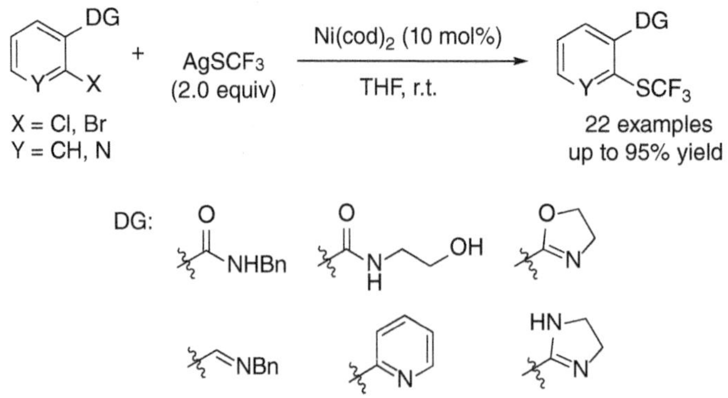

X = Cl, Br
Y = CH, N

DG:

Scheme 4.10 Ligandless nickel-catalyzed *ortho*-selective directed trifluoromethylthiolation of aryl chlorides and bromides using AgSCF$_3$.

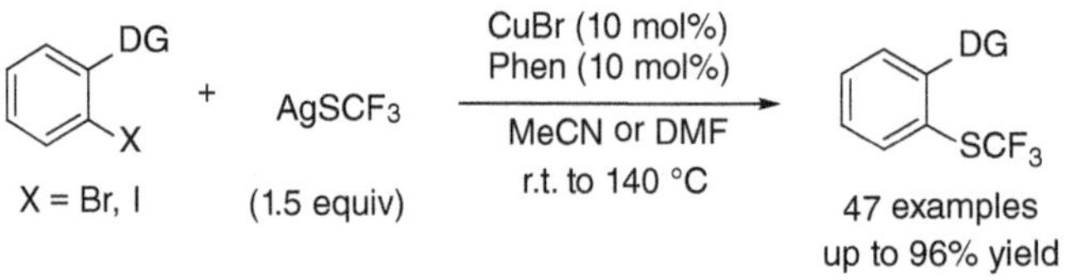

X = Br, I

Scheme 4.11 Copper-catalyzed trifluoromethylthiolation of aryl halides with directing groups using AgSCF$_3$.

Hypervalent iodide reagents were better electrophiles than aryl iodides in cross-coupling. Based on this phenomenon, Anbarasan and coworker reported a Cu-catalyzed trifluoromethylthiolation reaction using di(hetero)aryl-λ^3-iodanes as the electrophiles [24]. The reaction exhibited a broad substrate scope, and various common functional groups such as nitrile, enolizable ketone, ester, nitro, and even free carboxylic acid were compatible (Scheme 4.12). Mechanistic studies showed that [(SPhos)CuSCF$_3$] is initially generated upon mixing CuI, SPhos (SPhos = 2-dicyclohexylphosphino-2′,6′-dimethoxybiphenyl), and AgSCF$_3$. This species then reacts with Ar$_2$IOTf in the presence of a silver salt to generate an ionic copper(III) complex [(SPhos)Cu(SCF$_3$)(Ar)]$^+$X$^-$, which then reductively eliminates to give the corresponding trifluoromethylthiolated arene. The silver salt plays an important role since, in the absence of silver salt, reaction of [(SPhos)CuSCF$_3$] with Ar$_2$IOTf does not take place at all.

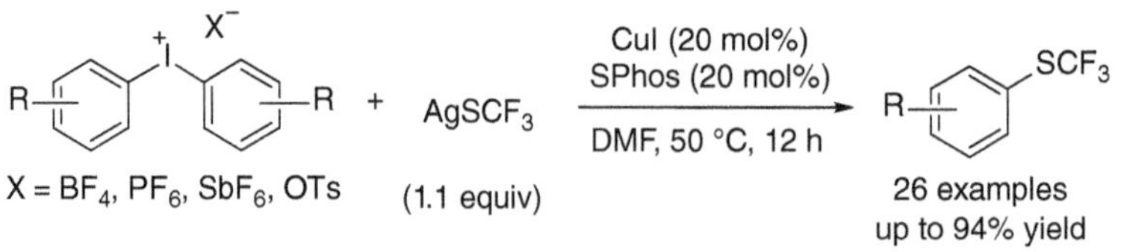

X = BF$_4$, PF$_6$, SbF$_6$, OTs

Scheme 4.12 Copper-catalyzed trifluoromethylthiolation of hypervalent diaryliodonium salts with AgSCF$_3$.

4.2.2.4 Oxidative Trifluoromethylthiolation

Copper-catalyzed oxidative functionalization of aryl boronic acids such as Chan–Lam–Evans reaction [25] has emerged in the past two decades as a powerful tool for C—X bond formation, and has found wide applications in organic synthesis because of its mild reaction conditions. Such a reaction has also been extended into aryl trifluoromethylthioether preparation. The first example of oxidative trifluoromethylthiolation was reported by Qing and coworkers in 2012 [26]. In this work, L_nCuSAr was initially generated *in situ* by the reaction of CuSCN, phen, elemental sulfur, and aryl boronic acid. This species then reacted with Ruppert–Prakash reagent TMSCF$_3$ in the presence of oxidant Ag$_2$CO$_3$ to give the corresponding aryl trifluoromethylthioethers at room temperature. The mild reaction allowed a broad substrate scope of aryl boronic acids with many common functional groups (Scheme 4.13). Nevertheless, heteroaryl boronic acids were not compatible with the reaction conditions. Later on, this protocol was successfully extended to metal-free oxidative trifluoromethylthiolation of alkynes [27] (Scheme 4.13). Interestingly, in 2017, Zhao et al. discovered that [(bpy)CuSCF$_3$], which was prepared from CuF$_2$, TMSCF$_3$, and sulfur in MeCN, a procedure described by Weng and coworkers, reacted with a variety of aryl boronic acids in good to excellent yields using oxygen as the oxidant [28].

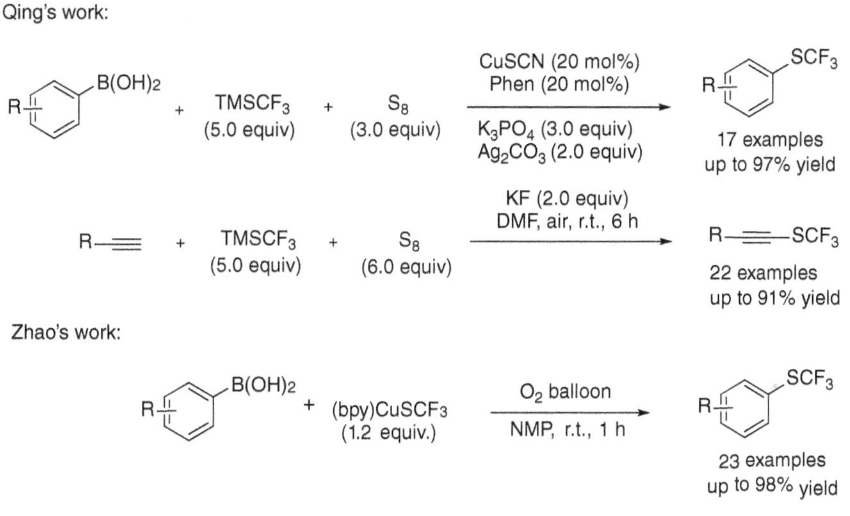

Qing's work:

Zhao's work:

Scheme 4.13 Oxidative trifluoromethylthiolation of aryl boronic acid or terminal alkynes with TMSCF$_3$ and elemental sulfur.

Shortly after, Vicic and coworker developed a similar reaction using Me$_4$NSCF$_3$ as the nucleophilic trifluoromethylthiolating reagent [29]. In this reaction, aryl boronic acids with both electron-withdrawing and -donating groups reacted smoothly to give the corresponding aryl trifluoromethylthioethers in high yields. Two examples of heteroaryl boronic acids such as 3-thiopheyl boronic acid and dibenzo[b,d]furan-4-ylboronic acid were also successfully trifluoromethylthiolated in good yields. The protocol was further extended to oxidatively trifluoromethylthiolate *trans*-vinyl boronic acids with retention of the configuration in moderate yields (Scheme 4.14).

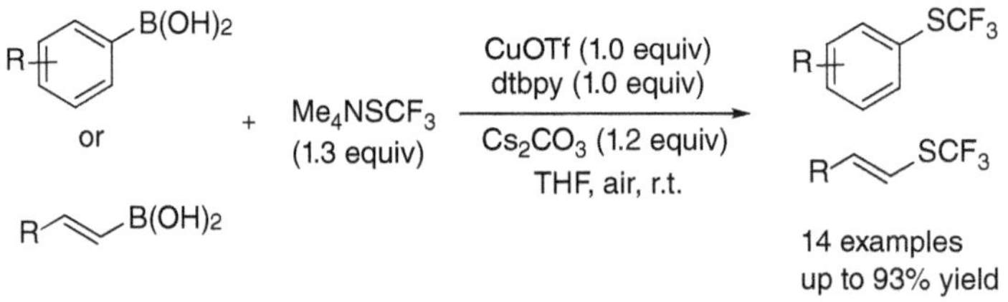

Scheme 4.14   Copper-mediated trifluoromethylthiolation of aryl boronic acids with Me$_4$NSCF$_3$.

In 2013, Duan and coworkers reported oxidative trifluoromethylthiolation of aryl boronic acids utilizing a combination of sulfur and CF$_3$CO$_2$Na as the nucleophilic trifluoromethylthiolating source [30]. Nevertheless, high temperature was required due to the difficulty in decarboxylation of CF$_3$CO$_2$Na (Scheme 4.15). Notably, one example of heteroaryl boronic acid 4-pyridyl boronic acid was reported to react to give 4-trifluoromethylpyridylthioether, albeit in low yield (30%).

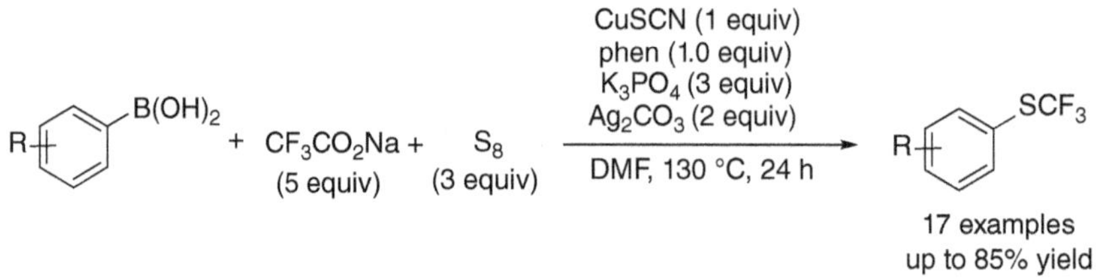

Scheme 4.15   Oxidative trifluoromethylthiolation of boronic acids with CF$_3$CO$_2$Na.

The above-mentioned oxidative trifluoromethylthiolating protocols typically required stoichiometric copper salt or nucleophilic trifluoromethylthiolating copper reagents. The first copper-catalyzed oxidative trifluoromethylthiolating method was reported in 2017 by Cao and coworkers [31]. In this protocol, a combination of 10 mol% CuI and 20 mol% of bpy was used as the catalyst, and 1.5 equiv of AgSCF$_3$ was used as the nucleophilic trifluoromethylthiolating reagent. A broad scope of aryl boronic acids were investigated to give the desired aryl trifluoromethylthioethers in moderate to good yields (Scheme 4.16). Specifically, two heteroaryl boronic acids, benzo[*b*]thiophen-2-ylboronic acid and pyrimidin-5-ylboronic acid, reacted to give the corresponding aryl trifluoromethylthioethers in 50% and 45% yields, respectively.

4.2.2.5   Transition Metal-Catalyzed Trifluoromethylthiolation of Arenes via C–H Activation

In the past two decades, transition metal-catalyzed C–H functionalization has developed into a general protocol for the late-stage functionalization of target molecules

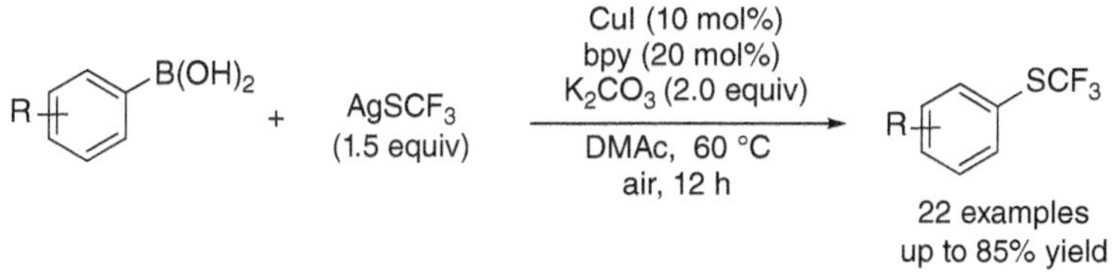

Scheme 4.16 Copper-catalyzed oxidative trifluoromethylthiolation of boronic acids with AgSCF₃.

because it circumvents the use of prefunctionalized aryl halides and avoids the generation of halogenated salt waste. Not surprisingly, transition metal-catalyzed C–H trifluoromethylthiolation of arenes has also been investigated and two such protocols have been reported in the literature.

In 2014, Huang and coworkers reported an example of *ortho*-pyridyl-directed palladium-catalyzed trifluoromethylthiolation of aromatic rings [32]. In this work, 10 mol% Pd(OAc)$_2$ was used as the catalyst and no ligand was required. AgSCF$_3$ was used as the nucleophilic trifluoromethylthiolating reagent and Select-fluor was used as the oxidant (Scheme 4.17). Under the optimized conditions, many pyridyl-directed arenes with different substituents at different positions of arenes were successfully trifluoromethylthiolated in high yields. Inter- and intramolecular H/D kinetic isotope effects (KIEs) were studied and the large kinetic isotope effect (KIE) suggests that the irreversible C–H palladation step is the rate-determining step.

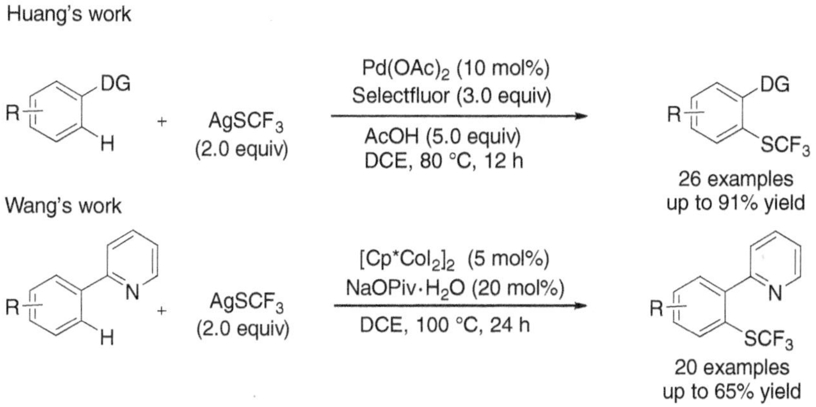

Huang's work

Wang's work

Scheme 4.17 Oxidative trifluoromethylthiolation of C–H bonds.

A similar C–H trifluoromethylthiolation catalyzed by a cobalt catalyst was reported in 2017 by Wang and coworkers [33]. With the assistance of a pyridyl directing group, a variety of trifluoromethylthiolated aromatic rings were synthesized in moderate yields under air (Scheme 4.17). A primary kinetic isotope effect of $k_H/k_D = 2.2$ was observed. In addition, a deuterated substrate was subjected to the reaction conditions and a significant amount of H/D exchange at the *ortho* positions of both starting material and product took place. The authors proposed that the C–H activation is reversible and is the rate-determining step of the reaction.

4.2.2.6 Miscellaneous Methods for the Formation or Aryl Trifluoromethylthioethers via Nucleophilic Trifluoromethylthiolating Reagents

Aryne is an important intermediate in organic chemistry due to its high reactivity toward nucleophilic species. By taking advantage of the aryne's high reactivity, while at the same time utilizing the nucleophilicity of $AgSCF_3$, Lee and coworkers reported in 2013 a regioselective cascade trifluoromethylthiolative cyclization reaction from the *in situ*-generated aryne intermediate from multi-alkyne substrates [34]. The reaction was proposed to be initiated via a Ag(OTf)-catalyzed Diels–Alder cyclization of bis-1,3-diyne to give an aryne intermediate, which might be stabilized by the silver cation. This reactive species is then attacked by $AgSCF_3$ to give arylated silver, which is then pronated to give the final product. With this protocol, multifunctionalized aryl trifluoromethylthioethers could be generated in good yields (Scheme 4.18). Further studies showed that not only 1,3-diyne but also allene-enynes could generate the highly reactive aryne intermediate. It was proposed that an Alder-ene type of reaction took place for the substrate bearing a propargyl hydrogen, producing the aryne intermediate, which was then trapped by $AgSCF_3$ to give aryltrifluoromethylthioethers in moderate yields [35] (Scheme 4.18).

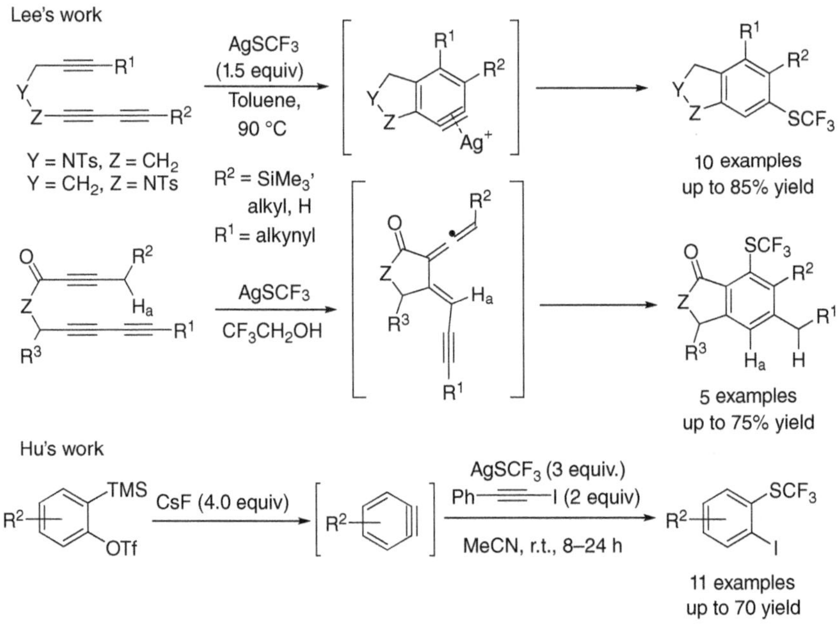

Scheme 4.18 Trifluoromethylthiolation of aryne and allene-enyne intermediates.

Similarly, Hu and coworker developed a trifluoromethylthiolative functionalization of arynes under mild conditions [36]. In this protocol, the *in situ*-generated aryne from easily available 2-(trimethylsilyl)phenyl trifluoromethanesulfonate underwent Ag—S bond insertion to give an arylated silver intermediate, which was then trapped by 1-iodophenylacetylene to give 2-iodophenyl trifluoromethylthioether in good yield. The choice of a suitable trapping electrophile was crucial for this transformation since it should be compatible with the strong basic conditions for the generation of aryne (Scheme 4.18).

Diaryliodonium salts are another type of highly reactive electrophiles that are able to react with a wide range of nucleophiles. In 2016, Rueping and coworkers reported that CuSCF$_3$ reacted with unsymmetric (aryl)(mesityl)iodonium triflates to give aryltrifluoromethylthioethers in good yields [37]. Notably, nitrogen-containing heteroarylated hypervalent iodonium triflates could also be generated and reacted under optimized conditions to give the desired products in good yields (Scheme 4.19). It is worth mentioning that the side product mesityl iodide was recovered in 91% yield and could be reused in the preparation of hypervalent iodide substrates. The same method can be applied in trifluoromethylthiolation of alkynyl(mesityl)iodonium triflates, which was reported by Zhang and coworkers [38] (Scheme 4.19).

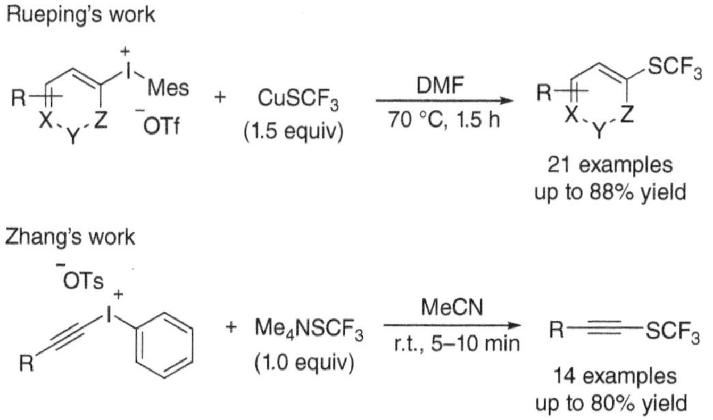

Rueping's work

Zhang's work

Scheme 4.19 Trifluoromethylthiolation of unsymmetrical λ^3-iodane derivatives.

In 2014, Wu and coworkers reported a method for the preparation of trifluoromethylthiolated isoquinoline derivatives from 2-alkynylbenzaldoxime substrates [39]. It was proposed that in the presence of AgOTf, 2-alkynylbenzaldoxime cyclizes to give an isoquinolinoxide intermediate, which is then activated by arylsulfonyl chloride. Nucleophilic attack of the activated species by AgSCF$_3$ followed by base-promoted dehydrosulfonylation generates 1-[(trifluoromethyl)thio]isoquinoline derivatives in good yields (Scheme 4.20).

Based on a similar mechanistic assumption, Qing and coworkers found that heteroaryl N-oxides could react with AgSCF$_3$ when a combination of Ts$_2$O and ^{n}Bu$_4$NI was used as the activator [40] (Scheme 4.20). In addition, Kuninobu's group also reported a similar protocol [41]. Rather than using Ts$_2$O/^{n}Bu$_4$NI as the activator, 2,4-dinitrobenzenesulfonyl chloride was used as the activator to deliver the same products in good yields (Scheme 4.20).

Wu's work:

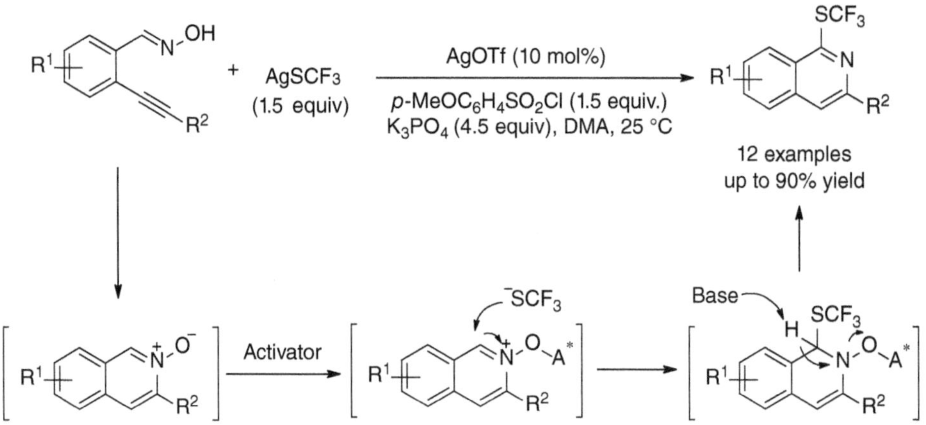

Qing's work:

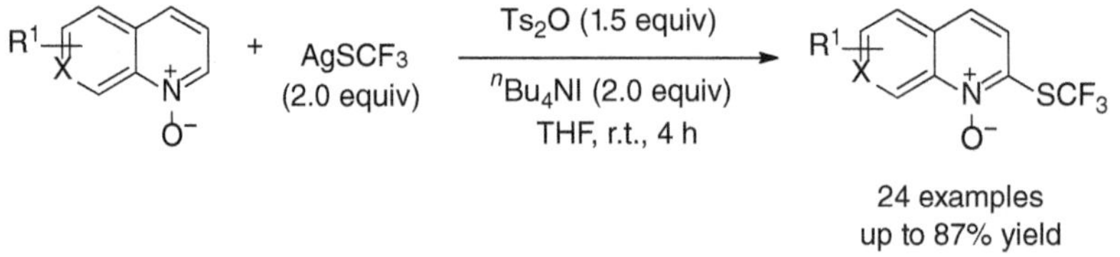

24 examples
up to 87% yield

Kuninobu's work:

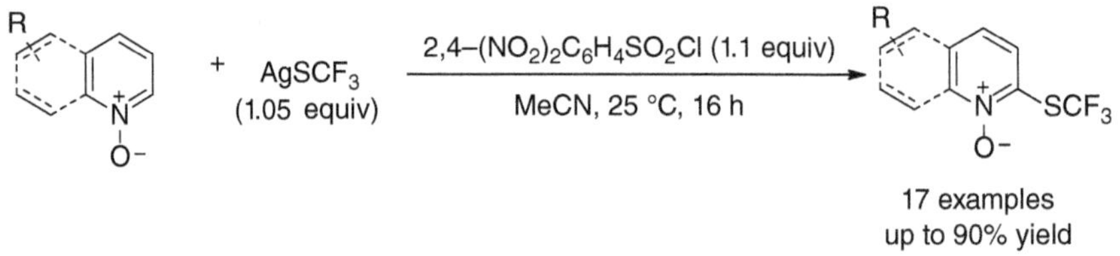

17 examples
up to 90% yield

Scheme 4.20 Trifluoromethylthiolation of *N*-oxides.

4.2.3 Formation of C(sp³)-SCF₃ by Nucleophilic Trifluoromethylthiolating Reagents

4.2.3.1 Reaction of CuSCF₃ with Activated Alkylated Halides

Direct nucleophilic trifluoromethylthiolation of an alkyl electrophile by a nucleophilic trifluoromethylthiolating reagent represents a general and straightforward method for the preparation of alkyl-trifluoromethylthioethers. Yet, with a few exceptions, most of the substrates in the reported reactions at early times were limited to activated electrophiles such as allylic halides, benzylic halides, or α-halo carbonyl compounds. For example, early studies by Muetterties and coworkers showed that reactions of Hg(SCF₃)₂ with allyl chloride afforded the trifluoromethylthioether in

23% yield [3] (Scheme 4.21). Kolomeitsev et al. reported in 2000 that reaction of $TADA^{2+}2SCF_3$ (TADA = tetrakis(dimethylamino)ethylene) with benzylic chloride or bromide in a mixed DMF/MeCN (DMF = *N*,*N*-dimethylformamide) solvent gave the corresponding trifluoromethylthioethers in almost quantitative yields [7a] (Scheme 4.21). Similarly, Weng and coworkers reported that the isolated stable complex [(bpy)$CuSCF_3$] reacted with a variety of benzylic bromides at room temperature in excellent yields [12b]. Furthermore, they showed that [(bpy)$CuSCF_3$] was able to trifluoromethylthiolate α-bromoketone derivatives in good yields [12c] (Scheme 4.21). Another example of such reaction was reported by Cahard and coworker [42]. They found that reaction of Morita–Baylis–Hillman carbonates with an *in situ*-generated nucleophilic trifluoromethylthiolating intermediate from $TMSCF_3$, elemental sulphur, and KF in the presence of 10 mol% DABCO (1,4-diazabicyclo[2.2.2]octane) as the catalyst afforded the corresponding trifluoromethylthioethers in good to excellent yields (Scheme 4.21).

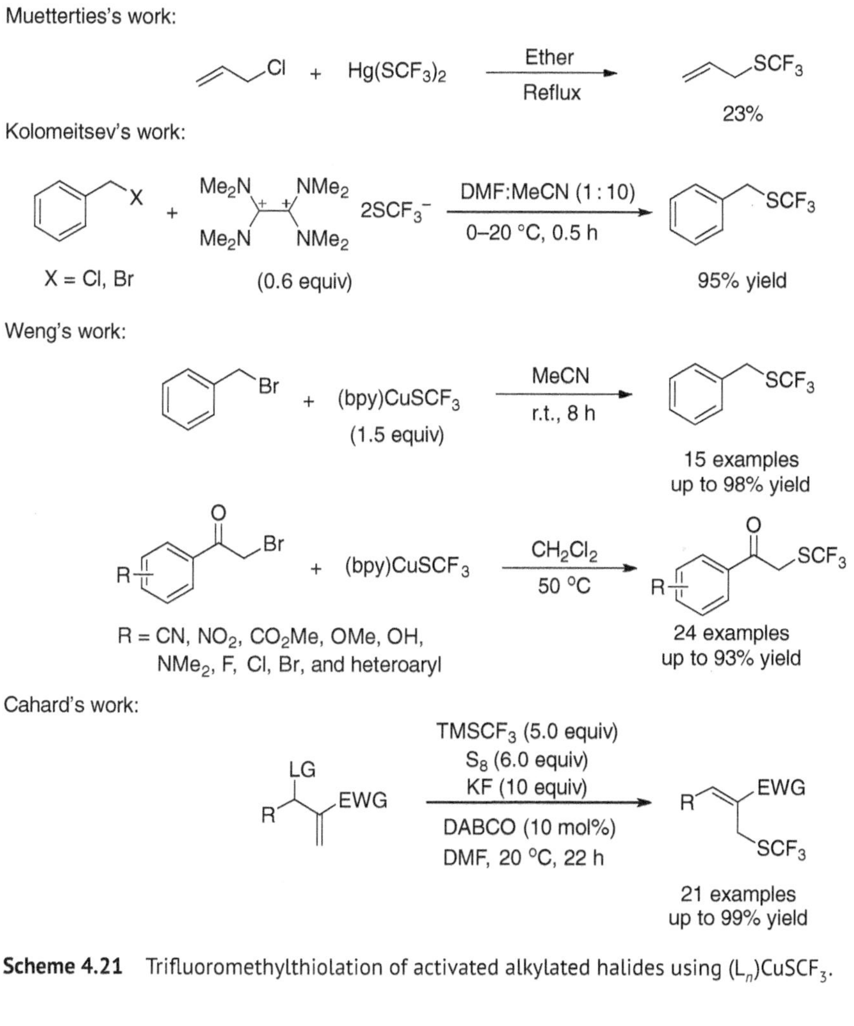

Muetterties's work:
Cl
+ Hg(SCF3)2
Ether
Reflux
SCF3
23%
Kolomeitsev's work:
X
+
Me2N
NMe2
Me2N
NMe2
2SCF3−
DMF:MeCN (1 : 10)
0–20 °C, 0.5 h
SCF3
X = Cl, Br
(0.6 equiv)
95% yield
Weng's work:
Br
+ (bpy)CuSCF3
(1.5 equiv)
MeCN
r.t., 8 h
SCF3
15 examples
up to 98% yield
O
Br
R
+ (bpy)CuSCF3
CH2Cl2
50 °C
SCF3
24 examples
up to 93% yield
R = CN, NO2, CO2Me, OMe, OH,
NMe2, F, Cl, Br, and heteroaryl
Cahard's work:
LG
EWG
TMSCF3 (5.0 equiv)
S8 (6.0 equiv)
KF (10 equiv)
DABCO (10 mol%)
DMF, 20 °C, 22 h
EWG
SCF3
21 examples
up to 99% yield

Scheme 4.21 Trifluoromethylthiolation of activated alkylated halides using $(L_n)CuSCF_3$.

4.2.3.2 Reaction of MSCF$_3$ with Unactivated Alkyl Halides

Nucleophilic trifluoromethylthiolation of unactivated alkyl halides is quite challenging since previous studies showed that MSCF$_3$ species were not good nucleophiles. To solve this problem, Weng and coworkers discovered that by employing KF as additive, [(bpy)CuSCF$_3$] was able to react with primary alkyl bromides in MeCN at elevated temperature (110 °C) to give the corresponding alkyl trifluoromethylthioethers in good to excellent yields [12d]. Under the same reaction conditions, reaction with alkyl chlorides or secondary bromides did not occur at all, while reactions of secondary iodides occurred in moderate yields (Scheme 4.22).

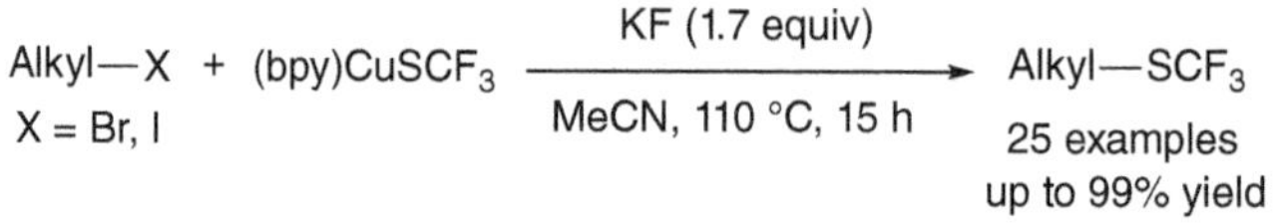

$$
\text{Alkyl—X} + \text{(bpy)CuSCF}_3 \xrightarrow[\text{MeCN, 110 °C, 15 h}]{\text{KF (1.7 equiv)}} \text{Alkyl—SCF}_3
$$

X = Br, I

25 examples
up to 99% yield

Scheme 4.22 Trifluoromethylthiolation of unactivated alkyl halides using (bpy)CuSCF$_3$.

Shen and coworkers adopted another approach to trifluoromethylthiolate alkyl electrophiles [43]. It was found that reaction of alkyl bromides with AgSCF$_3$ in the presence of 1.3 equiv of nBu$_4$NI in acetone after 24 hours at 80 °C afforded the corresponding alkyl trifluoromethylthioethers in excellent yields, while reactions of alkyl tosylates required the reaction temperature to be increased to 100 °C. Reactions of alkyl chlorides were more challenging and a combination of two different ammonia salts (nBu$_4$NI and nBu$_4$NBr) was necessary to promote the reaction to give the desired products in good yields (Scheme 4.23). Under these conditions, reactions of secondary alkyl halides or tosylates did not occur at all.

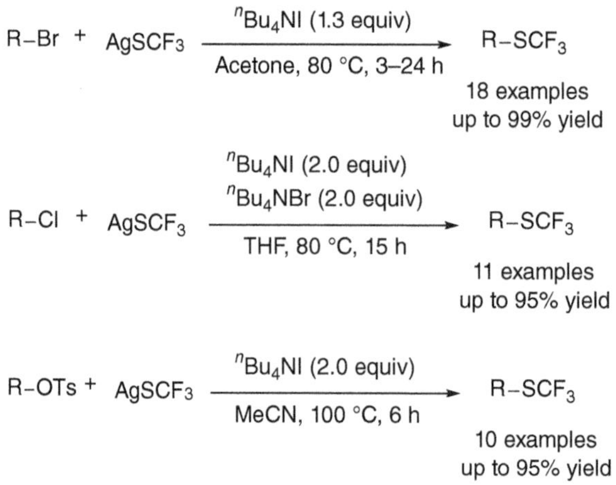

$$
\text{R–Br} + \text{AgSCF}_3 \xrightarrow[\text{Acetone, 80 °C, 3–24 h}]{^n\text{Bu}_4\text{NI (1.3 equiv)}} \text{R–SCF}_3
$$

18 examples
up to 99% yield

$$
\text{R–Cl} + \text{AgSCF}_3 \xrightarrow[\text{THF, 80 °C, 15 h}]{\substack{^n\text{Bu}_4\text{NI (2.0 equiv)} \\ ^n\text{Bu}_4\text{NBr (2.0 equiv)}}} \text{R–SCF}_3
$$

11 examples
up to 95% yield

$$
\text{R–OTs} + \text{AgSCF}_3 \xrightarrow[\text{MeCN, 100 °C, 6 h}]{^n\text{Bu}_4\text{NI (2.0 equiv)}} \text{R–SCF}_3
$$

10 examples
up to 95% yield

Scheme 4.23 Trifluoromethylthiolation of unactivated alkyl halides using AgSCF$_3$.

4.2.3.3 Nucleophilic Dehydroxytrifluoromethylthiolation of Alcohols

An alternative method for the preparation of alkyl trifluoromethylthioethers is dehydroxytrifluoromethylthiolation of alcohols. The first dehydroxytrifluoromethylthiolation was reported almost concurrently by Rueping, Qing, and their coworkers. In 2014, Rueping and coworkers reported that in the presence of a Lewis acid $BF_3 \cdot OEt_2$, allylic and benzylic alcohols reacted with $CuSCF_3$ to give the corresponding alkyl trifluoromethylthioethers in good yields [44]. When optically pure allyl or benzylic alcohols were used, racemic products were observed. Based on these observations, an S_N1-type mechanism was proposed (Scheme 4.24). Based on this discovery, Liang's group reported a silver-mediated cascade cyclization/trifluoromethylthiolation of propynols when $BF_3 \cdot OEt_2$ was employed as activator and $AgSCF_3$ as the trifluoromethylthio source [45] (Scheme 4.24). In 2018, Lebœuf's group found that the same protocol can be applied to trifluoromethylthiolate 3-hydroxyisoindolinone derivatives [46] (Scheme 4.24).

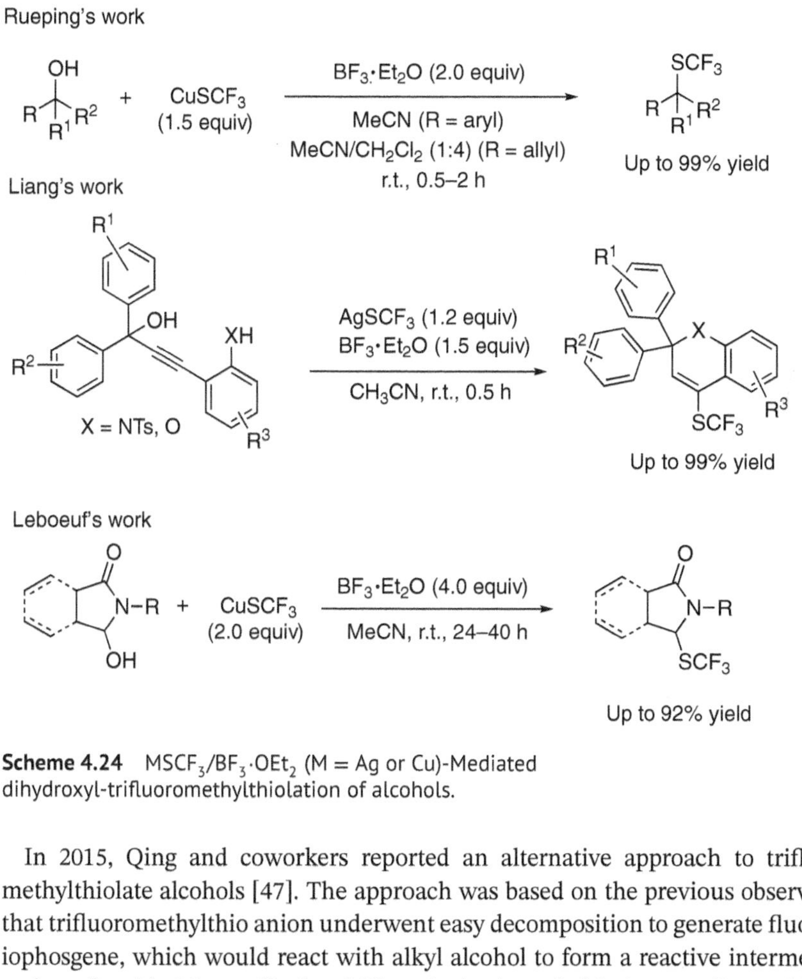

Rueping's work

Liang's work

Leboeuf's work

Scheme 4.24 $MSCF_3/BF_3 \cdot OEt_2$ (M = Ag or Cu)-Mediated dihydroxyl-trifluoromethylthiolation of alcohols.

In 2015, Qing and coworkers reported an alternative approach to trifluoromethylthiolate alcohols [47]. The approach was based on the previous observation that trifluoromethylthio anion underwent easy decomposition to generate fluorothiophosgene, which would react with alkyl alcohol to form a reactive intermediate carbonofluoridothioate. Nucleophilic substitution of this intermediate with the trifluoromethylthio anion generated the corresponding alkyl trifluoromethylthioether. In this approach, the use of nBu_4NI as additive was crucial. Not only primary alkyl alcohols but also secondary alkyl alcohols all reacted to give the desired products in moderate yields (Scheme 4.25). In addition, other substrates such as enols, allylic alcohols, and propargylic alcohols can also be converted to the corresponding products in high yields [48, 49]. Particularly, in the case of allylic alcohols and propargylic alcohols, S_N2'-type products were afforded (Scheme 4.26).

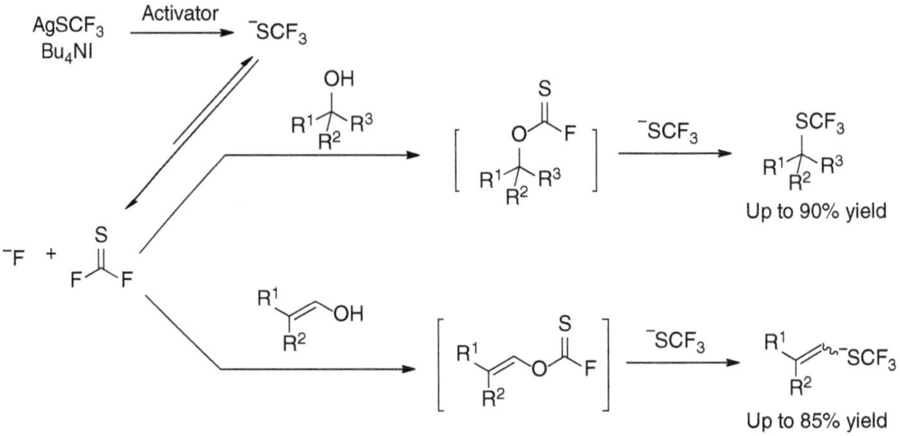

Scheme 4.25 Direct dehydeoxyltrifluoromethylthiolation of alkyl alcohols and enols.

4.2.3.4 Nucleophilic Trifluoromethylthiolation of Alcohol Derivatives

Another approach for the preparation of alkyl trifluoromethylthioethers is to trifluoromethylthiolate the activated alkyl electrophiles other than alkyl halides. For example, You and coworkers reported that allylic carbonic ester was successfully trifluoromethylthiolated in the presence of a ruthenium catalyst [50]. The reaction was highly selective to give linear allylic trifluoromethylthioethers, likely due to its thermostability (Scheme 4.27).

In 2017, Ma, Cahard, and coworkers reported that Me_4NSCF_3 was able to react with 1,2- and 1,3-sulfamidates under mild conditions [51]. The reaction was highly regio- and stereoselective and allowed to easily get access to optically active β- and γ-trifluoromethylthiolated amines and α-amino esters (Scheme 4.28).

4.2.3.5 Nucleophilic Trifluoromethylthiolation of α-Diazoesters

α-Diazoesters are easily available and high reactive electrophiles that can readily form electrophilic metal–carbene intermediate in the presence of a copper(I) or palladium(II) catalyst, which can be further applied in migratory insertion with different nucleophilic species to access a wide range of different functionalized compounds. Two approaches for the reactions of α-diazoesters with $CuSCF_3$ were reported under similar reaction conditions by two teams. The first team, led by Jinbo Hu, reported that reaction of α-diazoesters with $CuSCF_3$, which was *in situ* generated from CuCl and $AgSCF_3$, occurred after the solvent was mixed with certain amount of H_2O at room temperature. Under these conditions, α-trifluoromethylthiolated esters were synthesized in high yields [52]. The addition of water was crucial to high conversion since the yield decreased significantly in the absence of water. It

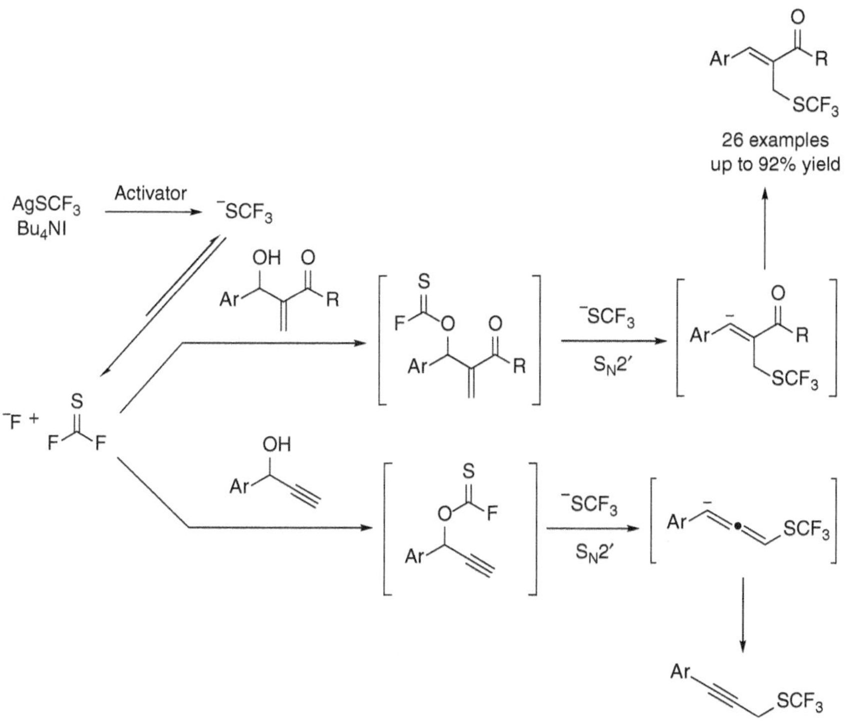

Scheme 4.26 S_N2'-type dehydeoxyltrifluoromethylthiolation of allylic and propargylic alcohols.

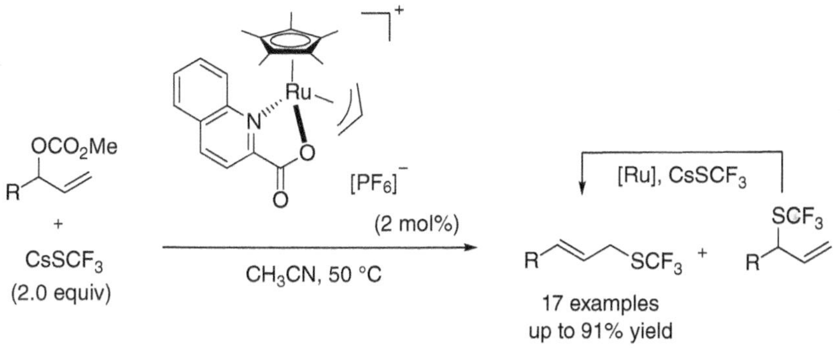

Scheme 4.27 Ruthenium-catalyzed regioselective allylic trifluoromethylthiolation.

was proposed that after migratory insertion, a copper enolate intermediate was formed. Quenching this species with water gave α-trifluoromethylthiolated ester. A second team led by Jianbo Wang developed the same reaction independently [53]. The only difference between the two approaches is that the reaction was quenched with an aqueous solution of NH_4Cl in Wang's approach (Scheme 4.29).

Interestingly, one month later, Rueping and coworkers reported the same reaction for the hydrotrifluoromethylthiolation of α-diazoesters [54]. Again, H_2O was added

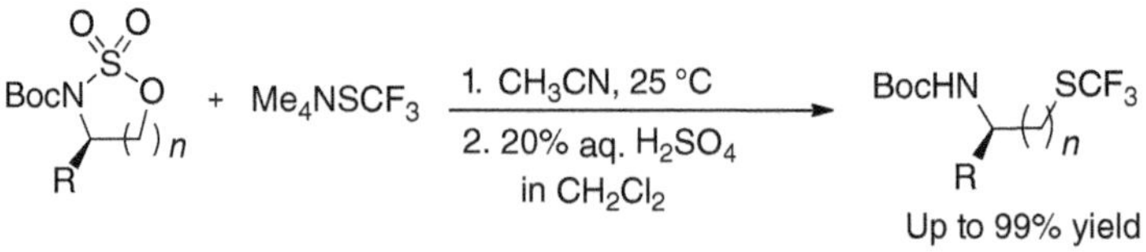

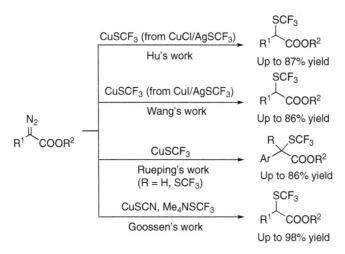

Scheme 4.28 Trifluoromethylthiolation of cyclic sulfamidates.

Scheme 4.29 Trifluoromethylthiolation of diazoesters via copper carbene migratory insertion.

as an additive to promote the reaction. In addition, bis-trifluoromethylthiolated esters could be obtained if N-trifluoromethylthiophthalimide was used to quench the enolate intermediate (Scheme 4.29).

Nevertheless, the above-mentioned methods required stoichiometric amount of CuSCF$_3$. The first copper-catalyzed trifluoromethylthiolation of α-diazoesters and α-diazoketones was reported by Goossen and coworkers in 2016 [55]. It was found that the reaction of α-diazoester with Me$_4$NSCF$_3$ in the presence of 10 mol% CuSCN occurred smoothly at room temperature to give α-trifluoromethylthiolated esters in high yield. Furthermore, switching the nucleophile from Me$_4$NSCF$_3$ to Me$_4$NSeCF$_3$ led to the formation of α-trifluoromethylselenolated esters and ketones in good yields (Scheme 4.29).

4.2.3.6 Formation or Alkyl Trifluoromethylthioethers via *In Situ* Generated Nucleophilic Trifluoromethylthiolating Reagent

Most of the nucleophilic trifluoromethylthiolating reagents discussed in Section 4.2.3 were stable and isolable reagents that could be prepared and stored for later use. Alternatively, nucleophilic trifluoromethylthiolating reagent can be generated *in situ* from a stable trifluoromethylthiolated precursor. In 2013, Zard and coworker reported that trifluoromethylthiolcarbonate reacted with KF in a mixed solvent THF/H$_2$O to generate trifluoromethylthio anion, which reacted with α-bromoketones or gramine derivatives to give the corresponding trifluoromethylthiolated compounds in excellent yields [56] (Scheme 4.30).

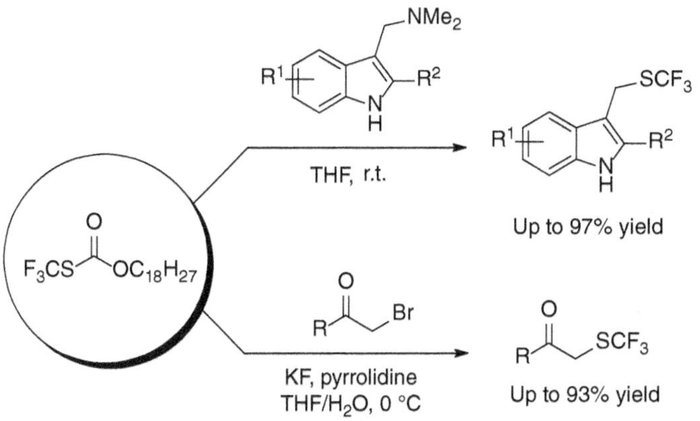

Scheme 4.30 Nucleophilic trifluoromethylthiolating reagent and its applications.

In 2015, Shi and coworkers applied this strategy for regioselective trifluoromethylthiolation of Morita–Baylis–Hillman carbonates [57]. Furthermore, in the presence of an optically tertiary amine catalyst, introduction of difluoromethylthiolated fragment was also achieved [58] (Scheme 4.31).

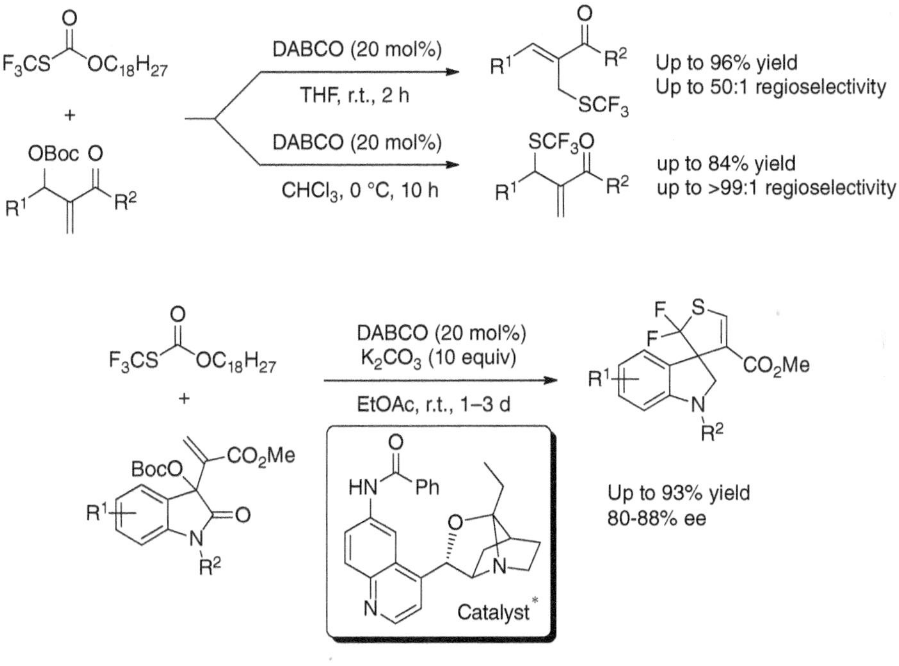

Scheme 4.31 Nucleophilic trifluoromethylthiolation of Morita–Baylis–Hillman carbonates.

4.2.3.7 Formation of Alkyl Trifluoromethylthioethers via C−H Bond Trifluoromethylthiolation

C–H activation and further functionalization represents a straightforward and atom-economic approach for the construction of C—C and C—X bond. Activation of aliphatic C—H bond for trifluoromethylthiolation was also feasible. Qing and coworkers reported a copper-catalyzed oxidative trifluoromethylthiolation of benzylic C—H bond in 2014 [59]. In this protocol, copper(I) thiophene-2-carboxylate (CuTc) was used as the catalyst and a peroxide ester *tert*-butyl 3-(trifluoromethyl)benzoperoxoate was used as the oxidant. The reaction occurred smoothly at 80 °C for 16 hours to give the benzylic C—H bond oxidative trifluoromethylthiolated products in moderate to good yields (Scheme 4.32). The trifluoromethylthiobenzylethers were further oxidized to sulfone derivatives under mild conditions. Several other methods involving radical type of alkane C–H trifluoromethylthiolation have also been developed, which will be discussed in Section 4.4.

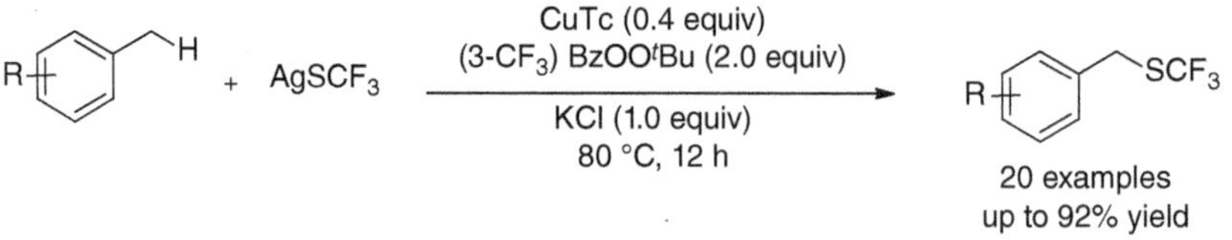

Scheme 4.32 Copper-catalyzed oxidative trifluoromethylthiolation of benzylic C−H bond.

4.3 Electrophilic Trifluoromethylthiolating Reagents

Reaction of a nucleophile with an electrophilic trifluoromethylthiolating reagent represents an alternative broadly applicable method for the construction of carbon—SCF_3 bond. The first two electrophilic trifluoromethylthiolating reagents CF_3SSCF_3 and CF_3SCl were reported in the 1950s. Yet, the high toxicity of both reagents limited their broad applications. In the next close to 50 years, no new highly reactive electrophilic trifluoromethylthiolating reagents were invented. Until 2008, Billard, Langlois, and their coworkers reported a highly reactive electrophilic trifluoromethylthiolating reagent $PhN(Me)SCF_3$ that could be readily prepared from diethylaminosulfur trifluoride (DAST). In the next 10 years, a variety of shelf-stable, scalable, and easy-to-handle electrophilic trifluoromethylthiolating reagents with even higher reactivities appeared and these developments are summarized in this section (Scheme 4.33).

4.3.1 CF₃SCl

Trifluoromethanesulfenyl chloride (CF_3SCl) was prepared by ultraviolet light irradiation of CF_3SSCF_3. An improved procedure for the preparation of CF_3SCl was reported by Emeléus and Nabi [60] in 1960 by reaction of trifluoromethanesulfenamide (CF_3SNH_2) with dry hydrogen chloride in a sealed

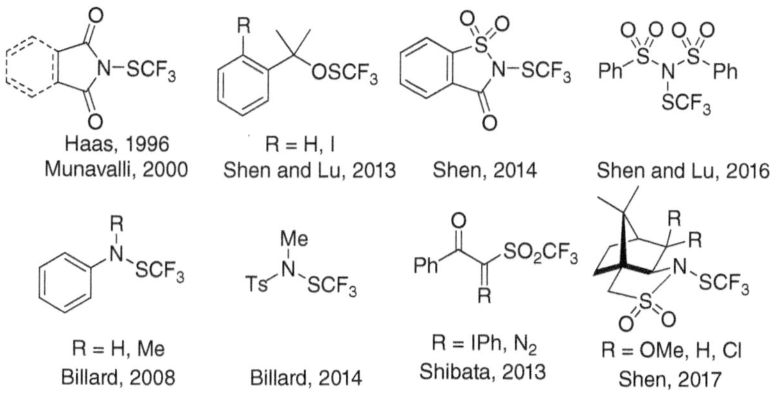

Scheme 4.33 Different types of stable electrophilic trifluoromethylthiolation reagents.

tube at room temperature to afford CF_3SCl in 97% yield. CF_3SCl was a highly reactive electrophilic trifluoromethylthiolating reagent that reacts with various nucleophiles, such as electron-rich (hetero)arenes, active methylene carbonylated compounds, organometallic reagent, silyl enol ether, and enamine [61] (Scheme 4.34). The reactions were generally conducted at low temperature to ensure chemoselectivity and prevent over-trifluoromethylthiolation. It was reported that CF_3SCl is highly toxic with an L(ct)50 of between 440 and 880 ppm/min [62], thus significantly limiting its broad applications in organic synthesis.

4.3.2 CF₃SSCF₃

CF_3SSCF_3 was first prepared by Haszeldine and Kidl in 1953 by the reaction of CS_2 and pure iodine pentafluoride (IF_5) at 60–200 °C [63]. Compared with CF_3SCl, CF_3SSCF_3 showed much lower electrophilicity. In the early years, a few limited examples for nucleophilic trifluoromethylthiolation of CF_3SSCF_3 with active carbon nucleophile as well as heteroaryl lithium reagent were reported [64]. With an L(ct)50 of 2200 ppm/min, CF_3SSCF_3 is highly toxic, and extreme care is required for its use [62].

In 2012, Daugulis and coworkers reported a copper-mediated directed arene *ortho* C—H bond trifluoromethylthiolation reaction. In this case, 8-aminoquinoline was found to be the directing group. Interestingly, 2,5-bis-trifluoromethylthiolated arene was obtained as the main product, while monotrifluoromethylthiolated product was not observed [65] (Scheme 4.35).

4.3.3 Haas Reagent

In 1996, Haas and Möller reported that reaction of silver succinimide with CF_3SCl generated *N*-trifluoromethylthiosuccinimide (Haas Reagent), even though the

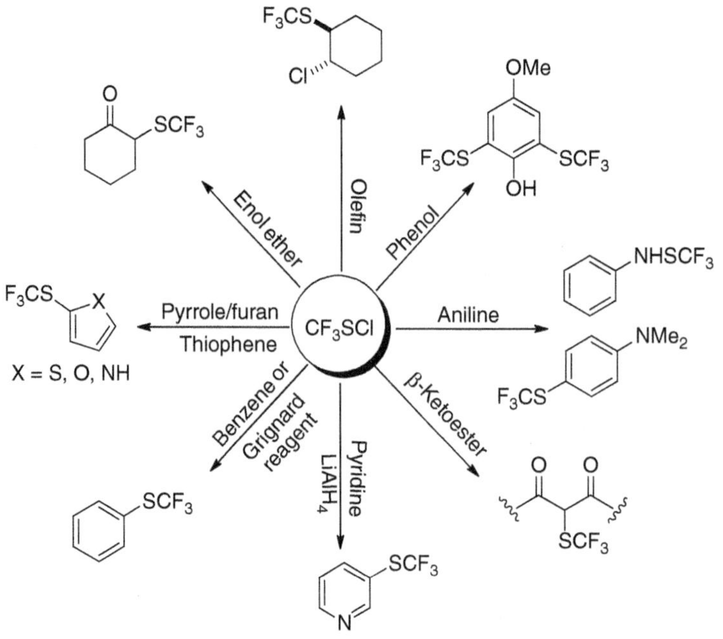

Scheme 4.34 Electrophilic trifluoromethylthiolation using CF₃SCl.

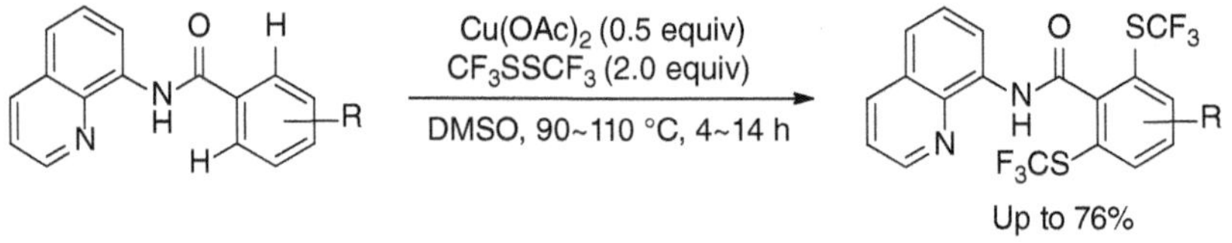

Scheme 4.35 Auxiliary-assisted trifluoromethylthiolation using CF₃SSCF₃.

reactivity of the reagent was not investigated, likely due to the use of toxic CF₃SCl for its preparation [66]. In 2014, Shen and coworkers reported an improved method for the preparation of *N*-trifluoromethylthiosuccinimide by reaction of AgSCF₃ with *N*-bromosuccinimide in acetonitrile at room temperature [67]. Furthermore, the same group demonstrated that *N*-trifluoromethylthiosuccinimide could be used in the palladium-catalyzed *ortho*-directed C–H activation and trifluoromethylthiolation of arenes when 10 mol% Pd(CH₃CN)₄(OTf)₂ was used as the catalyst [68] (Scheme 4.36).

Utilizing the Haas reagent as the electrophilic trifluoromethylthiolating source, Du and coworker [69] reported a bifunctional squaramide-catalyzed one-pot trifluoromethylthiolative Michael/aldol cascade cyclization reaction for the construction of trifluoromethylthio-substituted spirocyclopentanone–thiochromanes. The enantioselectivity of the reactions was up to 99% ee (Scheme 4.37).

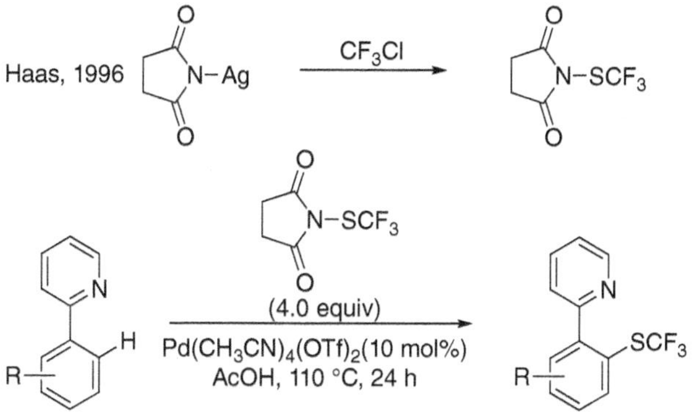

Haas, 1996

Scheme 4.36 Palladium-catalyzed trifluoromethylthiolation of aryl C—H bonds.

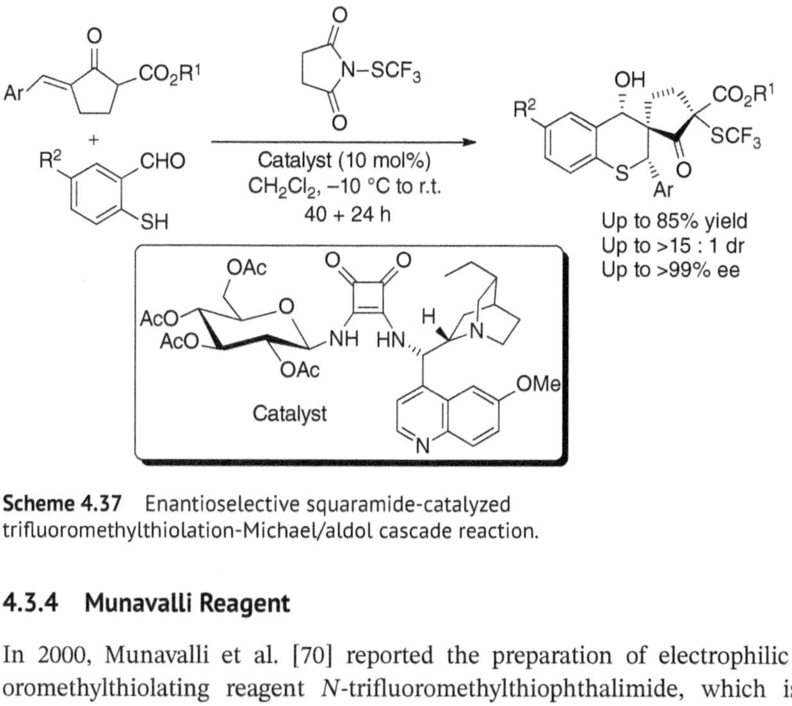

Up to 85% yield
Up to >15 : 1 dr
Up to >99% ee

Scheme 4.37 Enantioselective squaramide-catalyzed trifluoromethylthiolation-Michael/aldol cascade reaction.

4.3.4 Munavalli Reagent

In 2000, Munavalli et al. [70] reported the preparation of electrophilic trifluoromethylthiolating reagent N-trifluoromethylthiophthalimide, which is now called Munavalli reagent, by reaction of silver phthalimide with CF_3SCl. Initial studies showed that N-trifluoromethylthiophthalimide is a good electrophilic trifluoromethylthiolating reagent since it reacted with enamines to give the α-SCF$_3$-substituted carbonyl compounds in 88% yield. However, the reagent seems to have been forgotten by researchers in the next 10 years, largely due to, again, the use of toxic CF_3SCl in its preparation (Scheme 4.38).

In 2014, two new methods for the preparation of N-trifluoromethylthiophthalimide were reported almost at the same time. The first method was reported by Rueping and coworkers [71] by reaction of N-chlorosuccinimide (NCS) with $CuSCF_3$ in acetonitrile at room temperature, while the second method was reported by Shen

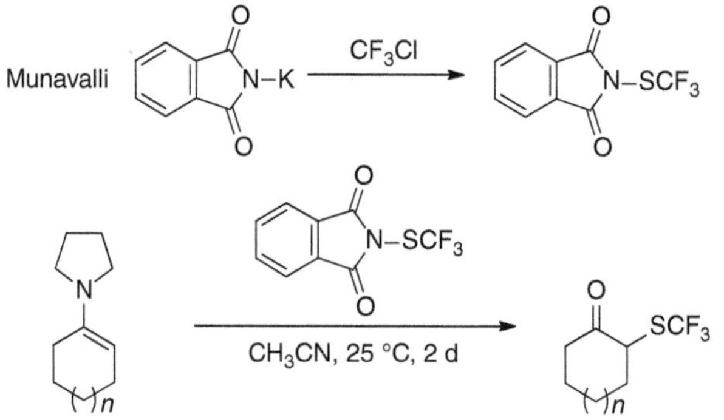

Scheme 4.38 Preliminary trifluoromethylthiolation attempts by Munavalli group.

and coworkers by reaction of *N*-bromophthalimide with $AgSCF_3$ in acetonitrile at room temperature [67]. In both cases, close to quantitative yields for the formation of *N*-trifluoromethylthiophthalimide were observed. The ease of preparation of the shelf-stable electrophilic reagent *N*-trifluoromethylthiophthalimide significantly boosts its use as electrophilic trifluoromethylthiolating reagent for the introduction of the trifluoromethylthio group (Scheme 4.39).

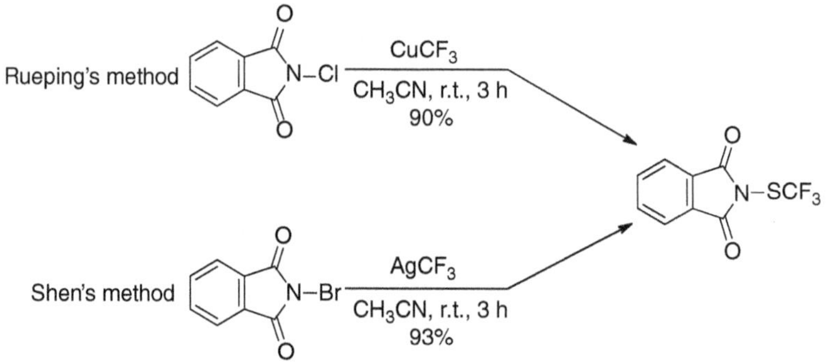

Scheme 4.39 Modified preparation of *N*-(trifluoromethylthio)phthalimide by Rueping and Shen group.

In 2013, Rueping and coworkers reported an organo-catalyzed asymmetric trifluoromethylthiolation of β-ketoesters with *N*-trifluoromethylthiophthalimide using quinidine as the catalyst. Under optimized conditions, various β-ketoesters derived from indanone, tetralone, or non-phenyl fused β-ketoesters could be trifluoromethylthiolated in high yields and excellent enantioselectivity [72]. Later on, the same group extended the asymmetric trifluoromethylthiolation reaction to 3-aryl oxindoles using $(DHQD)_2Pyr$ (2,5-diphenyl-4,6-bis(dihydroquinidine)-pyrimidine) as the catalyst [73]. The reaction occurred smoothly in toluene at $-10\,°C$ in excellent yields and enantioselectivity (Scheme 4.40).

In 2014, the research groups of Rueping [71] and Shen [67] independently reported the copper-catalyzed coupling reaction between aryl boronic acids and

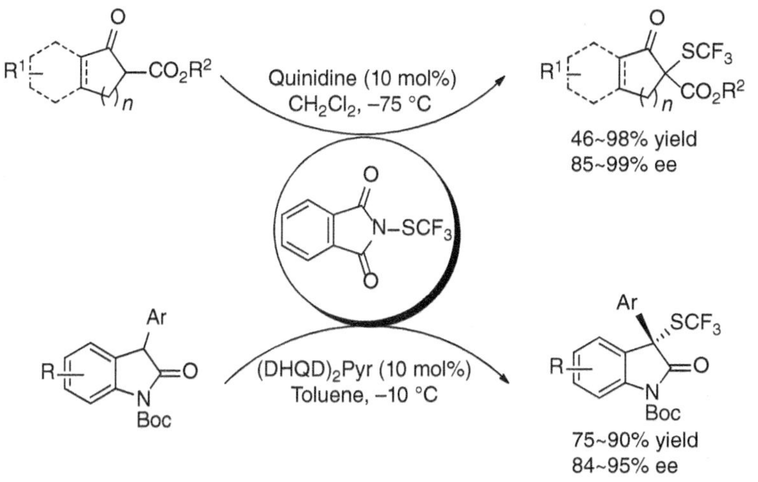

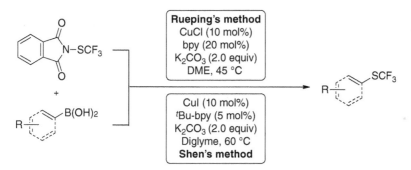

Scheme 4.40 Asymmetric trifluoromethylthiolation using *N*-(trifluoromethylthio)-phthalimide.

N-trifluoromethylthiophthalimide under similar reaction conditions. The reactions offered a general method to access a broad range of trifluoromethylthiolated arenes (Scheme 4.41). In 2019, Knochel and coworkers discovered that in the presence of catalytic amount of $Cu(OAc)_2 \cdot H_2O$, arylzinc chlorides were able to couple with *N*-trifluoromethylthiophthalimide to give trifluoromethylthiolated arenes in high yields [74].

Scheme 4.41 Cross-coupling reaction between boronic acids and *N*-(trifluoromethylthio)phthalimide.

Even though *N*-trifluoromethylthiophthalimide is a reactive electrophilic trifluoromethylthiolating reagent, Friedel–Crafts type of trifluoromethylthio-lation of electron-rich arene represents a challenging task. In 2015, Glorius and coworkers [75] discovered that when NaCl was used as an additive, Friedel–Crafts trifluoromethylthiolation of pyrroles, indoles, and azaindoles with *N*-trifluoromethylthiophthalimide was successfully realized. It was proposed that NaCl might first react with *N*-trifluoromethylthiophthalimide to *in situ* generate highly active CF_3SCl, which would then undergo Friedel–Crafts trifluoromethylthiolation with arene. Nevertheless, the presence of CF_3SCl in the reaction mixture was not detected and the role of NaCl remains elusive (Scheme 4.42).

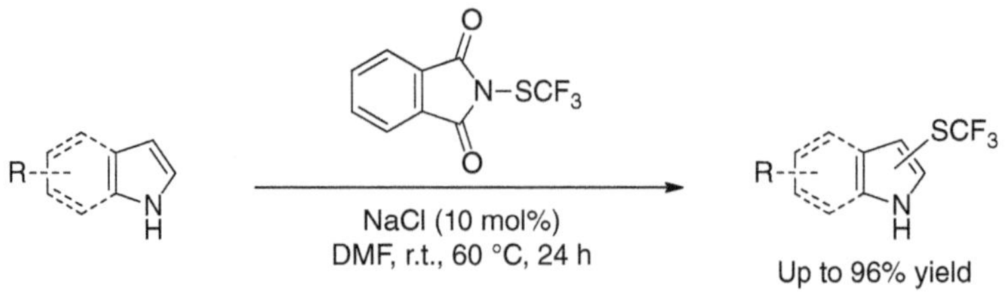

Scheme 4.42 NaCl-catalyzed trifluoromethylthiolation of N-heteroarenes.

Another application of N-trifluoromethylthiophthalimide for the formation of trifluoromethylthiolated arenes is the palladium-catalyzed C–H trifluoromethylthiolation. Besset and coworkers reported that in the presence of 10 mol% $PdCl_2$, arene with a 5-methoxy-quinolyl directing group underwent ortho-trifluoromethylthiolation with N-trifluoromethylthiophthalimide using air as the oxidant [76]. The same protocol could be applied to trifluoromethylthiolate 5-methoxy-aminoquinoline directed acylamide in excellent regio- and stereoselectivity (Scheme 4.43).

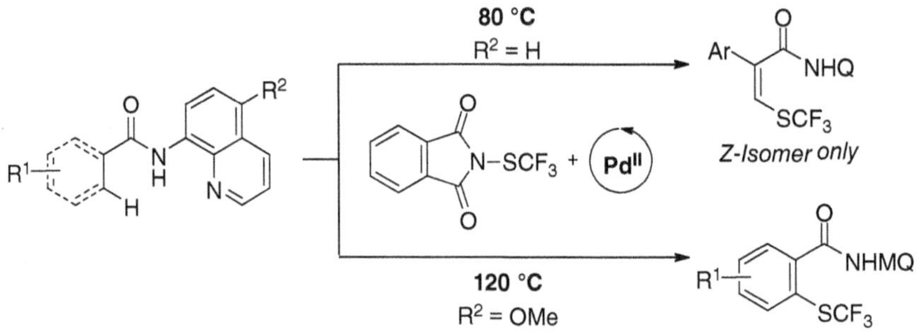

Scheme 4.43 Trifluoromethylthiolation of olefinic and aromatic derivatives.

Soft carbon nucleophiles such as enolates are known to react with a variety of electrophiles. Cahard and coworkers reported that enolate with Evans auxiliary reacted with N-trifluoromethylthiophthalimide to give α-trifluoromethylthiolated products in good to excellent diastereoselectivity [77] (Scheme 4.44). Likewise, Besset and coworkers found that using DABCO as the base, α-chloroaldehydes reacted with N-trifluoromethylthiophthalimide to afford α-trifluoromethylthio chloroaldehydes in moderate to good yields [78]. Very recently, Hu and coworkers were able to convert carboxylic acids to trifluoromethylthioesters when the carboxylic acids were treated with N-trifluoromethylthiophthalimide in the presence of a Lewis acid catalyst and stoichiometric amount of PPh_3 [79].

O
N–SCF3
O
R
O
O
N
O
Ph
LiHMDS (1.20 equiv)
THF, −78 °C, 7 h
R
O
O
N
O
SCF3
Ph

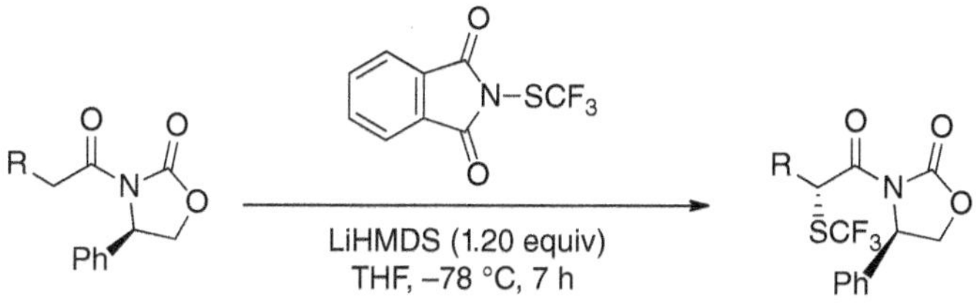

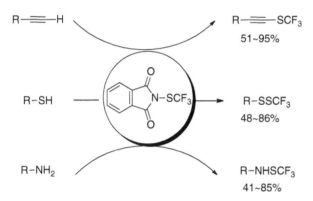

Scheme 4.44 Trifluoromethylthiolation of chiral oxazolidinones.

Several other nucleophiles including heteroatom nucleophiles could also react with *N*-trifluoromethylthiophthalimide to give the corresponding trifluoromethylthiolated compounds. For example, Rueping and coworkers reported that *N*-trifluoromethylthiophthalimide can react with alkynes in the presence of a copper catalyst to give alkynyl trifluoromethylthioethers in good yields [71], while reactions with thiols, thiophenols, and amines did not need the copper catalyst [80] (Scheme 4.45).

R—≡—H
R—≡—SCF3
51~95%
R–SH
O
N–SCF3
O
R–SSCF3
48~86%
R–NH2
R–NHSCF3
41~85%

Scheme 4.45 Other works by Rueping group.

Further studies showed that *N*-trifluoromethylthiophthalimide could also react with alkenes. In 2015, Wang and coworkers first demonstrated that *N*-trifluoromethylthiophthalimide reacted with α,β-unsaturated ketones or esters using DABCO as the base to give trifluoromethylthiolated bifunctionalized products in good yield [81]. Interestingly, when the nitro-substituted *N*-trifluoromethylthiophthalimide was used under the same reaction conditions, the yields of the reaction increased significantly (Scheme 4.46). Likewise, Yang and coworkers reported an iron-catalyzed chloro-trifluoromethylthiolation of alkenes with *N*-trifluoromethylthiophthalimide in the presence of 2.0 equiv of $SOCl_2$ [82]. The scope of the alkenes was broad enough to cover a wide range of electron-deficient, aromatic, and unactivated alkenes. Mechanistic studies showed that CF_3SCl was initially formed upon mixing *N*-trifluoromethylthiophthalimide with $SOCl_2$. In the presence of Fe_2O_3, a radical addition of CF_3SCl took place to give the corresponding addition products.

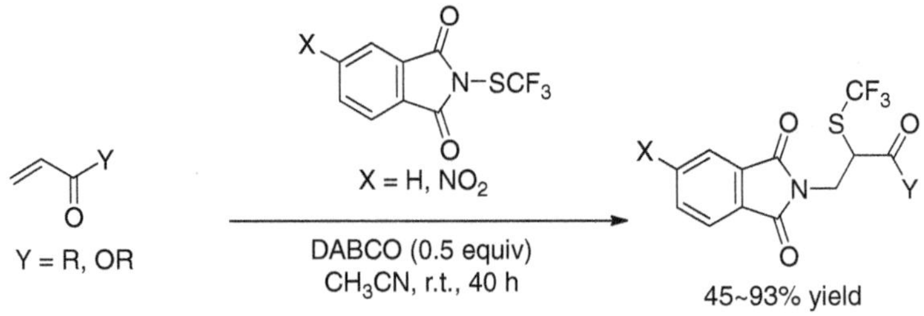

Scheme 4.46 DABCO-promoted aminotrifluoromethylthiolation of α,β-unsaturated carbonyl compounds.

N-Trifluoromethylthiophthalimide can act not only as electrophilic trifluoromethylthiolating reagent but also as radical trifluoromethylthio reagent; the related work is discussed in Section 4.4.

4.3.5 Billard Reagent

In 2008, Billard, Langlois, and coworkers reported two easily available, shelf-stable electrophilic trifluoromethylthiolating reagents PhNHSCF$_3$ and PhN(Me)SCF$_3$ that are now called Billard reagent. PhNHSCF$_3$ was prepared from the reaction of DAST and Ruppert–Prakash reagent using *N,N*-diisopropylethylamine (DIEA) as the base, followed by the addition of aniline [83]. PhNHSCF$_3$ was then further methylated with MeI to give PhN(Me)SCF$_3$. Initial studies showed that both are reactive trifluoromethylthiolating reagents toward highly reactive nucleophiles such as Grignard reagent and alkynyl lithium. On being activated by a Lewis or Brønsted acid, the first generation Billard reagent was able to trifluoromethylthiolate less nucleophilic substrates such as alkenes [84], alkynes [85], electron-rich arenes [86], and amines [87] (Scheme 4.47).

The choice of Lewis acid or Brønsted acid is crucial for the reaction of Billard reagent with the nucleophiles. In 2012, Qing and coworkers discovered that in the presence of TsOH·H$_2$O (Ts = Tosyl), reaction of tryptophan derivatives with PhNHSCF$_3$ gave tandem intramolecular cyclization trifluoromethylthiolation products, which introduced the trifluoromethylthio group at 4-position of indole moiety, while C-2 trifluoromethylthiolated product was obtained when BF$_3$·Et$_2$O was used as the activator [88]. In addition, the same group found that for allylic trifluoromethylthiolation of allylsilanes with PhN(Me)SCF$_3$, using acetyl chloride as the activator was beneficial [89]. Likewise, Liu and coworkers reported pyridine-directed Pd-catalyzed C–H trifluoromethylthiolation of arenes using Billard reagent [90]. In this case, benzoyl chloride was used as the activator. Similarly, Cao and coworkers adopted the same strategy of using acetyl chloride as the activator to achieve α-trifluoromethylthiolation of various acyclic and cyclic ketones [91]. Finally, Tu and coworkers applied this strategy for the trifluoromethylthiolative semi-pinacol rearrangement of allylic silyl ethers to give β-SCF$_3$-substituted carbonyl compounds in high yields [92] (Scheme 4.48).

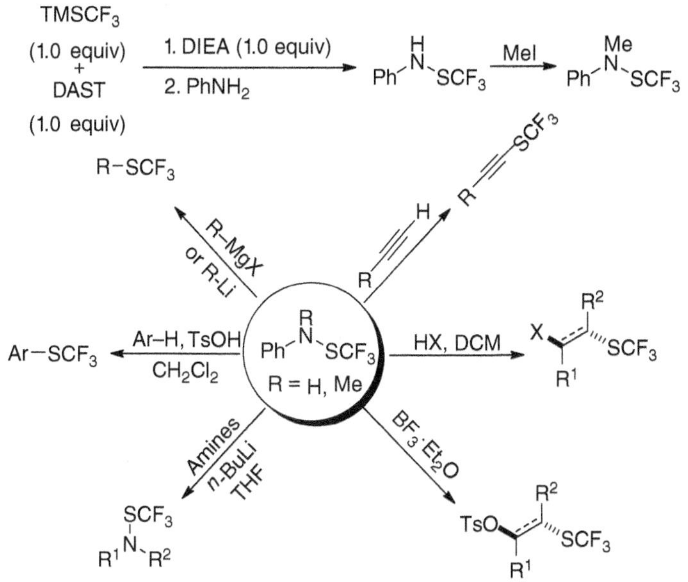

Scheme 4.47 Preparation and reactions of first generation of Billard reagent.

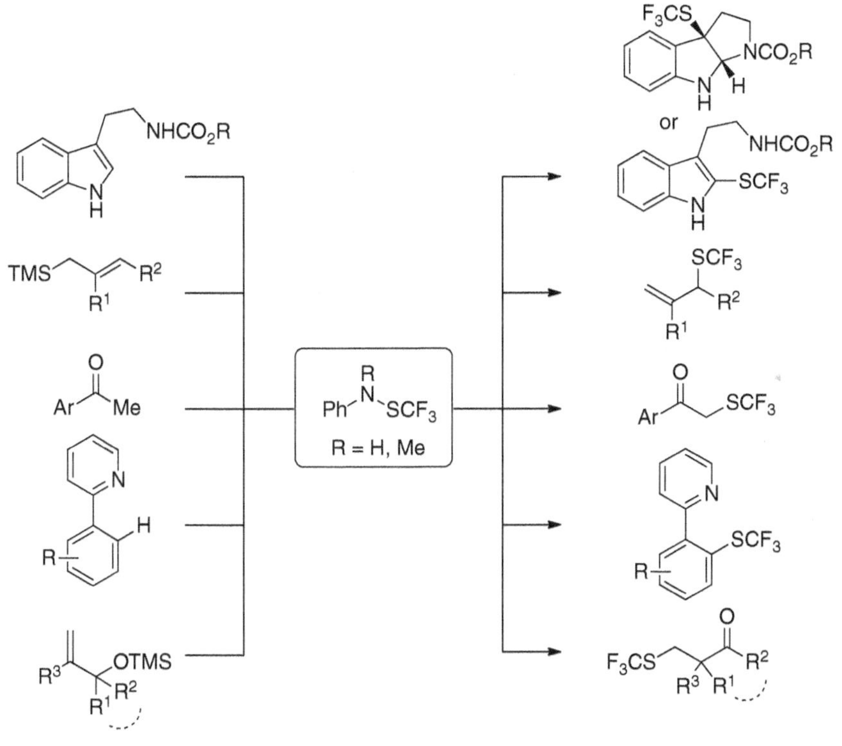

Scheme 4.48 Trifluoromethylthiolation under the activation of acid or acyl chloride using the first generation of Billard reagent.

In 2017, Qiu, Sheng, and coworkers reported that using 4-methylbenzenesulfonic acid (*p*-TsOH) as the activator, PhNHSCF$_3$ was able to react with *in situ*-generated sulfinates from sulfonyl chloride to give trifluoromethylthiolsulfonates in high yields [93] (Scheme 4.49).

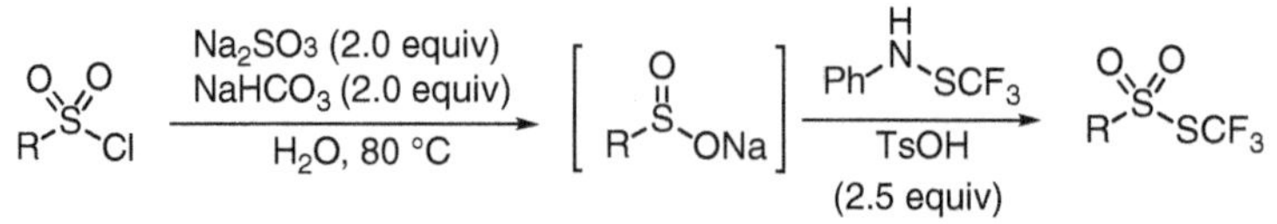

Scheme 4.49 One-pot reaction of sulfonyl chloride and Billard reagent.

Not only the commonly used Lewis or Brønsted acid, BiCl$_3$ can also be used as an activator to activate Billard reagent. Utilizing BiCl$_3$ as the activator, Wu and coworkers developed a series of tandem trifluoromethylthiolative cyclization reactions to construct various trifluoromethylthio-substituted scaffolds, such as benzo[*e*][1,2]thiazine 1,1-dioxide [94], indole [95], benzofuran [96], benzothiophene [96], and benzopyran derivatives [97]. In addition, alkynylsilicon substrate [98] can also be successfully trifluoromethylthiolated under these conditions (Scheme 4.50).

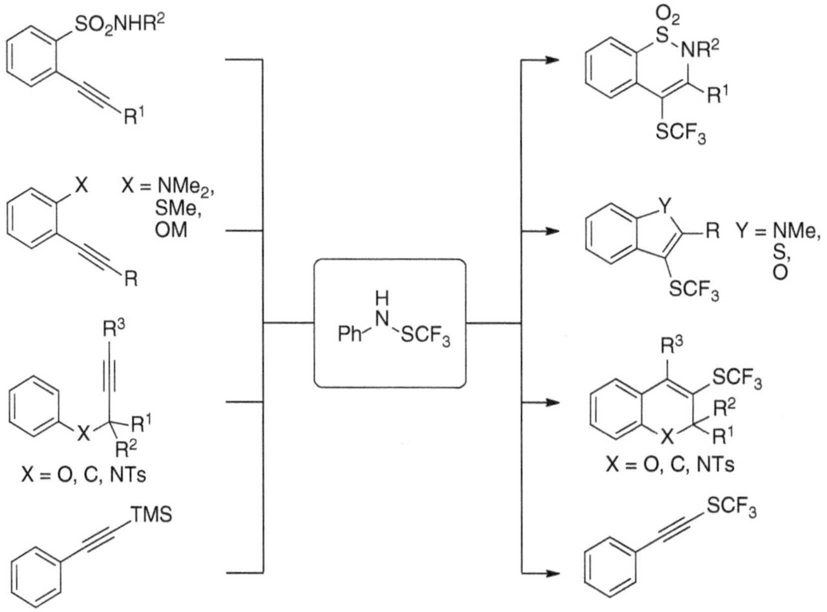

Scheme 4.50 Wu's work using the first generation of Billard reagent.

To further improve the electrophilicity of the reagent, in 2015, Billard and coworkers developed the second generation reagent PhN(Ts)SCF$_3$ by installing a tosyl group on the amino moiety of the reagent [99b]. Studies showed that PhN(Ts)SCF$_3$ is much more reactive than PhNHSCF$_3$ and PhN(Me)SCF$_3$ and the scope of the nucleophiles was expanded significantly to previously unreacted substrates such as enolates of aldehydes, ketones, esters or β-ketoesters, silyl enol ethers [100], electron-rich arenes [99], aryl boronic acids [101], thiols, alcohols [102], and alkynes [103] (Scheme 4.51).

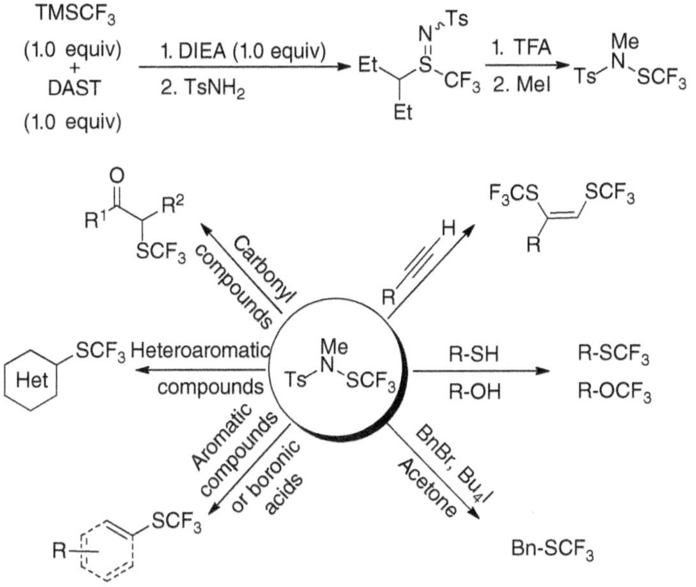

Scheme 4.51 Preparation and reactions of second generation of Billard reagent.

Two more reactions are worthy of further discussion. In the first reaction, Billard and coworkers discovered that the second generation reagent could be further activated by the addition of a super acid (HF/SbF$_5$). Nuclear magnetic resonance (^{19}F NMR) studies at −20 °C showed two new singlet peaks with chemical shifts at 15.53 and 6.55 ppm, which suggests that a more reactive diprotonated 4-nitrobenzensulfonamide intermediate was generated. Under these conditions, Billard and coworkers realized, for the first time, carbon-trifluoromethylthiolation of acetanilide derivatives [104] (Scheme 4.52).

In the second reaction, Billard and coworkers reported that in the presence of catalytic amount of nBu$_4$NI, PhN(Ts)SCF$_3$ underwent an "umpolung" process to become a nucleophilic trifluoromethylthiolating species that reacted with primary alkyl bromides or secondary benzyl bromides to give the corresponding alkyl trifluoromethylthioethers in good yields [105] (Scheme 4.53).

4.3.6 Shen Reagent

Inspired by the fact that the trifluoromethylthio group is a pseudohalogen group and with an aim to develop a more efficient electrophilic trifluoromethylthiolating reagent, Shen and coworkers designed and developed an

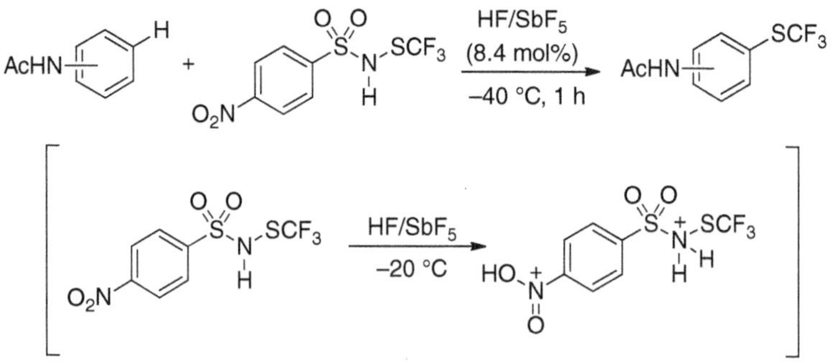

Scheme 4.52 Superacid-catalyzed trifluoromethylthiolation of aromatic amines.

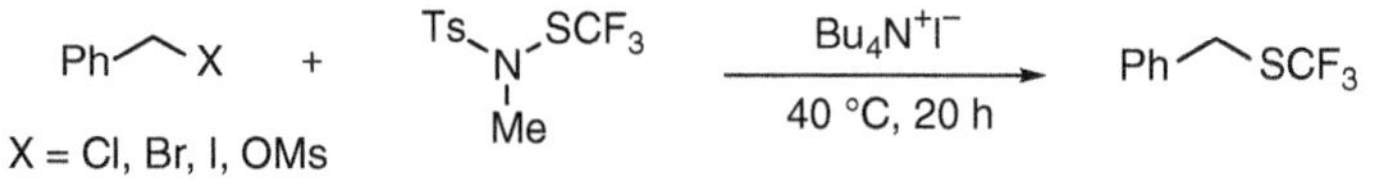

X = Cl, Br, I, OMs

Scheme 4.53 Trifluoromethylthiolation reaction of RX.

electrophilic trifluoromethylthiolating reagent based on saccharin skeleton [106]. *N*-Trifluoromethylthiosaccharin can be easily prepared from cheap and commercially available saccharin via a one-pot two-step process. Treatment of saccharin with ^{t}BuOCl in methanol gave *N*-chlorosaccharin in quantitative yield, which was used without further purification. Addition of AgSCF$_3$ to a solution of *N*-chlorosaccharin in acetonitrile at room temperature for 10 minutes gave *N*-trifluoromethylthiosaccharin in 84% yield [107]. The procedure can be easily scaled up to 16 g with 76–79% yield [108]. The cheap and easily available starting material, with ease of preparation, renders *N*-trifluoromethylthiosaccharin an ideal electrophilic trifluoromethylthiolating reagent (Scheme 4.54).

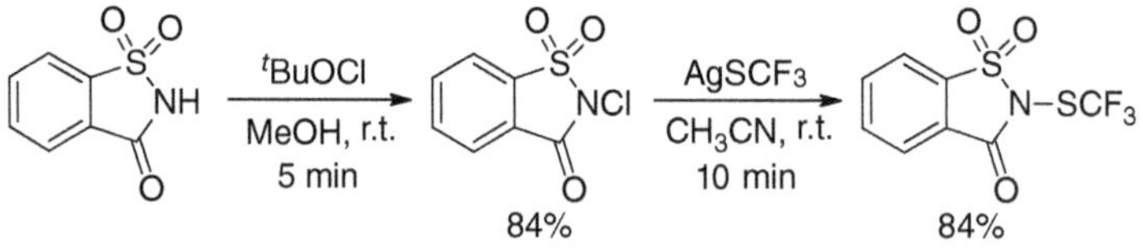

Scheme 4.54 Preparation of *N*-trifluoromethylthiosaccharin.

Extensive studies from Shen's group have demonstrated that *N*-trifluoromethylthiosaccharin is an electrophilic reagent with much higher reactivity than other electrophilic trifluoromethylthiolating reagents. It was found that *N*-trifluoromethylthiosaccharin reacted with various heteroatom nucleophiles such as alcohols, amines, thiols, and thiophenols to give the corresponding trifluoromethylthiolated products in excellent yields [106] (Scheme 4.55).

Utilizing *N*-trifluoromethylthiosaccharin as the electrophilic trifluoromethylthiolating reagent, for the very first time (except for CF$_3$SCl), Friedel–Crafts type of trifluoromethylthiolation occurred to a broad range of electron-rich arenes, when a

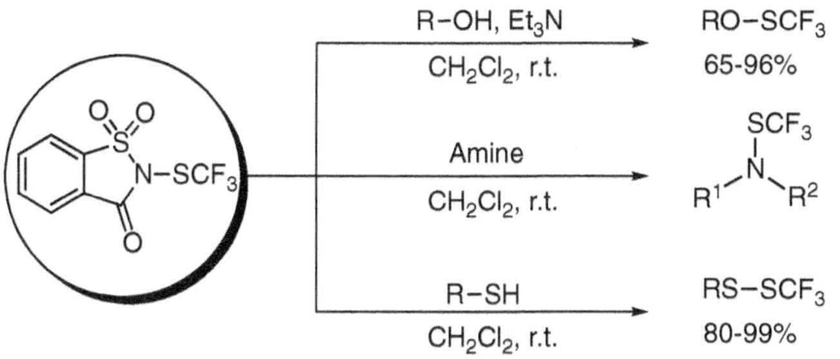

Scheme 4.55 Trifluoromethylthiolation reaction of various heteroatom nucleophiles.

Lewis acid TMSCl or triflic acid was used as a promoter. Similarly, Li and coworkers reported that other Lewis acids such as $FeCl_3$ or $AuCl_3$ were also able to promote Friedel–Crafts trifluoromethylthiolation of electron-rich arenes and heteroarenes [109]. Likewise, Gustafson and coworkers [110] reported that Friedel–Crafts trifluoromethylthiolation of electron-rich arenes with *N*-trifluoromethylthiosaccharin could be catalyzed by a combination of 10 mol% diarylselenide and 10 mol% TfOH as the catalyst. The reaction occurred smoothly in chloroform at room temperature to give the corresponding trifluoromethylthioarenes in high yields. Furthermore, Glorius and coworkers [111] discovered that NaCl was able to promote Friedel–Crafts trifluoromethylthiolation of *N*-trifluoromethylthiosaccharin with furans. Specifically, 2-trifluoromethylthiofuran was obtained as the major product. More recently, Shen and coworkers discovered that using 2,2,2-trifluoroethanol (TFE) as the solvent, a promoter-free Friedel–Crafts type of trifluoromethylthiolation of electron-rich arenes and heteroarenes with *N*-trifluoromethylthiosaccharin took place smoothly at 40 °C [112]. Interestingly, Xu and coworkers also reported a similar reaction using hexafluoroisopropanol (HFIP) as the solvent [113] (Scheme 4.56).

Trifluoromethylthiolated arenes could also be accessed via a C—H bond functionalization strategy. Li and coworkers reported that using $[Cp^*RhCl_2]_2$ as the catalyst, indoles with pyridyl directing group reacted with *N*-trifluoromethylthiosaccharin to give 2-trifluoromethylthiolated indole derivatives in high yields [114] (Scheme 4.57).

Studies from Shen's group also showed that soft carbon nucleophiles such as cyclic or acyclic β-ketoesters, malonates, aldehydes, or ketones were able to react with *N*-trifluoromethylthiosaccharin to give α-trifluoromethylthio-carbonyl compounds in excellent yields [106]. Cahard and coworkers reported that reaction of enolates with Evans auxiliary reacted with *N*-trifluoromethylthiosaccharin to give α-trifluoromethylthiolated amides with excellent diastereoselectivity [77]. In addition, under mild conditions, the auxiliary could be easily cleaved. Later, the same group reported that another type of enolate generated from β-ketocarboxylic acids via decarboxylation reacted with *N*-trifluoromethylthiosaccharin to give α-trifluoromethylthiolated ketones in high yield [115]. Likewise, Benaglia, Rossi, and their coworkers found that using tetrahydrothiophene as the catalyst, silyl enol ethers reacted with *N*-trifluoromethylthiosaccharin to afford

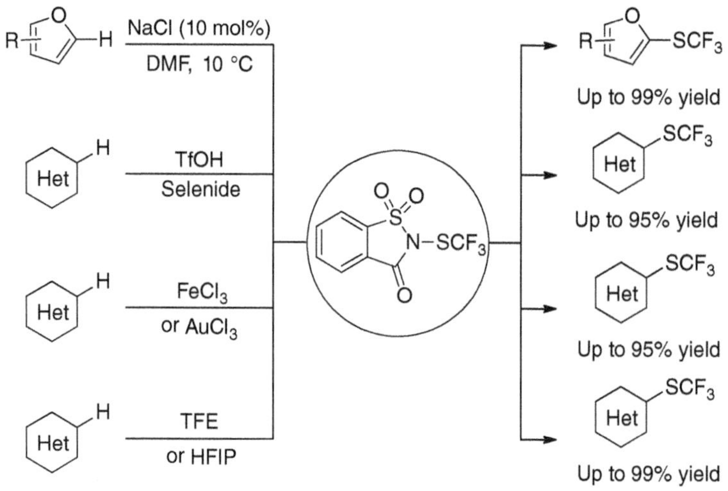

Scheme 4.56 Friedel–Crafts type of trifluoromethylthiolation.

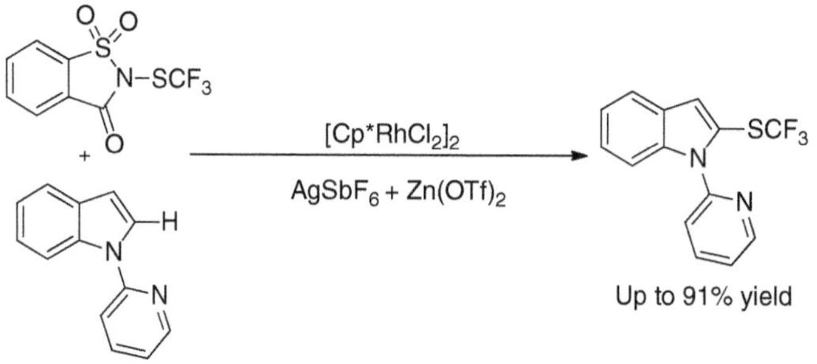

Scheme 4.57 Trifluoromethylthiolation of electron-rich arenes.

trifluoromethylthio-substituted ketone derivatives in high yields [116]. Furthermore, reaction of enolate derived from aldehyde with N-trifluoromethylthiosaccharin was reported by Wu, Sun, and coworkers in 2016 [117]. It was found that in the presence of a catalytic amount of pyrrolidine, aldehydes reacted with N-trifluoromethylthiosaccharin to give α-trifluoromethylthiolated aldehydes that were further reduced by NaBH$_4$ to give the corresponding β-trifluoromethylthiolated alcohol derivatives. Presumptively, pyrrolidine first reacts with aldehyde to give an enamine intermediate, which then reacts with N-trifluoromethylthiosaccharin to give α-trifluoromethylthiolated imine. Subsequent hydrolysis of imine gave the target α-trifluoromethylthiolated aldehyde (Scheme 4.58).

Owing to the high reactivity of N-trifluoromethylthiosaccharin, alkenes could also react with it to give trifluoromethylthiolated compounds. Shen and coworker reported that reactions of 2-(1-arylvinyl)-benzoic acid with N-trifluoromethylthiosaccharin gave trifluoromethylthiolated lactones in high yields. It was proposed that the alkene moiety of the substrate first reacts with

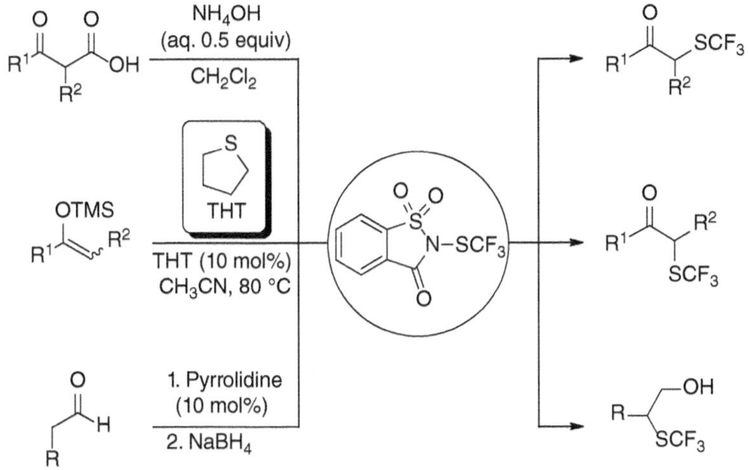

Scheme 4.58 Trifluoromethylthiolation of enolate nucleophiles.

N-trifluoromethylthiosaccharin to generate a trifluoromethyl-substituted thiira-nium ion intermediate, which is attacked by the oxygen of the carboxylic group to generate the final product. The reaction can also be extended to the synthesis of trifluoromethylthiolated lactams [107] (Scheme 4.59).

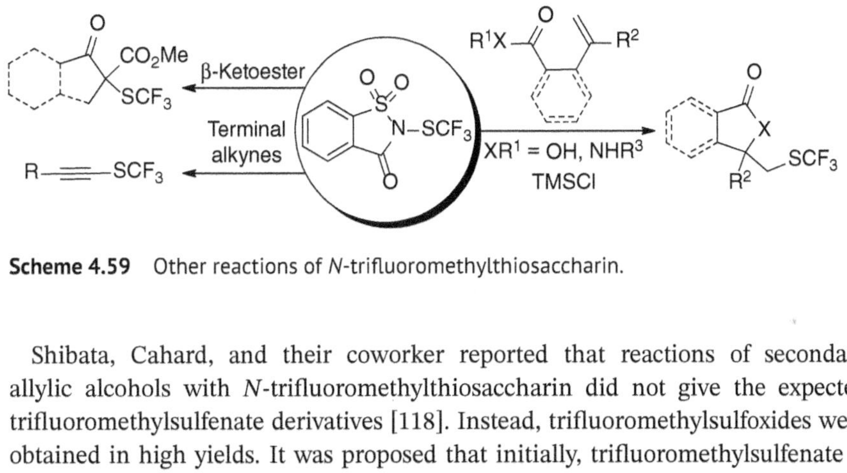

Scheme 4.59 Other reactions of N-trifluoromethylthiosaccharin.

Shibata, Cahard, and their coworker reported that reactions of secondary allylic alcohols with N-trifluoromethylthiosaccharin did not give the expected trifluoromethylsulfenate derivatives [118]. Instead, trifluoromethylsulfoxides were obtained in high yields. It was proposed that initially, trifluoromethylsulfenate is generated between the hydroxy group and N-trifluoromethylthiosaccharin. Yet, this species is not stable and quickly undergoes [2,3]-sigmatropic rearrangement to give thermodynamically more stable trifluoromethylsulfoxides (Scheme 4.60).

Ye and coworkers reported that reaction of glycals with N-trifluoromethylthios-accharin occurred smoothly to give trifluoromethylthio/chloro-difunctionalized glycose derivatives in high yields when TMSCl (trimethyl chlorosilane) was used as a Lewis acid activator. In the presence of DBU (1,8-diazabicyclo[5.4.0]undec-7-ene), elimination of HCl gave 2-trifluoromethylthioglycals in good yields [119] (Scheme 4.61).

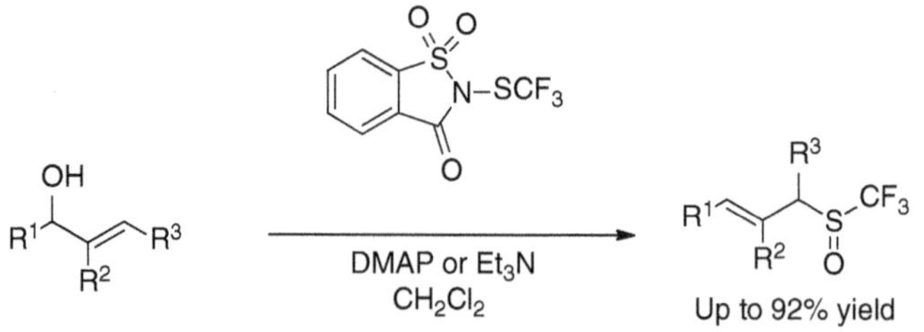

Scheme 4.60 Reactions between allylic alcohols and Shen reagent to form trifluoromethyl sulfoxides via [2,3]-sigmatropic rearrangement.

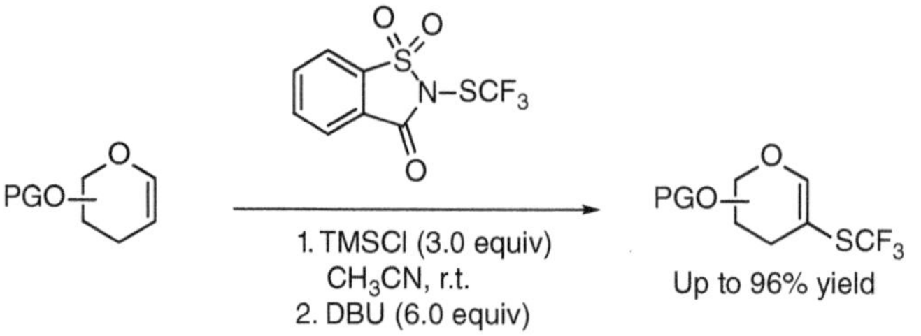

Scheme 4.61 2-Trifluoromethylthiolation of glycals.

Addition of a trifluoromethylthio group to alkene creates a chiral carbon center, which encourages organic chemists to develop an asymmetric addition reaction. Such a reaction was realized successfully by Zhao and coworkers [120]. Initially, Zhao and coworkers discovered that in the presence of a combination of diarylselenide and TfOH, reaction of an alkene with N-trifluoromethylthiosaccharin in acetonitrile occurred smoothly to give trifluoromethylthio-amino difunctionalized alkane derivatives in high yields. When the solvent was switched to nitromethane, hydroxy-trifluoromethylthio difunctionalized products were obtained in high yields. Later on, they discovered that when an optically pure diaryl selenide derived from indane was used as the catalyst, a highly enantioselective trifluoromethylthiolative lactonization occurred smoothly to give the desired products in excellent enantioselectivity (Scheme 4.62).

N-Trifluoromethylthiosaccharin can act not only as a highly reactive electrophilic trifluoromethylthiolating reagent but also as a trifluoromethylthio radical precursor. Related work for using N-trifluoromethylthiosaccharin as trifluoromethylthio radical precursor is discussed in Section 4.4.

4.3.7 Shen Reagent-II

During the development of electrophilic trifluoromethylthiolating reagent N-(trifluoromethylthio)saccharin, Shen and coworkers noticed that N-(trifluoromethylthio)saccharin is much more reactive than N-(trifluoromethylthio)phthalimide, while the only difference in the structures of two reagents is that one of the carbonyl groups in N-(trifluoromethylthio)phthalimide is replaced by a sulfonyl group in N-(trifluoromethylthio)saccharin. Consequently, Shen and coworkers hypothesized

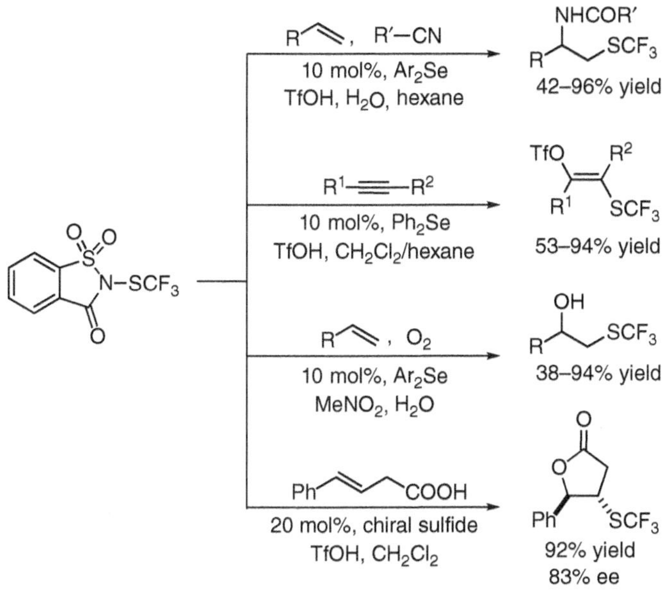

Scheme 4.62 Zhao's work for trifluoromethylthiolation of alkenes.

that the electrophilicity of the reagent can be further increased if both carbonyl groups are replaced by two sulfonyl groups. In 2016, Lu, Shen, and coworkers synthesized such an electrophilic trifluoromethylthiolating reagent $(PhSO_2)_2NSCF_3$ by first preparing $(PhSO_2)_2NCl$ from $(PhSO_2)_2NH$ by oxidation with ^tBuOCl in methanol, followed by replacement of the chloride with trifluoromethylthio using $AgSCF_3$ [121] (Scheme 4.63).

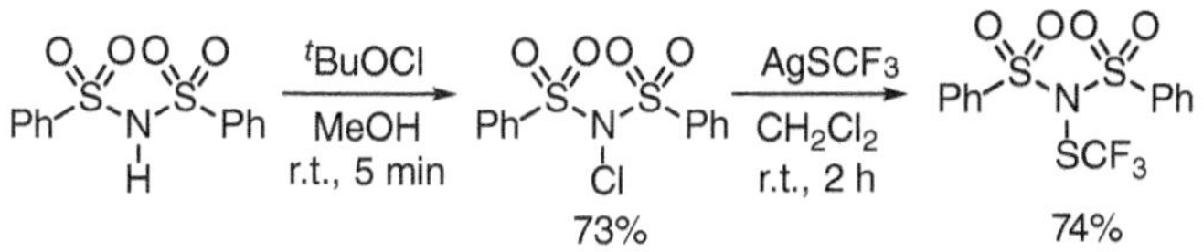

Scheme 4.63 Preparation of N-trifluoromethylthio-dibenzenesulfonimide.

Systematic investigation showed that, as expected, $(PhSO_2)_2NSCF_3$ exhibits even higher electrophilicity than N-(trifluoromethylthio)saccharin. In the absence of any Lewis acid or Brønsted acid, $(PhSO_2)_2NSCF_3$ reacted with a wide range of electron-rich arenes to give the corresponding trifluoromethylthiolated arenes in high yields when the reaction was conducted in DMF at 80 °C. Under the same conditions, electronic-rich heteroaromatic rings such as indoles, pyrroles, thiophenes, benzofurans, benzothiophenes, benzothiazoles, and imidazo[1,5-a]pyridines can react with $(PhSO_2)_2NSCF_3$ to give the corresponding trifluoromethylthiolated products in high yields. Interestingly, when 2-naphthol derivatives were used as the substrates, Friedel–Crafts type electrophilic trifluoromethylthiolation products were not produced. Instead, dearomatizative trifluoromethylthiolation products with the trifluoromethylthio group at 1-position were isolated in good yields [121] (Scheme 4.64).

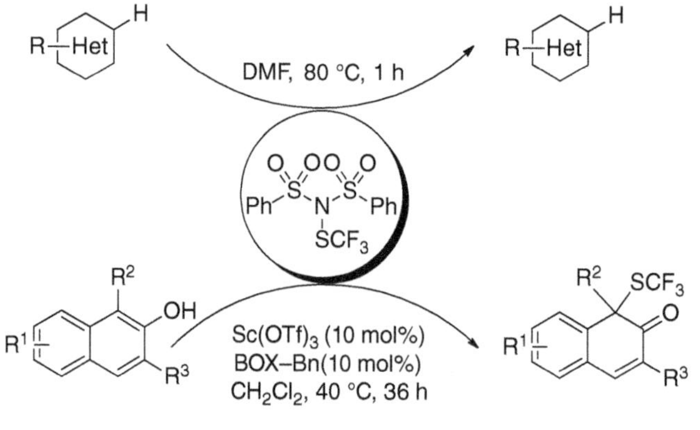

Scheme 4.64 Reactions between heteroarenes and naphthol derivatives with *N*-trifluoromethylthio-dibenzenesulfonimide.

Under the same reaction conditions, not only electron-rich arenes but also styrene derivatives reacted with $(PhSO_2)_2NSCF_3$ to give *trans*-trifluoromethylthiolated styrene derivatives in high yields. Interestingly, when the reaction was conducted at room temperature, a hydroxy-trifluoromethylthio difunctionalized product was isolated after aqueous workup. Mechanistic studies suggested that $(PhSO_2)_2NSCF_3$ is initially activated by DMF to generate trifluoromethylthio cation, which adds to styrene to form a benzylic cation. The cation is then attacked by DMF, followed by aqueous workup to form the final product. Following the same strategy, a variety of trifluoromethylthiolative difunctionalization products including formoxy, acetoxy, hydroxyl, and amino-trifluoromethylthio difunctionalized compounds were obtained by simply changing the reaction solvents [121] (Scheme 4.65).

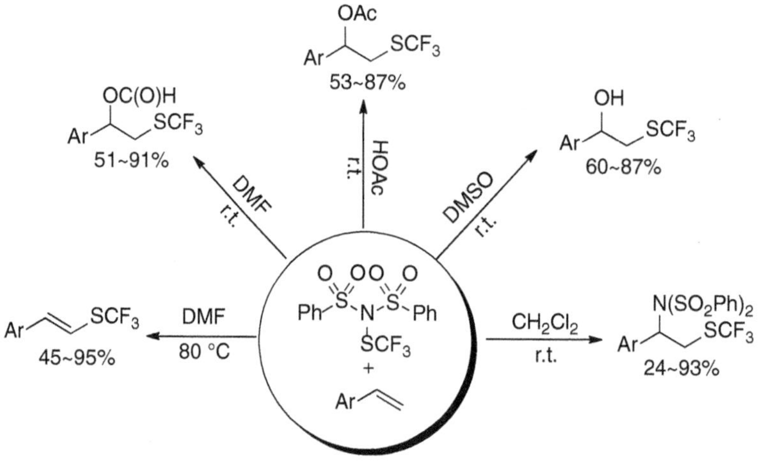

Scheme 4.65 Trifluoromethylthiolation of styrenes.

With the experience for the development of asymmetric trifluoromethylth-iolative addition to alkenes using *N*-trifluoromethylthiosaccharin, Zhao and coworkers discovered that in some cases, replacing *N*-trifluoromethylthiosaccharin with $(PhSO_2)_2NSCF_3$ resulted in better enantioselectivity. Consequently, a series of asymmetric trifluoromethylthiolative bifunctionalization of olefins with excellent enantioselectivity were developed, again using optically pure diarylselenide as the catalyst [120d, 122] (Scheme 4.66).

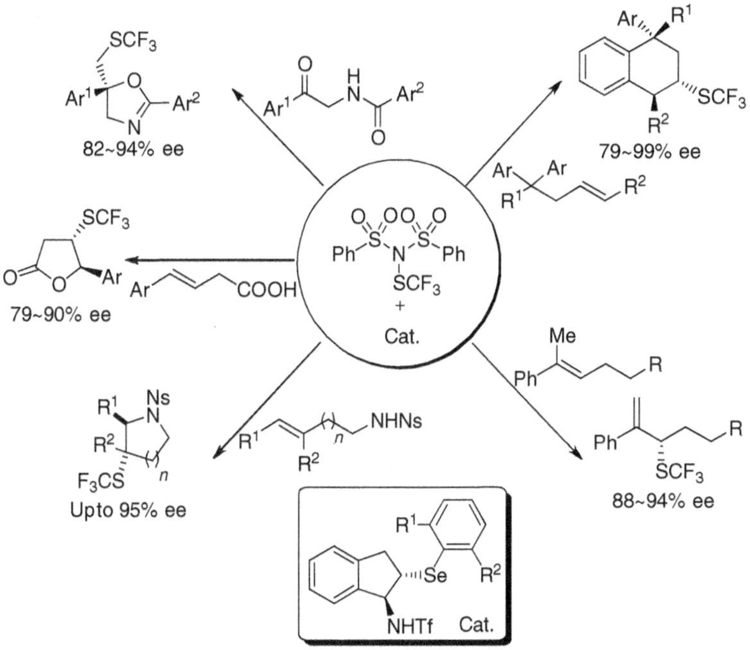

Scheme 4.66 Asymmetric trifluoromethylthiolation/cyclization bifunctionalization of olefins.

Inspired by Hooz's multicomponent reaction, Szabó and coworkers reported that $(PhSO_2)_2NSCF_3$ was able to react with α-diazocarbonyl compounds in the presence of a rhodium catalyst or Lewis acid. In 2017, they reported that when $Rh_2(OAc)_4$ was used as a catalyst, three-component reaction between α-diazoketone, $(PhSO_2)_2NSCF_3$, and an alcohol afforded the oxytrifluoromethylthiolated product in good yield. A variety of primary and secondary alcohols could be incorporated in the reaction. In addition, when an acyclic or cyclic ether was used instead of an alcohol, the C—O bond underwent cleavage to give oxytrifluoromethylthiolated products in good yields [123]. In 2019, the same group reported a Zn-catalyzed three-component reaction of α-diazoketone, $(PhSO_2)_2NSCF_3$, and tetraphenylborate to give α,α′-trifluoromethylthiolative phenylation product in high yields. Under the same conditions, reactions of various α-diazoesters or amides were also successful [124]; $(PhSO_2)_2NSCF_3$ could also be employed in transition metal-catalyzed directed arene C–H trifluoromethylthiolation or benzylic C–H trifluoromethylthiolation.

In 2017, Yoshino, Matsunaga, and coworkers reported that in the presence of 10 mol% $[Cp^*Co(CH_3CN)_3](SbF_6)_2$, pyridyl-, or purinyl-directed arenes reacted with $(PhSO_2)_2NSCF_3$ to give *ortho*-trifluoromethylthiolated arenes in good yields [125]. Likewise, in 2018, Britton and coworkers found that the benzylic C—H bond in alkylquinazolines and purines could be easily trifluoromethylthiolated when the substrate was allowed to react with $(PhSO_2)_2NSCF_3$ in acetonitrile at 75–125 °C when Li_2CO_3 was used as the base [126] (Scheme 4.67).

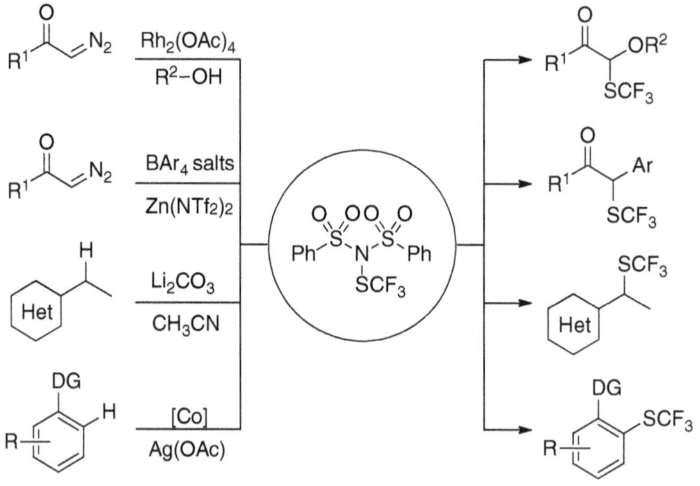

Scheme 4.67 Other applications of $(PhSO_2)_2NSCF_3$.

Further studies showed that like *N*-trifluoromethylthiosaccharin, $(PhSO_2)_2NSCF_3$ could also react with a variety of heteroatom nucleophiles such as aza-heterocycles, amines, thiols, and sulfinates under mild conditions. Typically, excellent yields of the corresponding trifluoromethylthiolated compounds could be obtained under mild conditions [121] (Scheme 4.68).

4.3.8 Optically Active Pure Trifluoromethylthiolation Reagents

In general, there are two different methods for the construction of trifluorom-ethylthio-substituted optically active compounds. The first method is to use transition metal catalyst or organocatalyst to enable the highly enantioselective asymmetric trifluoromethylthiolation of a substrate with an electrophilic trifluo-romethylthiolating reagent, which is also currently attracting much interest. An alternative, yet classic method for the building of the chiral trifluoromethylthi-olated carbon center is to trifluoromethylthiolate the substrate with an optically active trifluoromethylthiolating reagent, which could serve as a supplement to catalytic trifluoromethylthiolative method. In the past 10 years, a series of electrophilic trifluoromethylthiolating reagents have emerged, but the optically active trifluoromethylthiolating reagents have not been reported. Inspired by the optically active electrophilic fluorinating reagent [127], Shen and coworkers

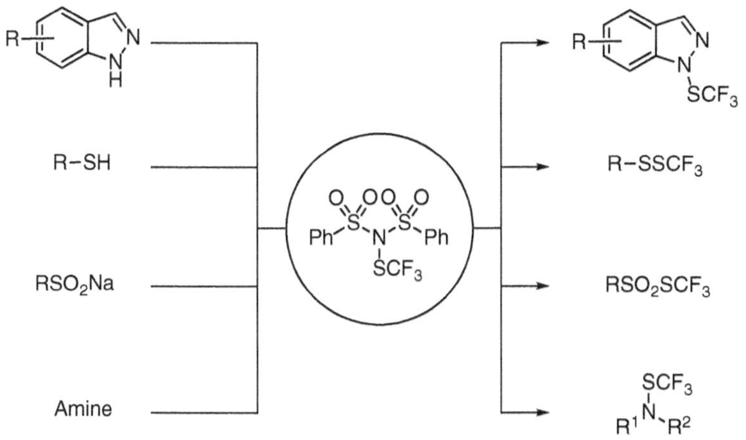

Scheme 4.68 Trifluoromethylthiolation of heteroatom nucleophiles..

successfully developed the first optically active trifluoromethylthiolating reagent based on camphorsultam framework [128], which reacted with β-ketoesters, oxindoles, benzofuran-2(3H)-ones to give the corresponding optically active trifluoromethylthiolated products with good to excellent enantioselectivity (Scheme 4.69).

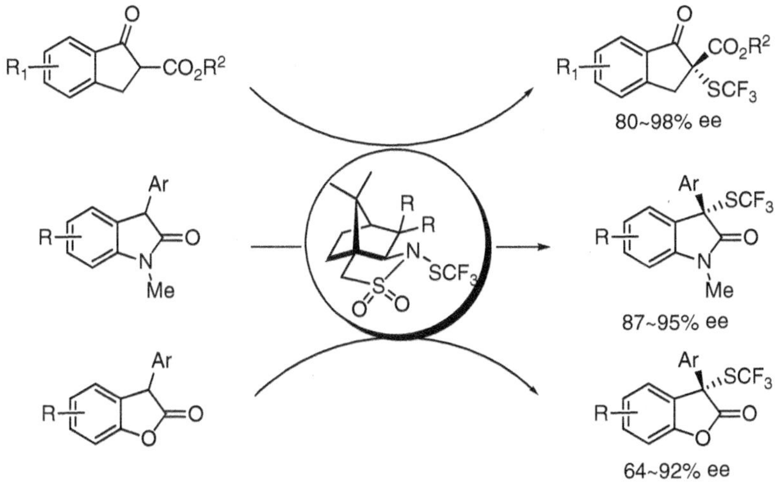

Scheme 4.69 Reactions of optically active pure trifluoromethylthiolation reagents.

4.3.9 Lu–Shen Reagent

Togni and coworkers's hypervalent iodine-based trifluoromethylating reagent [129] in 2013, Lu, Shen, and their coworkers designed a trifluoromethylthiolated analog, with the hope that it would be an excellent electrophilic trifluoromethylthiolating reagent [130]. It was found that reaction of 1-chloro-1,3-dihydro-3,3-dimethyl-1,2-

benziodoxole with AgSCF$_3$ gave a trifluoromethylthiolated compound. The structure of the compound was initially assigned as a trifluoromethylthiolated hypervalent iodine compound. However, this structure was revised by Buchwald and coworkers in 2014 to be a trifluoromethanesulfenate derivative by spectroscopic techniques and metal-organic-framework (MOF)-crystal sponge technology [131] (Scheme 4.70). In the revised structure, the trifluoromethylthio group is not connected to iodide. Instead, it is connected to the oxygen. In 2015, Lu, Shen, and coworkers conducted structure–reactivity studies and showed that the iodide of the structure was not so important for the reactivity of the reagent [132]. Thus, a secondary generation reagent PhC(Me$_2$)OSCF$_3$, which was much easier to be synthesized, was developed and serves as general electrophilic trifluoromethylthiolating reagent.

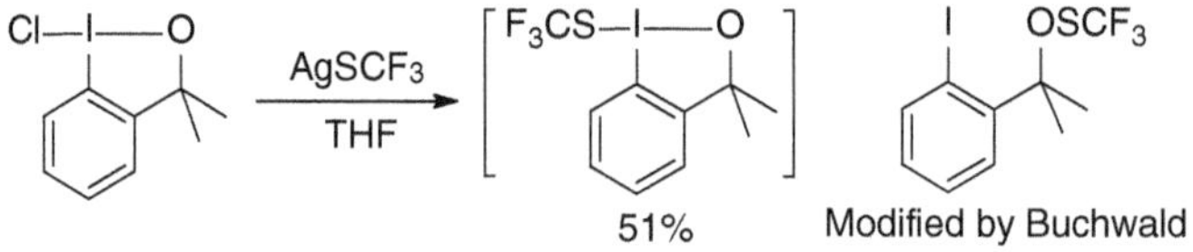

Scheme 4.70 Preparation and modification of Lu and Shen reagent.

Extensive studies from Lu, Shen, and their coworkers showed that this reagent is indeed a good electrophilic trifluoromethylthiolating reagent. In the presence of a copper catalyst, Lu–Shen reagent was able to couple with a variety of aryl boronic acids to give trifluoromethylthiolated arenes in high yields [133]. The coupling with alkyl boronic acids was more challenging. Nevertheless, increasing the reaction temperature to 120 °C and using a combination of 20 mol% CuCl$_2$·2H$_2$O and 40% bpy (bpy = 2,2′-bipyridine) as the catalyst, coupling of primary and secondary alkyl boronic acids with Lu–Shen reagent gave the corresponding alkyl trifluoromethylthioethers in good to excellent yields (Scheme 4.75).

In addition, in the presence of a copper catalyst, alkyne could also couple with Lu–Shen reagent to give alkynyl trifluoromethylthioethers in high yields. Nevertheless, the electrophilicity of Lu–Shen reagent is not as high as that of Shen reagent and Shen reagent-II, since Friedel–Crafts type of trifluoromethylthiolation with Lu–Shen reagent only worked for extremely electron-rich arenes such as indoles or pyrroles [134]. Recently, Lu, Shen, and coworkers described a transition metal-free method to trifluoromethylthiolate the *ipso*-carbon of lithium aryl boronates with trifluoromethanesulfenate under mild conditions. With this protocol, late-stage site-specific C–H and C–Cl trifluoromethylthiolation of biologically active molecules took place in high yields [135] (Scheme 4.71).

Soft carbon nucleophiles could also react with Lu–Shen reagent to give α-trifluoromethylthiolated carbonyls in high yields. Using *N,N*-dimethylamino-pyridine (DMAP) as the base, various cyclic β-ketoesters and a few acyclic β-ketoesters, aldehydes were successfully trifluoromethylthiolated in good to excellent yields [133]. In 2015, Dai and coworkers reported the first copper-catalyzed ring-opening electrophilic trifluoromethylation of cyclopropanols to synthesize

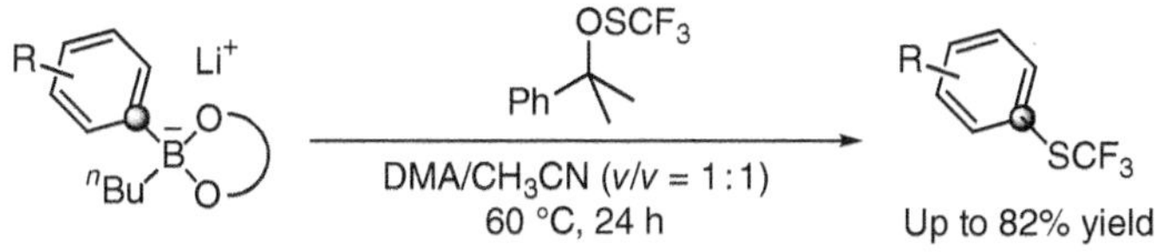

Scheme 4.71 Transition metal-free *ipso*-trifluoromethylthiolation of lithium aryl boronates.

β-SCF$_3$-substituted carbonyl compounds [136] (Scheme 4.72). Mechanistically, it was proposed that in the presence of CuSCF$_3$/bpy and a base, cyclopropanol undergoes ring-opening isomerization to give a β-alkylcopper intermediate, which then reacts with Lu–Shen reagent to give the final product.

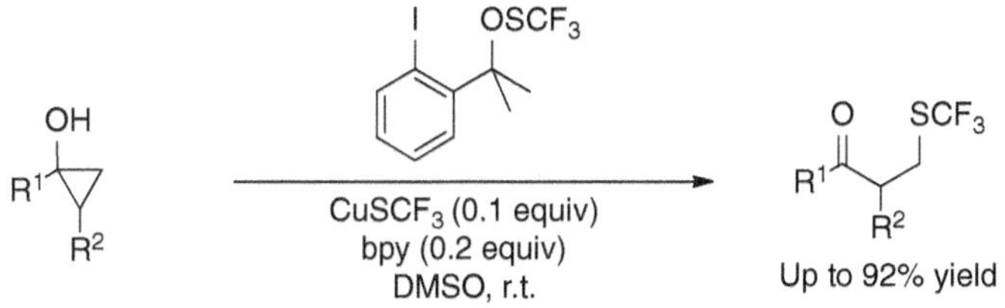

Scheme 4.72 Copper-catalyzed electrophilic ring-opening cross-coupling of cyclopropanols.

Asymmetric trifluoromethylthiolation of soft carbon nucleophiles and Lu–Shen reagent has also be reported. Shen and coworkers described that using quinine as the catalyst, five-membered β-ketoesters were able to react with Lu–Shen reagent to give the corresponding product in excellent enantioselectivity [137]. Surprisingly, under the same reaction conditions, six- or seven-membered β-ketoesters did not react at all. By switching the catalyst to a quinine-base phase transfer catalyst, six- or seven-membered β-ketoesters were able to react with Lu–Shen reagent with good to excellent enantioselectivity. Later, the quinine-catalyzed asymmetric trifluoromethylthiolation protocol was extended to substrates such as oxidoles with excellent enantioselectivity [138] (Scheme 4.73).

In 2014, Gade and coworkers reported a copper-catalyzed asymmetric trifluoromethylthiolation of β-ketoesters with Lu–Shen reagent [139]. In this protocol, a combination of 10 mol% Cu(OTf)$_2$ and 12 mol% Boxmi ligand was used as the catalyst. Both five- and six-membered β-ketoesters reacted to give the corresponding products in excellent enantioselectivity (Scheme 4.74).

Reaction of Lu–Shen reagent with heteroatom nucleophiles was difficult, mainly due to its relatively low electrophilicity. Nevertheless, Lu, Shen, and coworkers discovered that sodium arylsulfinates reacted with Lu–Shen reagent to give trifluoromethylthiosulfonates in high yields [132] (Scheme 4.75).

Lu–Shen reagent can not only be used as electrophilic trifluoromethylthiolating reagent but also, surprisingly, as a reagent for glycosidation. In 2014, Zhu and coworker [140] demonstrated that Lu–Shen reagent, in the presence of TMSOTf (trimethylsilyl triflate) as a Lewis acid, could activate the C—S bond of

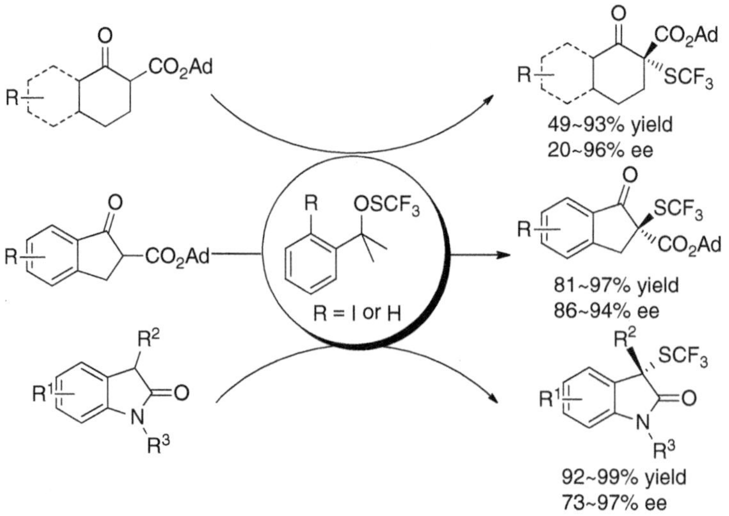

Scheme 4.73 Enantioselective electrophilic trifluoromethylthiolation using Lu–Shen reagent.

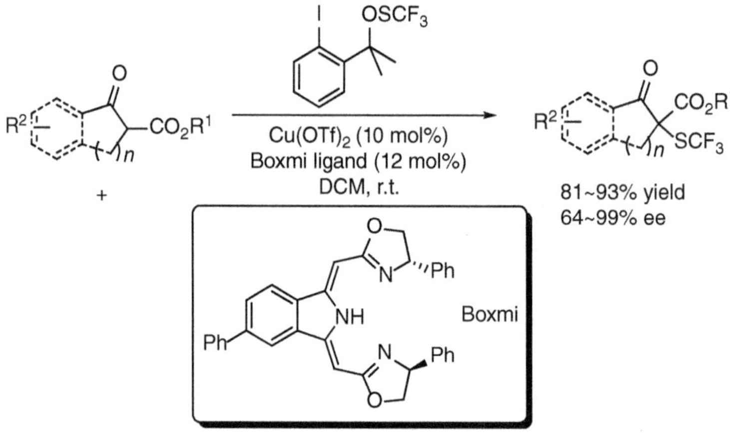

Scheme 4.74 Enantioselective electrophilic trifluoromethylthiolation using copper–boxmi complexes.

thioglycoside to realize the glycosidation of alcohols. Under these conditions, a variety of disaccharides were synthesized in excellent yields (Scheme 4.76).

In addition to the electrophilic trifluoromethylthiolation, Lu–Shen reagent could be used as the quenching agent of free radicals to realize the free radical trifluoromethylthiolation reaction, which is discussed in Section 4.2.3.

4.3.10 α-Cumyl Bromodifluoromethanesulfenate

Positron emission tomography (PET) is a leading noninvasive imaging technology that enables the *in vivo* study of physiological processes. Yet, a few methods to access

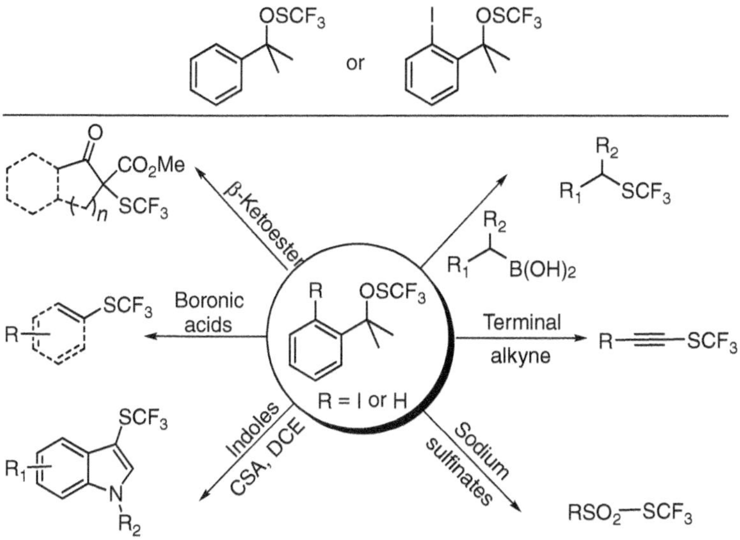

Scheme 4.75 Reactions of Lu–Shen reagent.

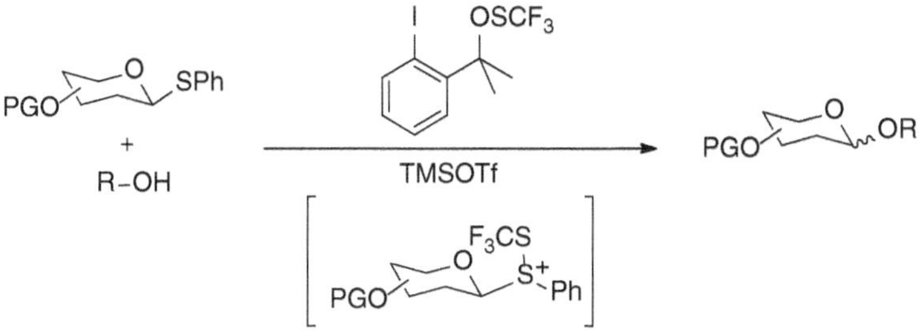

Scheme 4.76 Thioperoxide-mediated activation of thioglycoside donors.

[18]F-labeled [^{18}F]ArSCF$_3$ have been reported previously. Shen's group designed a new strategy for the preparation of [^{18}F]ArSCF$_3$. In this strategy, α-cumyl bromodifluoromethanesulfenate was first synthesized from easily available starting materials. In the presence of a copper catalyst, this reagent was able to couple with a variety of (hetero)aryl boronic pinacol esters to give bromodifluoromethylthiolated (hetero)arenes in excellent yields. The Shen group then collaborates with Gouverneur and coworkers to incorporate ^{18}F-labeling by a halogen-exchange fluorination with [^{18}F]KF with good radiochemical conversions. The strategy was then applied to synthesize [^{18}F]-labeled Tiflorex, a potent anorectic drug for the treatment of obesity [141] (Scheme 4.77).

4.3.11 Shibata Reagent

In 2013, Shibata and coworkers [142] developed an electrophilic trifluoromethylthiolating reagent based on the hypervalent iodonium ylide skeleton, which could be

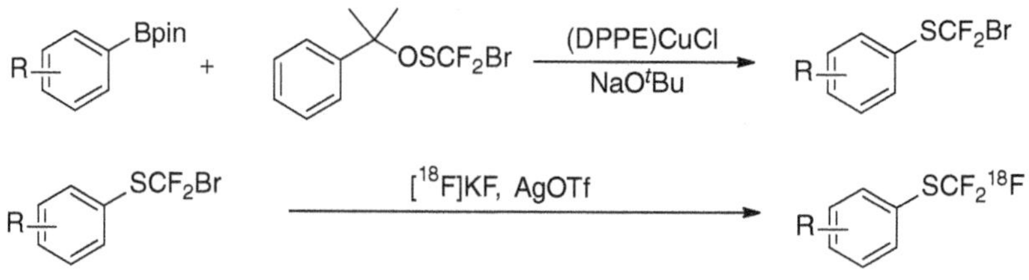

Scheme 4.77 Indirect trifluoromethylthiolation with α-cumyl bromodifluoromethanesulfenate.

synthesized with CF_3SO_2Na, 2-bromoacetophenone, and PIDA (iodobenzene diacetate) in high yield (Scheme 4.78).

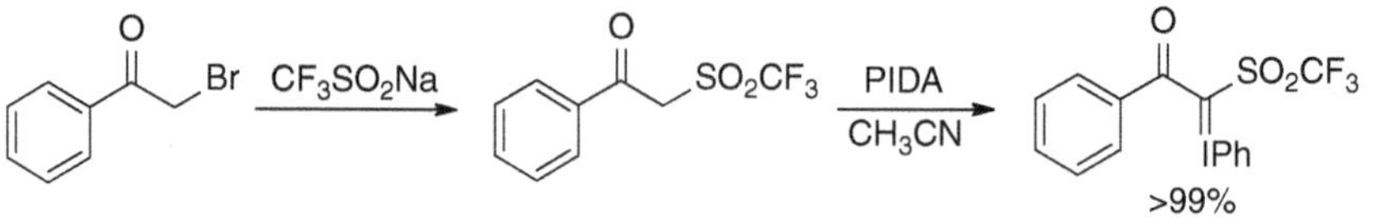

Scheme 4.78 Preparation of Shibata reagent.

In the presence of a copper catalyst, the reagent reacted with various nucleophiles including β-ketoesters, silyl enol ethers [143], aryl amines [144], indoles and pyrroles [145], enamines, aryl boronic acids [146], and allylsilanes [143] to give the corresponding trifluoromethylthiolated products in high yields (Scheme 4.79).

Mechanistically, it was proposed that the hypervalent iodonium ylide initially reacts with the cuprous salts to form a carbene intermediate, which undergoes successive rearrangement to give the active trifluoromethylthiolating species – trifluoromethylsulfenate. This species is then nucleophilically attacked by the substrate to give the final trifluoromethylthiolated product (Scheme 4.80).

4.3.12 *In Situ*-Generated Electrophilic Trifluoromethylthiolating Reagents

Most of the electrophilic trifluoromethylthiolating reagents developed in the past 20 years are shelf-stable compounds that are easily prepared from commercially available starting materials within three steps. Some of these reagents are now commercially available, thus paving the way for their further applications in medicinal chemistry. Alternatively, another complementary strategy for the electrophilic incorporation of the trifluoromethylthio group is to generate the target electrophilic trifluoromethylthiolating reagent *in situ* without further isolation and purification.

4.3.12.1 AgSCF$_3$ + TCCA

In 2014, Tan, Liu, and their coworkers described an asymmetric trifluoromethylthiolation of 3-aryl-2-indolinone substrates with $(DHQD)_2Pyr$ as the chiral catalyst [147]. In this approach, the electrophilic trifluoromethylthio species was *in situ* generated from $AgSCF_3$ and trichloroisocyanuric acid (TCCA). In their preliminary

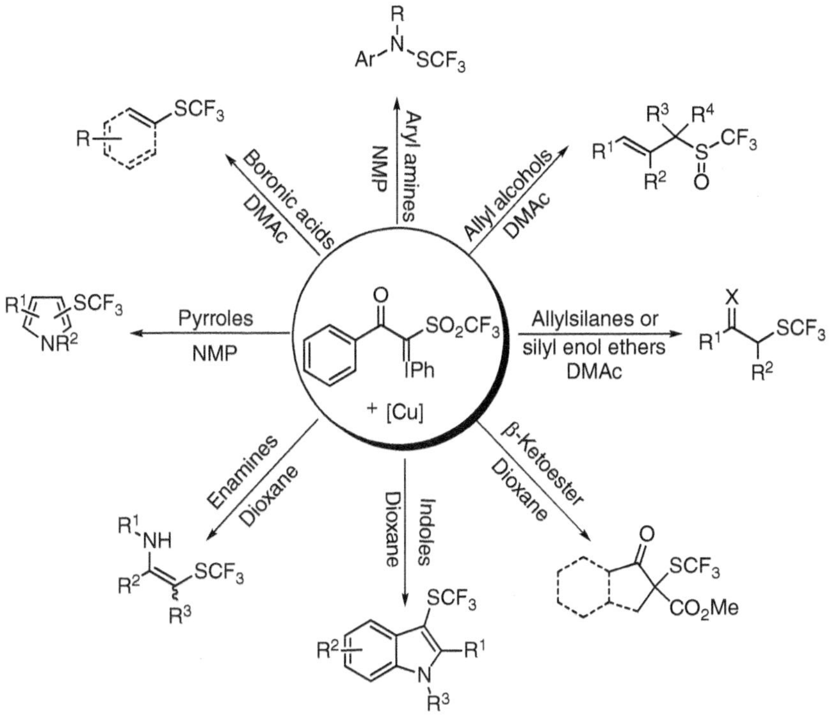

Scheme 4.79 Reactions of Shibata reagent.

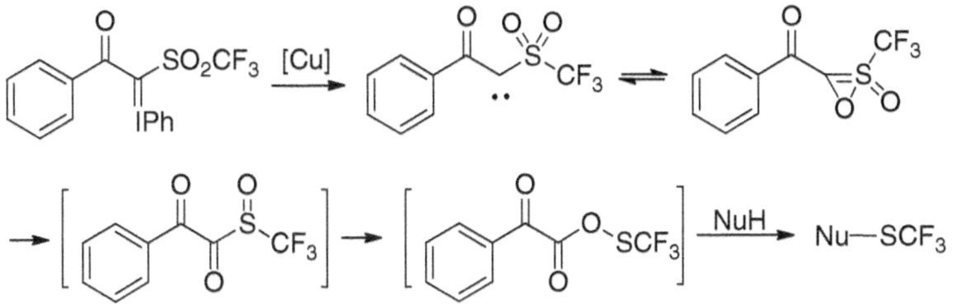

Scheme 4.80 Proposed mechanism for Cu-catalyzed trifluoromethylthiolation with Shibata reagent.

mechanism study, they proposed that CF_3SSCF_3 was one of the possible active species involved in the reaction (Scheme 4.81).

Yang and coworker adopted the same approach for the preparation of trifluoromethylthiolated chromone derivatives [148]. Mechanistically, it was proposed that the phenolic hydroxyl group first nucleophilically attacks the β-amino-α,β-unsaturated ketone; the resulting enolate then reacts with the *in situ*-generated electrophilic trifluoromethylthiolated species to give α-trifluoromethylthiolated intermediate. In the presence of a weak base, elimination of amine gives the final product (Scheme 4.82).

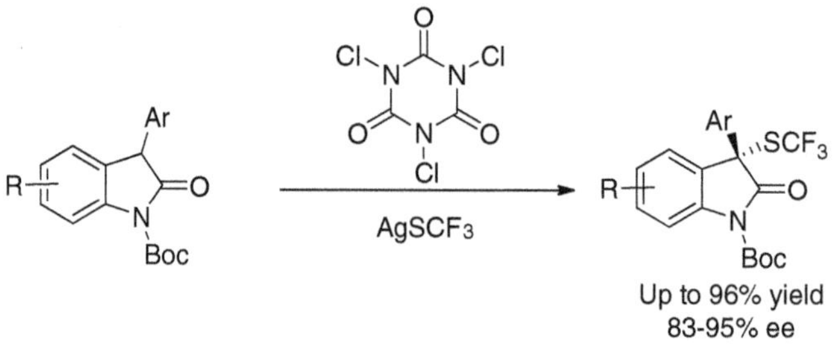

Scheme 4.81 *In situ* generation of electrophilic trifluoromethylthio reagents for enantioselective trifluoromethylthiolation of oxindoles.

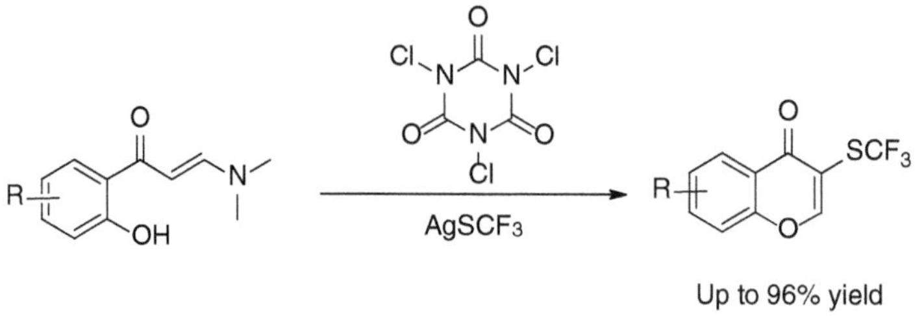

Scheme 4.82 *In situ*-generated electrophilic trifluoromethylthiolating reagent for the synthesis of 3-((trifluoromethyl)thio)-4*H*-chromen-4-one.

4.3.12.2 AgSCF$_3$ + NCS

In 2014, Qing and coworkers reported another approach for the *in situ* generation of the electrophilic trifluoromethylthiolating species [149]. In this approach, AgSCF$_3$ was allowed to react with NCS rather than TCCA to realize the trifluoromethylthiolation of terminal alkynes using potassium phosphate as the base. Interestingly, for this reaction, if the isolated electrophilic trifluoromethylthiolating reagent N-trifluoromethylthiosuccinimide was used, the formation of alkynyl trifluoromethylthioethers was not observed (Scheme 4.83).

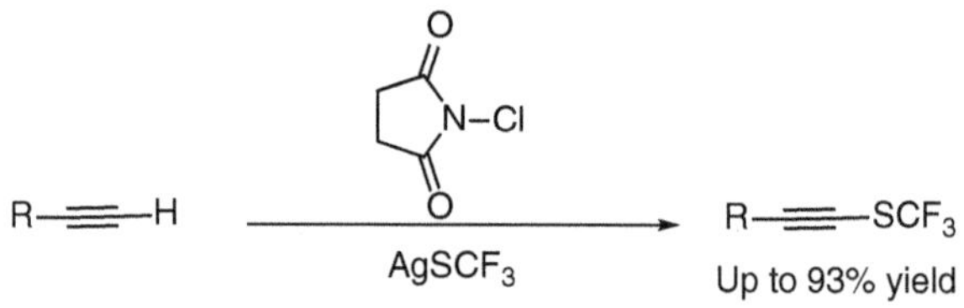

Scheme 4.83 Trifluoromethylthiolation of terminal alkynes using AgSCF$_3$ and NCS.

4.3.12.3 Langlois Reagent (CF$_3$SO$_2$Na) with Phosphorus Reductants

In 2015, Yi, Zhang, and their coworkers reported that (EtO)$_2$P(O)H was able to reduce CF$_3$SO$_2$Na to form CF$_3$S-SCF$_3$, which then reacted with CuCl to *in situ* generate an electrophilic trifluoromethylthiolating species [150] (Scheme 4.84).

This species was highly reactive and reacted with various electron-rich (hetero)arenes to give the corresponding trifluoromethylthio-substituted (hetero)arenes in good yields. Shortly after, Cai and coworkers discovered that PPh_3 was able to reduce CF_3SO_2Na to *in situ* generate CF_3SCl in the presence of 1.0 equiv of NCS [151] (Scheme 4.85). Friedel–Crafts type reaction of the *in situ*-generated CF_3SCl with indoles, pyrroles, and enamines occurred in good yields. In addition, it was reported that other phosphorus reductants such as PCl_3, $P(O)Cl_3$, or $P(O)Br_3$ could also reduce CF_3SO_2Na to generate active electrophilic trifluoromethylthiolating intermediate [152].

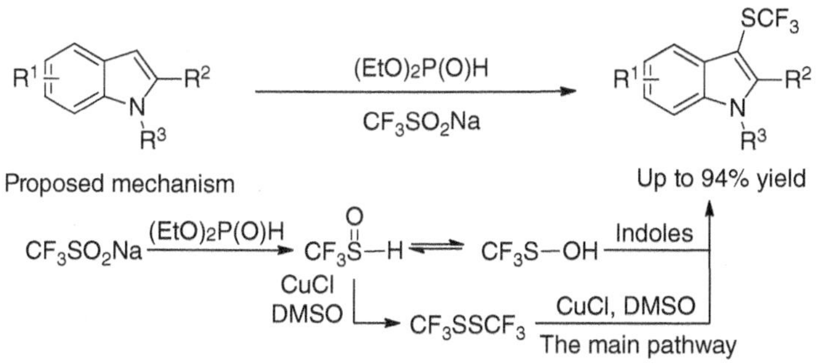

Scheme 4.84 Direct trifluoromethylthiolation of indoles with CF_3SO_2Na in the presence of $(EtO)_2P(O)H$.

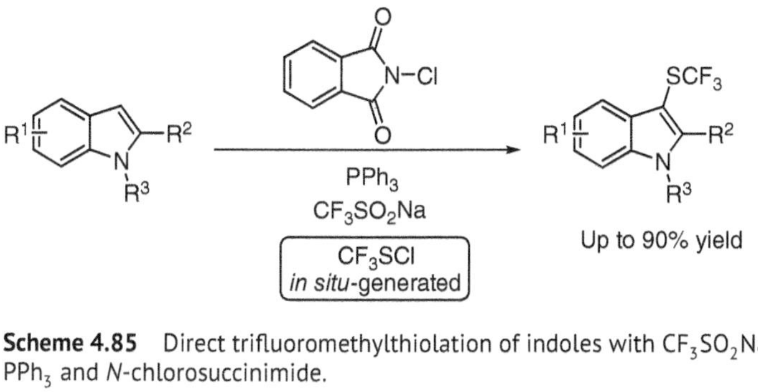

Scheme 4.85 Direct trifluoromethylthiolation of indoles with CF_3SO_2Na in the presence of PPh_3 and *N*-chlorosuccinimide.

4.3.12.4 Use of CF_3SO_2Cl with Phosphorus Reductants

Similar to the case of CF_3SO_2Na, CF_3SO_2Cl, which is an easily commercially available compound, could also be reduced by phosphorus reductants to generate active electrophilic trifluoromethylthiolating species. In 2016, Shibata, Cahard, and their coworkers adopted this strategy for the successful electrophilic trifluoromethythiolation of indoles, pyrroles, enamines, and other electron-rich substrates when PMe_3 was used as the reducing agent [153]. Preliminary mechanistic investigation suggested that CF_3SCl instead of CF_3SSCF_3 was the active species in the reaction. Zhao and coworkers discovered that addition of a catalytic amount of sodium iodide to the system could promote the reaction more effectively [154]. Likewise, Yi and coworkers reported based on their previous work on CF_3SO_2Na that phosphite was an efficient reductant to generate electrophilic trifluoromethylthiolating species from CF_3SO_2Cl for the Friedel–Crafts type of trifluoromethylthiolation of indole, pyrrole, and anisole derivatives [155] (Scheme 4.86).

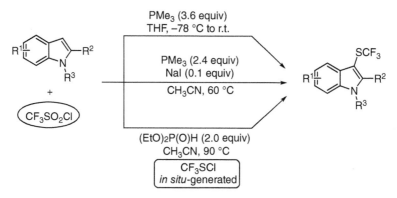

Scheme 4.86 Direct trifluoromethylthiolation of indoles with CF_3SO_2Cl.

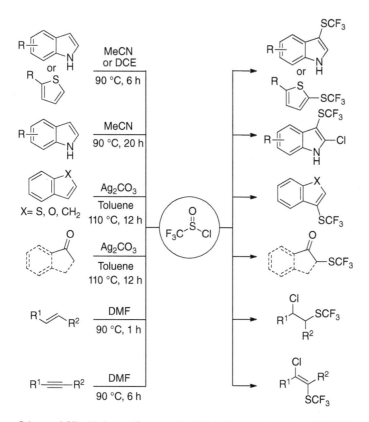

Scheme 4.87 Various trifluoromethylthiolation reactions with CF_3SOCl.

4.3.12.5 Reagent Based on CF₃SOCl and Phosphorus Reductants

Trifluoromethanesulfinyl chloride (CF$_3$SOCl), like its analog CF$_3$SO$_2$Cl, could be used to generate active trifluoromethylthiolative species. Yi, Zhang, and their coworkers discovered that CF$_3$SOCl could undergo self-disproportionation in the absence of any reductant to generate the active electrophilic trifluoromethylthiolating intermediate, which was highly reactive with electron-rich arenes, ketones, and thiols [156]. The same species could also be used for trifluoromethylthiolative bifunctionalization of alkenes and alkynes (Scheme 4.87).

Very recently, Liu and coworkers discovered that in the presence of pyridine, reaction of CF$_3$SOCl with 1,3-diketones occurred to give chloro-trifluoromethylthio difunctionalized 1,3-diketones in high yields [157]. Interestingly, if the substrate contains a hydroxyl or acetamide group, it would undergo amino-trifluoromethylthio bifunctionalization or Michael addition/trifluoromethylthiolation cascade reaction (Scheme 4.88).

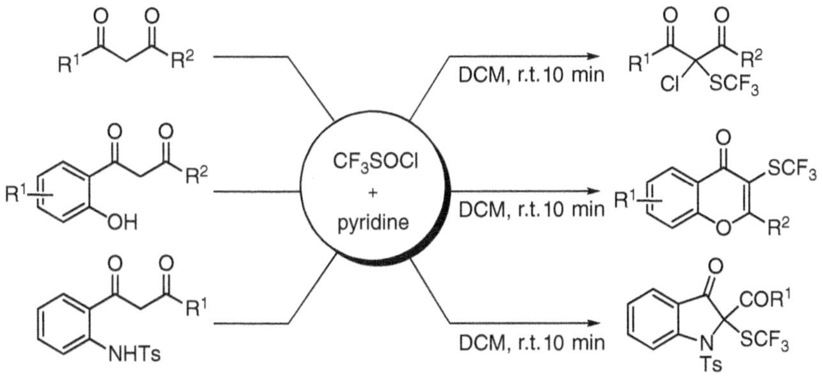

Scheme 4.88 Base-promoted trifluoromethylthiolation of 1,3-diketones with CF$_3$SOCl.

4.4 Radical Trifluoromethylthiolation

Free radical reaction often exhibits different regioselectivity from nucleophilic or electrophilic reaction. Thus, radical trifluoromethylthiolation may complement those nucleophilic or electrophilic trifluoromethylthiolative reactions. Early studies of radical trifluoromethylthiolation typically used CF$_3$SH, CF$_3$SCl, and CF$_3$SSCF$_3$ to generate trifluoromethylthio radical (·SCF$_3$) under the irradiation of ultraviolet light or X-ray. Subsequently, the highly reactive trifluoromethylthio radical reacted with alkane, alkenes, and other substrates. However, due to the toxicity of the reagents used, as well as the harsh reaction conditions, the development of radical trifluoromethylthiolation has been relatively slow. In the past 10 years, many easily available nucleophilic or electrophilic trifluoromethylthiolating reagents appeared, which stimulated the development of radical trifluoromethylthiolation chemistry.

As a result, numerous new trifluoromethylthiolative methods involving radical trifluoromethylthiolations have been reported.

4.4.1 Trifluoromethylthiolation by AgSCF$_3$/S$_2$O$_8^{2-}$

In 2014, Wang and coworker reported that a trifluoromethylthio radical could be easily generated from a combination of AgSCF$_3$ with K$_2$S$_2$O$_8$ in acetonitrile in the presence of substoichiometric amount of hexamethylphosphoramide (HMPA) [158]. Under these conditions, they designed a radical cascade reaction for the preparation of trifluoromethylthio-substituted oxindole derivatives from easily available starting material N-methyl-N-arylmethacrylamides. Detailed mechanistic studies suggest that AgSCF$_3$ might be initially oxidized by K$_2$S$_2$O$_8$ to form a Ag(II)SCF$_3$ species, which is unstable and easily undergoes Ag—S bond homolytic cleavage to generate ·SCF$_3$. Addition of ·SCF$_3$ to the terminal position of the alkene gives a secondary alkyl radical, which is then trapped by the arene (Scheme 4.89).

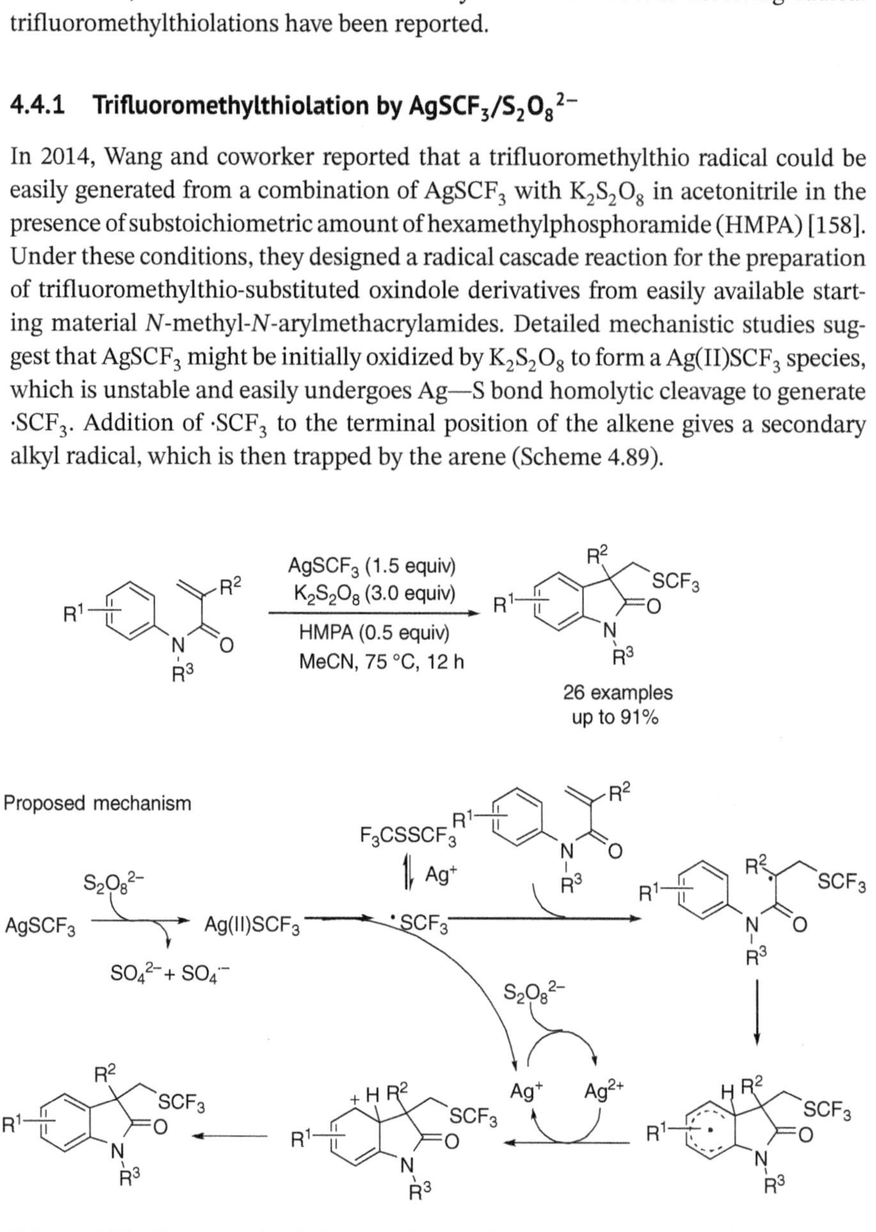

Scheme 4.89 First example of silver-mediated radical aryltrifluoromethlythiolation.

The discovery that ·SCF$_3$ could be easily generated from a combination of AgSCF$_3$ and K$_2$S$_2$O$_8$ paves the way for its wide application in radical cascade cyclization reactions. For example, following Wang's pioneer work, Nevado and coworkers reported a trifluoromethylthio radical-initiated addition/cyclization/desulfonylation cascade for the preparation of a variety of trifluoromethylthiolated heterocyclic

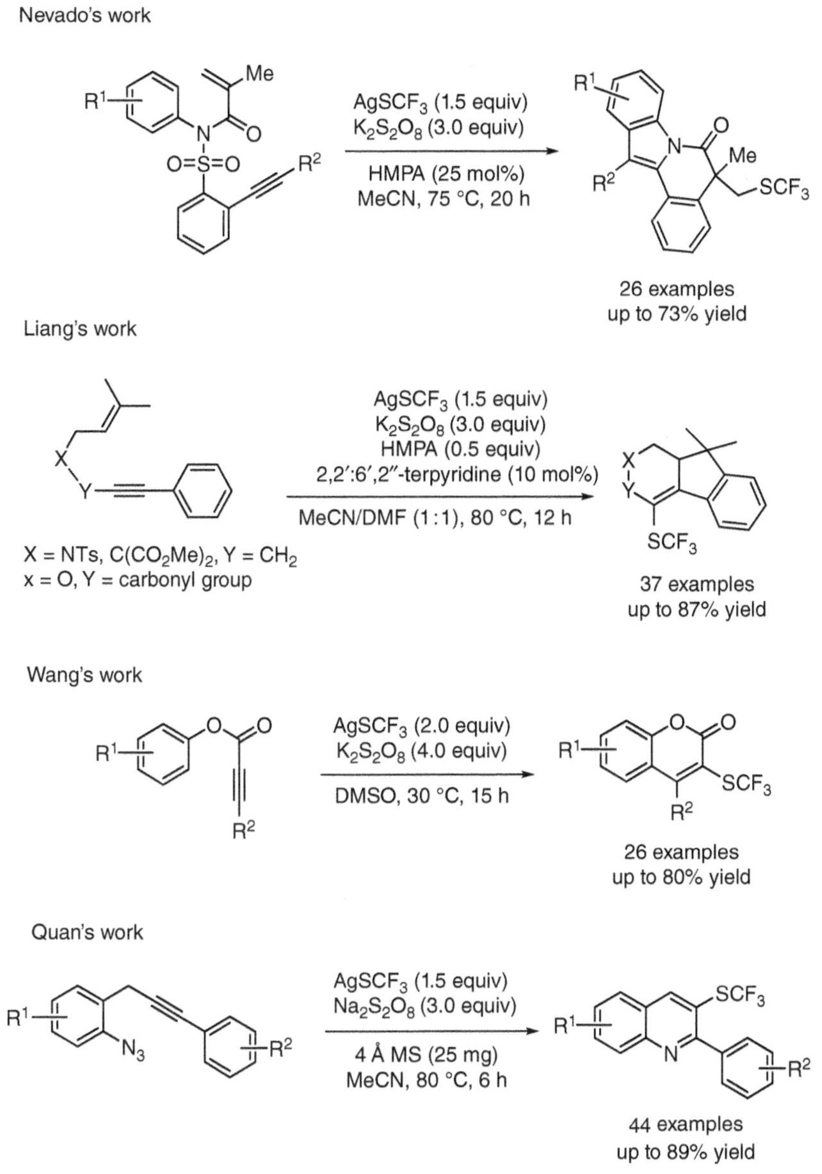

Nevado's work

AgSCF$_3$ (1.5 equiv)
K$_2$S$_2$O$_8$ (3.0 equiv)

HMPA (25 mol%)
MeCN, 75 °C, 20 h

26 examples
up to 73% yield

Liang's work

AgSCF$_3$ (1.5 equiv)
K$_2$S$_2$O$_8$ (3.0 equiv)
HMPA (0.5 equiv)
2,2′:6′,2″-terpyridine (10 mol%)

MeCN/DMF (1:1), 80 °C, 12 h

X = NTs, C(CO$_2$Me)$_2$, Y = CH$_2$
x = O, Y = carbonyl group

37 examples
up to 87% yield

Wang's work

AgSCF$_3$ (2.0 equiv)
K$_2$S$_2$O$_8$ (4.0 equiv)

DMSO, 30 °C, 15 h

26 examples
up to 80% yield

Quan's work

AgSCF$_3$ (1.5 equiv)
Na$_2$S$_2$O$_8$ (3.0 equiv)

4 Å MS (25 mg)
MeCN, 80 °C, 6 h

44 examples
up to 89% yield

Scheme 4.90 Other examples of silver-mediated radical trifluoromethylthiolation/cyclization.

compounds from substituted *N*-(arylsulfonyl)acrylamides substrates [159] (Scheme 4.90). Likewise, in 2015, Liang and coworkers developed a cascade trifluoromethylthiolation/radical cyclization of 1,6-enynes for the synthesis of various trifluoromethylthio-substituted polycyclic fluorene derivatives [160] (Scheme 4.90). Similarly, the same protocol has been successfully applied for the preparation of trifluoromethylthiolated coumarins, quinolines, and isoquinolines

[161, 162] (Scheme 4.90), while Shi and coworkers utilized this protocol for radical addition/ring opening/cyclization of methylenecyclopropanes to afford cyclic vinyl trifluoromethylthiolated products in high yields [163] (Scheme 4.91).

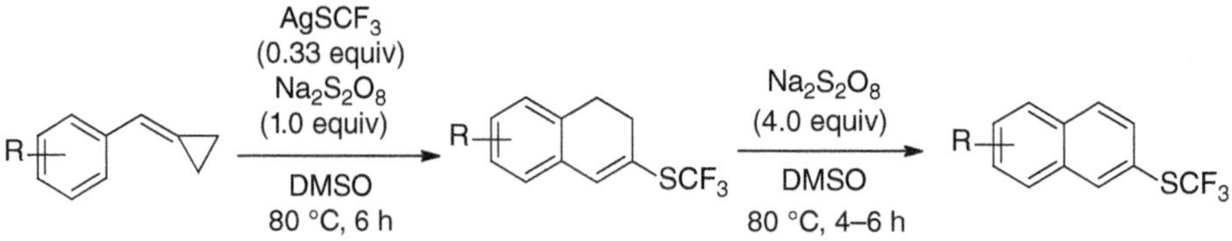

Scheme 4.91 Trifluoromethylthiolation of methylenecyclopropanes using $AgSCF_3/S_2O_8{}^{2-}$.

In 2014, Qing's group reported a copper-mediated reaction between a terminal alkene and $AgSCF_3/K_2S_2O_8$ [164]. Instead of giving a common radical addition product or radical-initiated polyolefin, an allylic trifluoromethylthioether was obtained as the main product. Mechanistically, it was proposed that the trifluoromethylthio radical is initially added to alkene to form a secondary radical, which is quickly oxidized by $Cu(OAc)_2$ to generate an alkyl cation. α-Hydrogen elimination affords the allylic trifluoromethylthioether. The yields of the reactions were typically good. Yet, the E/Z selectivity of the reaction was moderate, with a ratio range of 4.5 : 1–11 : 1 (Scheme 4.92).

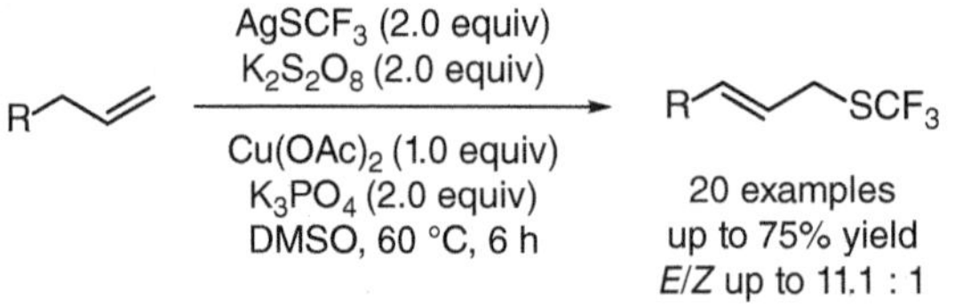

Scheme 4.92 Copper-mediated oxidative trifluoromethylthiolation of unactivated terminal alkenes.

In 2016, Cao and coworkers reported the application of $AgSCF_3/K_2S_2O_8$ protocol for hydro-trifluoromethylthiolation of alkynes [165]. Interestingly, *anti*-Markovnikov-selective hydro-trifluoromethylthiolation with moderate E/Z ratio occurred when the reaction was conducted with 50 mol% HMPA and 10 mol% phenanthroline (phen), while Markovnikov-selective hydro-trifluoromethylthiolation products were obtained when the reaction was conducted with 10 mol% CuCl and 3.0 equiv of H_2O (Scheme 4.93).

Liu and coworkers reported a trifluoromethylthio radical triggered migratory rearrangement of α,α-diaryl allylic alcohols to give β-trifluoromethylthiolated ketone derivatives in moderate yields [166] (Scheme 4.94).

In 2016, Qing and coworkers utilized the $AgSCF_3/K_2S_2O_8$ protocol for decarboxylative trifluoromethylthiolation of cinnamic acids [167]. Vinyl trifluoromethylthioethers and α-trifluoromethylthiolated ketones can be selectively synthesized by adjusting the reaction parameters (Scheme 4.95). Likewise, in

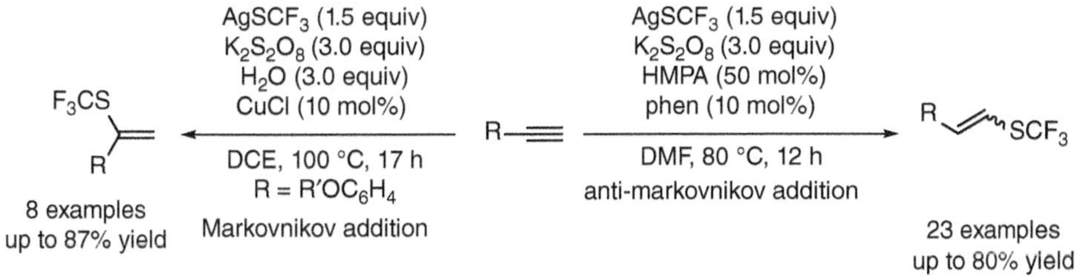

Scheme 4.93 showing reaction conditions:

Left side (toward F3CS vinyl product):
AgSCF₃ (1.5 equiv)
K₂S₂O₈ (3.0 equiv)
H₂O (3.0 equiv)
CuCl (10 mol%)
DCE, 100 °C, 17 h
R = R'OC₆H₄
Markovnikov addition
8 examples
up to 87% yield

Right side (toward SCF₃ product):
AgSCF₃ (1.5 equiv)
K₂S₂O₈ (3.0 equiv)
HMPA (50 mol%)
phen (10 mol%)
DMF, 80 °C, 12 h
anti-markovnikov addition
23 examples
up to 80% yield

Scheme 4.93 Selective hydro-trifluoromethylthiolation of alkynes using $AgSCF_3/S_2O_8^{2-}$.

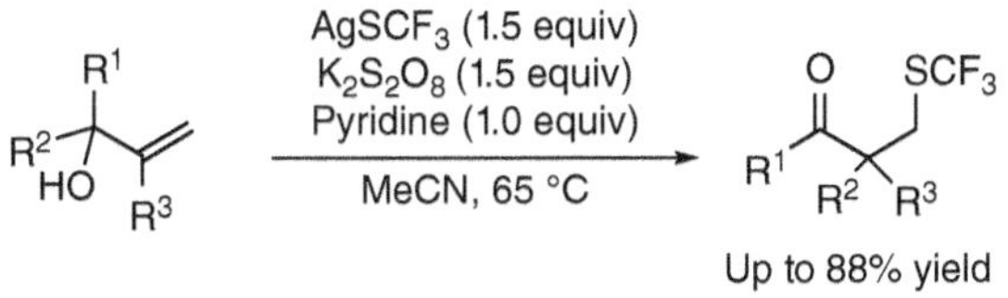

Scheme 4.94 conditions:
AgSCF₃ (1.5 equiv)
K₂S₂O₈ (1.5 equiv)
Pyridine (1.0 equiv)
MeCN, 65 °C
Up to 88% yield

Scheme 4.94 Trifluoromethylthiolation of diaryl allylic alcohols via radical neophyl rearrangement.

2018, Xu and coworkers improved the method for the preparation of vinyl trifluoromethylthioethers from cinnamic acids [168]. With the addition of $CuSO_4 \cdot 5H_2O$, the reaction conditions were simplified, and E/Z ratio of the alkenes was increased to up to 97 : 3. Furthermore, under the same reaction conditions, α-trifluoromethylthiolated ketones could be prepared from propiolic acids (Scheme 4.95). Finally, Hoover and coworkers applied the same strategy for the decarboxylative trifluoromethylthiolation of coumarin derivatives [169] (Scheme 4.95). In addition, decarboxylative trifluoromethylthiolation of alkyl carboxylic acids proceeded smoothly when Selectfluor was used as the oxidant [170].

Finally, the $AgSCF_3/K_2S_2O_8$ protocol could also be applied for alkane C—H bond trifluoromethylthiolation. In 2015, Chen and coworkers and Tang and coworkers reported back to back the first example of alkane C—H bond radical trifluoromethylthiolation in moderate yields [171, 172]. It was proposed that reaction of $AgSCF_3/K_2S_2O_8$ generates $Ag(II)SCF_3$ and $SO_4^{\cdot-}$. The $SO_4^{\cdot-}$ is able to abstract a hydrogen from alkane to give an alkyl radical, which is then trapped by $Ag(II)SCF_3$, CF_3SSCF_3, or $\cdot SCF_3$ (Scheme 4.96).

Several copper-mediated radical trifluoromethylthiolation reactions using $AgSCF_3$ as the trifluoromethylthio source have also been reported. For example, in 2018, Zhu and coworkers reported that in the presence of 20 mol% $Cu(OAc)_2$, O-benzoyl oxime with an alkene moiety reacted with $AgSCF_3$ to give trifluoromethylthio-substituted pyrrolines in high yields [173] (Scheme 4.97). It was proposed that a Cu(I) species initially reacts with $AgSCF_3$ to form $CuSCF_3$, which reductively cleaves the acyl oxime's N—O bond to generate an iminyl radical and a $Cu(II)SCF_3$ species. Subsequently, intramolecular 5-exo-trig cyclization of iminyl radical with the alkene moiety gives an alkyl radical, which is then trapped by the $Cu(II)SCF_3$ species to produce the desired product.

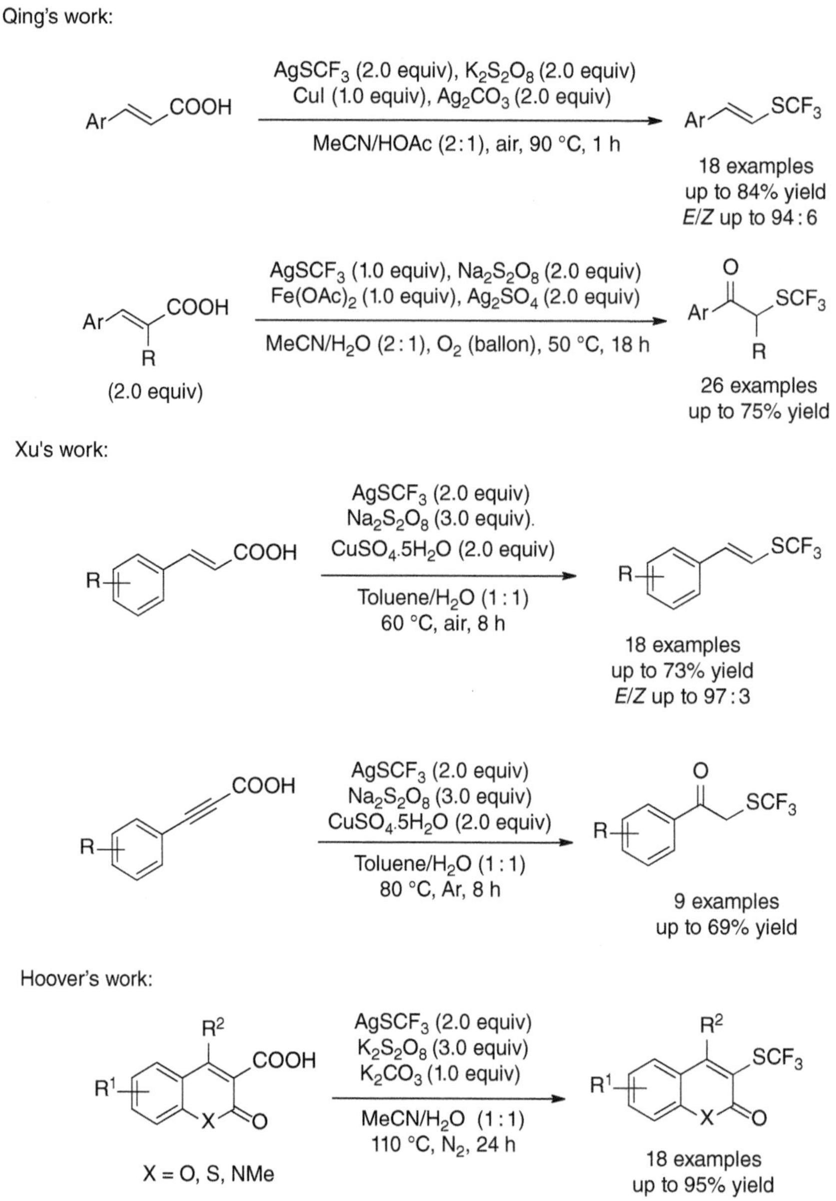

Qing's work:

AgSCF₃ (2.0 equiv), K₂S₂O₈ (2.0 equiv)
CuI (1.0 equiv), Ag₂CO₃ (2.0 equiv)

MeCN/HOAc (2:1), air, 90 °C, 1 h

18 examples
up to 84% yield
E/Z up to 94:6

AgSCF₃ (1.0 equiv), Na₂S₂O₈ (2.0 equiv)
Fe(OAc)₂ (1.0 equiv), Ag₂SO₄ (2.0 equiv)

MeCN/H₂O (2:1), O₂ (ballon), 50 °C, 18 h

26 examples
up to 75% yield

(2.0 equiv)

Xu's work:

AgSCF₃ (2.0 equiv)
Na₂S₂O₈ (3.0 equiv).
CuSO₄.5H₂O (2.0 equiv)

Toluene/H₂O (1:1)
60 °C, air, 8 h

18 examples
up to 73% yield
E/Z up to 97:3

AgSCF₃ (2.0 equiv)
Na₂S₂O₈ (3.0 equiv)
CuSO₄.5H₂O (2.0 equiv)

Toluene/H₂O (1:1)
80 °C, Ar, 8 h

9 examples
up to 69% yield

Hoover's work:

AgSCF₃ (2.0 equiv)
K₂S₂O₈ (3.0 equiv)
K₂CO₃ (1.0 equiv)

MeCN/H₂O (1:1)
110 °C, N₂, 24 h

18 examples
up to 95% yield

X = O, S, NMe

Scheme 4.95 Decarboxylative trifluoromethylthiolation using AgSCF₃/S₂O₈²⁻.

In 2019, Cook and coworkers reported a copper-catalyzed trifluoromethylthiolation of C(sp³)—H bond of an alkane using AgSCF₃ as the trifluoromethylthio source based on 1,5-hydrogen atom transfer (1,5-HAT) process [174] (Scheme 4.98). It was proposed that a copper(I) species first reacts with fluorosulfonamide in the substrate to produce a high-energy N-centered radical and a Cu(II) species. The nitrogen radical then abstracts a hydrogen through 1,5-HAT to generate an alkyl radical.

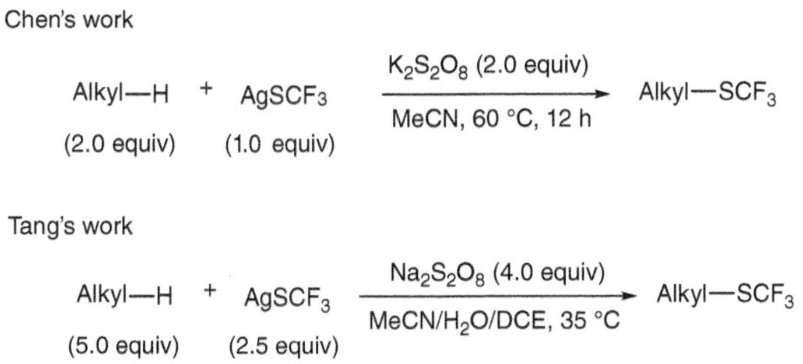

Chen's work

$$\text{Alkyl—H} \; + \; \text{AgSCF}_3 \xrightarrow[\text{MeCN, 60 °C, 12 h}]{\text{K}_2\text{S}_2\text{O}_8 \text{ (2.0 equiv)}} \text{Alkyl—SCF}_3$$

(2.0 equiv) (1.0 equiv)

Tang's work

$$\text{Alkyl—H} \; + \; \text{AgSCF}_3 \xrightarrow[\text{MeCN/H}_2\text{O/DCE, 35 °C}]{\text{Na}_2\text{S}_2\text{O}_8 \text{ (4.0 equiv)}} \text{Alkyl—SCF}_3$$

(5.0 equiv) (2.5 equiv)

Scheme 4.96 Trifluoromethylthiolation of unactivated $C(sp^3)$–H using $AgSCF_3/S_2O_8{}^{2-}$.

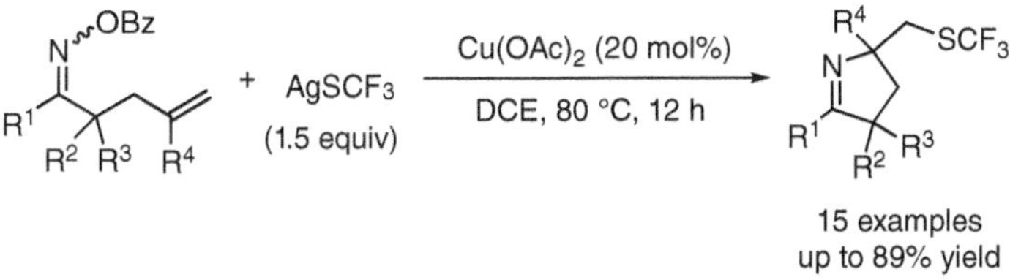

Scheme 4.97 Copper-catalyzed trifluoromethylthiolation/cyclization of unactivated alkenes.

Meanwhile, the Cu(II) species reacts with $AgSCF_3$ to form $Cu(II)SCF_3$, which is oxidized by the alkyl radical to give a Cu(III) intermediate with an alkyl group and a SCF_3. Reductive elimination from this intermediate produces the desired trifluoromethylthiolated compounds.

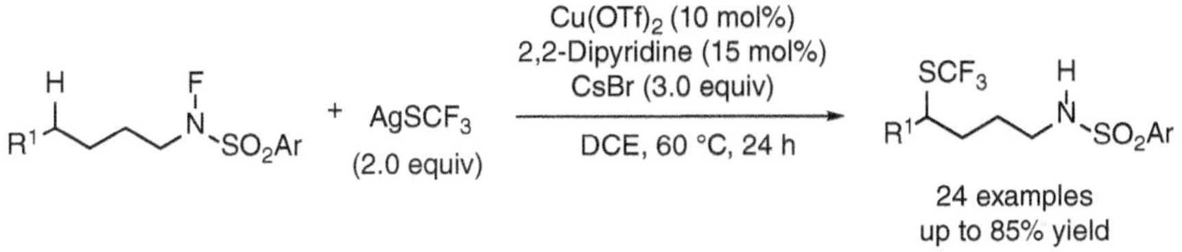

Scheme 4.98 Copper-catalyzed trifluoromethylthiolation of $C(sp^3)$–H bond based on 1,5-HAT.

Likewise, Xiao and coworkers reported in 2018 a silver-mediated trifluoromethylthiolation of alkane $C(sp^3)$—H bond using trifluoromethylsulfinyl chloride as the trifluoromethylthio source and triphenylphosphine as the reducing reagent [175]. In general, the tertiary C—H bonds of various alkanes were successfully trifluoromethylthiolated (Scheme 4.99). It was proposed that CF_3SOCl is reduced by PPh_3 to *in situ* generate CF_3SSCF_3. At the same time, electron transfer between Ag(I) and $K_2S_2O_8$ generates $SO_4{}^{\cdot-}$, which abstracts a hydrogen from the weakest tertiary C—H bonds of alkane to give an alkyl radical. Quenching the radical by CF_3SSCF_3 affords the desired product.

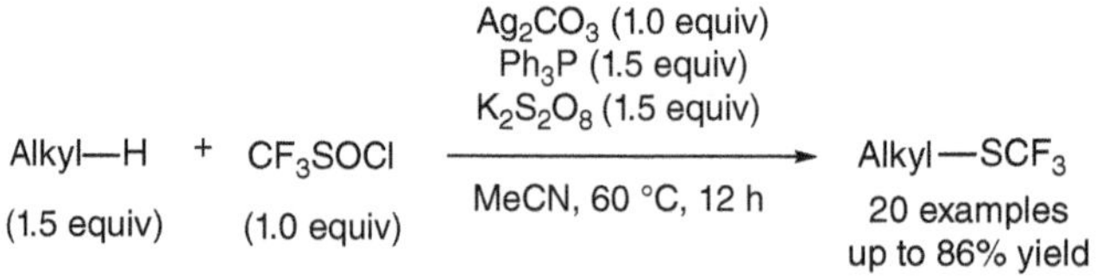
Ag2CO3 (1.0 equiv)
Ph3P (1.5 equiv)
K2S2O8 (1.5 equiv)
Alkyl—H
+
CF3SOCl
(1.5 equiv)
(1.0 equiv)
MeCN, 60 °C, 12 h
Alkyl—SCF3
20 examples
up to 86% yield

Scheme 4.99 Silver-mediated trifluoromethylthiolation of inert $C(sp^3)-H$ bond.

4.4.2 Electrophilic Reagents Involved in Radical Trifluoromethylthiolation

A secondary type of radical trifluoromethylthiolation reaction involves the participation of electrophilic trifluoromethylthiolating reagents. In this type of reaction, an alkyl radical is generated from the substrate, and subsequently quenched by an electrophilic trifluoromethylthiolating reagent to give the corresponding product.

Early in 2014, Lu, Shen, and their coworkers reported a silver-catalyzed decarboxylative trifluoromethylthiolation of alkyl carboxylic acids with Lu–Shen's reagent in a mixed solvent of CH_3CN/water [176]. A variety of secondary and tertiary alkyl carboxylic acids underwent decarboxylative trifluoromethylthiolation smoothly to give the corresponding alkyl trifluoromethylthioethers in high yields. It was found that a surfactant sodium dodecyl sulfate (SDS) played a crucial role in the reaction since it generates "oil-in-water" droplets. The silver-catalyzed decarboxylation of the alkyl carboxylic acid process took place in the water phase to give an alkyl radical, which penetrated into the oil phase to react with trifluoromethylsulfenate to give the corresponding product (Scheme 4.100).

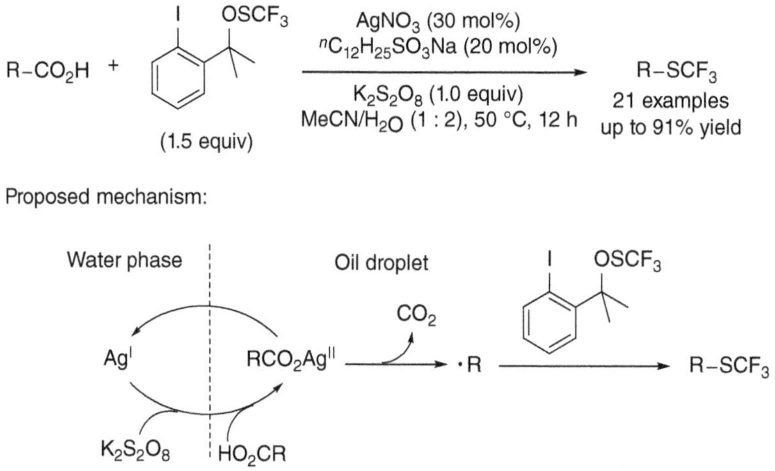
R–CO2H
+
I
OSCF3
(1.5 equiv)
AgNO3 (30 mol%)
nC12H25SO3Na (20 mol%)
K2S2O8 (1.0 equiv)
MeCN/H2O (1 : 2), 50 °C, 12 h
R–SCF3
21 examples
up to 91% yield
Proposed mechanism:
Water phase
Oil droplet
Ag^I
RCO2Ag^II
K2S2O8
HO2CR
CO2
•R
I
OSCF3
R–SCF3

Scheme 4.100 Decarboxyltrifluoromethylthiolation of alkyl carboxylic acid using Lu-Shen reagent as radical quencher.

In 2015, Yang and Shen developed an iron-mediated Markovnikov-selective hydro-trifluoromethylthiolation of unactivated alkenes with trifluoromethylsulfenate

(Lu–Shen reagent) [177]. Preliminary mechanistic studies suggests that an [Fe–H] species, which is *in situ* generated by iron(III) salt and BH_3, adds to the alkene to give an alkyl radical. The radical is then quenched by the electrophilic trifluoromethylthiolating reagent (Scheme 4.101). Very recently, a similar Co-catalyzed Markovnikov-selective hydro-trifluoromethylthiolation of unactivated alkenes with $PhSO_2SCF_3$ was reported [178] (Scheme 4.101).

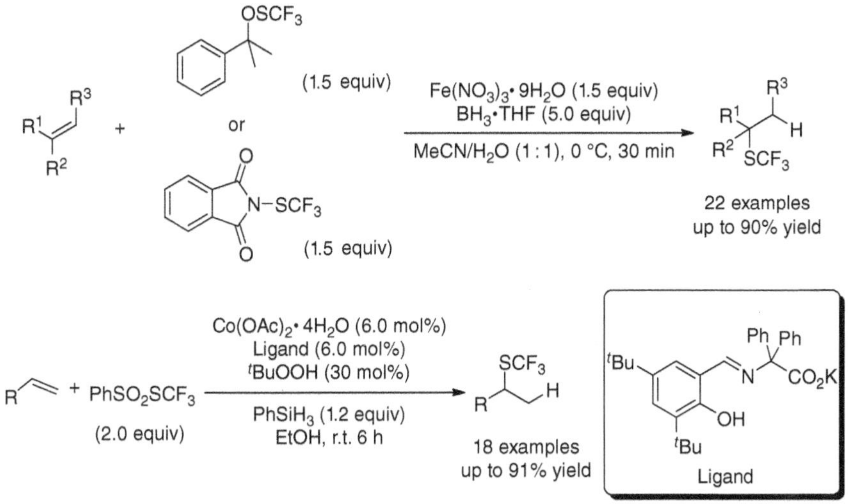

Scheme 4.101 Metal-mediated/catalyzed hydro-trifluoromethylthiolation using electrophilic reagents as radical quenchers.

4.4.3 Visible Light-Promoted Trifluoromethylthiolation by Using Electrophilic Reagents

In addition to the transition metal-promoted radical trifluoromethylthiolation methodologies, visible light-promoted radical trifluoromethylthiolation emerged recently. In general, visible light-induced radical trifluoromethylthiolation strategy can be divided into two different types of reaction modes. (i) The electrophilic trifluoromethylthiolating reagent is first reduced by the photocatalyst to generate $\cdot SCF_3$, which then gets involved in the subsequent reaction. (ii) The radical is initially generated under visible irradiation in the presence of photoredox catalyst and then quenched by an electrophilic trifluoromethylthiolating reagent.

In 2016, Glorius and coworkers reported the first visible light-promoted trifluoromethylthiolation of styrenes using *N*-(trifluoromethylthio)phthalimide as the trifluoromethylthio source in good yields [179]. It was proposed that assisted by the halide additives, *N*-(trifluoromethylthio)phthalimide is reduced by the photoredox catalysis to deliver $\cdot SCF_3$ in an oxidative quenching cycle. Addition of $\cdot SCF_3$ to styrene gives a benzylic radical, which is oxidized by the photoredox catalyst to become benzylic cation. α-Hydrogen elimination from this cation generates the corresponding product (Scheme 4.102).

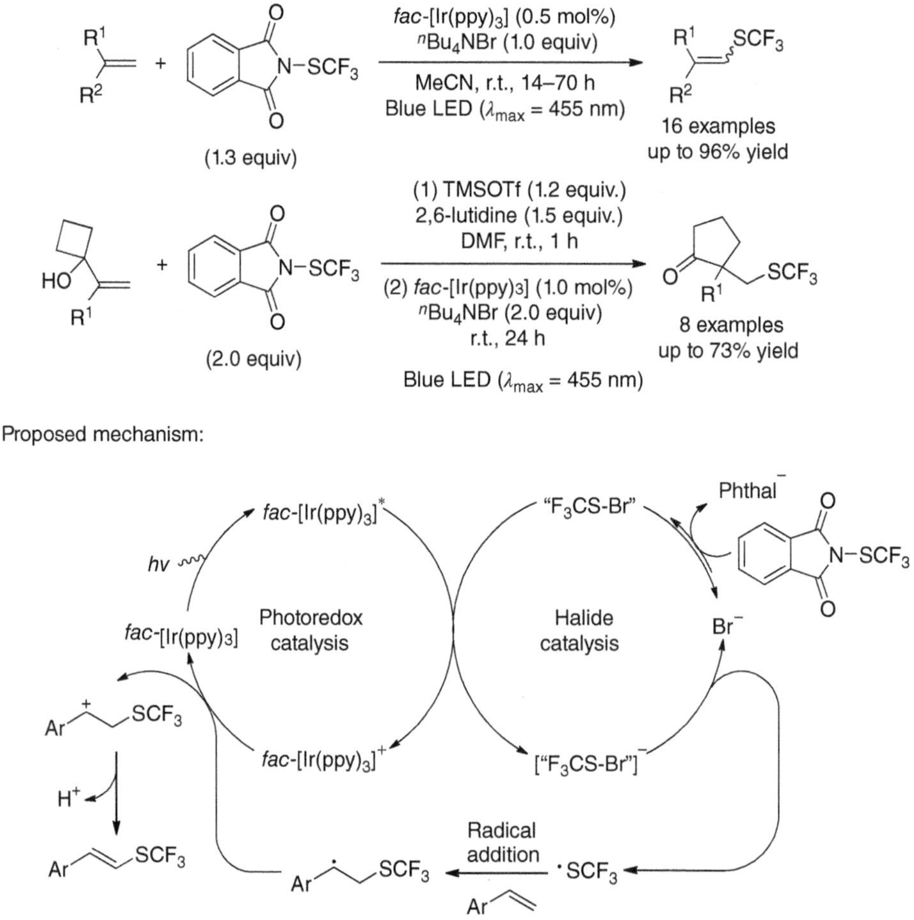

Scheme 4.102 Reaction and proposed mechanism of visible light-promoted trifluoromethylthiolation using *N*-(trifluoromethylthio)phthalimide as electrophilic reagents.

Density-functional theory (DFT) calculation by Cheng, Xue, and coworkers has established that electrophilicity and the bond strength of the N—SCF$_3$ bond of the electrophilic trifluoromethylthiolating reagents is correlated to the acidity of the corresponding amide moiety of the reagent [180]. Based on Glorius's work, it is reasonable to expect that more electrophilic trifluoromethylthiolating reagents such as *N*-trifluoromethylthiosaccharin (Shen reagent) should be easier to generate ·SCF$_3$ under similar reaction conditions.

In 2017, Magnier and coworkers reported that under photoredox catalysis, *N*-trifluoromethylthiosaccharin underwent homolytic N—S bond cleavage to give trifluoromethylthio radical, which further reacted with *N*-aryl acrylamide to give radical addition/cyclization products in high yields [181]. The same protocol could be extended to trifluoromethylthiolative difunctionalization of styrene derivatives (Scheme 4.103). Similarly, Fu and coworkers found that trifluoromethylthio radical could be generated from *N*-trifluoromethylthiosaccharin under similar conditions, which was then added to olefinic amides or *N*-(*o*-cyanobiaryl)acrylamide to give benzoxazine, oxazoline, or ring-fused phenanthridine derivatives in high yields [182, 183] (Scheme 4.103). Likewise, under similar conditions, *N*-trifluoromethylthiosuccinimide could also be used as trifluoromethylthio radical precursor for difunctionalization of styrenes [184] (Scheme 4.104).

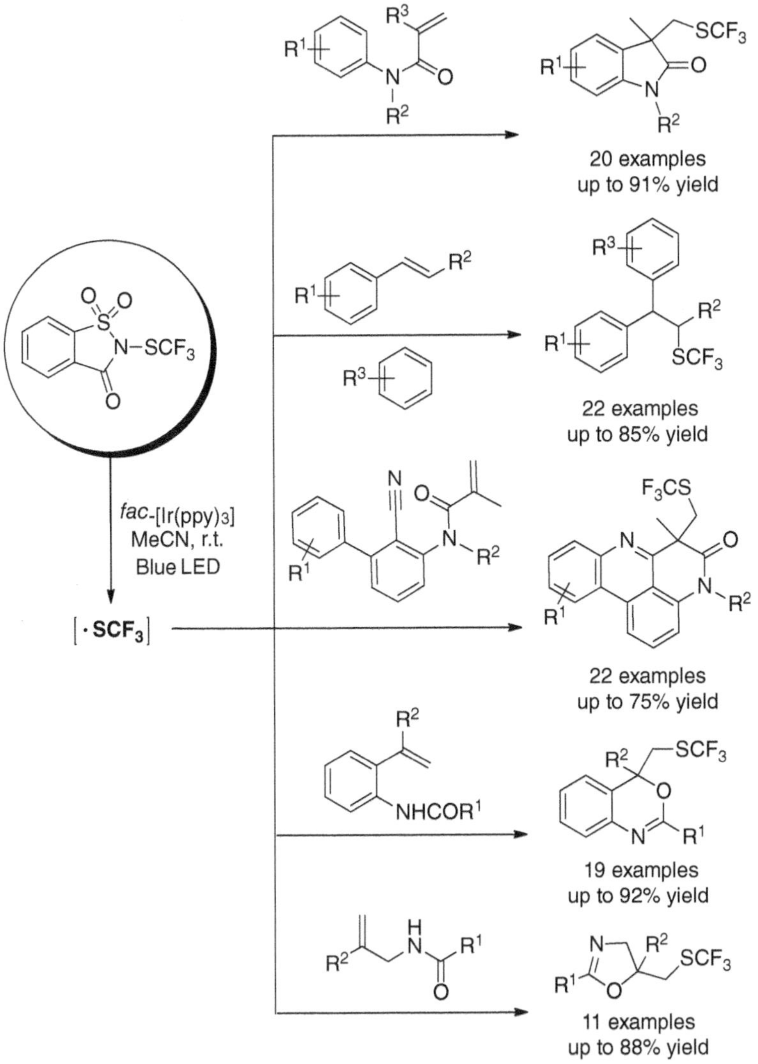

Scheme 4.103 Visible light-promoted trifluoromethylthiolation using *N*-trifluoromethylthiosaccharin as electrophilic reagent.

The second type of radical trifluoromethylthiolation model was also reported by Glorius and coworkers for visible light-promoted decarboxylative trifluoromethylthiolation of aliphatic carboxylic acids [185]. In this work, rather than

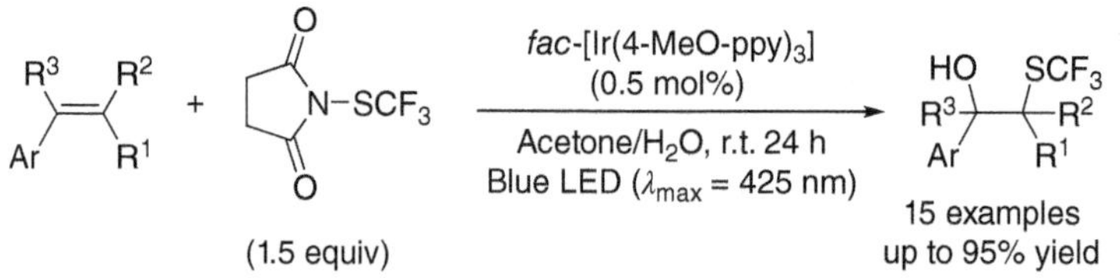

Scheme 4.104 Visible light-promoted trifluoromethylthiolation using 1-(trifluoromethylthio)-pyrrolidine-2,5-dione as electrophilic reagent.

in situ generating ·SCF$_3$ from N-(trifluoromethylthio)phthalimide via photoredox catalysis, an alkyl radical was initially generated from the corresponding alkyl carboxylic acid under photocatalytic cycle, and subsequently quenching of the alkyl radical by the electrophilic reagent N-(trifluoromethylthio)phthalimide gave alkyl trifluoromethylthioethers (Scheme 4.105).

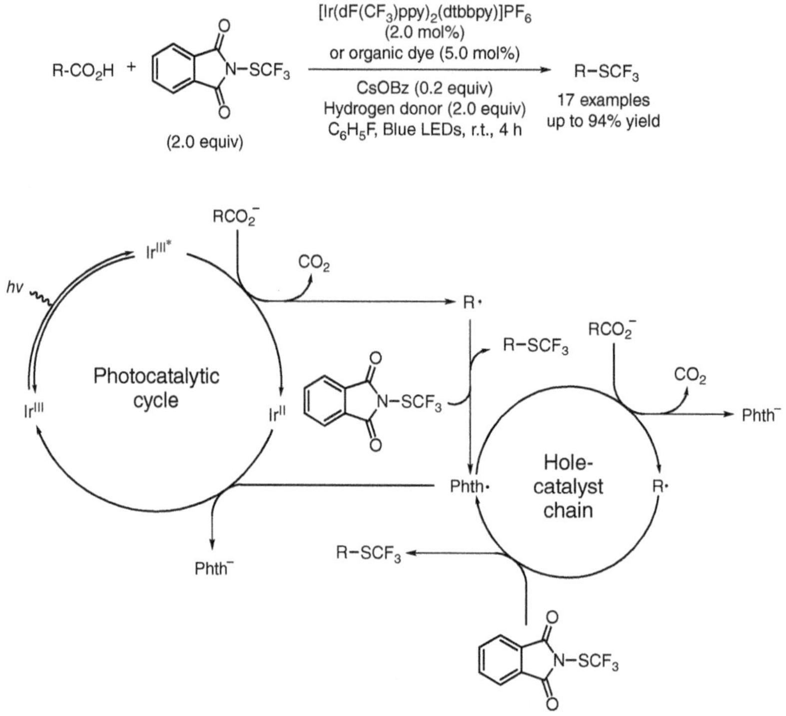

Scheme 4.105 Visible light-promoted trifluoromethylthiolation involving radical quenching by electrophilic reagent.

Inspired by the success in trifluoromethylthiolation of aliphatic carboxylic acids, the same group reasoned that if an alkyl radical could be generated through hydrogen atom transfer (HAT) process under photoredox catalysis, a more straightforward method for the preparation of alkyl trifluoromethylthioether could be developed. In 2016, Glorius and coworkers realized such a reaction by employing sodium benzoate as the HAT reagent [186]. Selective trifluoromethylthiolation of several alkyl compounds can be accomplished with high selectivity in good to excellent yields (Scheme 4.106). Similarly, Zhu's group developed an umpolung trifluoromethylthiolation of unactivated tertiary alkyl ethers [187]. In this reaction, the thiyl radical generated under photoredox catalysis abstracts the weakest hydridic α-C–H group of the ether to generate an alkoxyl radical. Homolytic cleavage of the C—O bond forms the tertiary alkyl radical, which is then quenched by electrophilic trifluoromethylthiolating reagent to give the product (Scheme 4.106). Applying the same strategy to benzylic C—H bonds led the same group to achieve efficient trifluoromethylthiolation of benzylic C—H bonds [188] (Scheme 4.107).

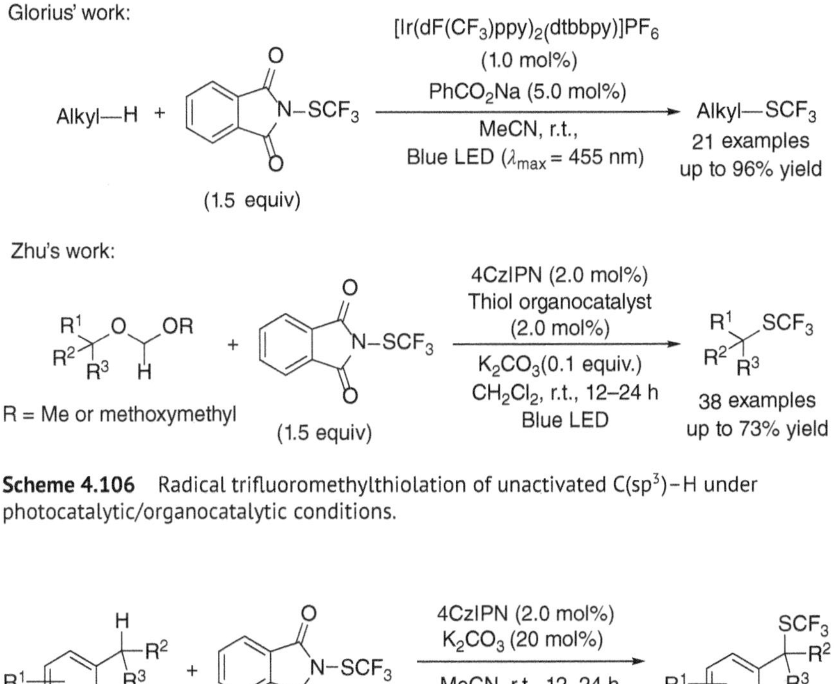

Glorius' work:

Zhu's work:

Scheme 4.106 Radical trifluoromethylthiolation of unactivated C(sp³)–H under photocatalytic/organocatalytic conditions.

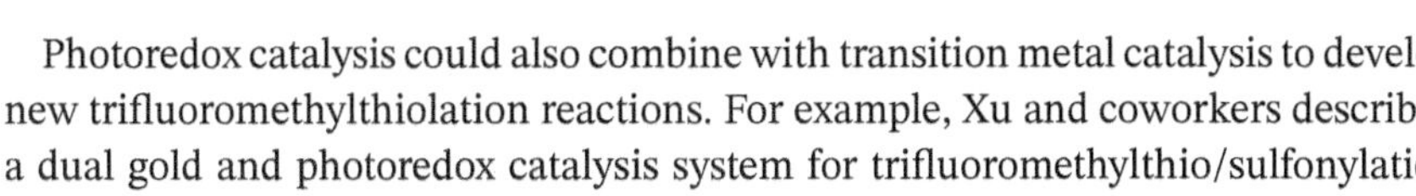

Scheme 4.107 Visible light-promoted trifluoromethylthiolation of benzylic C–H bonds.

Photoredox catalysis could also combine with transition metal catalysis to develop new trifluoromethylthiolation reactions. For example, Xu and coworkers described a dual gold and photoredox catalysis system for trifluoromethylthio/sulfonylation dual-functionalization of alkenes and alkynes [189, 190]. Particularly, arylsulfonyl anion, which is formed from the electrophilic reagent, can be utilized as the nucleophile in this reaction (Scheme 4.108).

Finally, recent studies from Qing's group showed that not only the electrophilic trifluoromethylthiolating reagents could be engaged in photoredox-promoted

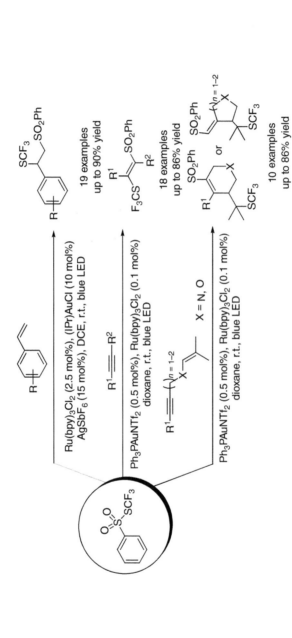

Scheme 4.108 Visible light-promoted dual gold and photoredox-catalyzed trifluoromethylthiolation of alkenes and alkynes.

radical trifluoromethylthiolation reaction, trifluoromethanesulfonic anhydride (Tf_2O), which is a cheap, commercially available reagent, could also be used as trifluoromethylthio source. In 2019, Qing and coworkers reported such a reaction for anti-Markovnikov selective hydro-trifluoromethylthiolation of unactivated alkenes and alkynes [191]. The addition of a phosphine Ph_2MeP as the reducing reagent was important for the reaction (Scheme 4.109). It was proposed that reaction of Ph_2MeP with Tf_2O generates $[R_3P\text{-}OTf]$, which is further reduced by the phosphine to give $[Ph_3SPMePh_2]^+OTf^-$. This intermediate can then be easily reduced by the Ir^{III} complex to generate $\cdot SCF_3$. Addition of the trifluoromethylthio radical to alkene or alkyne, followed by reducing the resulting alkyl or alkenyl radical by the Ir^{IV} species, gives the corresponding alkyl or alkenyl anion. The anion is then quenched by H_2O to give the final product.

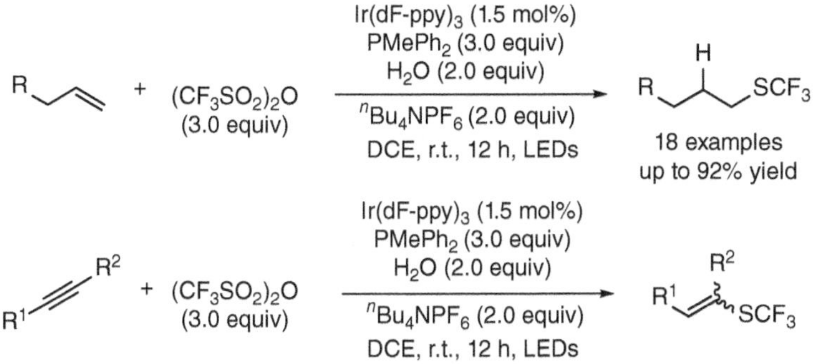

Scheme 4.109 Visible light-promoted hydro-trifluoromethylthiolation of unactivated alkenes and alkynes using trifluoromethanesulfonic anhydride as reagent.

4.5 Summary and Prospect

Owing to the special properties of trifluoromethylthio group and the increasing demands of the trifluoromethylthiolated molecules in the field of pharmaceutical and agrochemical industry, the past 10 years have witnessed the explosive development of methodologies for the efficient incorporation of the trifluoromethylthio group, including transition metal-catalyzed trifluoromethylthiolating reactions, electrophilic trifluoromethylthiolating reactions via highly reactive electrophilic trifluoromethylthiolating reagents, and radical trifluoromethylthiolation. These methods have provided useful tools for the introduction of the trifluoromethylthio group that could potentially facilitate the new drug discovery. Nevertheless, high-efficiency, low-cost trifluoromethylthiolating methodologies are still urgent to be actively explored.

References

1 (a) Leo, A., Hansch, C., and Elkins, D. (1971). *Chem. Rev.* 71: 525–616.
  (b) Hansch, C., Leo, A., and Taft, R.W. (1991). *Chem. Rev.* 91: 165–195.

2 (a) Boiko, V.N. (2010). *Beilstein J. Org. Chem.* 6: 880–921. (b) Liang, T., Neumann, C.N., and Ritter, T. (2013). *Angew. Chem. Int. Ed.* 52: 8214–8264. (c) He, W. and Weng, Z. (2013). *Prog. Chem.* 25: 1071–1078. (in Chinese). (d) Fang, L. (2014). *Huaxue Tongbao* 77: 1058–1063. (in Chinese). (e) Toulgoat, F., Alazet, S., and Billard, T. (2014). *Eur. J. Org. Chem.*: 2415–2428. (f) Chu, L. and Qing, F. (2014). *Acc. Chem. Res.* 47: 1513–1522. (g) Shao, X., Xu, C., Lu, L., and Shen, Q. (2015). *Acc. Chem. Res.* 48: 1227–1236. (h) Xu, X., Matsuzaki, K., and Shibata, N. (2015). *Chem. Rev.* 115: 731–764. (i) Zhang, K., Xu, X., and Qing, F. (2015). *Chin. J. Org. Chem.* 35: 556–569. (in Chinese). (j) Yang, X., Wu, T., Phipps, R.J., and Toste, F.D. (2015). *Chem. Rev.* 115: 826–870. (k) Lin, J., Ji, Y., and Xiao, J. (2015). *Curr. Org. Chem.* 19: 1541. (l) Chachignon, H. and Cahard, D. (2016). *Chin. J. Chem.* 34: 445–454. (m) Zheng, H., Huang, Y., and Weng, Z. (2016). *Tetrahedron Lett.* 57: 1397–1409. (n) Guo, Y., Huang, M., Fu, X. et al. (2017). *Chin. Chem. Lett.* 28: 719–728. (o) Barata-Vallejo, S., Bonesi, S., and Postigo, A. (2016). *Org. Biomol. Chem.* 14: 7150–7182. (p) Zeng, Y. and Hu, J. (2016). *Synthesis* 48: 2137–2150. (q) Zhang, C. (2017). *Chem. Sci.* 129: 1795–1805. (r) Zhang, P., Lu, L., and Shen, Q. (2017). *Acta Chim. Sinica* 75: 744–769. (s) Barthelemy, A., Magnier, E., and Dagousset, G. (2018). *Synthesis* 50: 4765–4776. (t) Li, S. and Wang, J. (2018). *Acta Chim. Sinica* 76: 913–924. (u) Rossi, S., Puglisi, A., Raimondi, L., and Benaglia, M. (2018). *ChemCatChem* 10: 2717–2733. (v) Ghiazza, C., Billard, T., and Tlili, A. (2019). *Chem. Eur. J.* 25: 6482–6495. (w) Hamzehloo, M., Hosseinian, A., Ebrahimiasl, S. et al. (2019). *J. Fluorine Chem.* 224: 52–60. (x) Hardy, M.A., Chachignon, H., and Cahard, D. (2019). *Asian J. Org. Chem.* 8: 591–609.

3 Man, E.H., Coffman, D.D., and Muetterties, E.L. (1959). *J. Am. Chem. Soc.* 81: 3575–3577.

4 Emeleus, H.J. and MacDuffie, D.E. (1961). *J. Chem. Soc. (Resumed)* 83: 2597–2599.

5 Yagupolskii, L.M., Kondratenko, N.V., and Sambur, V.P. (1975). *Synthesis* 1975: 721–723.

6 Chen, Q. and Duan, J. (1993). *J. Chem. Soc., Chem. Commun.*: 918–919.

7 (a) Kolomeitsev, A., Medebielle, M., Kirsch, P. et al. (2000). *J. Chem. Soc., Perkin Trans.* 1: 2183–2185. (b) Dmowski, W. and Haas, A. (1987). *J. Chem. Soc., Perkin Trans.* 1: 2119–2124. (c) Clark, J.H. and Taverner, S.J. (1997). *J. Fluorine Chem.* 85: 169–172. (d) Adams, D.J., Clark, J.H., Heath, P.A. et al. (2000). *J. Fluorine Chem.* 101: 187–191. (e) Taverner, S.J., Adams, D.J., and Clark, J.H. (1999). *J. Fluorine Chem.* 95: 171–176.

8 Tyrra, W., Naumann, D., Hoge, B., and Yagupolskii, Y.L. (2003). *J. Fluorine Chem.* 119: 101–107.

9 Munavalli, S., Rossman, D.I., Rohrbaugh, D.K. et al. (1992). *Heteroat. Chem.* 3: 189–192.

10 Weng, Z., He, W., Chen, C. et al. (2013). *Angew. Chem. Int. Ed.* 52: 1548–1552.

11 Yang, Y., Xu, L., Yu, S. et al. (2016). *Chem. Eur. J.* 22: 858–863.

12 (a) Zhang, M. and Weng, Z. (2016). *Adv. Synth. Catal.* 358: 386–394. (b) Kong, D., Jiang, Z., Xin, S. et al. (2013). *Tetrahedron* 69: 6046–6050. (c) Huang, Y.,

He, X., Li, H., and Weng, Z. (2014). *Eur. J. Org. Chem.*: 7324–7328. (d) Lin, Q., Chen, L., Huang, Y. et al. (2014). *Org. Biomol. Chem.* 12: 5500–5508. (e) Zhu, P., He, X., Chen, X. et al. (2014). *Tetrahedron* 70: 672–677. (f) Huang, Y., Ding, J., Wu, C. et al. (2015). *J. Org. Chem.* 80: 2912–2917. (g) Zhang, M., Chen, J., Chen, Z., and Weng, Z. (2016). *Tetrahedron* 72: 3525–3530.

13 Kondratenko, N.V. and Sambur, V.P. (1975). *Ukr. Khim. Zh. (Russ. Ed.)* 41: 516–519.

14 Adams, D.J., Goddard, A., Clark, J.H., and Macquarrie, D.J. (2000). *Chem. Commun.*: 987–988.

15 Danoun, G., Bayarmagnai, B., Gruenberg, M.F., and Goossen, L.J. (2014). *Chem. Sci.* 5: 1312–1316.

16 Bertoli, G., Exner, B., Evers, M.V. et al. (2018). *J. Fluorine Chem.* 210: 132–136.

17 Teveroskiy, G., Surry, D.S., and Buchwald, S.L. (2011). *Angew. Chem. Int. Ed.* 50: 7312–7314.

18 Yin, G., Englert, U., and Schoenebeck, F. (2015). *Angew. Chem. Int. Ed.* 54: 6809–6813.

19 Zhang, C. and Vicic, D.A. (2012). *J. Am. Chem. Soc.* 134: 183–185.

20 Yin, G., Kalvet, I., Englert, U., and Schoenebeck, F. (2015). *J. Am. Chem. Soc.* 137: 4164–4172.

21 Dürr, A.B., Yin, G., Kalvet, I. et al. (2016). *Chem. Sci.* 7: 1076–1081.

22 Nguyen, T., Chiu, W., Wang, X. et al. (2016). *Org. Lett.* 18: 5492–5495.

23 Xu, J., Mu, X., Chen, P. et al. (2014). *Org. Lett.* 16: 3942–3945.

24 Saravanan, P. and Anbarasan, P. (2015). *Adv. Synth. Catal.* 357: 3521–3528.

25 West, M.J., Fyfe, J.W.B., Vantourout, J.C., and Watson, A.J.B. (2019). *Chem. Rev.* 119: 12491–12523.

26 Chen, C., Xie, Y., Chu, L. et al. (2012). *Angew. Chem. Int. Ed.* 51: 2492–2495.

27 Chen, C., Chu, L., and Qing, F. (2012). *J. Am. Chem. Soc.* 134: 12454–12457.

28 Zhao, M., Zhao, X., Zheng, P., and Tian, Y. (2017). *J. Fluorine Chem.* 194: 73–79.

29 Zhang, C. and Vicic, D.A. (2012). *Chem. Asian J.* 7: 1756–1758.

30 Zhai, L., Li, Y., Yin, J. et al. (2013). *Tetrahedron* 69: 10262–10266.

31 Wu, W., Wang, B., Ji, X., and Cao, S. (2017). *Org. Chem. Front.* 4: 1299–1303.

32 Yin, W., Wang, Z., and Huang, Y. (2014). *Adv. Synth. Catal.* 356: 2998–3006.

33 Liu, X., Li, Q., and Wang, H. (2017). *Adv. Synth. Catal.* 359: 1942–1946.

34 Wang, K., Yun, S., Mamidipalli, P., and Lee, D. (2013). *Chem. Sci.* 4: 3205–3211.

35 Karmakar, R., Mamidipalli, P., Salzman, R.M. et al. (2016). *Org. Lett.* 18: 3530–3533.

36 Zeng, Y. and Hu, J. (2016). *Org. Lett.* 18: 856–859.

37 Nikolaienko, P., Yildiz, T., and Rueping, M. (2016). *Eur. J. Org. Chem.*: 1091–1094.

38 Fang, W., Dong, T., Han, J. et al. (2016). *Org. Biomol. Chem.* 14: 11502–11509.

39 Xiao, Q., Sheng, J., Ding, Q., and Wu, J. (2014). *Eur. J. Org. Chem.*: 217–221.

40 Zhang, S., Xu, X., and Qing, F. (2019). *J. Fluorine Chem.* 227: 109367. https://doi.org/10.1016/j.jfluchem.2019.109367.

41 Muta, R., Torigoe, T., and Kuninobu, Y. (2019). *Org. Lett.* 21: 4289–4292.

42 Dai, X. and Cahard, D. (2015). *Synlett* 26: 40–44.

43 Xu, C., Chen, Q., and Shen, Q. (2016). *Chin. J. Chem.* 34: 495–504.

44 Nikolaienko, P., Pluta, R., and Rueping, M. (2014). *Chem. Eur. J.* 20: 9867–9870.

45 Qiu, Y., Song, X., Li, M. et al. (2016). *Org. Lett.* 18: 1514–1517.

46 Maury, J., Force, G., Darses, B., and Lebœuf, D. (2018). *Adv. Synth. Catal.* 360: 2752–2756.

47 Liu, J., Xu, X., Chen, Z., and Qing, F. (2015). *Angew. Chem. Int. Ed.* 54: 897–900.

48 Liu, Y., Xu, X., and Qing, F. (2018). *J. Fluorine Chem.* 74: 5827–5832.

49 Liu, Y., Xu, X., and Qing, F. (2019). *J. Fluorine Chem.* 60: 953–956.

50 Ye, K., Zhang, X., Dai, L., and You, S. (2014). *J. Org. Chem.* 79: 12106–12110.

51 Zeng, J., Chachignon, H., Ma, J., and Cahard, D. (2017). *Org. Lett.* 19: 1974–1977.

52 Hu, M., Rong, J., Miao, W. et al. (2014). *Org. Lett.* 16: 2030–2033.

53 Wang, X., Zhou, Y., Ji, G. et al. (2014). *Eur. J. Org. Chem.*: 3093–3096.

54 Lefebvre, Q., Fava, E., Nikolaienko, P., and Rueping, M. (2014). *Chem. Commun.* 50: 6617–6619.

55 Matheis, C., Krause, T., Bragoni, V., and Goossen, L.J. (2016). *Chem. Eur. J.* 22: 12270–12273.

56 Li, S. and Zard, S. (2013). *Org. Lett.* 15: 5898–5901.

57 Yang, H., Fan, X., Wei, Y., and Shi, M. (2015). *Org. Chem. Front.* 2: 1088–1093.

58 Fan, X., Yang, H., and Shi, M. (2017). *Adv. Synth. Catal.* 359: 49–57.

59 Chen, C., Xu, X., Yang, B., and Qing, F. (2014). *Org. Lett.* 16: 3372–3375.

60 Eméleus, H.J. and Nabi, S.N. (1960). *J. Chem. Soc.*: 1103–1108.

61 (a) Andreades, S., Harris, J.F., and Sheppard, W.A. (1964). *J. Org. Chem.* 29: 898–900. (b) Sheppard, W.A. (1964). *J. Org. Chem.* 29: 895–898. (c) Scribner, R.M. (1966). *J. Org. Chem.* 31: 3671–3682. (d) Bayreuther, H. and Haas, A. (1973). *Chem. Ber.* 106: 1418–1422. (e) Croft, T.S. and McBrady, J.J. (1975). *J. Heterocycl. Chem.* 12: 845–849. (f) Haas, A. and Hellwig, V. (1976). *Chem. Ber.* 109: 2475–2484. (g) Haas, A. and Niemann, U. (1977). *Chem. Ber.* 110: 67–77. (h) Haas, A., Lieb, M., and Zhang, Y. (1985). *J. Fluorine Chem.* 29: 311–322. (i) Bogdanowicz-Szwed, K., Kawalek, B., and Lieb, M. (1987). *J. Fluorine Chem.* 35: 317–327. (j) Rossman, D.I., Muller, A.J., and Lewis, E.O. (1991). *J. Fluorine Chem.* 55: 221–224.

62 *Chem. Eng. News*, 1967, 45: 44, 42–44. http://dx.doi.org/10.1021/cen-v045n044.p042a.

63 Haszeldine, R.N. and Kidl, J.M. (1953). *J. Chem. Soc.*: 3219–3225.

64 (a) Sharpe, T.R., Cherkofsky, S.C., Hewes, W.E. et al. (1985). *J. Med. Chem.* 28: 1188–1194. (b) South, M.S. and Van Sant, K.A. (1991). *J. Heterocycl. Chem.* 28: 1017–1024. (c) Boese, R., Haas, A., Lieb, M., and Roeske, U. (1994). *Chem. Ber.* 127: 449–455.

65 Tran, L.D., Popov, I., and Daugulis, O. (2012). *J. Am. Chem. Soc.* 134: 18237–18240.

66 Haas, A. and Möller, G. (1996). *Chem. Ber.* 129: 1383–1388.

67 Kang, K., Xu, C., and Shen, Q. (2014). *Org. Chem. Front.* 1: 294–297.

68 Xu, C. and Shen, Q. (2014). *Org. Lett.* 16: 2046–2049.

69 Zhao, B. and Du, D. (2017). *Org. Lett.* 19: 1036–1039.

70 Munavalli, S., Rohrbaugh, D.K., Rossman, D.I. et al. (2000). *Synth. Commun.* 30: 2847–2854.

71 Pluta, R., Nikolaienko, P., and Rueping, M. (2014). *Angew. Chem. Int. Ed.* 53: 1650–1653.

72 Bootwicha, T., Liu, X., Pluta, R. et al. (2013). *Angew. Chem. Int. Ed.* 52: 12856–12859.

73 Rueping, M., Liu, X., Bootwicha, T. et al. (2014). *Chem. Commun.* 50: 2508–2511.

74 Graßl, S., Hamze, C., Koller, T.J., and Knochel, P. (2019). *Chem. Eur. J.* 25: 3752–3755.

75 Honeker, R., Ernst, J.B., and Glorius, F. (2015). *Chem. Eur. J.* 21: 8047–8051.

76 Zhao, Q., Chen, M., Poisson, T. et al. (2018). *Eur. J. Org. Chem.*: 6167–6175.

77 Chachignon, H., Kondrashov, E.V., and Cahard, D. (2018). *Adv. Synth. Catal.* 360: 965–971.

78 Gelat, F., Poisson, T., Biju, A.T. et al. (2018). *Eur. J. Org. Chem.*: 3693–3696.

79 Mao, R., Bera, S., Cheseaux, A., and Hu, X. (2019). *Chem. Sci.* 10: 9555–9559.

80 Pluta, R. and Rueping, M. (2014). *Chem. Eur. J.* 20: 17315–17318.

81 Xiao, Q., He, Q., Li, J., and Wang, J. (2015). *Org. Lett.* 17: 6090–6093.

82 Jia, Y., Qin, H., Wang, N. et al. (2018). *J. Org. Chem.* 83: 2808–2817.

83 Ferry, A., Billard, T., Langlois, B.R., and Bacque, E. (2008). *J. Org. Chem.* 73: 9362–9365.

84 Ferry, A., Billard, T., Langlois, B.R., and Bacque, E. (2009). *Angew. Chem. Int. Ed.* 48: 8551–8555.

85 Alazet, S., Zimmer, L., and Billard, T. (2013). *Angew. Chem. Int. Ed.* 52: 10814–10817.

86 Ferry, A., Billard, T., Bacqué, E., and Langlois, B.R. (2012). *J. Fluorine Chem.* 134: 160–163.

87 Alazet, S., Ollivier, K., and Billard, T. (2013). *Beilstein J. Org. Chem.* 9: 2354–2357.

88 Yang, Y., Jiang, X., and Qing, F. (2012). *J. Org. Chem.* 77: 7538–7547.

89 Liu, J., Chu, L., and Qing, F. (2013). *Org. Lett.* 15: 894–897.

90 Xu, J., Chen, P., Ye, J., and Liu, G. (2015). *Acta Chim. Sinica* 73: 1294–1297.

91 Wu, W., Zhang, X., Liang, F., and Cao, S. (2015). *Org. Biomol. Chem.* 13: 6992–6999.

92 Xi, C., Chen, Z., Zhang, S., and Tu, Y. (2018). *Org. Lett.* 20: 4227–4230.

93 Liu, Y., Qiu, G., Wang, H., and Sheng, J. (2017). *Tetrahedron Lett.* 58: 690–693.

94 Xiao, Q., Sheng, J., Chen, Z., and Wu, J. (2013). *Chem. Commun.* 49: 8647–8649.

95 Sheng, J., Fan, C., and Wu, J. (2014). *Chem. Commun.* 50: 5494–5496.

96 Sheng, J., Li, S., and Wu, J. (2014). *Chem. Commun.* 50: 578–580.

97 Liu, T., Qiu, G., Ding, Q., and Wu, J. (2016). *Tetrahedron* 72: 1472–1476.

98 Sheng, J. and Wu, J. (2014). *Org. Biomol. Chem.* 12: 7629–7633.

99 (a) Alazet, S. and Billard, T. (2015). *Synlett* 26: 76–78. (b) Alazet, S., Zimmer, L., and Billard, T. (2015). *J. Fluorine Chem.* 171: 78–81.

100 (a) Alazet, S., Zimmer, L., and Billard, T. (2014). *Chem. Eur. J.* 20: 8589–8593. (b) Alazet, S., Ismalaj, E., Glenadel, Q. et al. (2015). *Eur. J. Org. Chem.*: 4607–4610.

101 Glenadel, Q., Alazet, S., Tlili, A., and Billard, T. (2015). *Chem. Eur. J.* 21: 14694–14698.

102 Glenadel, Q. and Billard, T. (2016). *Chin. J. Chem.* 34: 455–458.

103 Tlili, A., Alazet, S., Glenadel, Q., and Billard, T. (2016). *Chem. Eur. J.* 22: 10230–10234.

104 Bonazaba Milandou, L.J.C., Carreyre, H., Alazet, S. et al. (2017). *Angew. Chem. Int. Ed.* 56: 169–172.

105 Glenadel, Q., Bordy, M., Alazet, S. et al. (2016). *Asian J. Org. Chem.* 5: 428–433.

106 Xu, C., Ma, B., and Shen, Q. (2014). *Angew. Chem. Int. Ed.* 53: 9316–9320.

107 Xu, C. and Shen, Q. (2015). *Org. Lett.* 17: 4561–4563.

108 Zhu, J., Xu, C., Xu, C., and Shen, Q. (2017). *Org. Synth.* 94: 217–233.

109 Wang, Q., Qi, Z., Xie, F., and Li, X. (2015). *Adv. Synth. Catal.* 357: 355–360.

110 Nalbandian, C.J., Brown, Z.E., Alvarez, E., and Gustafson, J.L. (2018). *Org. Lett.* 20: 3211–3214.

111 Ernst, J.B., Rakers, L., and Glorius, F. (2017). *Synthesis* 49: 260–268.

112 Lu, S., Chen, W., and Shen, Q. (2019). *Chin. Chem. Lett.* 30: 2279–2281.

113 Liu, S., Zeng, X., and Xu, B. (2019). *Asian J. Org. Chem.* 8: 1372–1375.

114 Wang, Q., Xie, F., and Li, X. (2015). *J. Org. Chem.* 80: 8361–8366.

115 Guyon, H., Chachignon, H., Tognetti, V. et al. (2018). *Eur. J. Org. Chem.*: 3756–3763.

116 Abubakar, S.S., Benaglia, M., Rossi, S., and Annunziata, R. (2018). *Catal. Today* 308: 94–101.

117 Hu, L., Wu, M., Wan, H. et al. (2016). *New J. Chem.* 40: 6550–6553.

118 Maeno, M., Shibata, N., and Cahard, D. (2015). *Org. Lett.* 17: 1990–1993.

119 Yu, Y., Xiong, D., and Ye, X. (2016). *Org. Biomol. Chem.* 14: 6403–6406.

120 (a) Luo, J., Zhu, Z., Liu, Y., and Zhao, X. (2015). *Org. Lett.* 17: 3620–3623. (b) Wu, J., Xu, J., and Zhao, X. (2016). *Chem. Eur. J.* 22: 15265–15269. (c) Zhu, Z., Luo, J., and Zhao, X. (2017). *Org. Lett.* 19: 4940–4943. (d) Liu, X., An, R., Zhang, X. et al. (2016). *Angew. Chem. Int. Ed.* 55: 5846–5850.

121 Zhang, P., Li, M., Xue, X. et al. (2016). *J. Org. Chem.* 81: 7486–7509.

122 (a) Luo, J., Liu, Y., and Zhao, X. (2017). *Org. Lett.* 19: 3434–3437; (b) Luo, J., Cao, Q., Cao, X., and Zhao, X. (2018). *Nat. Commun.* 9: 527. https://doi.org/10.1038/s41467-018-02955-0. (c) Liu, X., Liang, Y., Ji, J. et al. (2018). *J. Am. Chem. Soc.* 140: 4782–4786. (d) Qin, T., Jiang, Q., Ji, J. et al. (2019). *Org. Biomol. Chem.* 17: 1763–1766.

123 (a) Mai, B.K., Szabó, K., and Himo, F. (2018). *Org. Lett.* 20: 6646–6649. (b) Lübcke, M., Yuan, W., and Szabó, K. (2017). *Org. Lett.* 19: 4548–4551.

124 Lübcke, M., Bezhan, D., and Szabó, K. (2019). *Chem. Sci.* 10: 5990–5995.

125 Yoshida, M., Kawai, K., Tanaka, R. et al. (2017). *Chem. Commun.* 53: 5974–5977.

126 Meanwell, M., Adluri, B.S., Yuan, Z. et al. (2018). *Chem. Sci.* 9: 5608–5613.

127 Differding, E. and Lang, R. (1988). *Tetrahedron Lett.* 29: 6087–6090.

128 Zhang, H., Leng, X., Wan, X., and Shen, Q. (2017). *Org. Chem. Front.* 4: 1051–1057.

129 Charpentier, J., Fruh, N., and Togni, A. (2015). *Chem. Rev.* 115: 650–682.

130 Shao, X., Wang, X., Yang, T. et al. (2013). *Angew. Chem. Int. Ed.* 52: 3457–3460.

131 Vinogradova, E., Muller, P., and Buchwald, S. (2014). *Angew. Chem. Int. Ed.* 53: 3125–3128.

132 Shao, X., Xu, C., Lu, L., and Shen, Q. (2015). *J. Org. Chem.* 80: 3012–3021.

133 Shao, X., Liu, T., Lu, L., and Shen, Q. (2014). *Org. Lett.* 16: 4738–4741.

134 Ma, B., Shao, X., and Shen, Q. (2015). *J. Fluorine Chem.* 171: 73–77.

135 Shen, F., Zheng, H., Xue, X. et al. (2019). *Org. Lett.* 21: 6347–6351.

136 Li, Y., Ye, Z., Bellman, T.M. et al. (2015). *Org. Lett.* 17: 2186–2189.

137 Wang, X., Yang, T., Cheng, X., and Shen, Q. (2013). *Angew. Chem. Int. Ed.* 52: 12860–12864.

138 Yang, T., Shen, Q., and Lu, L. (2014). *Chin. J. Chem.* 32: 678–680.

139 Deng, Q.H., Rettenmeier, C., Wadepohl, H., and Gade, L.H. (2014). *Chem. Eur. J.* 20: 93–97.

140 He, H. and Zhu, X. (2014). *Org. Lett.* 16: 3102–3105.

141 Wu, J., Zhao, Q., Wilson, T.C. et al. (2019). *Angew. Chem. Int. Ed.* 58: 2413–2417.

142 Yang, Y., Azuma, A., Tokunaga, E. et al. (2013). *J. Am. Chem. Soc.* 135: 8782–8785.

143 Arimori, S., Takada, M., and Shibata, N. (2015). *Org. Lett.* 17: 1063–1065.

144 Huang, Z., Yang, Y.-D., Tokunaga, E., and Shibata, N. (2015). *Asian J. Org. Chem.* 4: 525–527.

145 Huang, Z., Yang, Y.-D., Tokunaga, E., and Shibata, N. (2015). *Org. Lett.* 17: 1094–1097.

146 Arimori, S., Takada, M., and Shibata, N. (2015). *Dalton Trans.* 44: 19456–19459.

147 Zhu, X., Xu, J., Cheng, D. et al. (2014). *Org. Lett.* 16: 2192–2195.

148 Xiang, H. and Yang, C. (2014). *Org. Lett.* 16: 5686–5689.

149 Zhu, S., Xu, X., and Qing, F. (2014). *Eur. J. Org. Chem.*: 4453–4456.

150 Jiang, L., Qian, J., Yi, W. et al. (2015). *Angew. Chem. Int. Ed.* 54: 14965–14969.

151 Bu, M., Lu, G., and Cai, C. (2017). *Org. Chem. Front.* 4: 266–270.

152 (a) Sun, D., Jiang, X., Jiang, M. et al. (2017). *Eur. J. Org. Chem.*: 3505–3511. (b) Liu, J., Zhao, X., Jiang, L., and Yi, W. (2018). *Adv. Synth. Catal.* 360: 4012–4016. (c) Sun, D., Jiang, X., Jiang, M. et al. (2018). *Eur. J. Org. Chem.* 83: 2078–2081.

153 Chachignon, H., Maeno, M., Kondo, H. et al. (2016). *Org. Lett.* 18: 2467–2470.

154 Lu, K., Deng, Z., Li, M. et al. (2017). *Org. Biomol. Chem.* 15: 1254–1260.

155 Jiang, L., Yi, W., and Liu, Q. (2016). *Adv. Synth. Catal.* 358: 3700–3705.

156 Jiang, L., Yan, Q., Wang, R. et al. (2018). *Chem. Eur. J.* 24: 18749–18756.

157 Sun, D., Jiang, M., and Liu, J. (2019). *Chem. Eur. J.* 25: 10797–10802.

158 Yin, F. and Wang, X. (2014). *Org. Lett.* 16: 1128–1131.

159 Fuentes, N., Kong, W., Fernández-Sánchez, L. et al. (2015). *J. Am. Chem. Soc.* 137: 964–973.

160 Qiu, Y., Zhu, X., Li, Y. et al. (2015). *Org. Lett.* 17: 3694–3697.

161 Zeng, Y., Tan, D., Chen, Y. et al. (2015). *Org. Chem. Front.* 2: 1511–1515.

162 Qiu, Y., Niu, Y., Wei, X. et al. (2019). *J. Org. Chem.* 84: 4165–4178.

163 Chen, M., Tang, X., and Shi, M. (2017). *Org. Chem. Front.* 4: 86–90.

164 Zhang, K., Liu, J., and Qing, F. (2014). *Chem. Commun.* 50: 14157–14160.

165 Wu, W., Dai, W., Ji, X., and Cao, S. (2016). *Org. Lett.* 18: 2918–2921.

166 Liu, K., Jin, Q., Chen, S., and Liu, P. (2017). *RSC Adv.* 7: 1546–1552.

167 Pan, S., Huang, Y., and Qing, F. (2016). *Chem. Asian J.* 11: 2854–2858.

168 Cheng, Z., Tao, T., Feng, Y. et al. (2018). *J. Org. Chem.* 83: 499–504.

169 Li, M., Petersen, J.L., and Hoover, J.M. (2017). *Org. Lett.* 19: 638–641.

170 He, B., Xiao, Z., Wu, H. et al. (2017). *RSC Adv.* 7: 880–883.

171 Wu, H., Xiao, Z., Wu, J. et al. (2015). *Angew. Chem. Int. Ed.* 54: 4070–4074.

172 Guo, S., Zhang, X., and Tang, P. (2015). *Angew. Chem. Int. Ed.* 54: 4065–4069.

173 Guo, K., Zhang, H., Cao, S. et al. (2018). *Org. Lett.* 20: 2261–2264.

174 Modak, A., Pinter, E.N., and Cook, S.P. (2019). *J. Am. Chem. Soc.* 141: 18405–18410.

175 Zhao, Y., Lin, J., Hang, X., and Xiao, J. (2018). *J. Org. Chem.* 83: 14120–14125.

176 Hu, F., Shao, X., Zhu, D. et al. (2014). *Angew. Chem. Int. Ed.* 53: 6105–6109.

177 Yang, T. and Shen, Q. (2015). *Chem. Commun.* 51: 5479–5481.

178 Shao, X., Hong, X., Lu, L., and Shen, Q. (2019). *Tetrahedron* 75: 4156–4166.

179 Honeker, R., Garza-Sanchez, R.A., Hopkinson, M.N., and Glorius, F. (2016). *Chem. Eur. J.* 22: 4395–4399.

180 Li, M., Guo, J., Xue, X., and Cheng, J. (2017). *Org. Lett.* 19: 2098.

181 Dagousset, G., Simon, C., Anselmi, E. et al. (2017). *Chem. Eur. J.* 23: 4282–4286.

182 Zhu, M., Fu, W., Guo, W. et al. (2019). *Org. Biomol. Chem.* 17: 3374–3380.

183 Zhu, M., Li, R., You, Q. et al. (2019). *Asian J. Org. Chem.* 8: 2002–2005.

184 Li, Y., Koike, T., and Akita, M. (2017). *Asian J. Org. Chem.* 6: 445–448.

185 Candish, L., Pitzer, L., Gómez-Suárez, A., and Glorius, F. (2016). *Chem. Eur. J.* 22: 4753–4756.

186 Mukherjee, S., Maji, B., Tlahuext-Aca, A., and Glorius, F. (2016). *J. Am. Chem. Soc.* 138: 16200–16203.

187 Xu, W., Ma, J., Yuan, X. et al. (2018). *Angew. Chem. Int. Ed.* 57: 10357–10361.

188 Xu, W., Wang, W., Liu, T. et al. (2019). *Nat. Commun.* 10: 4687. https://doi.org/10.1038/s41467-019-12844-9.

189 Li, H., Shan, C., Tung, C., and Xu, Z. (2017). *Chem. Sci.* 8: 2610–2615.

190 Li, H., Cheng, Z., Tung, C., and Xu, Z. (2018). *ACS Catal.* 8: 8237–8243.

191 Ouyang, Y., Xu, X., and Qing, F. (2019). *Angew. Chem. Int. Ed.* 58: 18508–18512.

5

Bifunctionalization-Based Catalytic Fluorination and Trifluoromethylation

Pinhong Chen and Guosheng Liu

Chinese Academy of Sciences, Shanghai Institute of Organic Chinese Academy of Sciences, State Key Laboratory of Organometallic Chemistry, 345 Lingling Road, Shanghai 200032, China

5.1 Introduction

The vicinal bifunctionalization of alkenes is an attractive transformation that converts feedstock olefins into valuable molecules for application in pharmaceutical, agrochemical, medical, and material sciences. Among the methodologies developed, transition metal-catalyzed fluorination and trifluoromethylation of organic molecules has received much attention. Especially, during the recent decade, palladium and copper were demonstrated to be excellent metals for fluorination and related reactions. In this chapter, we mainly focus on two aspects: one is the palladium-catalyzed fluorination and related reactions, such as trifluoromethylation and trifluoromethoxylation. The other is the copper-catalyzed radical trifluoromethylation of alkenes and the corresponding enantioselective versions, which are mainly initiated by addition of trifluoromethyl radical to alkenes. Those reactions initiated by nucleocupration of alkenes are not discussed here.

5.2 Palladium-Catalyzed Fluorination, Trifluoromethylation, and Trifluoromethoxylation of Alkenes

Owing to the strong negative property of fluorine anion and stability of the fluorine-containing anion, it is challenging to bring about reductive elimination from the metal center. Recently, a high-valent palladium strategy was applied for the synthesis of fluorine-containing organic molecules that are typically inaccessible by other methods [1]. Early literature showed that high-valent palladium (III or IV) has different property than palladium(II) species [2]. Reductive elimination from high-valent palladium is favored over β-hydride elimination, which facilitates the C—C and C—X bonds formation.

Organofluorine Chemistry: Synthesis, Modeling, and Applications,
First Edition. Edited by Kálmán J. Szabó and Nicklas Selander.
© 2021 WILEY-VCH GmbH. Published 2021 by WILEY-VCH GmbH.

5.2.1 Palladium-Catalyzed Fluorination of Alkenes

In Sanford's work, electrophilic fluorinating reagents were employed as both the fluorine source and oxidant to realize C–H fluorination reaction, in which a high-valent PdF species was proposed and later proved by others [3]. However, the employment of nucleophilic fluorinating reagents was rarely reported at that moment. Liu and coworkers reported a palladium-catalyzed regioselective aminofluorination of alkenes using PhI(OPiv)$_2$/AgF (Scheme 5.1) [4]. It is worth noting that inorganic fluoride was first used as the fluorine source to conduct oxidative fluorination of organopalladium complex. A series of β-fluoropiperidines can be synthesized in good to excellent yields with high regioselectivities. This catalytic system was further extended to the synthesis of fluorinated cyclic sulfamide derivatives [5]. Mechanistic studies indicated that this reaction proceeds via *trans*-aminopalladation of alkenes with attack at the terminal carbon, generating a secondary alkyl–Pd(II) complex, followed by oxidation with PhI(OPiv)$_2$/AgF and finally reductive elimination from Pd(IV) intermediate leading to the β-fluoropiperidines.

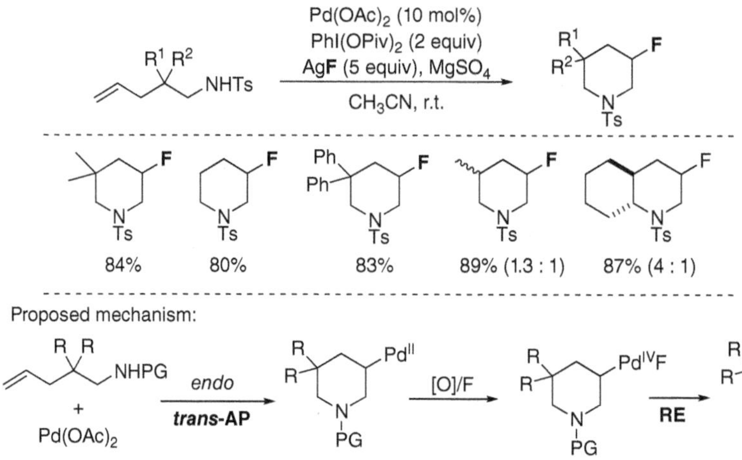

Scheme 5.1 Palladium-catalyzed 6-*endo* aminofluorination of alkenes.

Similar chlorination reaction suggested that a reversible aminopalladation was involved, and the oxidation of a secondary alkyl–Pd(II) complex was much faster than that of a primary alkyl–Pd(II) complex [6]. Thus, the regioselectivity of aminofluorination can be completely switched from 6-*endo* to 5-*exo* cyclizations by installing a chelating group on nitrogen. Later, Liu and coworkers achieved the 5-*exo* aminofluorination of alkenes to give the challenging primary C—F bonds (Scheme 5.2) [7]. Both monofluoromethylated pyrrolidines and imidazolines were obtained in moderate to good yields.

The intermolecular aminofluorination reaction of styrenes was further studied (Scheme 5.3) [8]. Screening of the nucleophiles indicated that sulfonylamides were efficient nucleophiles for this reaction.

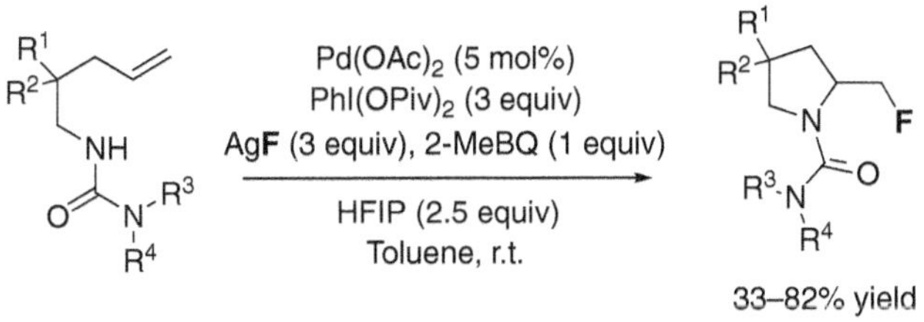
R1
R2
NH
O
N
R3
R4
Pd(OAc)2 (5 mol%)
PhI(OPiv)2 (3 equiv)
AgF (3 equiv), 2-MeBQ (1 equiv)
HFIP (2.5 equiv)
Toluene, r.t.
F
33–82% yield

Scheme 5.2 Palladium-catalyzed 5-*exo* aminofluorination of alkenes.

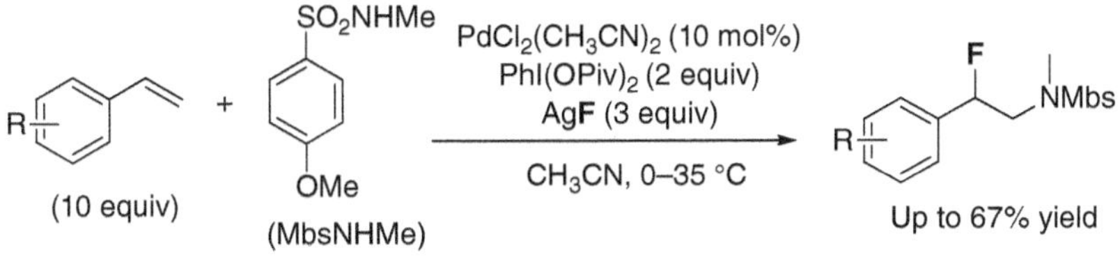
SO2NHMe
R
+
(10 equiv)
OMe
(MbsNHMe)
PdCl2(CH3CN)2 (10 mol%)
PhI(OPiv)2 (2 equiv)
AgF (3 equiv)
CH3CN, 0–35 °C
F
NMbs
Up to 67% yield

Scheme 5.3 Palladium-catalyzed intermolecular aminofluorination of alkenes.

Electrophilic fluorination reagents can be employed as both fluorine source and oxidant. Gouverneur and coworkers developed a *cis*-hydrofluorination of alkenylarenes with Selectfluor (Scheme 5.4) [9]. The addition of hydrogen and fluorine across the double bond is *syn* stereospecific; thus, the reaction of *E*-trisubstituted alkenes gives the *anti*-hydrofluorination products, while the *Z*-ones give the *syn*-products. Mechanistic studies indicate that the reaction is initiated with a reversible *syn* hydropalladation, and then the C—F bond is formed from high-valent palladium center with the retention of configuration.

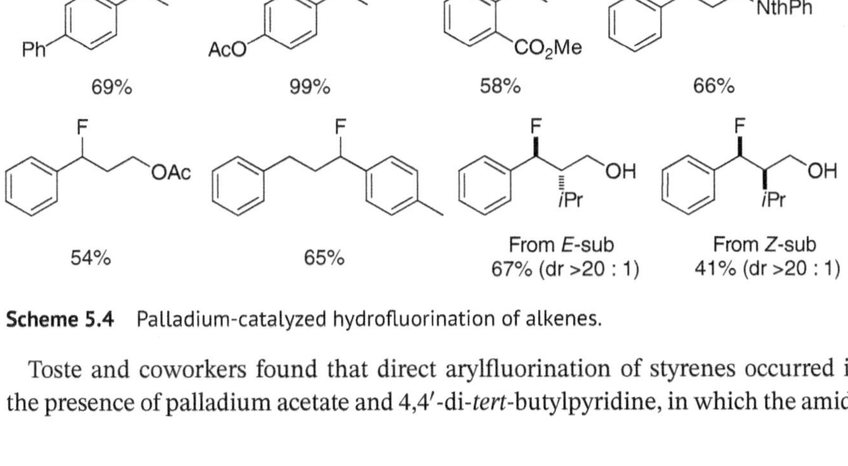
Pd(PPh3)4 (10 mol%)
Et3SiH (3 equiv)
Selectfluor (3 equiv)
MeCN, 0 °C, 2 h
R1
R2
R3
F
H
Ph
69%
AcO
99%
CO2Me
58%
NthPh
66%
OAc
54%
65%
OH
iPr
From E-sub
67% (dr >20 : 1)
From Z-sub
41% (dr >20 : 1)

Scheme 5.4 Palladium-catalyzed hydrofluorination of alkenes.

Toste and coworkers found that direct arylfluorination of styrenes occurred in the presence of palladium acetate and 4,4′-di-*tert*-butylpyridine, in which the amide group served as a chelating group to stabilize the high-valent palladium intermediate and avoid the oxidative Heck reaction (Scheme 5.5) [10]. When 8-aminoquinoline was employed as the directing group, the intermolecular arylfluorination products were formed in moderate to good yield with different types of arylboronic acids.

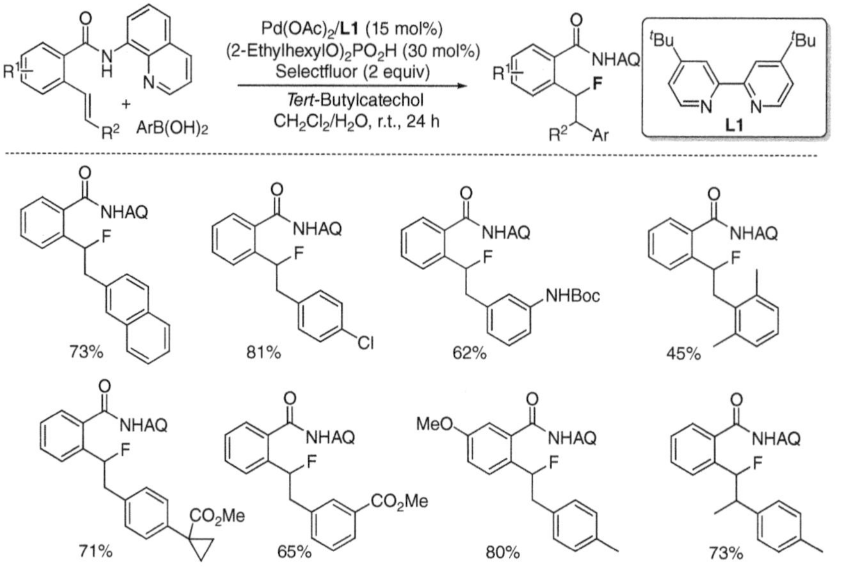

Scheme 5.5 Palladium-catalyzed 1,2-arylfluorination of alkenes.

The enantioenriched benzylic fluorides were obtained if bipyridine was switched to chiral pyridyloxazoline **L2** (Scheme 5.6). The reactions provide various enantioenriched benzylic fluorides in good yield.

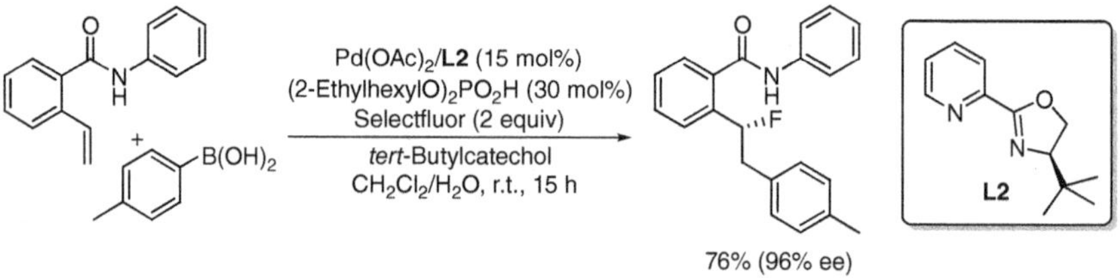

76% (96% ee)

Scheme 5.6 Palladium-catalyzed enantioselective 1,2-arylfluorination of alkenes.

As mentioned in the above reaction, the directing group is important for inhibiting β–H elimination reaction. When allylamines are subjected to the reactions instead of styrenes, the reactions afford the 1,1-arylfluorination rather than 1,2-arylfluorination products (Scheme 5.7) [11]. The lack of efficient coordination of amine to palladium species leads to β–H elimination and reinsertion step to form a more stable benzylic palladium intermediate. Related asymmetric reaction was achieved with Box ligand **L3** to give chiral benzylic fluorides in good to excellent enantioselectivities.

Condition A: $Pd(OAc)_2$ (5 mol%), *tert*-butylbipy (6 mol%), Selectfluor (2 equiv), $CH_2Cl_2/H_2O/MeCN$ (10 : 2 : 1), r.t., 18 h
Condition B: $Pd(MeCN)_2Cl_2$ (5 mol%), **L3** (6 mol%), Selectfluor (3 equiv), benzene/H_2O/BnCN (8 : 8 : 1), 4 °C, 18 h

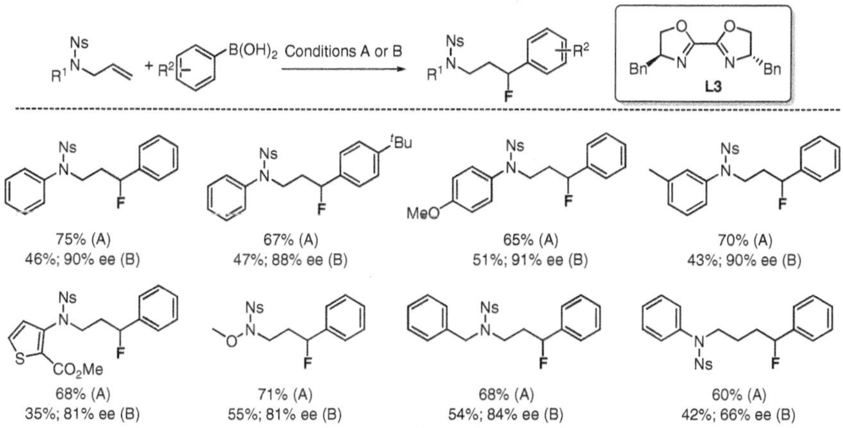

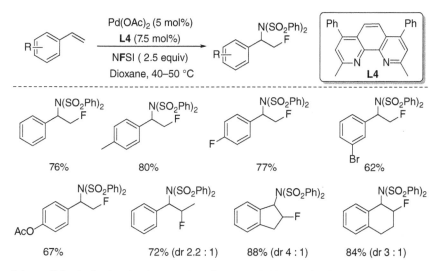

Scheme 5.7 Palladium-catalyzed 1,1-arylfluorination of alkenes.

Compared to aminopalladation, the related fluoropalladation has not been reported due to the special property of fluorine element, such as weak nucleophilicity and weak bonding with the soft transition metal catalysts. Liu and coworkers developed a regioselective palladium-catalyzed intermolecular aminofluorination of styrenes, in which a stoichiometric amount of *N*-fluorobenzenesulfonmide (NFSI) was used as the fluorine and amine source (Scheme 5.8) [12]. Diverse aminofluorination products from styrenes were obtained in moderate to good yields under mild condition. Markovnikov regioselectivity is observed in these products, which is complementary to their abovementioned aminofluorination reactions, although with different mechanisms.

Scheme 5.8 Aminofluorination reaction of styrenes initiated with fluoropalladation.

Preliminary studies indicated that fluoropalladation of styrenes may be involved. First, a F–Pd(II)–NZ$_2$ species **int.-II** is formed from the oxidation of Pd(0) complex **int.-I** by NFSI (Scheme 5.9). Then fluoropalladation of styrene gives the Pd(II) species **int.-III**, which is oxidized by NFSI to give the Pd(IV) intermediate **int.-IV**, and the following reductive elimination forms the C—N bond. The authors considered that the fluoropalladation process will open a new pathway for constructing the C—F bond. With the same reaction pattern, the intramolecular aminofluorination reaction of styrenes afforded a collection of fluorinated pyrrolidines [13].

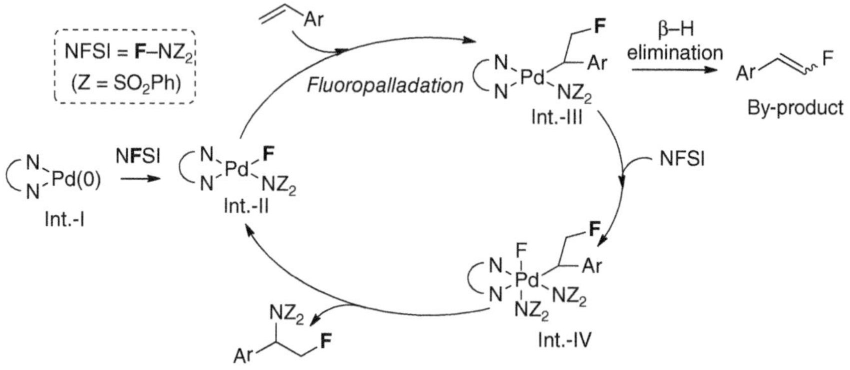

Scheme 5.9 Proposed mechanism of palladium-catalyzed aminofluorination.

This reaction mode was further applied to the intermolecular fluoroesterification of styrenes (Scheme 5.10) [14]. The ligand exchange between carboxylic acids and N(SO$_2$Ph)$_2$ anion is facilitated when the acid is of weak nucleophilicity but strong acidity, such as CF$_3$CO$_2$H and CCl$_3$CO$_2$H. When trifluoroacetic acid was used as the nucleophile, the related ester products were not stable enough and were easily hydrolyzed by pyridine to give fluoroalcohols. However, the trichloroacetate esters

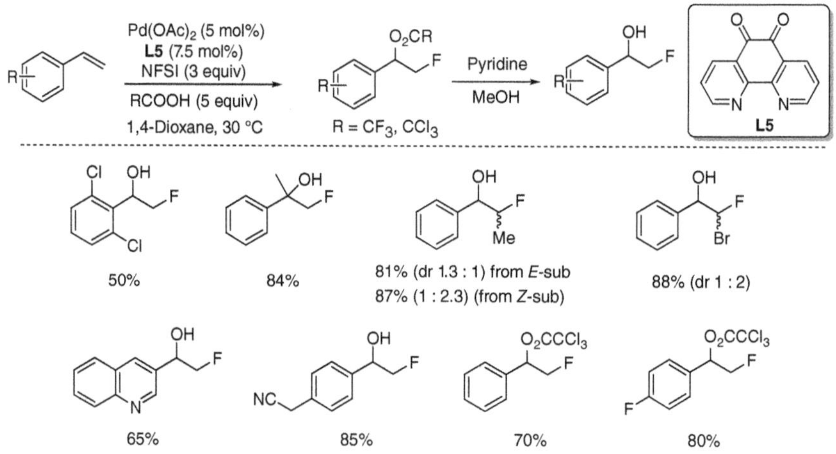

Scheme 5.10 Palladium-catalyzed fluoroesterification of styrenes.

were stable enough for isolation. In contrast, the reactions of weak acids with strong nucleophilicity, such as HOAc and BzOH, still formed aminofluorination products, rather than oxyfluorination products. The reaction presented an efficient synthetic pathway to afford a series of monofluoromethylbenzyl alcohols and esters in good to excellent yields. This reaction was further extended to intramolecular reaction with a tethered alcohol, giving the fluorinated tetrahydrofuran derivatives in good yields [15].

Liu and coworkers discovered that a sulfonyl radical species can be formed by the oxidation of NFSI and palladium catalyst [16]. A novel Pd-catalyzed intermolecular regio- and diastereoselective fluorosulfonylation of styrenes was developed. This reaction tolerates a wide range of functional groups in the styrene and arylsulfinic acid substrates to afford various β-fluorosulfones (Scheme 5.11). The reactions of both *E*- and *Z*-alkenes give anti-specific fluorosulfonylation products in good yields.

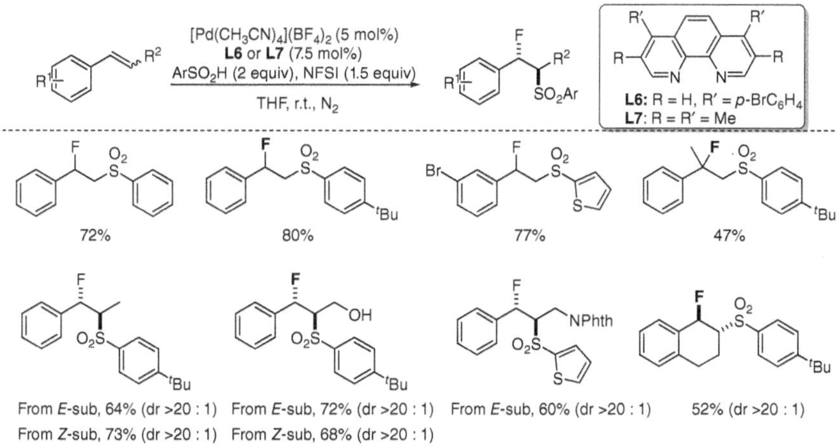

Scheme 5.11 Palladium-catalyzed fluorosulfonylation of styrenes.

A preliminary mechanistic study revealed that a high-valent palladium species was formed from the reaction of NFSI and palladium(II) salt, which acted as an oxidant to generate a sulfonyl radical species (Scheme 5.12). The sulfonyl radical was added to alkenes generating a benzyl radical, but the C—F bond formation was significantly different from the previous radical fluorination process. Owing to the high diasteroselectivities observed, it was more likely that a high-valent LPd^{III}F species site selectively occurred with a benzylic radical to deliver a C—F bond. Similar radical fluorinations could be achieved by other metals such as silver, iron, and so on [17].

5.2.2 Palladium-Catalyzed Trifluoromethylation of Alkenes

Inspired by the oxidative fluorination studies, Liu and coworkers further studied the combination of hypervalent iodine reagent and the Ruppert–Prakash reagent

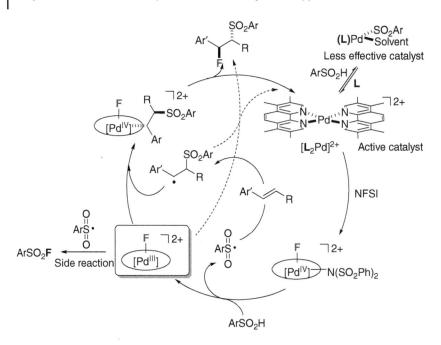

Scheme 5.12 Proposed mechanism of palladium-catalyzed fluorosulfonylation.

(TMSCF$_3$) as a nucleophilic trifluoromethylation reagent. The oxidative aryltrifluoromethylation reaction of alkenes proceeded efficiently to provide various trifluoromethylated oxindoles under mild conditions (Scheme 5.13) [18]. Other electrophilic [CF$_3{}^+$] reagents, such as Togni's reagent and Umemoto's reagent, were ineffective for the transformation. The addition of catalytic amount of Yb(OTf)$_3$ was beneficial for improvement of the yields. An achiral Box ligand was employed in this reaction; however, no enantioselectivity was observed with a chiral Box ligand. It is important to note that diene substrate can be converted to a single diastereomer in 62% yield, which indicated the involvement of a C(sp^3)–Pd intermediate generated via arylpalladation of alkenes (Scheme 5.14).

Further electrospray ionization-mass spectrometry (ESI–MS) experiments observed fragment signals at $m/z = 435$, 678, and 668 that were consistent with the proposed fragment mass of Pd(II) complex **int. I** [I – OAc]$^+$ and Pd(IV) complexes **int. IV** [IV – OAc]$^+$ and [IV – CF$_3$]$^+$. In addition, the signal at −27.1 ppm in the ^{19}F NMR spectrum was indicative of an alkyl–Pd^{IV}–CF$_3$ complex. Based on these observations, the proposed reaction mechanism was provided in Scheme 5.14. The final sp^3 C—CF^3 bond was derived from the reductive elimination of the Pd(IV) center.

5.2.3 Palladium-Catalyzed Trifluoromethoxylation of Alkenes

Compared to fluorination and trifluoromethylation reactions, direct trifluoromethoxylation is less developed, due to the instability of the OCF$_3$ anion, and

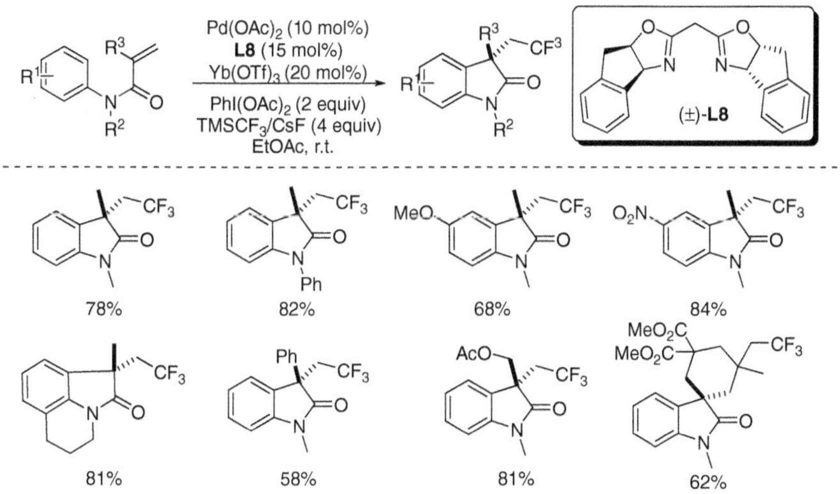

Scheme 5.13 Palladium-catalyzed aryltrifluoromethylation of alkenes.

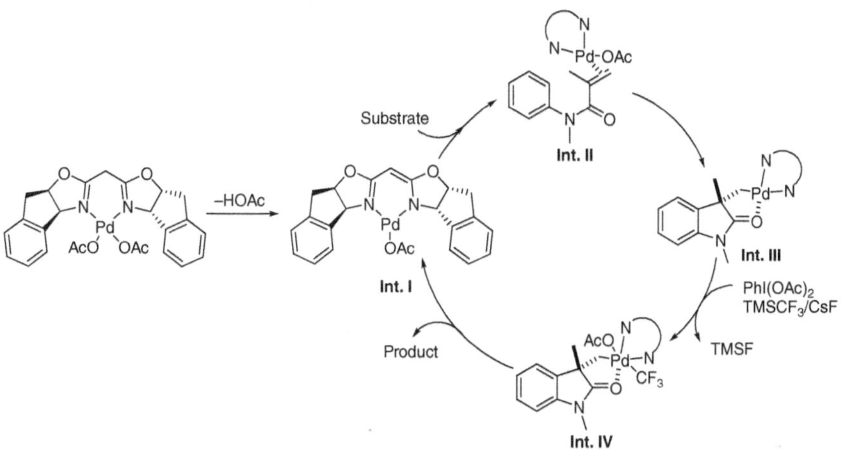

Scheme 5.14 Proposed mechanism of aryltrifluoromethylation of alkenes.

limited trifluoromethoxide reagents. More importantly, the facile β-fluoride elimination of the CF_3O-ligated transition metal complex ($MOCF_3$) further impedes discovery of the metal-catalyzed trifluoromethoxylation reactions. Liu and coworkers envisioned that the facile reductive elimination of high-valent palladium center might deliver the desired C—OCF_3 bond. After screening of the oxidants, they found that Selectfluor was a more suitable oxidant than I(III) reagents, providing higher yields with better reproducibility. The aminotrifluoromethoxylation reaction underwent 6-*endo* cyclization to yield β-CF_3O substituted piperidines in good yields (Scheme 5.15) [19]. It is important to note that this reaction was the first catalytic reaction that incorporated the OCF_3 group into organic molecules.

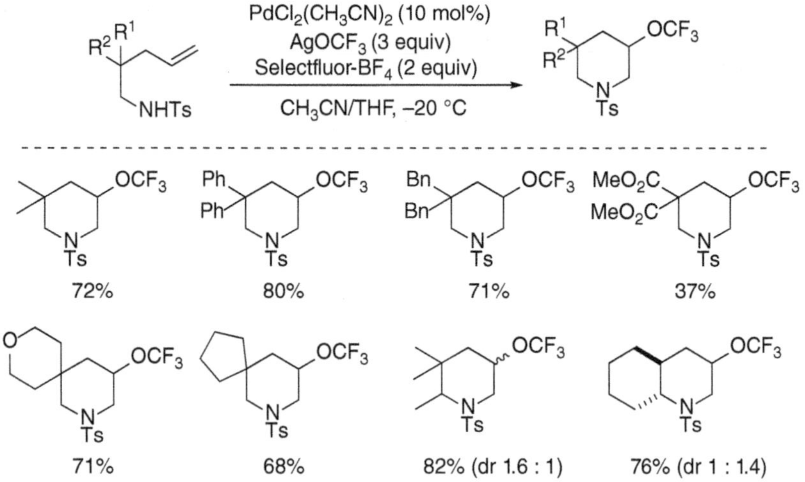

Scheme 5.15 Palladium-catalyzed aminotrifluoromethoxylation of alkenes.

The mechanism of C—OCF$_3$ bond formation was carefully studied. First, Pd(IV) fluoride complex **2** was obtained quantitatively with the oxidation of Selectfluor and reacted rapidly with AgOCF$_3$ to afford a new palladium complex **3** at 0 °C, which was observed by ^{19}F NMR and ^{1}H NMR. Upon warming the solution to room temperature, the sp^3 C—OCF$_3$ bond was formed in Pd(II) complex **4** via the reductive elimination of Pd(IV) trifluoromethoxide complex **3**. The structure of Pd(II) complex **4** was unambiguously determined by following derivatization with hydrogenation and ligand exchanges (Scheme 5.16).

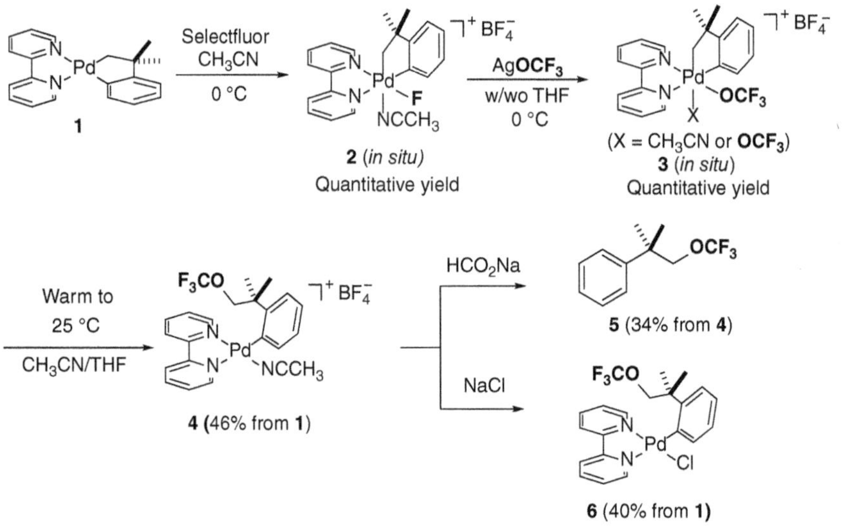

Scheme 5.16 Trifluoromethoxylation of palladium complex.

Furthermore, an enantioselective version of this reaction was developed with the stable $CsOCF_3$ as the trifluoromethoxide source (Scheme 5.17) [20]. $AgOCF_3$ was replaced due to the competitive coordination of silver and palladium with chiral ligands. A sterically bulky Pyox ligand was critical to the success of this reaction. Reaction monitoring indicated that the reaction using $CsOCF_3$ was much slower than that using $AgOCF_3$; however, it was remarkably accelerated by adding the sterically bulky Pyox ligand.

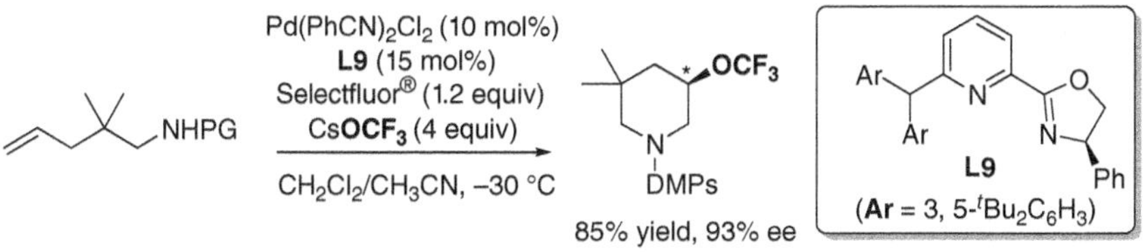

Scheme 5.17 Enantioselective palladium-catalyzed aminotrifluoromethoxylation of alkenes.

The palladium-catalyzed dioxygenations of alkenes were thoroughly studied, which were initiated through nucleopalladation, and followed by a sequential oxidative cleavage. Liu and coworkers reported a novel palladium-catalyzed intermolecular ditrifluoromethoxylation of unactivated alkenes (Scheme 5.18) [21]. Higher oxidative potential was required in this reaction, and the designed Selectfluor^{CN}-OTf was the best one. This reaction afforded ditrifluoromethoxylated products in moderate to good yields, along with the ineradicable Wacker-type products. Based on the stereochemistry of this reaction, they proposed that the reaction was possibly initiated by a high-valent Pd(IV) catalyst possessing strong Lewis acidity, which underwent a *cis*-trifluoromethoxypalladation (FOP) process to form an alkyl–Pd(IV) intermediate, which could undergo reductive elimination to generate the major product.

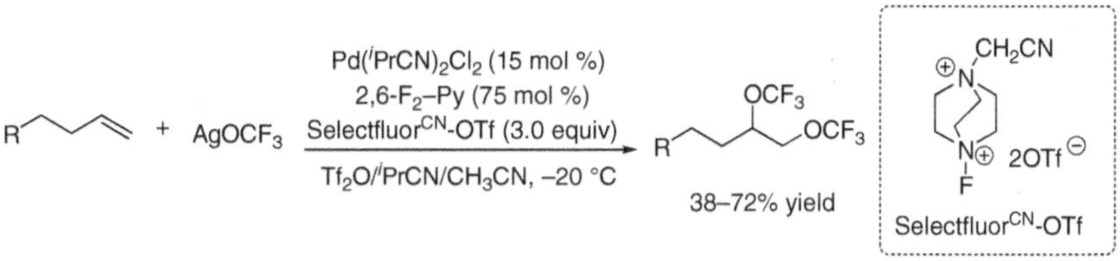

Scheme 5.18 Palladium-catalyzed ditrifluoromethoxylation of alkenes.

5.3 Copper-Catalyzed Trifluoromethylative Functionalization of Alkenes

Early studies on trifluoromethyl radical focused on the reactions of aromatic rings. Since 2011, the copper-catalyzed intermolecular trifluoromethylation of allylic C—H bonds with electrophilic trifluoromethyl reagents were independently developed by the groups of Buchwald, Wang, Liu, and coworkers [22]. Since then, radical trifluoromethylation of alkenes has received increasing attention. A plausible mechanism involved a radical process in which CF_3 radical was initially generated from the reaction between electrophilic trifluoromethylation reagent and copper(I) via single electron transfer (SET), followed by its addition to alkenes to form an alkyl radical intermediate. The resulting alkyl radical intermediate might be further oxidized to a carbocation, or combined with copper to form alkylcopper(III) species. The introduction of other nucleophiles could achieve the trifluoromethylative functionalization of alkenes (Scheme 5.19).

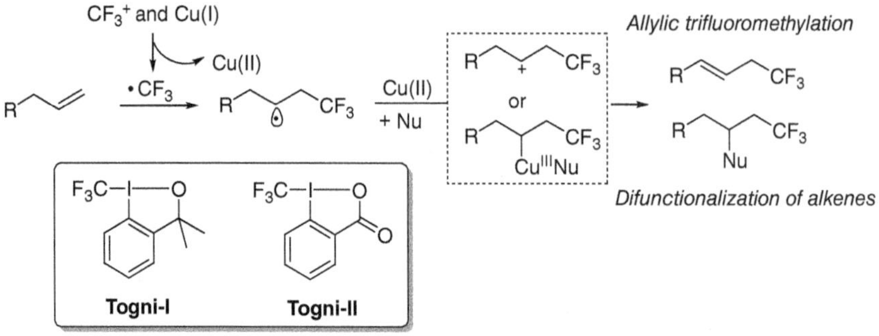

Scheme 5.19 Trifluoromethylation of alkenes.

5.3.1 Copper-Catalyzed Trifluoromethylamination of Alkenes

In 2013, Sodeoka and coworkers disclosed a trifluoromethylamination reaction of allylamines (Scheme 5.20) [23]. This reaction afforded a series of β-trifluoromethylamine derivatives with high efficiency. Substrates bearing different *N*-aryl rings were compatible with this reaction, and the benzylamine derivative also gave the desired product under the reaction conditions. The authors suggested that this *N*-migratory oxytrifluoromethylation reaction proceeded via an aziridine intermediate, which would be attacked by various nucleophiles, such as 2-iodobenzoate, a common by-product in the generation of CF_3 from the Togni-II

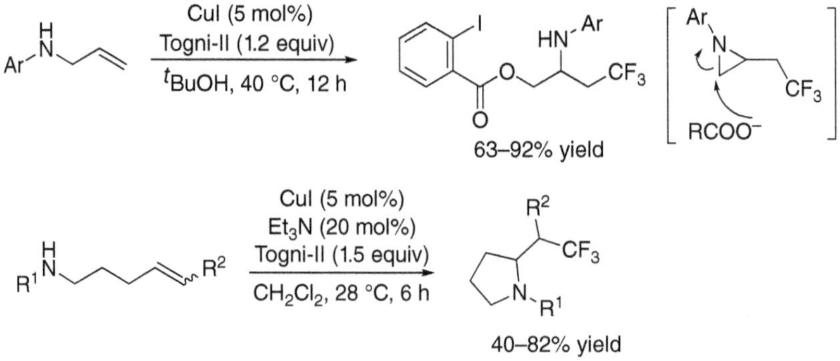

63–92% yield

40–82% yield

Scheme 5.20 Trifluoromethylamination reaction of allylamines.

reagent. They also found 4-methylthiophenol, decanethiol, aniline, ^{n}BuOH, phenol, and an indole derivative to be good nucleophiles for the reaction in the presence of additional Lewis acids. This reaction was further applied to the synthesis of trifluoromethylated pyrrolidines [24].

At the same time, Tan, Liu, and coworkers reported a copper-catalyzed intramolecular trifluoromethylamination of alkenes with diverse nitrogen-based nucleophiles (Scheme 5.21) [25]. For instance, primary aliphatic and aromatic amines, sulfonamides, carbonates, and ureas can be employed in this reaction to construct trifluoromethylated five-membered azaheterocycles. However, they proposed that the reaction underwent CF_3 radical addition and single-electron oxidation, and the resultant carbocation was trapped by nitrogen nucleophiles. Wang and coworker also reported a similar reaction to synthesize trifluoromethylated lactams [26].

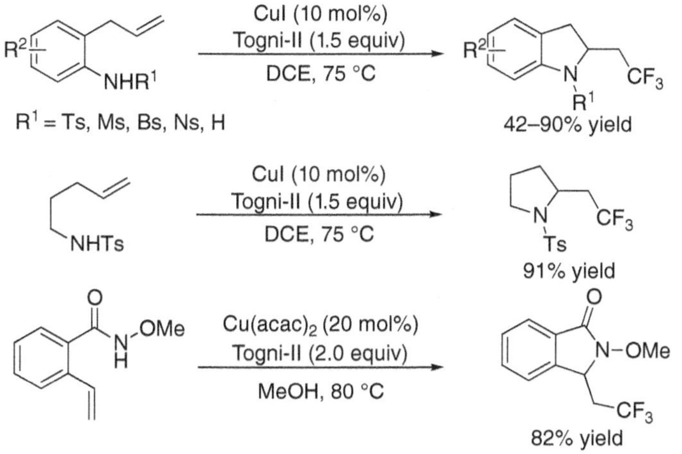

Scheme 5.21 Intermolecular trifluoromethylamination reaction of alkenes.

In 2014, Liu and coworkers developed a copper-catalyzed intermolecular azidotrifluoromethylation of alkenes using the less reactive Togni-I reagent as the CF_3 source and $TMSN_3$ as the nucleophile (Scheme 5.22) [27]. The reaction exhibited a broad substrate scope. For example, styrenes, mono- and disubstituted aliphatic alkenes, electron-deficient alkenes, and cyclic alkenes were suitable for the reaction. The reaction showed excellent diastereoselectivity for cyclic alkenes. Notably, the silyl reagents are crucial to these reactions, and other azido sources, such as TsN_3 and NaN_3, are ineffective. In addition, trifluoromethylated alkyl isocyanides could also be obtained with trimethylsilyl isothiocyanate (TMSNCS) as the nucleophilic reagent under modified conditions [28].

5.3.2 Copper-Catalyzed Trifluoromethyloxygenation of Alkenes

In 2012, a few cases of catalytic trifluoromethyloxygenation of styrenes with Togni reagent were independently reported by Sodeoka and Szabó [29], enabling direct

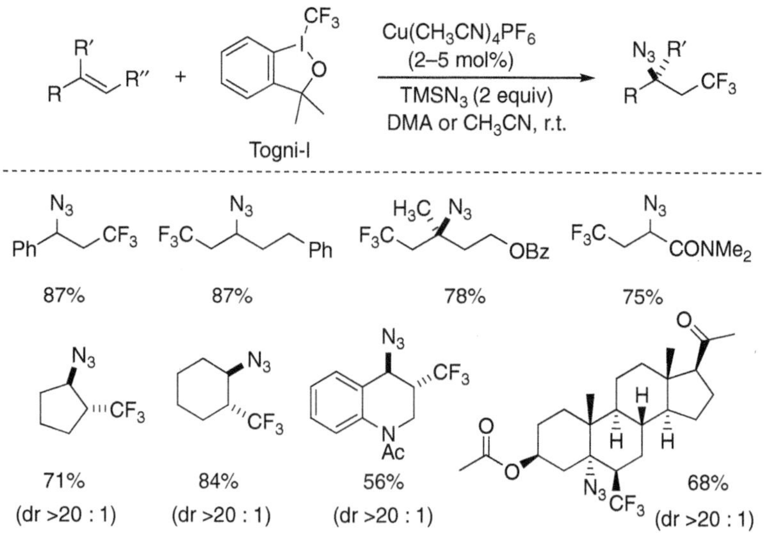

Scheme 5.22 Intermolecular azidotrifluoromethylation of alkenes.

introduction of 2-iodobenzoate. Loh and Feng reported the copper-catalyzed trifluoromethyloxygenation of enamides in methanol as the solvent (Scheme 5.23). In this transformation, the imine intermediate was attacked by methanol to form the β-trifluoromethyl aminal products.

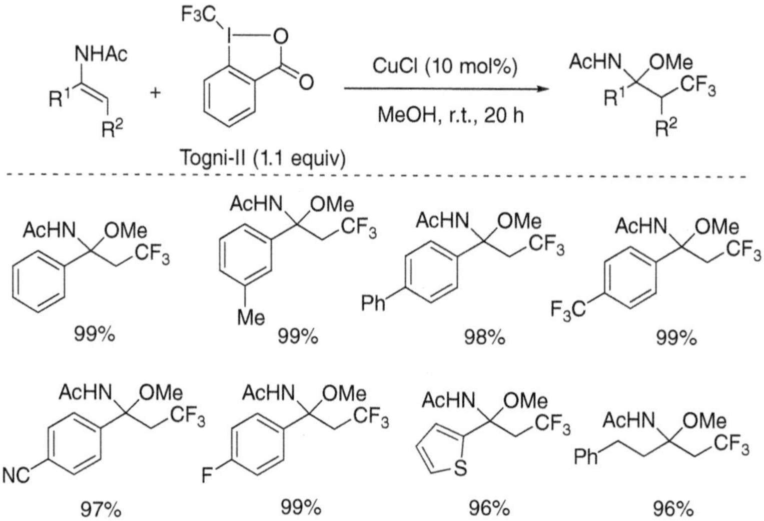

Scheme 5.23 Trifluoromethyloxygenation of enamides.

At the same time, Buchwald and coworker reported a mild copper-catalyzed trifluoromethyloxygenation of unactivated alkenes (Scheme 5.24) [30]. Carboxylic acids, alcohols, and phenols could serve as suitable nucleophiles to give cyclic products ranging from 3- to 6-membered rings. In their study, they found that bidentate nitrogen ligands could assist the C—O bond formation, which paved the way for further testing of the asymmetric version of this reaction with chiral Box ligands.

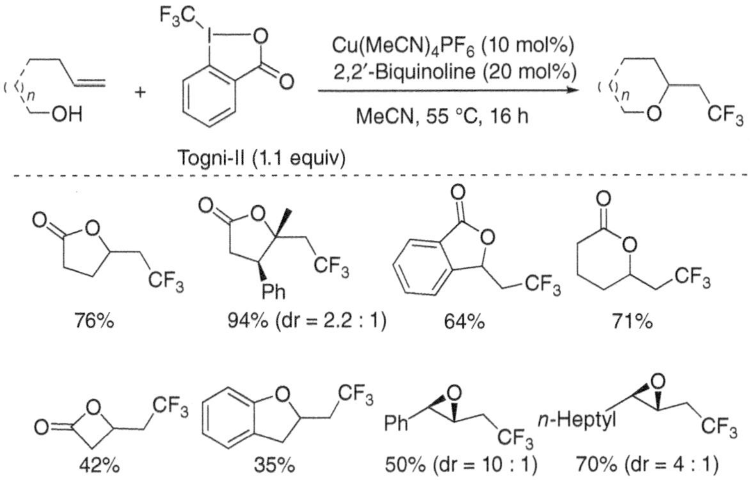

Scheme 5.24 Intermolecular trifluoromethyloxygenation of alkenes.

Yu and coworkers reported the copper-catalyzed trifluoromethyloxygenation of allylamines, which employed CO_2 in the construction of important heterocycles, thus leading to important CF_3-containing 2-oxazolidones (Scheme 5.25) [31]. Both primary and secondary alkyl groups were feasible substituents on nitrogen, while aryl, strongly electron-withdrawing substituents were not favored. CO_2 plays a vital role in tuning the reaction from amino- to oxy-trifluoromethylation with excellent chemoselectivity, regioselectivity, and diastereoselectivity.

5.3.3 Copper-Catalyzed Trifluoromethylcarbonation of Alkenes

Compared to the oxygen and nitrogen nucleophiles, carbon nucleophiles have much weaker nucleophilicity. Therefore, it is more difficult to form a C—C bond in the introduction of a second functional group. On the basis of their trifluoromethylazidation reaction of alkenes [27], Liu and coworkers further achieved the trifluoromethylcyanation of alkenes by using trimethylsilyl cyanide (TMSCN) as a nucleophilic reagent (Scheme 5.26) [32]. This reaction afforded a series of trifluoromethylated alkyl nitriles in excellent yields. Meanwhile, Liang and coworkers reported the same reaction with copper triflate as the catalyst [33]. Furthermore, this catalytic system could also be used for the trifluoromethylation of alkynes, allenes, and enynes [34].

Later, Liu and coworkers developed the first intermolecular trifluoromethylarylation of alkenes with arylboronic acids as the aryl sources (Scheme 5.27) [35]. Diverse alkenes were compatible with these conditions, including aryl- and alkyl-substituted alkenes, as well as multisubstituted alkenes. Moreover, reactions of cyclic alkenes

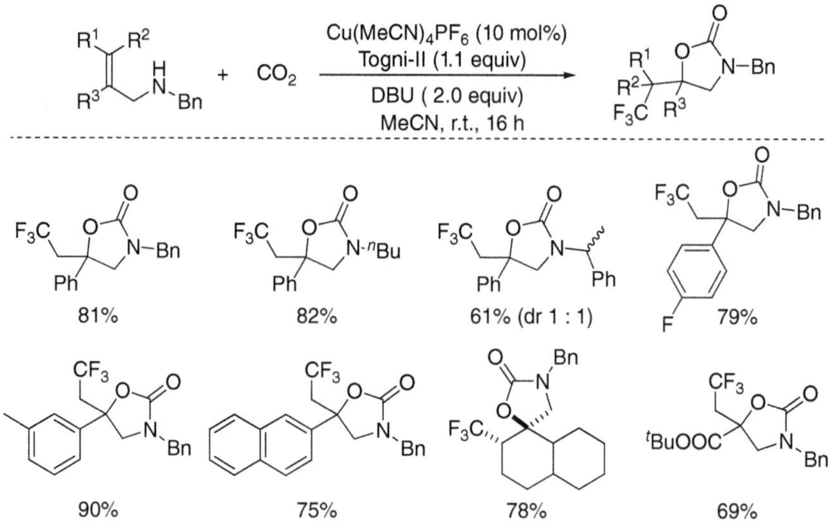

Scheme 5.25 Trifluoromethyloxygenation of allylamines.

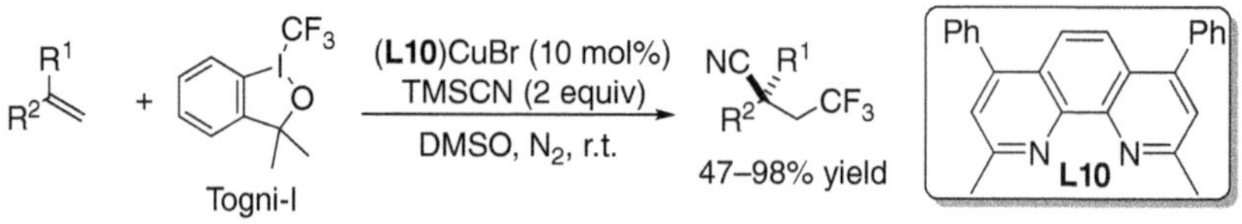

Scheme 5.26 Trifluoromethylcyanation of alkenes.

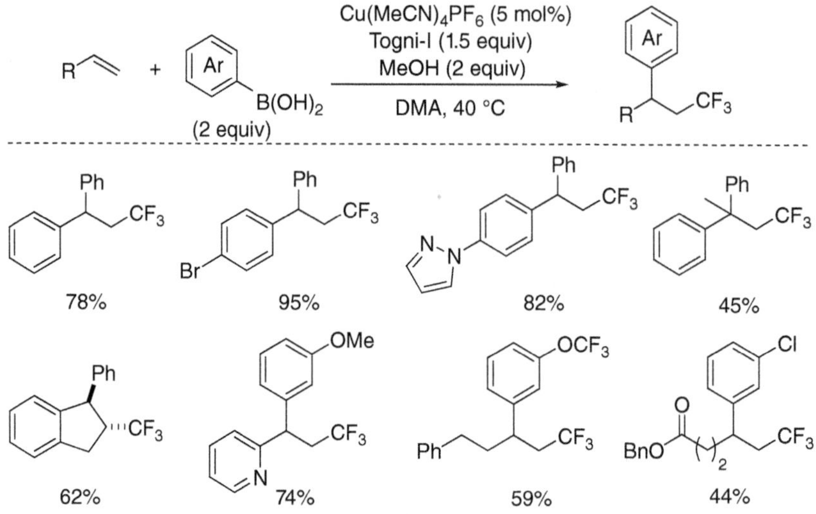

Scheme 5.27 Trifluoromethylarylation of alkenes with aryl boronic acid.

gave the desired products with excellent diastereoselectivities (>20 : 1). The reaction featured a broad substrate scope of arylboronic acids.

Notably, only boronic acid was suitable for this reaction, while other arylboron reagents (e.g. ArBpin and $ArBF_3$ salts) and arylmetallic reagents (e.g. $ArSnBu_3$ and $ArSiMe_3$) failed to afford the desired products. Moreover, Togni-I was the only feasible CF_3 reagent. A mutual activation between arylboronic acids and the Togni-I was proposed in the reaction, which was further indicated by ^{19}F NMR analysis of the mixture of Togni-I and boronic acids (Scheme 5.28). Finally, the mechanism was proposed as follows: with the treatment of activated CF_3^+ species **int. II**, a CF_3 radical was released and quickly added to the alkene, giving an alkyl radical intermediate (**int. III**); meanwhile, the alkoxy anion generated from the reduced Togni-I reagent acted as a strong base to activate $ArB(OH)_2$, promoting transmetalation to deliver the ArCu(II) species. These two species were then combined to provide Cu(III) species **int. IV**, and reductive elimination from this species led to the final products. This catalytic cycle was supported by kinetic studies, Hammett plot, and competitive reactions.

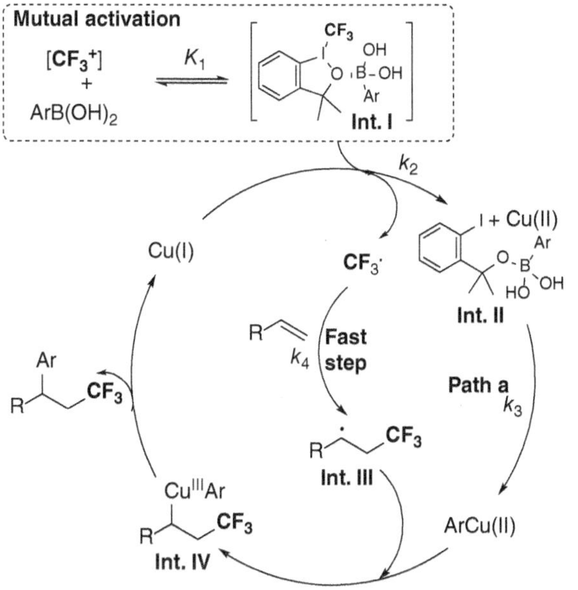

Scheme 5.28 Proposed mechanism of trifluoromethylarylation of alkenes.

Recently, Li and coworkers disclosed a copper-catalyzed oxidative trifluoromethylarylation of styrenes with indoles as the nucleophilic aryl sources (Scheme 5.29) [36]. Cheap $NaSO_2CF_3$ was used as the CF_3 source with the oxidation of *tert*-butyl peroxybenzoate (TBPB), $K_2S_2O_8$, and copper catalyst. A wide range of styrenes including 1,1-diarylethylene were compatible with the reaction conditions to afford the 1,2-trifluoromethylarylation products, while the reaction of electron-deficient styrenes resulted in low yields. In the plausible reaction mechanism, a carbocation intermediate was formed with the aid of Cu(II) species, and the electrophilic alkylation of indole C—H bonds afforded the final products.

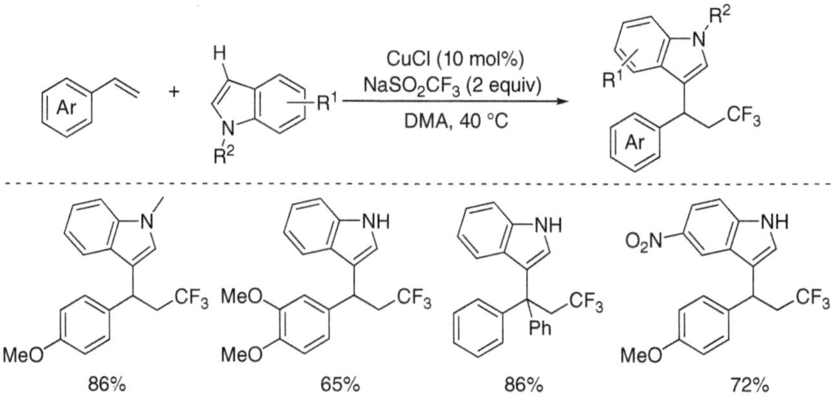

Scheme 5.29 Trifluoromethylarylation of styrenes with indoles.

5.3.4 Enantioselective Copper-Catalyzed Trifluoromethylation of Alkenes

Although great endeavors have been devoted to various racemic versions of copper-catalyzed radical trifluoromethylation of alkenes, the development of catalytic asymmetric methods remains a formidable challenge. Recently, mainly two catalytic chiral systems proved to be efficient for copper-catalyzed radical trifluoromethylation of alkenes [37]. One is Cu/chiral bidentate system, typically using chiral bisoxazoline (Box) ligand, and the other is Cu/chiral phosphoric acid (CPA) [38].

In 2013, Buchwald and coworker developed a pioneering method for the efficient enantioselective trifluoromethyloxygenation of alkenes using a copper-chiral Box catalyst system (Scheme 5.30) [39]. This method delivered a set of enantioenriched 5- and 6-member lactones with good functional group compatibility. Mechanistic studies evidenced a redox radical addition mechanism, in which a C—O bond is enantioselectively formed via a carbon radical intermediate in the presence of Cu(II) species.

Further studies suggested that bisligated complex L_2Cu is an off-cycle species, while monoligated complex LCu is the reactive species and is somewhat unstable (Scheme 5.31). Thus a slight excess of ligand helps to keep the catalyst stable via a balance between stability and reactivity [40]. More importantly, they proposed a model for the transition state of the enantiodetermining step, involving a distorted square planar alkyl–Cu(III) carboxylate intermediate formed via radical addition to Cu(II). A subsequent reductive elimination affords the chiral products with retention of the configuration.

During the studies of the trifluoromethylarylation of alkenes, Liu and coworkers speculated a Cu(I/II/III) catalytic cycle in which the carbon radical reacts with

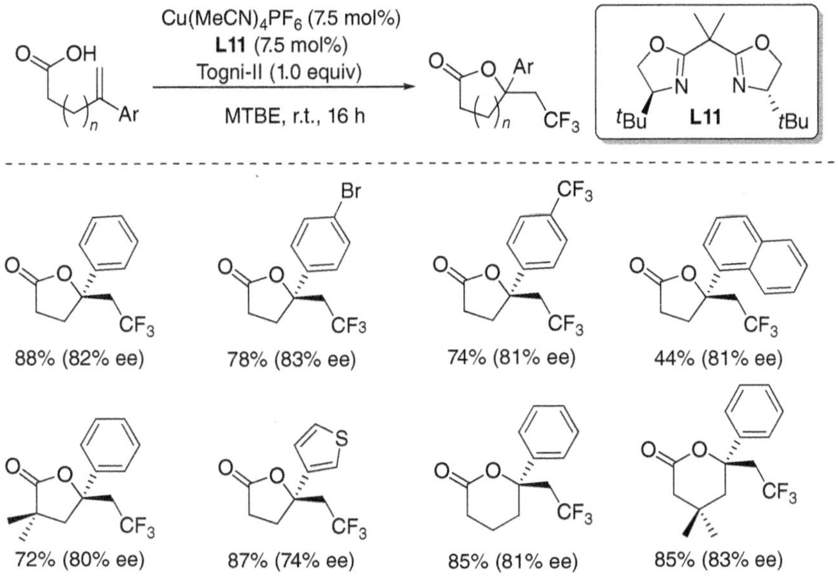

Scheme 5.30 Enantioselective trifluoromethyloxygenation of alkenes.

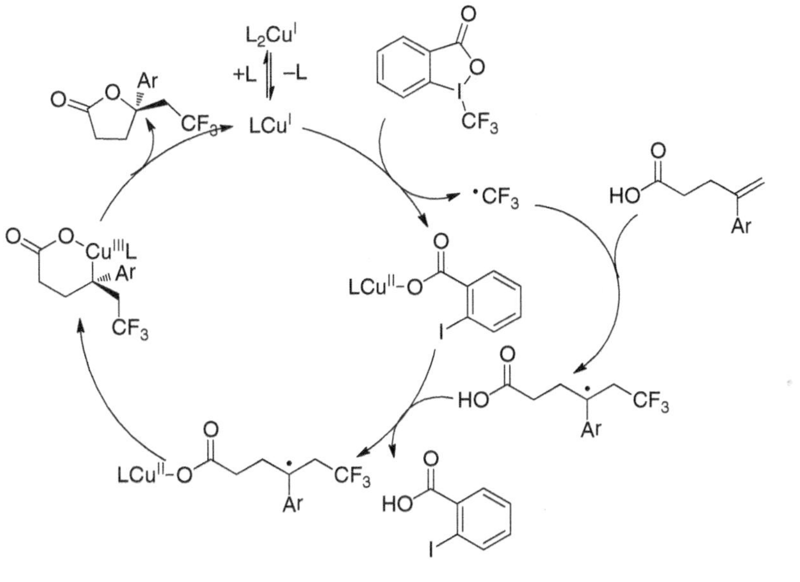

Scheme 5.31 Proposed mechanism of enantioselective trifluoromethyloxygenation of alkenes.

the Cu(II) species to form a highly reactive Cu(III) intermediate and its reductive elimination contributes to the final bond formation [35]. Inspired by this, they explored asymmetric radical transformations by introducing chiral ligands, and first achieved the enantioselective trifluoromethylcyanation of styrenes in the presence of chiral Cu(I)/Box catalysts [41]. This protocol exhibited a broad substrate scope and functional group compatibility, and provided efficient access to enantioenriched benzylic nitriles (Scheme 5.32). Internal alkenes could also be employed as substrates, and the reactions provided the desired products with excellent enantioselectivities. Notably, the configuration of the internal alkene (*E* or *Z*) did not affect the stereochemistry of products, which suggested a CF_3 radical addition pathway to form the benzylic radical non-selectively.

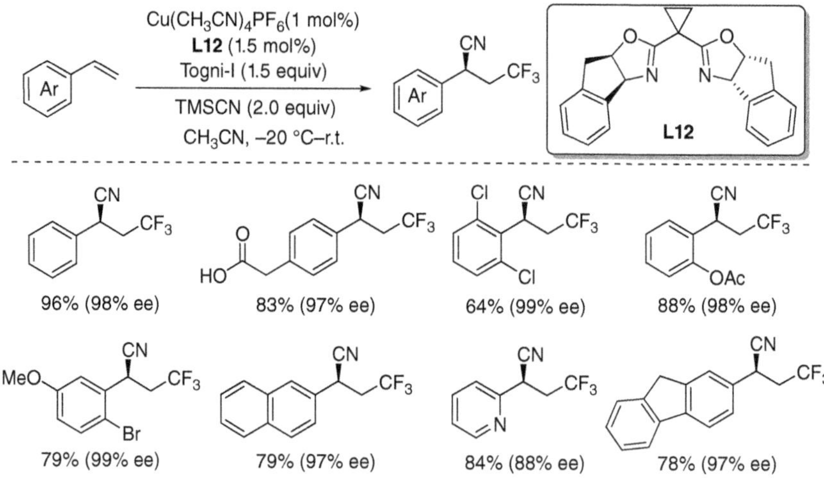

Scheme 5.32 Enantioselective trifluoromethylcyanation of styrenes.

Mechanistic studies revealed that a low concentration of cyanide in the reaction system was vital for excellent enantioselective control and the mutual activation between Togni-I and TMSCN was a good mode to release cyanide slowly. However, a slight amount of cyanide could shorten the induction period, so $(L12)Cu^I(CN)$ was proposed to be the active catalytic species. A possible reaction mechanism was described in Scheme 5.33: the mutual activation promoted the single-electron oxidation of $(L12)Cu^I CN$ by the activated Togni-I reagent. The released CF_3 radical then added to styrene, generating the benzylic radical **int. II** that was trapped by the chiral $(L12)Cu^{II}(CN)_2$ species to form chiral Cu(III) species **int. III**. Eventually, reductive elimination from the Cu(III) center afforded optically pure benzylic nitriles, which was the enantiodetermining step [42]. Notably, a ligand acceleration effect was observed, and excellent enantioselectivity was still obtained when the concentration of the ligand was less than that of the copper catalyst. This enantioselective cyanation of benzylic radicals was further applied to a series of asymmetric cyanofunctionalizations of styrenes [43]. A similar reaction was also reported by Wang, Xu, and coworkers using fluoroalkyl iodides under photoredox conditions [44].

Liu and coworkers further explored the enantioselective trifluoromethylarylation of styrenes [45]. The best ligand in the cyanation reaction only provided moderate enantiomeric excess. Interestingly, a clear correlation between the enantiomeric excess of the arylated product and the bite angle of the Box ligand

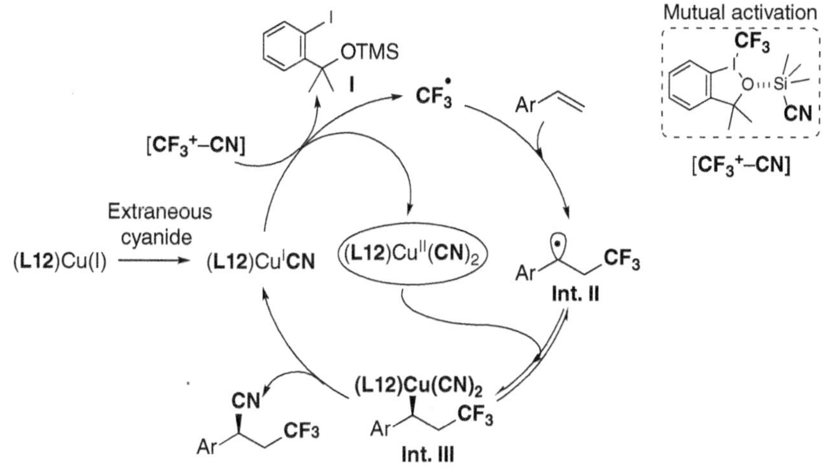

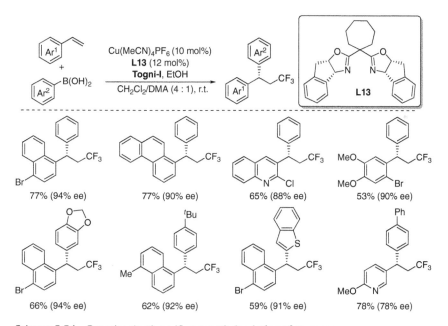

Scheme 5.33 Proposed mechanism of enantioselective trifluoromethylcyanation.

was observed. Ligand **L12** with a small bite angle gave the best enantiomeric excess (Scheme 5.34). And further increasing the steric hindrance of Box ligand dramatically decreased the enantioselectivity, which indicated that proper steric effect was vital for the bulky chiral (**L13**)Cu^{II}Ar species to trap benzylic radicals. The enantioselectivity was further improved by lowering the reaction temperature and adding ethanol as an additive. This reaction had a broad substrate scope for both styrenes and ArB(OH)$_2$, yielding various 1,1-diaryl-3-trifluoropropanes with

Scheme 5.34 Enantioselective trifluoromethylarylation of styrenes.

good to excellent enantioselectivities. Furthermore, heteroaryl frameworks in both the vinyl arene and ArB(OH)$_2$ substrates were well tolerated. Based on this work, they further developed an aminoarylation of alkenes and arylation of benzylic C—H bonds [46].

This trifluoromethylation reaction could be extended from styrenes to α-substituted acrylamides, providing a straightforward and efficient access to chiral quaternary all-carbon centers [47]. Initial trials on α-methylstyrene failed to give the arylation product, probably due to the increased steric hindrance or weaker oxidative ability of tertiary benzylic radical, which impedes the interaction of benzylic radical with L*Cu(II)Ar species. After screening various carbonyl groups, an acrylamidyl (CONHAr) group adjacent to the tertiary carbon radical was the best choice. They also identified that the *gem*-dibenzyl group in the Box ligand had a significant effect on the reactivity and enantioselectivity. The reaction exhibited broad substrate scope, and excellent functional group compatibility, and both aryl and alkyl substitutions on α-position were feasible. However, β-substituted internal alkenes, and alkenes without substituent or bearing a steric bulky alkyl group exhibited low efficiency (Scheme 5.35).

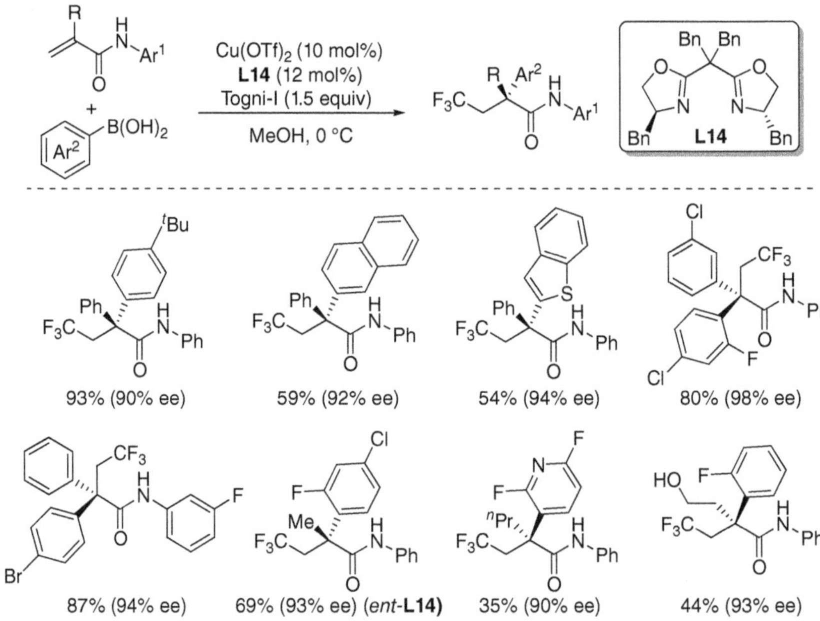

Scheme 5.35 Enantioselective arylation of tertiary carbon radical.

Liu and coworkers further explored the enantioselective trifluoromethylalkynylation of styrenes via a similar strategy [48]. Suitable alkynylating reagents were important for the success of this reaction. Only the reaction with alkynyl–Si(OMe)$_3$ afforded the desired products, while other alkynylating reagents such as alkynyl–trimethylsilyl (TMS), alkynyl boron, and terminal alkyne were inert to this reaction. This reaction had wide substrate scope and high functional group compatibilities, which provided an easy access to the structurally diverse and enantiomerically enriched CF_3-containing propargylic products (Scheme 5.36).

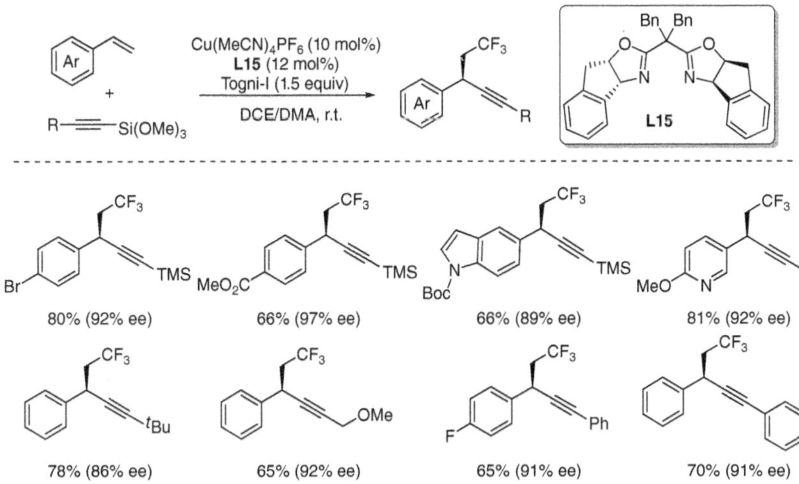

Scheme 5.36 Enantioselective trifluoromethylalkynylation of styrenes.

Since 2014, Liu and coworkers have been dedicated to designing a Cu/CPA catalytic system for asymmetric radical involved alkene difunctionalizations [49]. In 2016, they disclosed a novel copper-catalyzed trifluoromethylamination of alkenes with CPA (Scheme 5.37) [50]. The reaction of alkenyl amines with Togni I under Cu/CPA dual-catalytic system afforded an efficient access to CF_3-containing chiral pyrrolidines bearing a tertiary stereocenter with excellent enantioselectivity. Notably, two acidic N—H bonds in the urea group were vital for the success of this reaction, because urea acted as both the nucleophile and the directing group. Indeed, the reaction of N-methylated substrate did not take place. Similar reactions were further extended to other perfluoroalkyl groups, such as β-perfluorobutyl, trifluoromethyl, and so on, with fluoroalkylsulfonyl chlorides as the radical source [51].

Based on the mechanistic studies, a plausible mechanism was proposed as listed in Scheme 5.38. A CF_3 radical was first generated via a single-electron transfer between CuI and the CPA-activated Togni's reagent, accompanied by a chiral copper(II) phosphate. The CF_3 radical selectively added to the terminal alkene to afford a tertiary benzylic radical intermediate **int. I**, which would be trapped by Cu^{II}/CPA to generate a Cu^{III} complex **int. II**. The subsequent C—N bond formation delivered the final products while releasing Cu^{I} and CPA. The chiral phosphate could control the selectivity via both hydrogen-bonding interactions with the N—H bond adjacent to the aryl group. However, a carbocation intermediate **int. III** through ion pairing cannot be ruled out.

Liu and coworkers tried to extend this Cu^{I}/CPA dual-catalyst system to trifluoromethyloxygenation reaction by changing urea to alcohol; unfortunately, poor enantioselectivity was observed (<21% ee). The reason might be the poor hydrogen bonding property of the hydroxyl group, which was inefficient for anchoring a

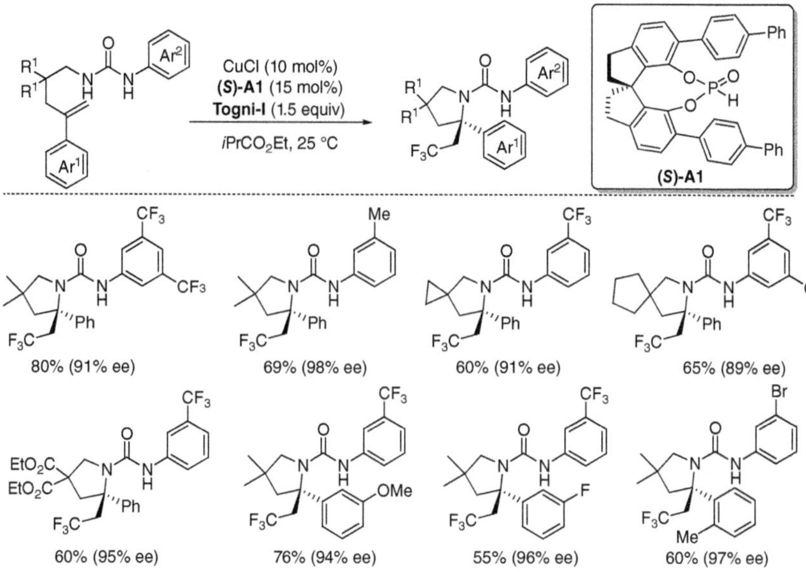

Scheme 5.37 Enantioselective trifluoromethylamination of alkenes.

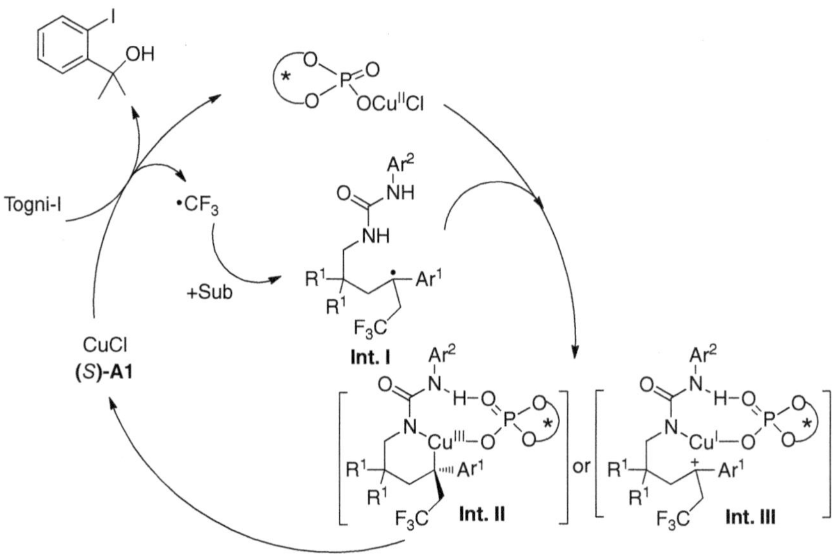

Scheme 5.38 Proposed mechanism of enantioselective trifluoromethylamination of alkenes.

CPA-ligated copper catalyst. They introduced an achiral Lewis base to realize this asymmetric trifluoromethyloxygenation of alkenes (Scheme 5.39) [52]. This reaction provided a direct approach to CF$_3$-containing chiral tetrahydrofurans with excellent enantioselectivity. However, Thorpe–Ingold effect was observed in this reaction, and the reaction of substrate without *gem*-disubstituents only gave moderate enantioselectivity. Mechanistic studies indicated that *N,N*-diethylnicotinamide might act as a good coordinative ligand to stabilize the loosely alcohol-bounded Cu/CPA complex.

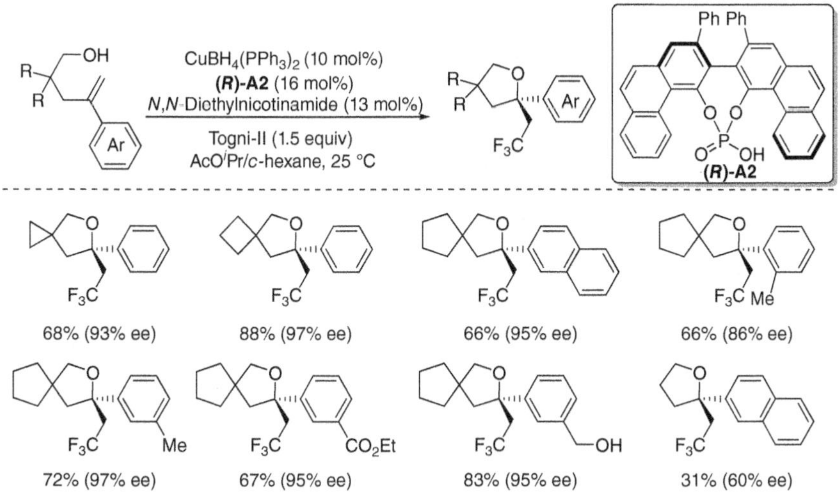

Scheme 5.39 Enantioselective trifluoromethyloxygenation of alkenes.

5.4 Summary and Conclusions

During the past decade, great progress has been made in the transition metal-catalyzed bifunctionalization-based fluorination and trifluoromethylation reactions. Notably, the addition of chiral ligands enables the synthesis of chiral fluorine-containing organic compounds, which are much more valuable for medicinal chemistry and drug discovery. However, due to the limitation of fluorinating reagents, the catalytic reaction for installing fluorine-containing groups, such as OCF$_3$, is still underdeveloped. More stable and easily available fluorinating reagents are highly in demand. Furthermore, the combination of radical reactions with copper catalysis exhibited an efficient way for the synthesis of trifluoromethylated alkanes, which has robust reactivity, outstanding chemoselectivity, and high functional group compatibility. Recently, the related enantioselective reactions have attracted extensive attention. However, most reactions were only suitable for the enantioselective control of benzylic carbon radicals; the enantioselective control of other carbon radicals is still highly challenging, especially carbon radicals generated from radical addition of unactivated alkenes. Thus, new catalytic systems need to be designed for the reaction of unactivated alkenes, and alkenes in complex molecular.

References

1 Yin, G., Mu, X., and Liu, G. (2016). *Acc. Chem. Res.* 49: 2413–2423.
2 Canty, A.J. (2009). *Dalton Trans.*: 10409–10417.

3 (a) Hull, K.L., Anani, W.Q., and Sanford, M.S. (2006). *J. Am. Chem. Soc.* 128: 7134–7135. (b) Wang, X., Mei, T.-S., and Yu, J.-Q. (2009). *J. Am. Chem. Soc.* 131: 7520–7521. (c) Ball, N.D. and Sanford, M.S. (2009). *J. Am. Chem. Soc.* 131: 3796–3797. (d) Furuya, T. and Ritter, T. (2008). *J. Am. Chem. Soc.* 130: 10060–10061.

4 Wu, T., Yin, G., and Liu, G. (2009). *J. Am. Chem. Soc.* 131: 16354–16355.

5 Cheng, J., Chen, P., and Liu, G. (2015). *Chin. J. Catal.* 36: 40–47.

6 Yin, G., Wu, T., and Liu, G. (2012). *Chem. Eur. J.* 18: 451–455.

7 Wu, T., Cheng, J., Chen, P., and Liu, G. (2013). *Chem. Commun.* 49: 8707–8709.

8 Zhu, H. and Liu, G. (2012). *Acta Chim. Sinica* 70: 2404–2407.

9 Emer, E., Pfeifer, L., Brown, J.M., and Gouverneur, V. (2014). *Angew. Chem. Int. Ed.* 53: 4181–4185.

10 Talbot, E.P.A., Fernandes, T.A., McKenna, J.M., and Toste, F.D. (2014). *J. Am. Chem. Soc.* 136: 4101–4104.

11 He, Y., Yang, Z., Thornbury, R.T., and Toste, F.D. (2015). *J. Am. Chem. Soc.* 137: 12207–12210.

12 Qiu, S., Xu, T., Zhou, J. et al. (2010). *J. Am. Chem. Soc.* 132: 2856–2857.

13 Xu, T., Qiu, S., and Liu, G. (2011). *Chin. J. Chem.* 29: 2785–2790.

14 Peng, H., Yuan, Z., Wang, H.-y. et al. (2013). *Chem. Sci.* 4: 3172–3178.

15 Yuan, Z., Peng, H., and Liu, G. (2013). *Chin. J. Chem.* 31: 908–914.

16 Yuan, Z., Wang, H.-y., Mu, X. et al. (2015). *J. Am. Chem. Soc.* 137: 2468–2471.

17 (a) Baker, T.J. and Boger, D.L. (2012). *J. Am. Chem. Soc.* 134: 13588–13591. (b) Shigehisa, H., Nishi, E., Fujisawa, M., and Hiroya, K. (2013). *Org. Lett.* 15: 5158–5161. (c) Li, Z., Song, L., and Li, C. (2013). *J. Am. Chem. Soc.* 135: 4640–4643. (d) Zhang, H., Song, Y., Zhao, J. et al. (2014). *Angew. Chem. Int. Ed.* 53: 11079–11083. (e) Lu, D.-F., Liu, G.-S., Zhu, C.-L. et al. (2014). *Org. Lett.* 16: 2912–2915. (f) Lu, D.-F., Zhu, C.-L., Sears, J.D., and Xu, H. (2016). *J. Am. Chem. Soc.* 138: 11360–11367.

18 Mu, X., Wu, T., Wang, H.-Y. et al. (2012). *J. Am. Chem. Soc.* 134: 878–881.

19 Chen, C., Chen, P., and Liu, G. (2015). *J. Am. Chem. Soc.* 137: 15648–15651.

20 Chen, C., Pflüger, P.M., Chen, P., and Liu, G. (2019). *Angew. Chem. Int. Ed.* 58: 2392–2396.

21 Chen, C., Luo, Y., Fu, L. et al. (2018). *J. Am. Chem. Soc.* 140: 1207–1210.

22 (a) Parsons, A.T. and Buchwald, S.L. (2011). *Angew. Chem. Int. Ed.* 50: 9120–9123. (b) Xu, J., Fu, Y., Luo, D.-F. et al. (2011). *J. Am. Chem. Soc.* 133: 15300–15303. (c) Wang, X., Ye, Y., Zhang, S. et al. (2011). *J. Am. Chem. Soc.* 133: 16410–16413.

23 Egami, H., Kawamura, S., Miyazaki, A., and Sodeoka, M. (2013). *Angew. Chem. Int. Ed.* 52: 7841–7844.

24 (a) Kawamura, S., Egami, H., and Sodeoka, M. (2015). *J. Am. Chem. Soc.* 137: 4865–4873. (b) Kawamura, S., Dosei, K., Valverde, E. et al. (2017). *J. Org. Chem.* 82: 12539–12553.

25 Lin, J.-S., Xiong, Y.-P., Ma, C.-L. et al. (2014). *Chem. Eur. J.* 20: 1332–1340.

26 Shen, K. and Wang, Q. (2016). *Org. Chem. Front.* 3: 222–226.

27 Wang, F., Qi, X., Liang, Z. et al. (2014). *Angew. Chem. Int. Ed.* 53: 1881–1886.

28 Liang, Z., Wang, F., Chen, P., and Liu, G. (2015). *Org. Lett.* 17: 2438–2441.

29 (a) Egami, H., Shimizu, R., and Sodeoka, M. (2012). *Tetrahedron Lett.* 53: 5503–5506. (b) Janson, P.G., Ghoneim, I., Ilchenko, N.O., and Szabó, K.J. (2012). *Org. Lett.* 14: 2882–2885. (c) Feng, C. and Loh, T.-P. (2012). *Chem. Sci.* 3: 3458–3462.

30 Zhu, R. and Buchwald, S.L. (2012). *J. Am. Chem. Soc.* 134: 12462–12465.

31 Ye, J.-H., Song, L., Zhou, W.-J. et al. (2016). *Angew. Chem. Int. Ed.* 55: 10022–10026.

32 Liang, Z., Wang, F., Chen, P., and Liu, G. (2014). *J. Fluorine Chem.* 167: 55–60.

33 He, Y.-T., Li, L.-H., Yang, Y.-F. et al. (2014). *Org. Lett.* 16: 270–273.

34 (a) Zhu, N., Wang, F., Chen, P. et al. (2015). *Org. Lett.* 17: 3580–3583. (b) Wang, F., Zhu, N., Chen, P. et al. (2015). *Angew. Chem. Int. Ed.* 54: 9356–9360. (c) Wang, F., Wang, D., Zhou, Y. et al. (2018). *Angew. Chem. Int. Ed.* 57: 7140–7145.

35 Wang, F., Wang, D., Mu, X. et al. (2014). *J. Am. Chem. Soc.* 136: 10202–10205.

36 Min, M.-Y., Song, R.-J., Ouyang, X.-H., and Li, J.-H. (2019). *Chem. Commun.* 55: 3646–3649.

37 Wang, F., Chen, P., and Liu, G. (2018). *Acc. Chem. Res.* 51: 2036–2046.

38 (a) Li, Z.-L., Fang, G.-C., Gu, Q.-S., and Liu, X.-Y. (2020). *Chem. Soc. Rev.* 49: 32–48. (b) Gu, Q.-S., Li, Z.-L., and Liu, X.-Y. (2020). *Acc. Chem. Res.* 53: 170–181.

39 Zhu, R. and Buchwald, S.L. (2013). *Angew. Chem. Int. Ed.* 52: 12655–12658.

40 Zhu, R. and Buchwald, S.L. (2015). *J. Am. Chem. Soc.* 137: 8069–8077.

41 Wang, F., Wang, D., Wan, X. et al. (2016). *J. Am. Chem. Soc.* 138: 15547–15550.

42 Zhang, W., Wang, F., McCann, S.D. et al. (2016). *Science* 353: 1014–1018.

43 (a) Wang, D., Wang, F., Chen, P. et al. (2017). *Angew. Chem. Int. Ed.* 56: 2054–2058. (b) Wang, D., Zhu, N., Chen, P. et al. (2017). *J. Am. Chem. Soc.* 139: 15632–15635. (c) Zhang, G., Fu, L., Chen, P. et al. (2019). *Org. Lett.* 21: 5015–5020.

44 Guo, Q., Wang, M., Peng, Q. et al. (2019). *ACS Catal.* 9: 4470–4476.

45 Wu, L., Wang, F., Wan, X. et al. (2017). *J. Am. Chem. Soc.* 139: 2904–2907.

46 (a) Wang, D., Wu, L., Wang, F. et al. (2017). *J. Am. Chem. Soc.* 139: 6811–6814. (b) Zhang, W., Chen, P., and Liu, G. (2017). *J. Am. Chem. Soc.* 139: 7709–7712. (c) Zhang, W., Wu, L., Chen, P., and Liu, G. (2019). *Angew. Chem. Int. Ed.* 58: 6425–6429.

47 Wu, L., Wang, F., Chen, P., and Liu, G. (2019). *J. Am. Chem. Soc.* 141: 1887–1892.

48 Fu, L., Zhou, S., Wan, X. et al. (2018). *J. Am. Chem. Soc.* 140: 10965–10969.

49 Yu, P., Lin, J.-S., Li, L. et al. (2014). *Angew. Chem. Int. Ed.* 53: 11890–11894.

50 Lin, J.-S., Dong, X.-Y., Li, T.-T. et al. (2016). *J. Am. Chem. Soc.* 138: 9357–9360.

51 Lin, J.-S., Wang, F.-L., Dong, X.-Y. et al. (2017). *Nat. Commun.* 8: 14841.

52 Cheng, Y.-F., Dong, X.-Y., Gu, Q.-S. et al. (2017). *Angew. Chem. Int. Ed.* 56: 8883–8886.

6

Fluorination, Trifluoromethylation, and Trifluoromethylthiolation of Alkenes, Cyclopropanes, and Diazo Compounds

Kálmán J. Szabó

Stockholm University, Department of Organic Chemistry, Arrhenius Laboratory, Stockholm SE-10691, Sweden

6.1 Introduction

Although, fluorine is the 13th most abundant element in the earth crust, very few natural products contain carbon–fluorine (C—F) bond [1]. The low abundance of organofluorines in natural products is a consequence of the physical properties of the naturally occurring fluorine sources, which makes the biosynthetic pathways cumbersome. Such properties involve the relatively low solubility of the naturally occurring fluorine salts, the low reactivity of fluoride ion in water solution, and the difficulties to oxidize fluorine in biological systems [1]. The low abundance of C—F bonds in natural products is in very sharp contrast to the broad application of organofluorines in life sciences. For example, 20% of all marketed drugs and over 50% of the recently registered innovative medicines contain at least one C—F bond [2–5]. Agrochemical products are another important area of application of organofluorines [6, 7]. Except for the natural isotope, even the unnatural fluorine-18 is applied in medical diagnostics [8, 9]. Fluorine-18 is a positron emitting nucleus, and therefore it can be used as a tracer in bioactive organic molecules in positron emission tomography (PET) (see Section 6.2.3).

As mentioned above, very few fluorine-containing natural products are identified, and therefore all practically useful organofluorines are obtained by organic synthesis. Selective synthesis of organic compounds is still a challenging task for organic chemists. The nucleophilic fluorine sources have usually low reactivity because of the high solvation energy of the fluoride ion, while electrophilic fluorination suffers from the high reactivity and low selectivity of the electron-deficient fluorine atoms. However, in the past decade a number of new reagents and catalytic procedures have appeared to control the reactivity and selectivity issues in the synthesis of the organic fluorine compounds [3, 10–16].

This chapter is mainly focused on application of electrophilic reagents for transfer of fluorine, trifluoromethyl, and trifluoromethylation functionalities. The targeted substrates in these processes include alkenes, alkynes, and diazocarbonyl compounds. The most important reagents considered in this chapter are hypervalent

Organofluorine Chemistry: Synthesis, Modeling, and Applications,
First Edition. Edited by Kálmán J. Szabó and Nicklas Selander.
© 2021 WILEY-VCH GmbH. Published 2021 by WILEY-VCH GmbH.

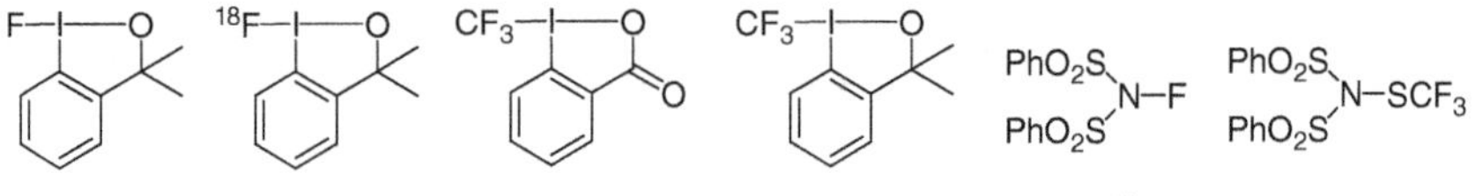

Figure 6.1 Main fluorine (**1**), trifluoromethyl (**2**), and trifluoromethylthiol (**3**) transfer reagents discussed in this chapter.

iodines **1** and **2** for fluorination and trifluoromethylation and dibenzenesulfonimide reagents **3a,b** for introduction of the F and SCF_3 groups (Figure 6.1). The selection of the reagents is mainly based on structural reasons. Reagents **1, 2** are based on hypervalent iodines, while reagent **3** is based on polarized N—X bonds. The primary interest was to compare and assess the hypervalent iodine reagents for synthesis of organofluorines. However, the hypervalent iodine-based SCF_3 transfer reagent cannot be isolated (probably because of stability issues) [17]. Therefore a potent dibenzenesulfonimide reagent **3b** was used in the studies. In addition, the reactivity of **3b** was compared to the popular fluorinating reagent NFSI, **3a.**

6.2 Fluorination of Alkenes, Cyclopropanes, and Diazocarbonyl Compounds

Fluorination of alkenes is one of the simplest and widely employed methods for creation of carbon–fluorine bonds. In principle, all widely used electrophilic fluorine transfer reagents, such as Selectfluor, NFSI (**3a**), and various hypervalent iodines (such as **1**) can be used for selective C—F bond formation using alkene substrates [3, 10, 12, 14, 18]. Although there are excellent studies on application of various hypervalent iodines for the formation of C—F bonds [19–21], our interest was focused on application of fluoro-benziodoxole **1** [15, 16, 22, 23]. The two main reasons for the choice of this reagent are based on its structural similarity (Section 6.4) to the CF_3-transfer reagents **2a,b** (Togni reagents [11]) and the promising application area of the fluorine-18 analog of **1** for labeling of PET tracers (Section 6.2.3).

An overview of the fluorination reactions carried out with fluoro-benziodoxole (**1**) discussed in this chapter is shown in Figure 6.2. These reactions involve various fluorination-based difunctionalization reactions involving alkene (Section 6.2.1), cyclopropane (Section 6.2.2), and diazocarbonyl (Section 6.2.3) compounds.

6.2.1 Application of Fluoro-Benziodoxole for Fluorination of Alkenes

Fluoro-benziodoxole **1** is particularly useful for fluorination of alkenes (Figure 6.2a–c) [15, 24, 25, 30–33]. Reagent **1** was reported first in 2012 by Legault and Prévost [34]. Subsequently, a modified procedure for the synthesis and purification of **1** was reported by the groups of Togni [22] and Stuart [23]. This reagent is an air-stable, odorless crystalline substance, which is easy to handle. In most fluorination reactions **1** has to be activated by metal, or Lewis or Brønsted acid catalysts.

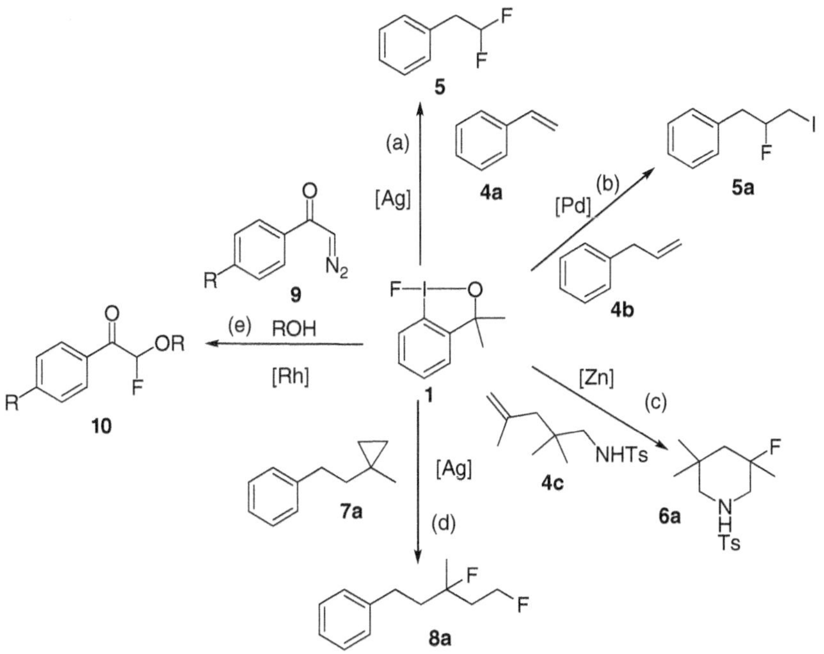

Figure 6.2 Reactions with fluoro-benziodoxole reviewed in this chapter. (a) Difluorination of styrene derivatives (Section 6.2.1.1) [24, 25]. (b) Palladium-catalyzed iodofluorination of alkenes (Section 6.2.1.2) [26]. (c) Fluorocyclization (Section 6.2.1.3) [27]. (d) 1,3-Difluorination (Section 6.2.2) [28]. (e) Rhodium-catalyzed difunctionalization of diazocarbonyl compounds [29].

6.2.1.1 Geminal Difluorination of Styrene Derivatives

The first fluorination reaction of alkenes by fluoro-benziodoxole **1** was reported by the Szabó group (Figure 6.2a) [24]. In these reactions styrenes (such as **4d,e**) were used as substrates with reagent **1** and AgBF$_4$ (Figure 6.3). The reaction proceeded with difluorination of the terminal position of styrenes affording difluoromethyl derivatives, such as **5c,d**. Interestingly, one of the fluorine atoms arose from fluoro-benziodoxole **1** and the other one from the BF$_4$ counterion of AgBF$_4$. Thus, formally an F$_2$ molecule was introduced to styrene under mild neutral reaction conditions.

When methyl-styrenes (such as **4e**) were used as substrates, the methyl group underwent 1,2-migration from the benzylic to the terminal position of the substrate (Figure 6.3b). Further interesting mechanistic features were revealed using di-deuterated styrene derivative **4f-d$_2$** (Figure 6.4). It was found that both deuterium atoms are dislocated from the terminal (**4f-d$_2$**) to the benzylic position (**5e-d$_2$**) under the difluorination process. However, despite a formal migration of two deuterium atoms, the deuterium isotope effect in the reaction was insignificantly small.

The above interesting rearrangement processes in the fluorination reaction are rationalized by the intermediacy of a phenonium intermediate in the reaction (Figure 6.5). Accordingly, AgBF$_4$ activated **1** by breaking of the I—O bond

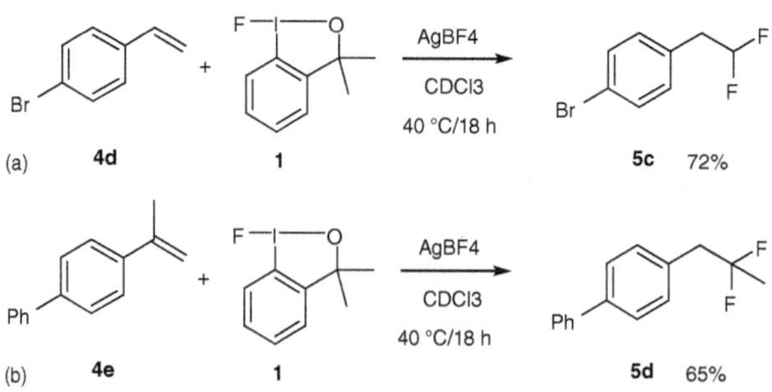

Figure 6.3 Examples for geminal difluorination of styrene (a) and methyl styrene (b) with **1** and AgBF$_4$. Source: Based on Ilchenko et al. [24].

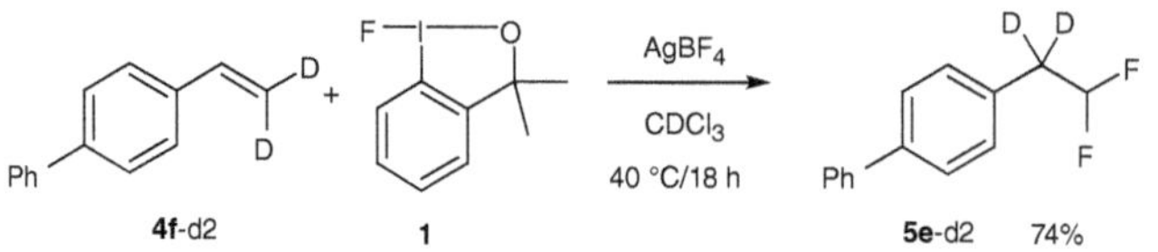

Figure 6.4 Mechanistic study with di-deuterated styrene derivative.

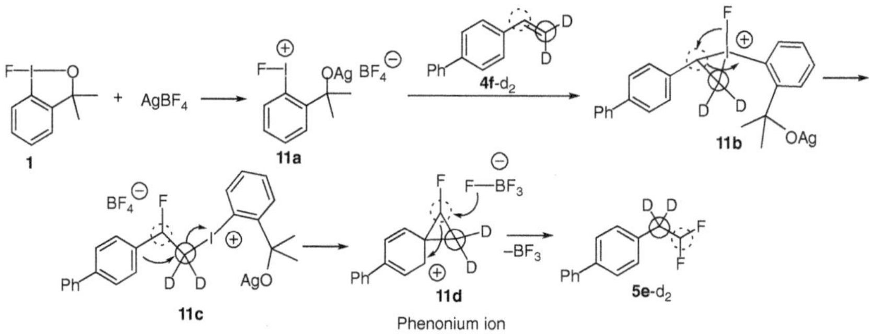

Figure 6.5 Proposed mechanism of the geminal difluorination of styrene derivatives.

to generate a strongly electrophilic I–F reagent **11a**. This reagent undergoes electrophilic addition to the double bond of **4f-d$_2$** affording **11c** via hypervalent iodine **11b**. Donation of a π-electron pair from the aromatic ring leads to the formation of phenonium ion **11d**, which is allowed by the superb leaving group ability of the iodo-arene moiety in **11c**. Phenonium ion **11d** can be opened by attack in two different positions by nucleophiles. In the fluorination by BF$_4^-$ the fluorinated carbon is attacked with high site selectivity, probably since it is more electron deficient than the other one, the deuterated/protonated carbon atom. This type of opening of the cyclopropane ring leads to formation of **5e-d$_2$** product. Thus, the dislocation of the two deuterium atoms of **4f-d$_2$** takes place without C—D bond cleavage. Instead, the Cα and Cβ carbon atoms of **4f-d$_2$** change place via phenonium ion intermediate **11d** affording **5e-d$_2$** product. DFT modeling studies of this reaction reported by Xia-Song Xue and coworkers [35] verified a phenonium ion mechanism.

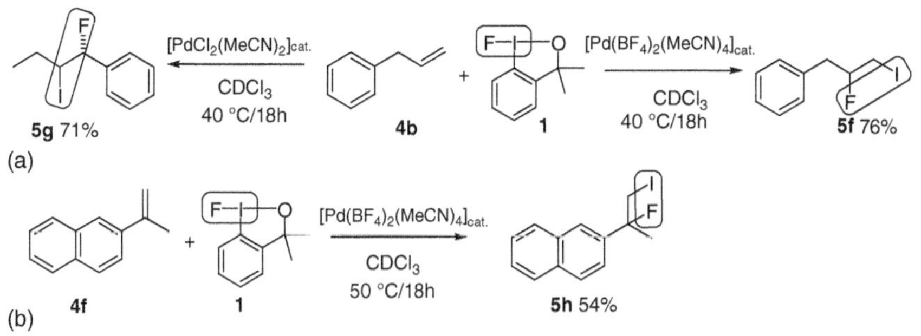

Figure 6.6 Palladium-catalyzed iodofluorination of alkenes with fluoro-benziodoxole **1**.

6.2.1.2 Iodofluorination of Alkenes

As mentioned above, styrenes and **1** undergo a geminal difluorination reaction in the presence of AgBF$_4$. Several other metal tetrafluoroborates can be applied to perform a similar process [24]. However, when the reaction is performed by catalytic amounts of Pd(BF$_4$)$_2$(MeCN)$_4$, formation of geminal difluorinated products (such as **5a–c**) was not observed [26]. Instead of difluorination, selective iodofluorination of the alkene substrate occurred (Figure 6.6). The site selectivity of the reaction was dependent on the applied palladium catalyst (Figure 6.6a). In the case of catalytic amounts of Pd(BF$_4$)$_2$(MeCN)$_4$, the fluoroiodination occurred at the original position of the carbon–carbon double bond. Both the fluorine and iodine atoms arose from fluoro-benziodoxole **1**. When PdCl$_2$(MeCN)$_2$ was used as catalyst different carbon atoms were iodofluorinated (Figure 6.6a). A possible explanation is a double bond migration prior to the iodofluorination using PdCl$_2$(MeCN)$_2$ as catalyst.

When methyl-styrene derivative **4f** was used as substrate the reaction proceeded without rearrangement of the methyl group (c.f. Figures 6.3b and 6.6b), indicating that the reaction of the alkenes with **1** occurs by different mechanisms by using AgBF$_4$ or Pd(BF$_4$)$_2$(MeCN)$_4$ catalysts. The exact mechanism of this interesting reaction is not known. However, a fluoro-iodoalkene intermediate, such as **11c**, is probably involved in this reaction. In the case of palladium catalysis, a cleavage of the aromatic carbon–iodine bond of **11c** occurs, leading to the formation of an iodofluorinated product.

6.2.1.3 Fluorocyclization with C−N, C−O, and C−C Bond Formation

Another class of reactions of alkenes with fluoro-benziodoxole **1** involves fluorocyclization processes. In these bifunctionalization reactions **1** is the source of an electrophilic fluorine substituent followed by an internal attack by a nitrogen, oxygen, or carbon nucleophile (Figure 6.7) [27]. In most of the studies, α-disubstituted alkenes were used as substrates, which gave cyclic quaternary fluorides [3] under mild conditions in the presence of zinc or copper catalysts. The aminocyclization reactions took place with high stereoselectivity (Figure 6.7a,b) affording fluoro-piperidines, such as **6b,c** in good yield. The related oxyfluorination reactions also proceeded efficiently, leading to fluoro-tetrahydropyrans and oxepanes (such as **12**). Interestingly, even carbocyclization could be carried out using **1** as fluorine source. In these

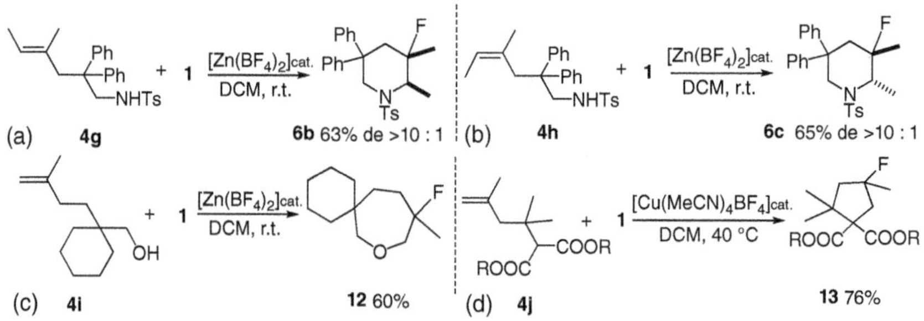

Figure 6.7 Examples for fluorocyclization of alkenes using fluoro-benziodoxole **1**.

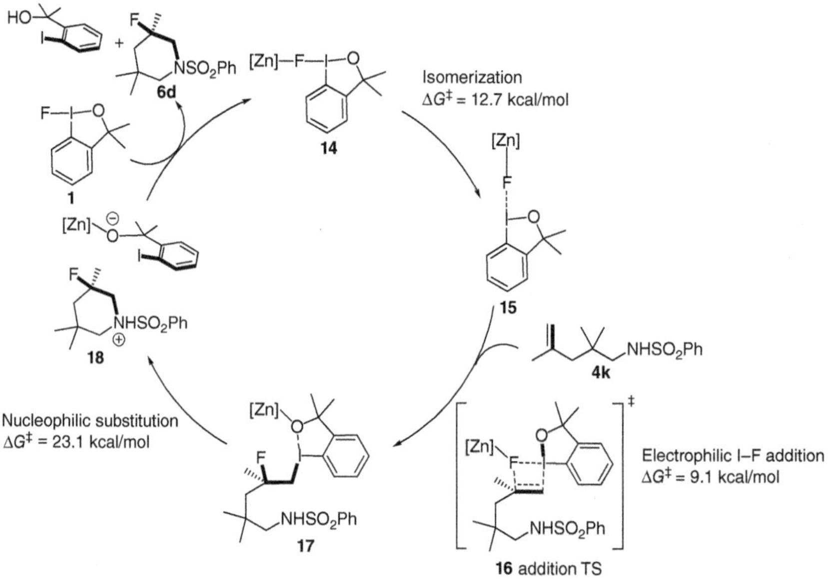

Figure 6.8 Mechanism of the aminofluorination reaction based on DFT modeling studies.

reactions, copper catalysis was used with slightly elevated temperature to furnish fluoro-cyclopentane derivatives, such as **13**.

The reaction mechanism of the aminofluorination reaction was studied by DFT modeling (Figure 6.8) [36]. These studies indicated that the zinc catalyst coordinated more preferentially to the fluorine than to the oxygen atom of **1**. The electrophilic reactivity of **1** could be increased by isomerization of the zinc coordinated complex **14** to **15**. Activation is also possible without the assistance of zinc but the activation energy is much higher. The activated reagent **15** undergoes electrophilic addition with disubstituted alkene **4k** affording **17** with a quaternary C–F center. The reaction proceeds via a four-membered ring TS **16** with surprisingly low activation barrier. In intermediate **17** the carbon–hypervalent iodine bond is very weak; thus aryl iodine is a very good leaving group. The internal attack by nitrogen gives **18**, which after deprotonation provides the fluoro piperidine product **6d** (Figure 6.8).

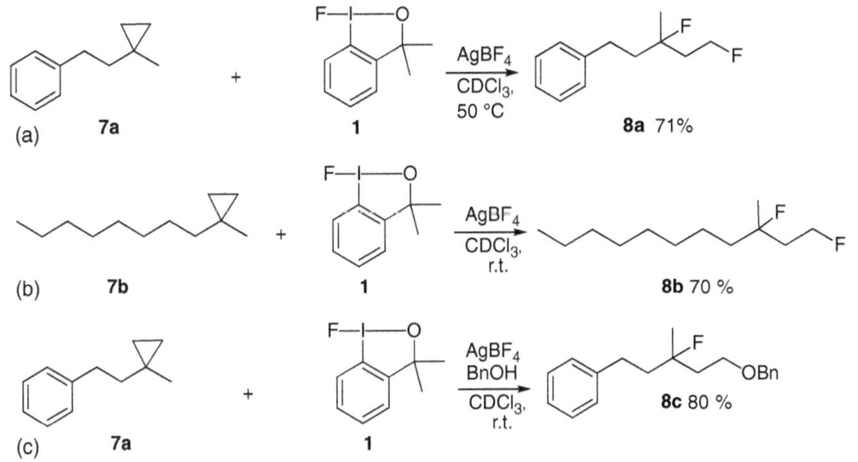

Figure 6.9 Opening of cyclopropane for 1,3-difluorination (a, b) and 1,3-oxyfluorination (c).

Other fluorocyclization reactions using **1** were also reported by the groups of Stuart [37], Gulder [30], and Lu [38].

6.2.2 Fluorinative Cyclopropane Opening

The above reactions based on electrophilic additions of fluoro-benziodoxole (**1**) to alkenes gave the idea to attempt a selective electrophilic ring opening of cyclopropane derivatives (Figure 6.9). This reaction could be used for extension of the fluorination-based geminal (Section 6.2.1.1) and vicinal difunctionalization reactions (Sections 6.2.1.2 and 6.2.1.3) to 1,3-difunctionalization [28]. Similarly to styrene derivatives (Section 6.2.1.1), disubstituted cyclopropane derivatives, such as **7a,b**, underwent 1,3-difluorination reactions using stoichiometric amounts of $AgBF_4$ (Figure 6.9a,b). Using excess of benzyl alcohol, the reaction led to the formation of 1,3-oxyfluorinated product **8c**.

According to the plausible mechanism, the reaction starts with activation of **1** (see also Figure 6.5) to give **11a** (Figure 6.10) [28]. Species **11a** breaks one of the carbon–carbon bonds of cyclopropane (such as **7a,b**) to give fluorinated carbocation **20**, which subsequently undergoes nucleophilic fluorination by the fluoride from the BF_4^- counterion of silver.

6.2.3 Fluorine-18 Labeling with Fluorobenziodoxole

As mentioned above, fluorine-18-labeled organic substances are employed as tracers in PET [8, 39, 40]. PET is a noninvasive, very efficient medical diagnostic method for *in vivo* imaging of biochemical processes. As fluorine-18 is a positron emitting nucleus, labeled organo-fluorine-18 substances are suitable tracers for this diagnostic method. In principle, all different types of PET imaging studies require a specific fluorine-18-labeled tracer. Accordingly, there is a large demand for adaptation of the modern methodology in organofluorine chemistry to produce a very

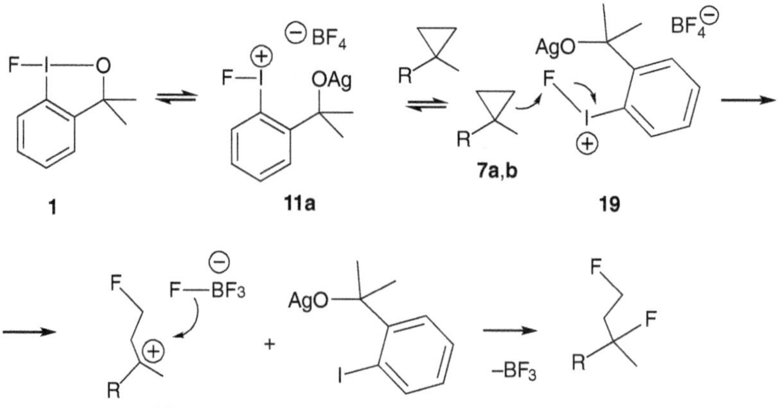

Figure 6.10 Plausible mechanism for ring opening of cyclopropanes with fluoro-benziodoxole (**1**).

broad variety of fluorine-18-labeled compounds. Selective fluorine-18 labeling is very challenging synthetically due to the relatively short half-life of the fluorine-18 (109.8 minutes) isotope and the extremely small amounts of fluorine-18 precursors generated by on-site hospital cyclotrons. These cyclotrons can relatively easily generate fluorine-18-labeled HF, which can be converted to nucleophilic fluorinating reagents. However, the synthesis of electrophilic fluorinating reagents, which are often more reactive than the nucleophilic ones, is complicated by changing the reactivity of the nucleophilic fluoride ion by the so called "umpolung" process [41–43]. As mentioned above electrophilic fluorinating reagent fluoro-benziodoxole **1** can be easily generated from fluoride ion sources, while other reagents, such as NFSI (**3a**) or Selectfluor, are synthetized from fluorine gas. Accordingly, [^{18}F]**1** has been considered as a promising reagent for electrophilic fluorine-18 labeling of PET tracers.

It was shown [44] that synthesis of [^{18}F]**1** can be accomplished from tosyl-benziodoxole **22** using fluorine-18-labeled TBAF, which is easily obtained from cyclotron-generated [^{18}F]HF (Figure 6.11). Fluoro-benziodoxole **1** has a very low polarity and therefore it is sparingly soluble in hexane. This fortuitous property could be exploited for the purification of [^{18}F]**1**, as the other components of the reaction (Figure 6.11), such as **22** and the rest of the products of TBAF, are insoluble in hexane. Thus, extraction of the reaction mixture with warm hexane led to purified [^{18}F]**1**, which could be used for fluorine-18 labeling studies [44, 45] (see also Section 6.3.2).

The reaction was employed for fluorocyclization of *o*-styrylamides [30] (such as **23a,b**) to obtain fluorine-18-labeled fluoro-benzoxazepines bearing a quaternary fluoride [44] (Figure 6.12). The reaction temperatures and times were optimized to give high radiochemical yields (RCYs) up to 74%. The RCY [46] is the amount of activity in the product expressed as a percentage of the activity used in a process (previously this parameter was often referred to as radiochemical conversion, RCC). In a

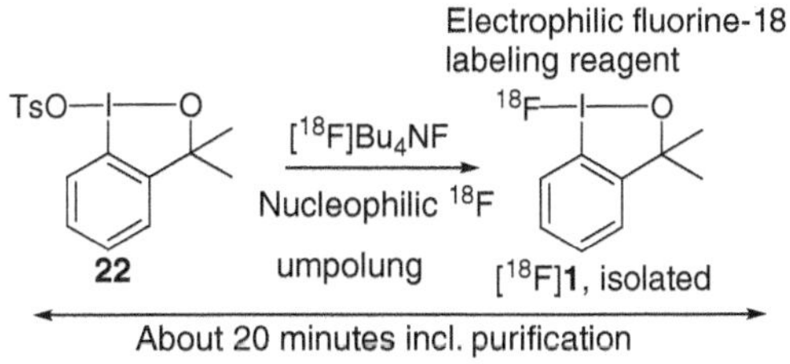

Figure 6.11 Synthesis of fluorine-18-labeled fluoro-benzodioxole.

Electrophilic fluorine-18 labeling reagent

About 20 minutes incl. purification

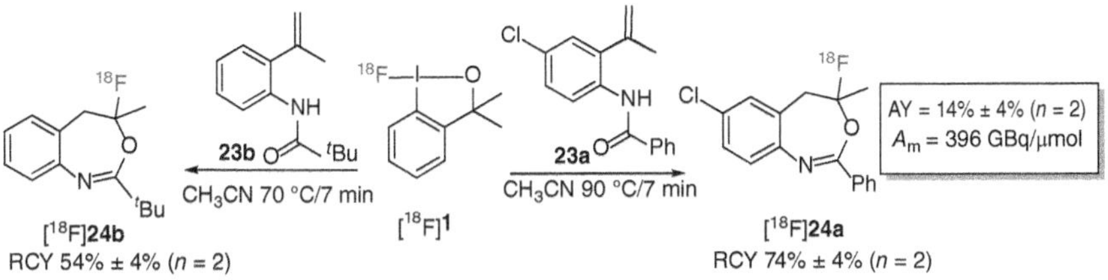

Figure 6.12 Examples for fluorocyclization with fluorine-18-labeled **1**.

separate, fairly large-scale experiment fluoro-benzoxazepine product [^{18}F]**24a** was isolated with an activity yield (AY) of 14% (Figure 6.12). The obtained sample showed an excellent level of molar activity (A_m) of 396 GBq/µmol. Molar activity (A_m) refers to the measured amount of activity per mole of a compound [46]. A high fluorine-18 molar activity indicates that the contamination of a labeled product with the natural isotope is relatively low. A high molar activity value is very important for PET saturation studies, in which the *in vivo* uptake of fluorinated drug or other bioactive substances can be measured.

6.3 Fluorination-Based Bifunctionalization of Diazocarbonyl Compounds

Diazo compounds are excellent reagents for the introduction of new functionalities [47–54] to organic molecules. An interesting approach for geminal difunctionalization-based fluorination [19, 55–59] involves the use of diazocarbonyl compounds.

6.3.1 Rhodium-Catalyzed Geminal Oxyfluorination Reactions

In difunctionalization reactions often a nucleophile and electrophile are used to introduce two functional groups at the same carbon by replacing the diazo group of diazo carbonyl compounds, which are regarded as safe, stabilized diazo reagents. We devised a rhodium (**25a**)-catalyzed process for geminal oxyfluorination of diazoketones (such as **9a–c**) using alcohols, such as **26a** (Figure 6.13a), phenols, such as **26b** (Figure 6.13b), and carboxylic acids, such as **27** (Figure 6.13c), as nucleophiles in combination with fluoro-benzodioxole (**1**) as electrophilic fluorine source [29].

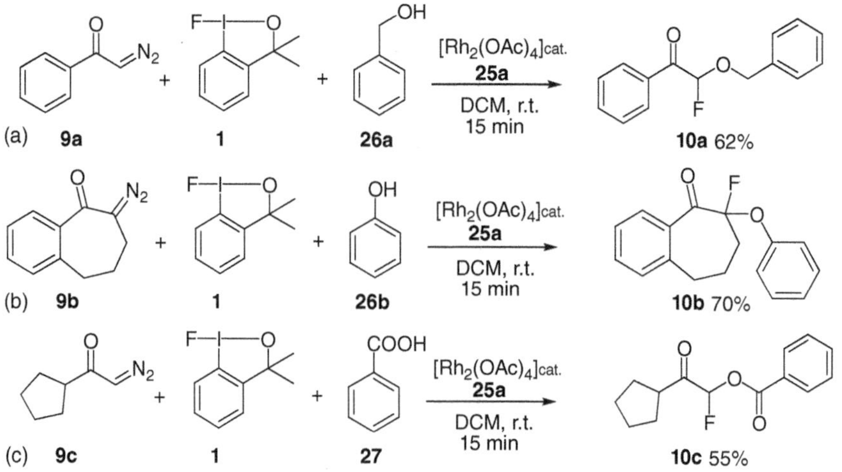

Figure 6.13 Examples for rhodium-catalyzed geminal oxyfluorination of diazoketones.

These multicomponent bifunctionalization reactions can be carried out rapidly at room temperature. The reaction has a very broad synthetic scope and it is suitable for one-step synthesis of α-fluorinated ethers (such as **10a–c**), which is an important class of bioactive motives [7]. The mechanism of the reaction was studied [60] by DFT modeling (Figure 6.14). The active catalyst (**28**) is formed from Rh$_2$(OAc)$_4$ (**25a**) by coordination of two fluoro-benziodoxole (**1**) molecule. The first step is the formation of rhodium-carbene **30** from intermediate **29** via an activation barrier of 22.6 kcal/mol. Rhodium carbene **30** undergoes a fast reaction (8.1 kcal/mol) with nucleophiles such as benzyl alcohol **26a** to give oxonium ylide **31**. Proton transfer leads to formation of a rhodium-stabilized enolate **32**. This enolate intermediate (**32**) is the key intermediate of the reaction. Previous mechanistic studies indicated (see Figure 6.8) that **1** smoothly undergoes electrophilic addition to carbon–carbon double bonds. The TS of the electrophilic addition (**33**) occurs with an activation energy of 20.3 kcal/mol to give intermediate **34**. After a couple of low-energy isomerization reactions via intermediates **35** and **36**, the fluorine is transferred from the hypervalent iodine to the carbon affording intermediate **37** via a low barrier of 9.8 kcal/mol. Dissociation of the rhodium complex leads to formation of the α-fluorinated ether product **10a** and regeneration of the catalyst. In the reaction, the rhodium catalyst has a dual role, carbene (**30**) formation and stabilization of the enolate intermediate **32**. Interestingly, rhodium does not undergo redox reactions, and thus the catalytic process is redox neutral.

6.3.2 [¹⁸F]Fluorobenziodoxole for Synthesis of α-Fluoro Ethers

As mentioned in Section 6.2.3 one of the most important challenges in synthetic organofluorine chemistry is adaptation [61], the so-called translation, of modern synthetic methodology to fluorine-18 labeling to develop new routes to PET imaging ligands. Considering this, we developed a new fluorine-18 labeling method by

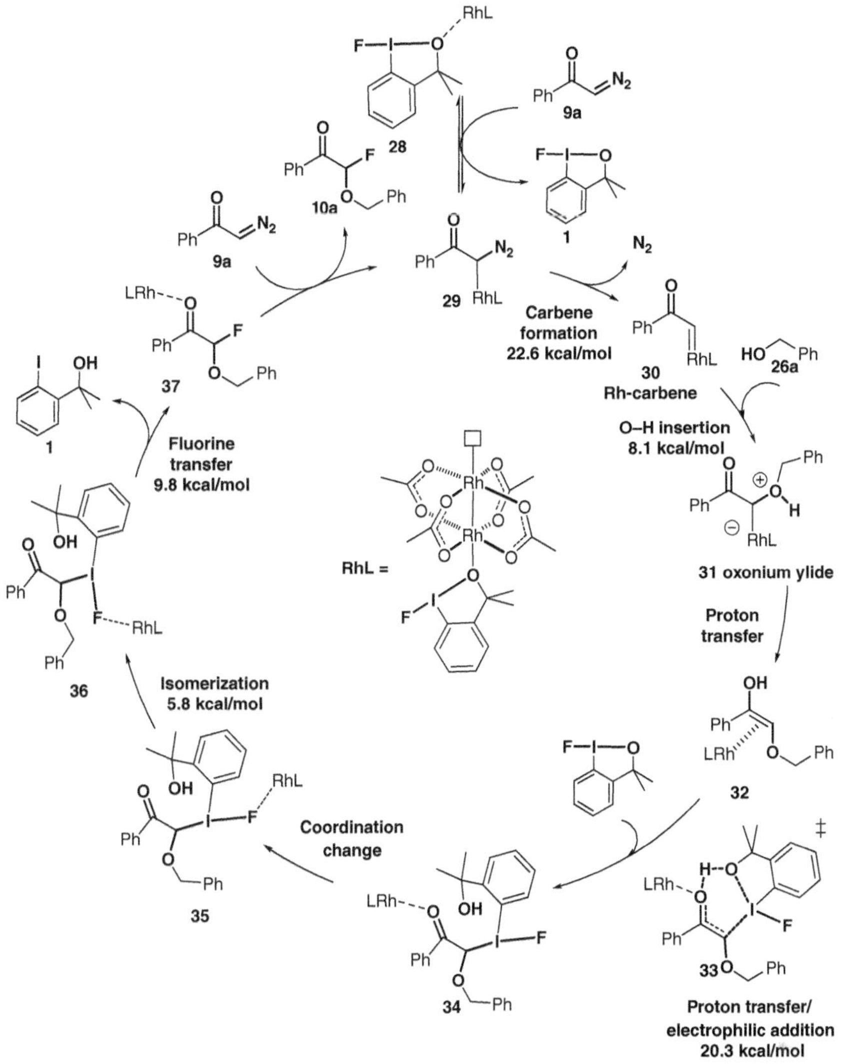

Figure 6.14 Mechanism of the rhodium-catalyzed oxyfluorination of diazocarbonyl compounds. The catalytic cycle is exemplified with the reaction of **9a**, **26a**, and **1** (see Figure 6.13a).

adapting the above (Section 6.3.1) rhodium-catalyzed oxyfluorination of diazocarbonyl compounds [45].

When the multicomponent oxyfluorination reaction was attempted (Figure 6.15) using fluorine-18-labeled fluoro-benziodoxole [^{18}F]**1** (Figure 6.11) and benzyl alcohol (**26a**) as nucleophile, the outcome of the reaction was different from the process with the natural fluorine isotope (Figure 6.13a). Using minute amounts of [^{18}F]**1**, formation of the oxyfluorinated product [^{18}F]**10a** was not observed. As mentioned above, the hospital cyclotrons produce very limited amounts of fluorine-18 isotopes; therefore only very small amounts of fluorine-18 reagents such

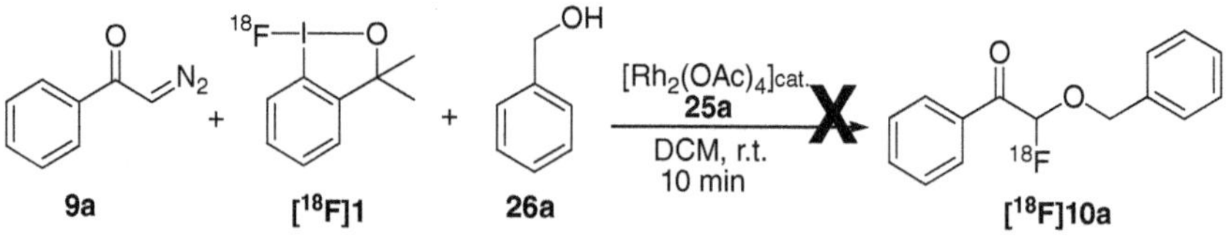

Figure 6.15 Attempted multicomponent fluorine-18 labeling of diazoketone **25a** with benzyl alcohol (c.f. Figure 6.13a).

as [^{18}F]**1** can be applied, which leads to challenging downscaling problems. In the above reaction an important downscaling problem is that stoichiometric amounts of benzylic alcohol **26a** solvolyze minute amounts of [^{18}F]**1** before the fluorine transfer to **9a** occurs. Such a solvolysis reaction is not significantly important, when stoichiometric amounts of **1**, **9a**, and benzylic alcohol **26a** are employed (Figure 6.13a) but prevent the formation of [^{18}F]**10a** under the reaction conditions of fluorine-18 labeling.

A possible solution for the above downsizing issue is application of trimethyl orthoformate **37** as oxygen nucleophile instead of alcohols (Figure 6.16a–c). Trimethyl orthoformate **37** is a much weaker nucleophile than alcohols, such as benzylic alcohol **26a** or MeOH, and therefore it does not solvolyze the small amounts of [^{18}F]**1** before the electrophilic fluorine-18 transfer happens [45]. The weak nucleophilicity of **37** may lead to very slow reaction with rhodium carbenes (such as **30**); therefore, **37** was employed as solvent in the oxyfluorination reaction. As Rh$_2$(OAc)$_4$ **25a** has a limited solubility in neat **37**, we replaced it with Rh$_2$(OPiv)$_4$ **25b**. Using these conditions, aromatic diazoketones such as **9a,d** could be converted to the corresponding fluorine-18 α-fluoro ethers, such as [^{18}F]**10d,e**, with a high RCY (Figure 6.16a,b). Even diazo amide **9e** could be used as substrate albeit the oxyfluorinated product [^{18}F]**10f** formed with poor RCY (Figure 6.16c). Product [^{18}F]**10e** is a particularly interesting synthon, as the aromatic bromide functionality can be used in subsequent Suzuki–Miyaura coupling as prosthetic group [9, 39]. After isolation by semi-preparative HPLC we obtained a pure sample of [^{18}F]**10e** with 9% AY, which showed a molar activity (A_m) of 216 GBq/μmol, determined 110 minutes after the end of the bombardment. This high A_m value confirmed our previous observation (Figure 6.12) [44] that application of [^{18}F]**1** leads to high fluorine-18 isotope purity (expressed by A_m), which is a very useful property of fluorine-18-labeled PET tracers.

6.4 Trifluoromethylation of Alkenes, Alkynes, and Diazocarbonyl Compounds with the Togni Reagent

As mentioned in the introduction we were interested in comparing the synthetic features of the benziodoxole based fluorine (**1**) and trifluoromethyl (**2**) transfer reagents. There are two different types of hypervalent iodide reagents, which are widely used in organic synthesis, benziodoxolone-based reagent **2a** and the benziodoxole-based

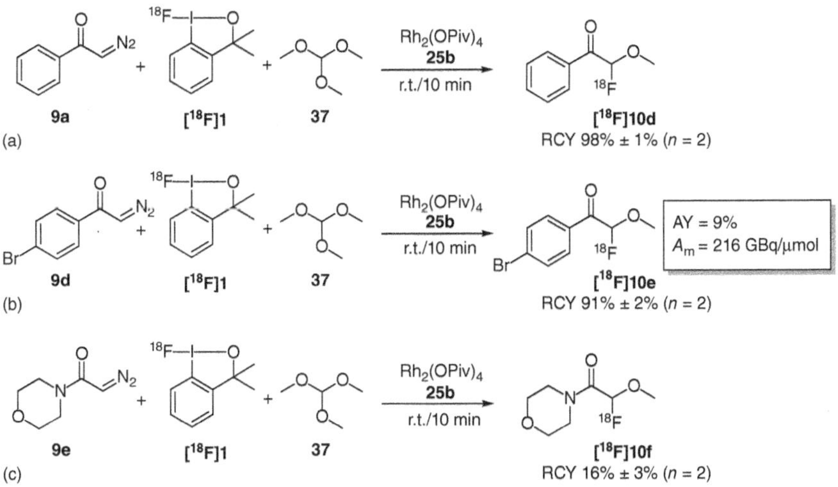

Figure 6.16 Fluorine-18 oxyfluorination of diazocarbonyl compounds.

reagent **2b**. These reagents are also referred to as the Togni I and II reagents [11]. Our studies indicated a higher reactivity for **2a**, and therefore most of the trifluoromethylation reactions were performed with this reagent (Figure 6.17).

6.4.1 Bifunctionalization of C—C Multiple Bonds

Bifunctionalization-based methods are very efficient for obtaining trifluoromethyl group functionalized products. The CF_3 group is one of the chemically most inert functional groups. The high oxidation stability makes the CF_3 group very stable in metabolic processes, which is one of the most important reasons for application of trifluoromethylated compounds in drugs and agrochemical products. In a trifluoromethylation-based bifunctionalization the introduction of the inert CF_3 group is usually accompanied by introduction of another reactive functional group (handle), most frequently with a new C—O or C—N bond formation.

6.4.1.1 Oxytrifluoromethylation of Alkenes and Alkynes

Oxytrifluoromethylation of alkenes is one of the most widely applied methods for formation of a C—CF_3 bond by reagent **2a** (e.g. Figure 6.17) and other CF_3 transfer reagents [11, 65]. The Sodeoka group [66] and our group [62] published the first studies on oxytrifluoromethylation of alkenes and alkynes.

We have found [62] that both styrene derivatives (such as **4k**) and non-aromatic alkenes (such as **4j**) underwent oxytrifluoromethylation with **2a** to give bifunctionalized products, such as **38a,b** (Figure 6.18a,b). The trifluoromethyl group was introduced at the terminal position of the double bond, while the iodobenzoyl group (arising from reagent **2a**) was introduced at the vicinal position. The iodobenzoyl group of **38a** could be easily hydrolyzed by K_2CO_3 in MeOH to obtain the alcohol derivative. It is important to point out that the outcome of this reaction is sharply different from the analogous fluorination process (see Figure 6.3 in Section 6.2.1.1).

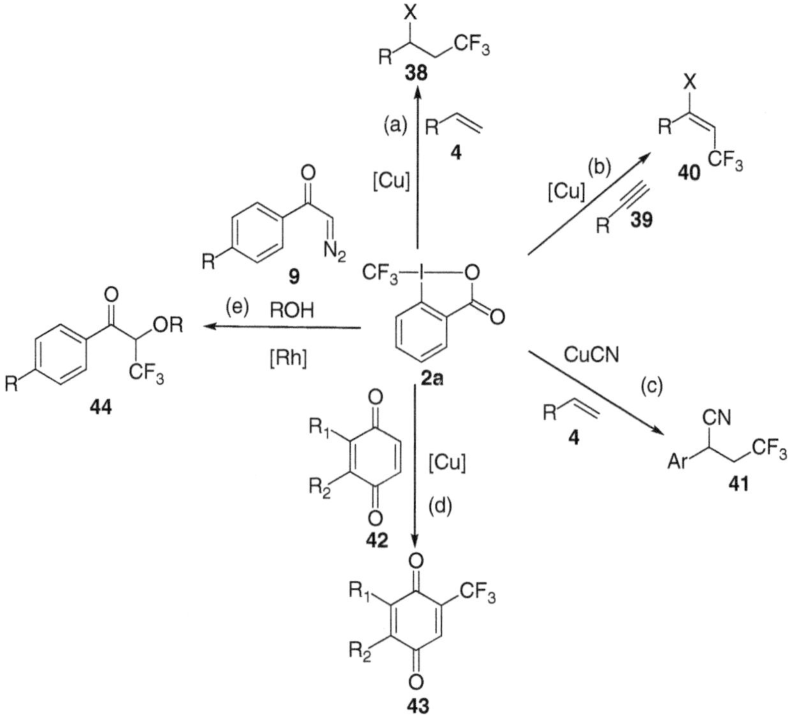

Figure 6.17 Trifluoromethylation with **2a**. (a, b) Oxytrifluoromethylation of alkenes and alkynes (Section 6.4.1.1). (c) Cyanotrifluoromethylation of alkenes (Section 6.4.1.2). (d) Direct trifluoromethylation of quinones (Section 6.4.1.3). (e) Oxytrifluoromethylation of diazocarbonyl compounds (Section 6.4.2). Source: (a, b) Janson et al. [62]; (c) Ilchenko et al. [63]; (d) Ilchenko et al. [64]; (e) Yuan et al. [29].

In case of using fluoro benziodoxole (**1**) with a styrene substrate in the presence of AgBF$_4$, a geminal difluorination process occurs. Apparently, hypervalent iodines **1** and **2a** react with styrenes by substantially different reaction mechanisms. Mono-substituted alkynes react also with **2a** in the presence of CuI as catalyst (Figure 6.18c) to give trifluoromethylated alkenes, such as **40a**. In the products the CF$_3$ and the iodo-benzoyl groups are in *trans* geometry, indicating that the insertion reaction did not take place by a concerted (*cis*) mechanism. The exact mechanism of this reaction is still unclear [67]. However, transfer of a radical CF$_3$ group is a possible mechanistic option [67].

6.4.1.2 Cyanotrifluoromethylation of Styrenes

The above example for oxy-trifluoromethylation of styrenes with **2a** resulted in iodo-benzoyl substituted products. In these reactions the iodo-benzoyl group arose from **2a**. We wished to develop a bifunctionalization reaction, in which another type of functionality can be introduced together with the CF$_3$ group. It was found [63] that application of stoichiometric amounts of CuCN leads to cyano-trifluoromethylation of styrenes (such as **4k–m**) using benziodoxolon

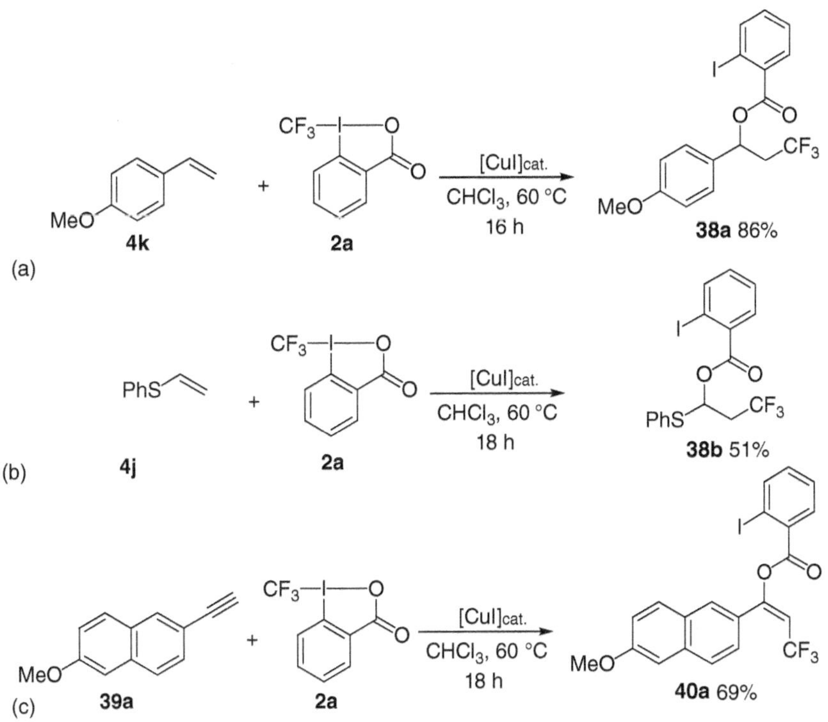

(a)

(b)

(c)

Figure 6.18 Examples for copper-catalyzed oxy-trifluoromethylation of alkenes and alkynes.

reagent **2a** (Figure 6.19). In order to get acceptable yields PCy_3 had to be added in catalytic amounts. Interestingly, application of bis(pinacolato)diboron (B_2Pin_2), which is a widely used borylation reagent, also showed a significant activating effect in the reaction. The activating effect of PCy_3 and B_2Pin_2 indicates that electron donating ligands (i.e. PCy_3 or Bpin) on copper may favorably influence the bifunctionalization reaction [67]. An important synthetic feature of the cyano-trifluoromethylation reaction is the introduction of two carbon–carbon bonds in a single reaction step. The reaction gave acceptable results for styrenes with both electron-donating (**4k,l**) and electron-withdrawing (**4m**) substituents to give the corresponding cyano-trifluoromethylated products, such as **41a–c**. Subsequently, excellent studies were published for cyano-trifluoromethylation of alkenes, in which the copper-catalyzed reactions were combined using TMSCN as a source of the cyano group [68].

6.4.1.3 C–H Trifluoromethylation of Benzoquinone Derivatives

As mentioned above, the oxy-trifluoromethylation and the analogous cyano-trifluoromethylation reactions probably involve a radical mechanism for the transfer of the CF_3 group. We hypothesized that formation of a stabilized radical intermediate can be the driving force of the copper-catalyzed trifluoromethylation

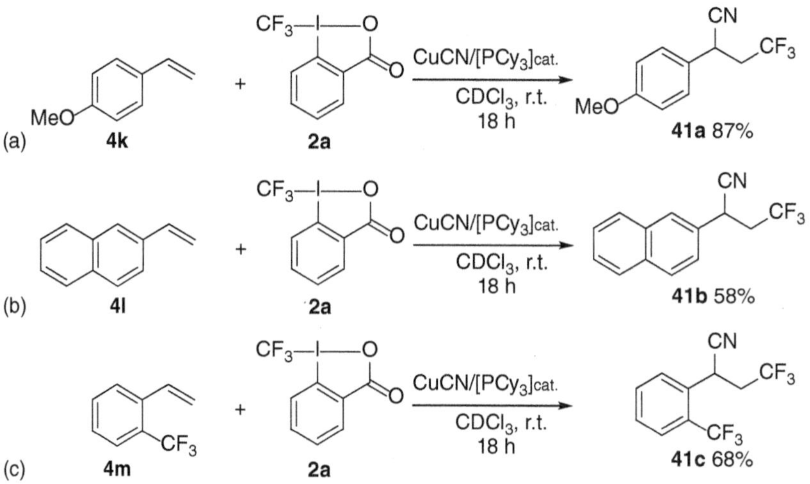

Figure 6.19 Examples for cyano-trifluoromethylation of styrenes.

reactions using **2a**. A possible proof of this concept could be trifluoromethylation of benzoquinone derivatives, which proceeds via radical quinoid intermediates. Indeed, trifluoromethylation of benzoquinone and its derivatives (**42a–c**) could be carried out [64] with **2a** in the presence of CuCN and catalytic amounts of B_2Pin_2 (Figure 6.20). The benzophenone derivatives are important bioactive compounds occurring in ubiquinone and in vitamin K. The CF_3 group may lend high metabolic stability for the corresponding drug substances. For example, trifluoro-menadione (**43b**) is a core motif in malaria drugs [69]. Davioud-Charvet and coworkers [69] described a four-step procedure for synthesis of **43b** from naphthoquinone **42b**. As shown in Figure 6.20b, this reaction can be performed in a single step using **2a** as a trifluoromethyl source. Compound **43c** is the trifluoromethyl analog of an apoptosis (programmed cell death) initiator, which could also be prepared in a single step from quinone **42c** (Figure 6.20c).

A plausible mechanism for the trifluoromethylation of benzoquinone is presented in Figure 6.21. Accordingly, the first step is supposed to be coordination of a Bpin ligand to the copper catalyst [70]. Bpin is one of the strongest σ-donor ligands [71], and therefore it can activate oxidative addition of **2a** with Cu(I) to give **44**. The Cu—CF_3 bond of **44** may undergo homolytic cleavage by transferring the CF_3 radical to **42a** to give semiquinone derivative **45**. A possible driving force of this homolytic cleavage is the formation of a Cu(II) intermediate and **45**, in which the unpaired electron can be delocalized on the quinoid system. Recombination of **45** and the (open-shell) Cu(II) intermediate leads to **46**, which after deprotonation provides the final product **43a**. The key step and a possible driving force of the reaction is the formation of a stabilized trifluoromethylated radical **45**. Studies on trifluoromethylation of benzoquinones were published by the groups of Wang [72] and Bi [73].

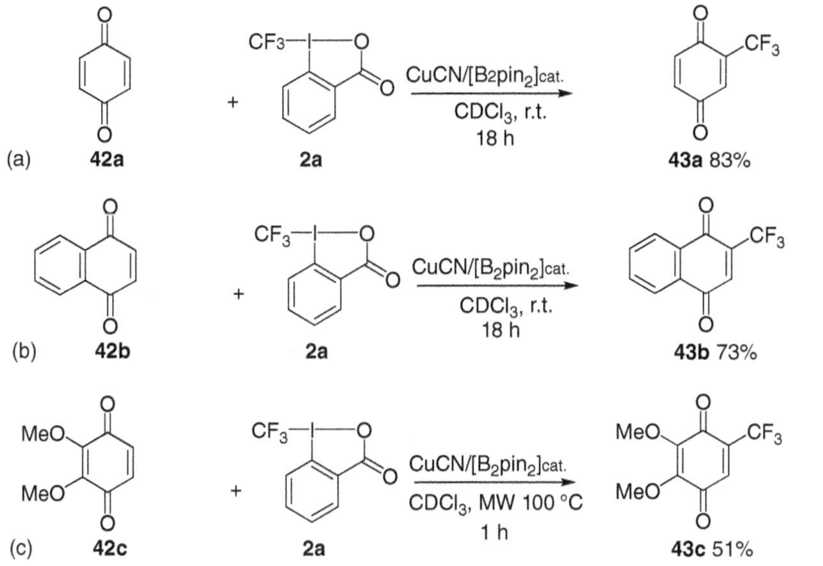

Figure 6.20 Examples for direct trifluoromethylation of benzoquinone derivatives.

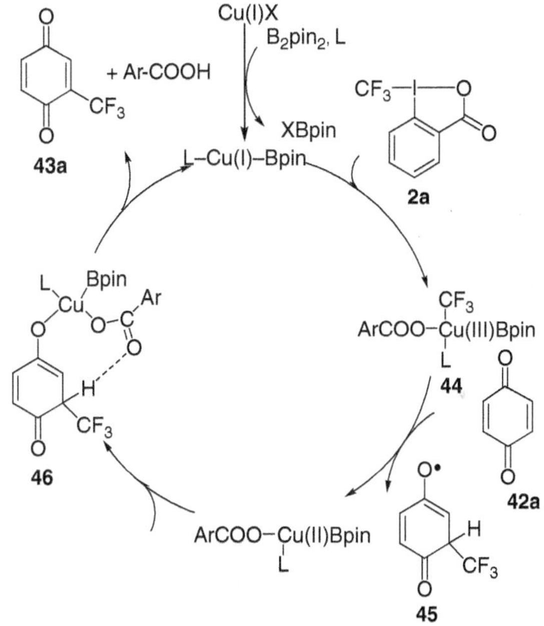

Figure 6.21 Plausible mechanism for the trifluoromethylation of benzoquinone with **2a**.

6.4.2 Geminal Oxytrifluoromethylation of Diazocarbonyl Compounds

Oxytrifluoromethylation of diazoketones (such as **9a,b,d**) can be performed using **2a** and various alcohols (Figure 6.22), such as benzylic alcohol **26a** [29]. The reaction proceeds under the same reaction conditions as the analogous fluorination of diazoketones (Section 6.3.1, Figure 6.13). Mechanistic studies indicate that the

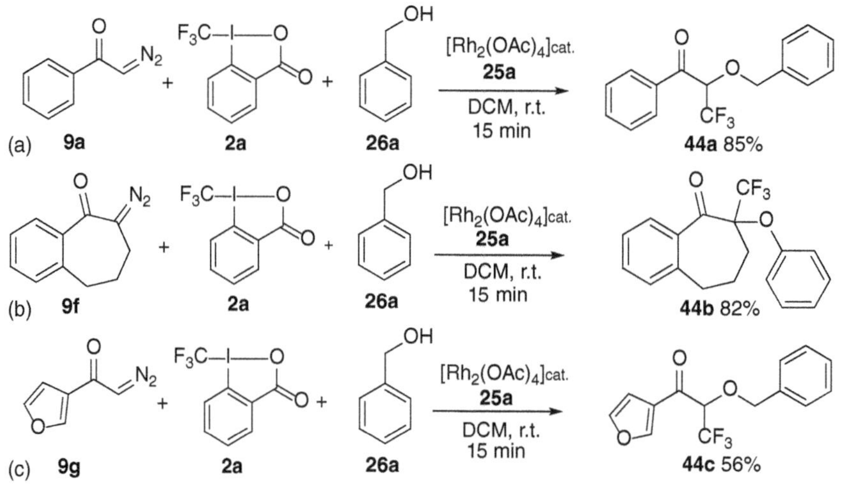

Figure 6.22 Examples for oxy-trifluoromethylation of diazoketones with **2a**.

oxyfluorination reaction with **1** is about twice as fast as the oxytrifluoromethylation with **2a**. DFT modeling studies show that the reaction mechanism of the rhodium-catalyzed oxy-trifluoromethylation (with **2a**) and oxyfluorination (with **1**) reactions proceeds with very similar mechanisms [60].

6.5 Bifunctionalization-Based Trifluoromethylthiolation of Diazocarbonyl Compounds

The above successful reactions for oxyfluorination and oxy-trifluoromethylation reactions encouraged us to develop analogous bifunctionalization reactions based on introduction of the SCF₃ group. The SCF₃ analog of hypervalent iodine reagents **1, 2** proved to be unstable, and thus it is not useful as a reagent for trifluoromethylthiolation reactions [17]. Therefore, we selected dibenzenesulfonimide reagent **3b** as the SCF₃ source, which is also an analog of the popular fluorinating reagent NFSI (**3a**). In fact, NFSI can also be used for oxy-fluorination reactions similarly to **1** [29, 74].

6.5.1 Multicomponent Approach for Geminal Oxy-Trifluormethylthiolation

Oxy-trifluoromethylthiolation of various diazocarbonyl compounds (such as **9h**) could be performed with **3e** and alcohols, such as **26a** (Figure 6.23a) [75]. The reaction was catalyzed by rhodium complex **25a**, and a NaOAc additive suppressed the competing protonation (instead of trifluoromethylthiolation) process. When cyclic ethers such as THF (**47**) were applied instead of alcohols, a trifunctionalization reaction occurred involving opening of the THF ring (Figure 6.23b). This interesting trifunctionalization reaction was also observed when the fluoro analog of **3b**, i.e.

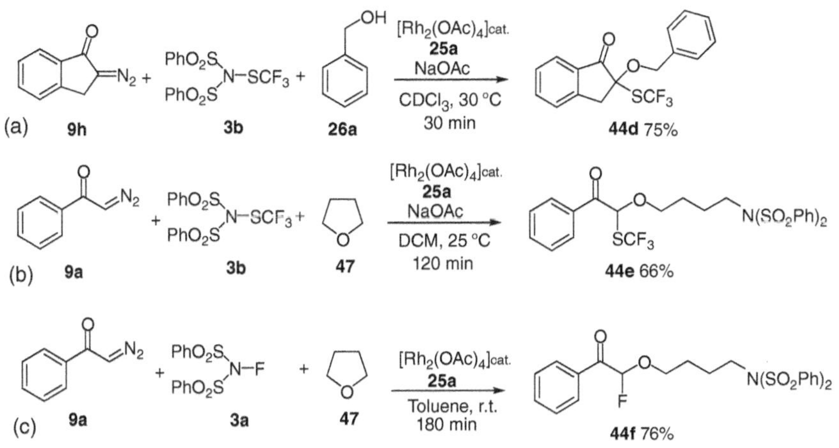

Figure 6.23 Trifunctionalization reactions with **3b** and its analog NFSI (**3a**).

NFSI **3a**, was used under similar conditions (Figure 6.23c) [74]. DFT modeling studies indicated a close analogy between the above (Figure 6.14) multicomponent fluorination and trifluoromethylthiolation reactions (such as Figure 6.23a,b) [76].

6.5.2 Simultaneous Formation of C−C and C−SCF₃ Bonds via Hooz-Type Reaction

In the above reactions (Figure 6.23a) an SCF_3 and new ether functionality are introduced in the diazo carbon of the substrates. We wished to find a new reaction in which the introduction of the SCF_3 functionality is coupled with carbon–carbon (instead of carbon–oxygen) bond formation. After several attempts, it was realized that several components of the oxy-trifluoromethylation reaction (Figure 6.23a) have to be replaced to get a bifunctionalization process based on C—SCF_3 and C—C bond formation. A particular problem in this type of multicomponent reactions is the undesired interaction between the nucleophilic (carbon source) and electrophilic (SCF_3 source) reagents. We have found that the so-called Hooz reaction (Figure 6.24) [77, 78] offers an attractive platform to use **3b** together with carbon nucleophiles to develop a new bifunctionalization process. This reaction is based on application of triaryl boranes, which are reacted with diazo compounds in the presence of an electrophile. The mechanism is based on the formation of a hypervalent boron intermediate with the diazocarbonyl compounds. These intermediates are known to undergo 1,2-migration by breaking of one of the carbon–boron bonds. The intermediate formed in this process rapidly undergoes a 1,3-borotropic rearrangement producing a boron enolate. Our mechanistic studies have shown that these types of enolate intermediates readily react with fluorinated electrophiles (c.f. Figure 6.14). This hints that the boron enolate intermediate (Figure 6.24) would react with SCF_3 transfer reagent **3b** as well.

Indeed, we have found that diazoketones such as **9a** and **9i** reacted with **48a** and **3b** affording trifluoromethylthiolation–phenylation products, such as **49a,b**

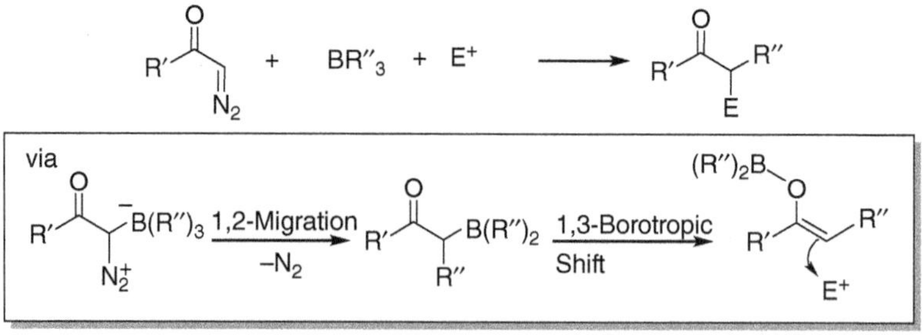

Figure 6.24 The Hooz reaction and its mechanism.

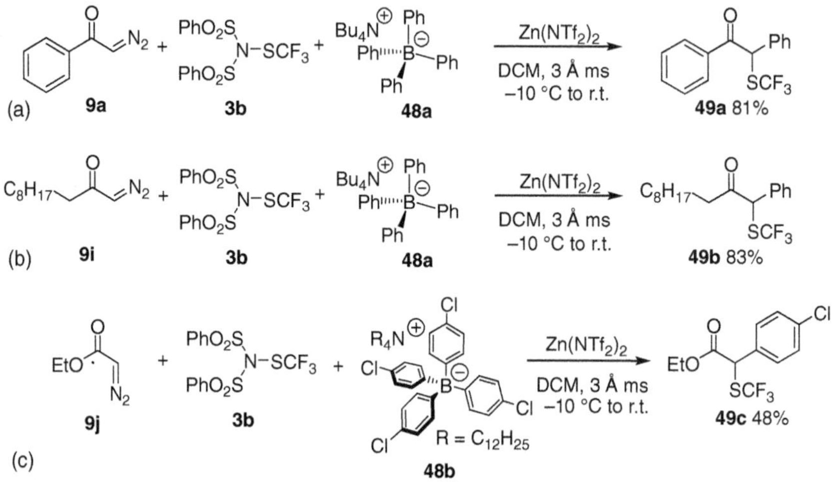

Figure 6.25 Examples for trifluoromethylthiolation-phenylation with **3b**.

(Figure 6.25a,b) [79]. The reactions proceeded with acceptable yields when $Zn(NTf_2)_2$ was used as additive in substoichiometric amounts. Even diazoesters, such as **9j**, or amides could be used in the reactions. Except tetraphenyl borate (**48a**) some other tetra aryl borates (**48b**) could also serve as aryl component for the carbon–carbon bond formation process (Figure 6.25c). The mechanistic studies verified that a Hooz-type reaction takes place in the geminal trifluoromethylthiolation–phenylation reactions.

6.6 Summary

This chapter demonstrates that hypervalent iodine-based fluoro and trifluoromethyl benziodoxole(on) reagents **1, 2** can be used for efficient bifunctionalization of alkene and diazocarbonyl substrates. The corresponding SCF_3 reagent is not stable but dibenzenesulfonimide reagent **3a** can be used for similar type of reactions as the fluoro- (**1**) and trifluoromethyl- (**2**) benziodoxole(on). The modeling studies pointed out the principal similarities between the fluorination (and trifluoromethylation) of alkenes and diazocarbonyl compounds (c.f. Figures 6.8 and 6.14). In both reaction types the key step is the addition of an electrophilic hypervalent iodine reagent to the π-bond of an alkene. In case of the diazocarbonyl substrates the π-system is the Rh-stabilized enolate formed from the Rh-carbene intermediate of the process. The rhodium-catalyzed F-, CF_3-, and SCF_3-based bifunctionalization reactions can be performed under similar conditions and the reactions afford similar type of α-ether products. An important application of the above chemistry involves efficient electrophilic fluorine-18 labeling of alkenes and diazocarbonyl compounds. This reaction could be realized by the very efficient radiosynthesis of fluorine-18-labeled benziodoxole [^{18}F]**1**.

Probably the most important challenge in future is extension of the above reactions to asymmetric fluorination, trifluoromethylation, and trifluoromethylthiolation methods. Another interesting area is the development of new type of bifunctionalization reactions involving C–F/C–CF$_3$/C–SCF$_3$ forming reactions with simultaneous formation of new carbon–carbon bonds. A further challenging area with very high impact on PET imaging-based diagnostics is the development of new efficient labeling reactions based on [^{18}F]**1** and its corresponding analogs for fluorine-18 CF$_3$ and SCF$_3$ labeling.

References

1 Murphy, C.D., Schaffrath, C., and O'Hagan, D. (2003). *Chemosphere* 52: 455–461.

2 Mei, H., Han, J., Fustero, S. et al. (2019). *Chem. Eur. J.* 25: 11797–11819.

3 Zhu, Y., Han, J., Wang, J. et al. (2018). *Chem. Rev.* 118: 3887–3964.

4 Zhou, Y., Wang, J., Gu, Z. et al. (2016). *Chem. Rev.* 116: 422–518.

5 Wang, J., Sánchez-Roselló, M., Aceña, J.L. et al. (2014). *Chem. Rev.* 114: 2432–2506.

6 Jeschke, P. (2004). *ChemBioChem* 5: 570–589.

7 Leroux, F., Jeschke, P., and Schlosser, M. (2005). *Chem. Rev.* 105: 827–856.

8 Deng, X., Rong, J., Wang, L. et al. (2019). *Angew. Chem. Int. Ed.* 58: 2580–2605.

9 Preshlock, S., Tredwell, M., and Gouverneur, V. (2016). *Chem. Rev.* 116: 719–766.

10 Liang, T., Neumann, C.N., and Ritter, T. (2013). *Angew. Chem. Int. Ed.* 52: 8214–8264.

11 Charpentier, J., Früh, N., and Togni, A. (2015). *Chem. Rev.* 115: 650–682.

12 Wolstenhulme, J.R. and Gouverneur, V. (2014). *Acc. Chem. Res.* 47: 3560–3570.

13 Tlili, A. and Billard, T. (2013). *Angew. Chem. Int. Ed.* 52: 6818–6819.

14 Cahard, D. and Bizet, V. (2014). *Chem. Soc. Rev.* 43: 135–147.

15 Kohlhepp, S.V. and Gulder, T. (2016). *Chem. Soc. Rev.* 45: 6270–6288.

16 Arnold, A.M., Ulmer, A., and Gulder, T. (2016). *Chem. Eur. J.* 22: 8728–8739.

17 Vinogradova, E.V., Müller, P., and Buchwald, S.L. (2014). *Angew. Chem. Int. Ed.* 53: 3125–3128.

18 Yang, X., Wu, T., Phipps, R.J., and Toste, F.D. (2015). *Chem. Rev.* 115: 826–870.

19 Tao, J., Tran, R., and Murphy, G.K. (2013). *J. Am. Chem. Soc.* 135: 16312–16315.

20 Molnár, I.G. and Gilmour, R. (2016). *J. Am. Chem. Soc.* 138: 5004–5007.

21 Scheidt, F., Schäfer, M., Sarie, J.C. et al. (2018). *Angew. Chem. Int. Ed.* 57: 16431–16435.

22 Matoušek, V., Pietrasiak, E., Schwenk, R., and Togni, A. (2013). *J. Org. Chem.* 78: 6763–6768.

23 Geary, G.C., Hope, E.G., Singh, K., and Stuart, A.M. (2013). *Chem. Commun.* 49: 9263–9265.

24 Ilchenko, N.O., Tasch, B.O.A., and Szabó, K.J. (2014). *Angew. Chem. Int. Ed.* 53: 12897–12901.

25 Ilchenko, N.O. and Szabó, K.J. (2017). *J. Fluorine Chem.* 203: 104–109.

26 Ilchenko, N.O., Cortés, M.A., and Szabo, K.J. (2016). *ACS Catal.* 6: 447–450.

27 Yuan, W. and Szabó, K.J. (2015). *Angew. Chem. Int. Ed.* 54: 8533–8537.

28 Ilchenko, N.O., Hedberg, M., and Szabo, K.J. (2017). *Chem. Sci.* 8: 1056–1061.

29 Yuan, W., Eriksson, L., and Szabó, K.J. (2016). *Angew. Chem. Int. Ed.* 55: 8410–8415.

30 Ulmer, A., Brunner, C., Arnold, A.M. et al. (2016). *Chem. Eur. J.* 22: 3660–3664.

31 Brunner, C., Andries-Ulmer, A., Kiefl, G.M., and Gulder, T. (2018). *Eur. J. Org. Chem.*: 2615–2621.

32 Walker, S.E., Jordan-Hore, J.A., Johnson, D.G. et al. (2014). *Angew. Chem. Int. Ed.* 53: 13876–13879.

33 Geary, G.C., Hope, E.G., Singh, K., and Stuart, A.M. (2015). *RSC Adv.* 5: 16501–16506.

34 Legault, C.Y. and Prévost, J. (2012). *Acta Crystallogr., Sect. E* 68: o1238.

35 Zhou, B., Yan, T., Xue, X.-S., and Cheng, J.-p. (2016). *Org. Lett.* 18: 6128–6131.

36 Zhang, J., Szabo, K.J., and Himo, F. (2017). *ACS Catal.* 7: 1093–1100.

37 Geary, G.C., Hope, E.G., and Stuart, A.M. (2015). *Angew. Chem. Int. Ed.* 54: 14911–14914.

38 Yang, B., Chansaenpak, K., Wu, H. et al. (2017). *Chem. Commun.* 53: 3497–3500.

39 van der Born, D., Pees, A., Poot, A.J. et al. (2017). *Chem. Soc. Rev.* 46: 4709–4773.

40 Tredwell, M. and Gouverneur, V. (2012). *Angew. Chem. Int. Ed.* 51: 11426–11437.

41 Teare, H., Robins, E.G., Kirjavainen, A. et al. (2010). *Angew. Chem. Int. Ed.* 49: 6821–6824.

42 Buckingham, F., Kirjavainen, A.K., Forsback, S. et al. (2015). *Angew. Chem. Int. Ed.* 54: 13366–13369.

43 Lee, E., Kamlet, A.S., Powers, D.C. et al. (2011). *Science* 334: 639–642.

44 Cortes Gonzalez, M.A., Nordeman, P., Bermejo Gomez, A. et al. (2018). *Chem. Commun.* 54: 4286–4289.

45 Cortés González, M.A., Jiang, X., Nordeman, P. et al. (2019). *Chem. Commun.* 55: 13358–13361.

46 Coenen, H.H., Gee, A.D., Adam, M. et al. (2017). *Nucl. Med. Biol.* 55: v–xi.

47 Ford, A., Miel, H., Ring, A. et al. (2015). *Chem. Rev.* 115: 9981–10080.

48 Guo, X. and Hu, W. (2013). *Acc. Chem. Res.* 46: 2427–2440.

49 Murphy, G.K., Stewart, C., and West, F.G. (2013). *Tetrahedron* 69: 2667–2686.

50 Davies, H.M.L. and Denton, J.R. (2009). *Chem. Soc. Rev.* 38: 3061–3071.

51 Xiao, Q., Zhang, Y., and Wang, J. (2013). *Acc. Chem. Res.* 46: 236–247.

52 Doyle, M.P., Duffy, R., Ratnikov, M., and Zhou, L. (2010). *Chem. Rev.* 110: 704–724.

53 DeAngelis, A., Panish, R., and Fox, J.M. (2016). *Acc. Chem. Res.* 49: 115–127.

54 Xia, Y., Zhang, Y., and Wang, J. (2013). *ACS Catal.* 3: 2586–2598.

55 Emer, E., Twilton, J., Tredwell, M. et al. (2014). *Org. Lett.* 16: 6004–6007.

56 Yasui, N., Mayne, C.G., and Katzenellenbogen, J.A. (2015). *Org. Lett.* 17: 5540–5543.

57 Hu, M., Ni, C., Li, L. et al. (2015). *J. Am. Chem. Soc.* 137: 14496–14501.

58 Chen, G., Song, J., Yu, Y. et al. (2016). *Chem. Sci.* 7: 1786–1790.

59 He, F., Pei, C., and Koenigs, R.M. (2020). *Chem. Commun.* 56: 599–602.

60 Mai, B.K., Szabó, K.J., and Himo, F. (2018). *ACS Catal.* 8: 4483–4492.

61 Campbell, M.G., Mercier, J., Genicot, C. et al. (2017). *Nat. Chem.* 9: 1–3.

62 Janson, P.G., Ghoneim, I., Ilchenko, N.O., and Szabó, K.J. (2012). *Org. Lett.* 14: 2882–2885.

63 Ilchenko, N.O., Janson, P.G., and Szabo, K.J. (2013). *J. Org. Chem.* 78: 11087–11091.

64 Ilchenko, N.O., Janson, P.G., and Szabó, K.J. (2013). *Chem. Commun.* 49: 6614–6616.

65 Egami, H. and Sodeoka, M. (2014). *Angew. Chem. Int. Ed.* 53: 8294–8308.

66 Egami, H., Shimizu, R., and Sodeoka, M. (2012). *Tetrahedron Lett.* 53: 5503–5506.

67 Janson, P.G., Ilchenko, N.O., Diez-Varga, A., and Szabó, K.J. (2015). *Tetrahedron* 71: 922–931.

68 Liang, Z., Wang, F., Chen, P., and Liu, G. (2014). *J. Fluorine Chem.* 167: 55–60.

69 Lanfranchi, D.A., Belorgey, D., Muller, T. et al. (2012). *Org. Biomol. Chem.* 10: 4795–4806.

70 Laitar, D.S., Müller, P., and Sadighi, J.P. (2005). *J. Am. Chem. Soc.* 127: 17196–17197.

71 Zhu, J., Lin, Z., and Marder, T.B. (2005). *Inorg. Chem.* 44: 9384–9390.

72 Wang, X., Ye, Y., Ji, G. et al. (2013). *Org. Lett.* 15: 3730–3733.

73 Fang, Z., Ning, Y., Mi, P. et al. (2014). *Org. Lett.* 16: 1522–1525.

74 Yuan, W. and Szabó, K.J. (2016). *ACS Catal.* 6: 6687–6691.

75 Lübcke, M., Yuan, W., and Szabó, K.J. (2017). *Org. Lett.* 19: 4548–4551.

76 Mai, B.K., Szabó, K.J., and Himo, F. (2018). *Org. Lett.* 20: 6646–6649.

77 Hooz, J. and Linke, S. (1968). *J. Am. Chem. Soc.* 90: 5936–5937.

78 He, Z., Zajdlik, A., and Yudin, A.K. (2014). *Dalton Trans.* 43: 11434–11451.

79 Lübcke, M., Bezhan, D., and Szabo, K.J. (2019). *Chem. Sci.* 10: 5990–5995.

7

Photoredox Catalysis in Fluorination and Trifluoromethylation Reactions

Takashi Koike and Munetaka Akita

Tokyo Institute of Technology, Institute of Innovative Research, Laboratory for Chemistry and Life Science, R1-27, 4259 Nagatsuta-cho, Midori-ku, Yokohama 226-8503, Japan

7.1 Introduction

For the past decade, photoredox catalysis with metal complexes such as $[Ru(bpy)_3]^{2+}$ (bpy: 2,2′-bipyridine) and organic dyes has emerged as a reliable strategy for radical reactions because visible light irradiation induces 1e-redox reactions below room temperature [1].

Light irradiation causes excitation of the photocatalyst (PC) in the ground state. The excited photocatalyst (*PC) may induce 1e-oxidation of electron-rich reagents (eD) to lead to the generation of radical species $(eD^{\cdot+})$ via a single electron transfer (SET) process and subsequent reduction of electron-deficient reagents (eA), with the resultant reduced catalyst (PC^-) generating another radical species $(eA^{\cdot-})$ accompanying regeneration of the photocatalyst (PC) (Scheme 7.1a). This reaction sequence is called "reductive quenching cycle," and the sequence following the initial reduction of eA by *PC giving $eA^{\cdot-}$ and PC^+, namely "oxidative quenching cycle," is also feasible. Thus, a pair of radical species, $eD^{\cdot+}$ and $eA^{\cdot-}$, can be generated by the action of photoredox catalysis. In principle, by taking redox potentials of photocatalysts and redox agents into account, design of organic radical reactions is attainable. On the other hand, when photocatalysts (PC) are carbonyl compounds, especially aromatic ketones, or some metal oxo species (M=O species), the excited photocatalyst (*PC) serves as a hydrogen abstractor (˙A) due to its highly electrophilic O-centered radical character and is amenable to homolytic cleavage of a $C(sp^3)$—H bond (R—H) via hydrogen atom transfer (HAT), leading to generation of alkyl radical (˙R) (Scheme 7.1b) [2]. PC is regenerated by a back-HAT step of the reduced H–PC species to one of the intermediates (X) formed in the reaction. This is not the genuine photoredox catalysis, but several seminal works on fluorination are also described in Section 7.2.1.

Radical species frequently play important roles in the construction of C—F and C—CF$_3$ bonds. In general, radical fluorination results from the reaction of organyl radicals with SOMOphilic fluorine sources. By contrast, radical trifluoromethylation can result either from (i) the reaction involving the trifluoromethyl radical or

Organofluorine Chemistry: Synthesis, Modeling, and Applications,
First Edition. Edited by Kálmán J. Szabó and Nicklas Selander.
© 2021 WILEY-VCH GmbH. Published 2021 by WILEY-VCH GmbH.

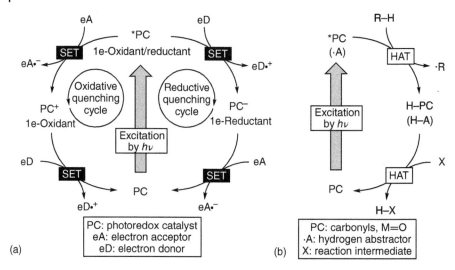

Scheme 7.1 (a) Photoredox catalysis and (b) photocatalytic HAT.

(ii) the reaction of organyl radicals with SOMOphilic CF_3 sources. Modern protocols are expected to exhibit site selectivity and high functional group compatibility. Thus, herein we highlight strategies for site-selective radical fluorination and trifluoromethylation by photocatalysis.

7.2 Fluorination

Late-stage functionalization, which requires high compatibility with various functional groups, is regarded as an ideal technology in the field of medicinal chemistry. In particular, development of late-stage fluorination can be connected to synthesis of ^{18}F-labeled drugs for PET (positron emission tomography) imaging [3]. In this section, photocatalytic strategies for site-selective introduction of fluorine atom(s) into organic skeletons are discussed. Functionalization of unactivated C—H bonds through HAT processes induced by the excited species of aryl ketones (˙A) has been studied more extensively compared to photoredox catalysis and thus will be discussed first.

7.2.1 Fluorination Through Direct HAT Process by Excited Photocatalyst

Fluorination of inert $C(sp^3)$—H bonds is highly attractive because alkyl C—H bonds are ubiquitous and prefunctionalization of substrates can be avoided. In 2013, the group of Chen reported on the mono- and difluorination of benzylic C—H bonds [4]. Monofluorinated products **2** were obtained by the photocatalytic reaction of alkylbenzenes **1** using 9-fluorenone and Selectfluor, as a catalyst and a SOMOphile, respectively. In contrast, the combination of xanthone as a catalyst and Selectfluor

II gave the difluorinated products **4** (Scheme 7.2). The monofluorination protocols could be applied to a gram-scale synthesis. The proposed mechanism is depicted in Scheme 7.2. The benzyl radical generated through photoinduced HAT with the triplet photoexcited catalyst (·A) reacts with Selectfluor or Selectfluor II to afford the fluorinated product **2** or **4**. The Frutos and Kappe group succeeded in significant shortening of the reaction time by application of a continuous flow system (residence time < 30 minutes) [5].

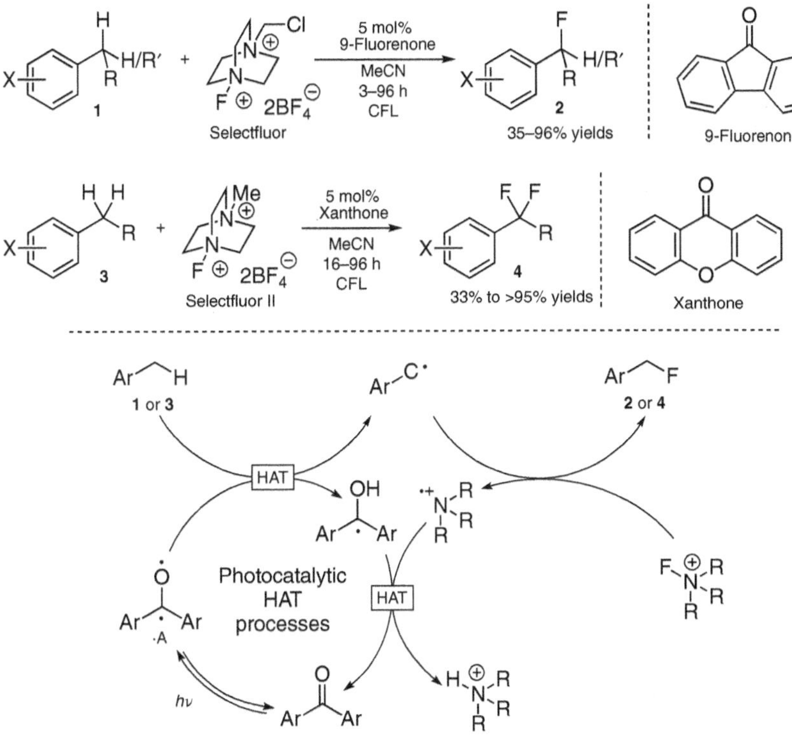

Scheme 7.2 Aryl ketone-catalyzed benzylic C–H fluorination (CFL: compact fluorescent lamp).

In 2014, the group of Chen extended the reaction system to fluorination of aliphatic compounds **5** [6]. In this system, acetophenone, an aryl alkyl ketone, served as an efficient photocatalyst (Scheme 7.3). While site selectivity of the acetophenone-catalyzed C–H fluorination was low, the reaction was operationally simple and utilized a cheap, readily available photocatalyst and light source, household CFL (compact fluorescence lamp). Almost at the same time, the group of Tan reported on anthraquinone (AQN)-catalyzed fluorination of aliphatic C—H bonds (Scheme 7.4) [7]. They proposed that the photoexcited AQN does not serve as a hydrogen abstractor (·A) of **5** but as an energy donor for Selectfluor, resulting in the formation of a cationic N-centered radical, which induced the HAT process.

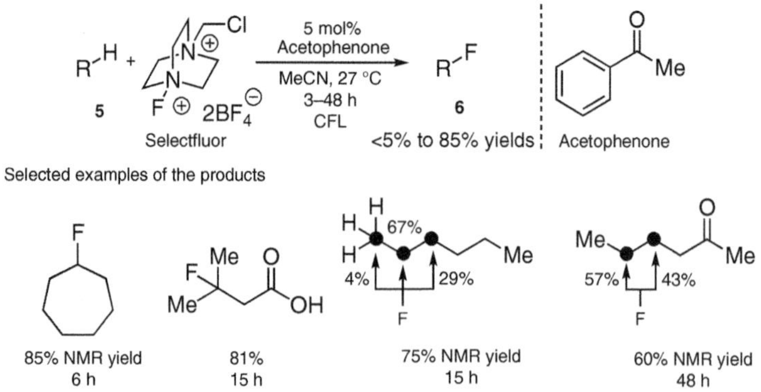

Scheme 7.3 Acetophenone-catalyzed aliphatic C–H fluorination.

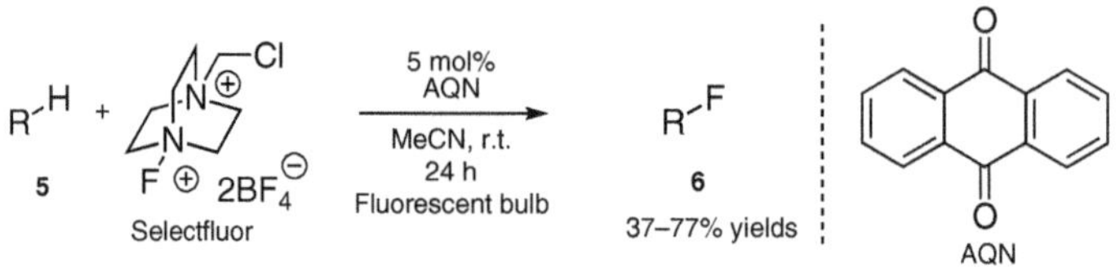

Scheme 7.4 AQN-catalyzed aliphatic C–H fluorination.

In 2014, the group of Britton developed fluorination of unactivated C—H bonds with NFSI (*N*-fluorobenzenesulfonimide), an electrophilic fluorinating reagent, using the tetrabutylammonium salt of decatungstate (TBADT) as the photocatalyst (Scheme 7.5) [8]. The proposed catalytic cycle is depicted in Scheme 7.5. The excited tungstate ($*[W_{10}O_{32}]^{4-}$) serves as a hydrogen abstractor ($\cdot A$) to give the organyl radicals ($\cdot R$) from alkanes **5** through HAT processes.

While there is room for improvement in the site selectivity, the present method could be extended to introduction of a ^{18}F atom because [^{18}F]-NFSI can be readily prepared [9].

The abovementioned catalysts, i.e. acetophenone and decatungstate, usually need near-UV-region energy to excite the photocatalyst. In 2016, the group of Sorensen reported on visible light fluorination of C—H bonds using the nitrate salt of uranyl cation photocatalyst, [UO_2](NO_3)$_2$, although the scope of the reaction is narrow compared to the abovementioned systems (Scheme 7.6) [10].

7.2.2 Fluorination Through Photoredox Processes

As described in Section 7.2.1, fluorination of unactivated C—H bonds is attractive as a modern synthetic technology. But high site selectivity of alkyl C(sp^3)—H bond activation still remains a challenging topic. In this section, several seminal works on selective fluorination by photoredox catalysis will be presented. The reactions are featured by the ingenious design of substrates in order that an appropriate functional group is utilized as a scaffold directing to the desired reaction site.

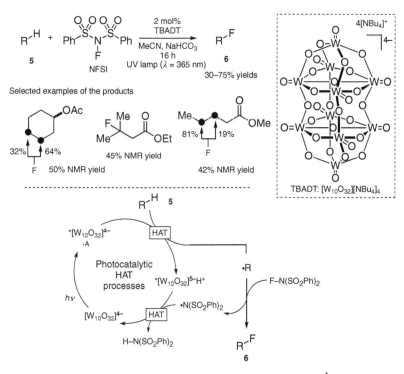

Scheme 7.5 Decatungstate salt-catalyzed aliphatic C–H fluorination.

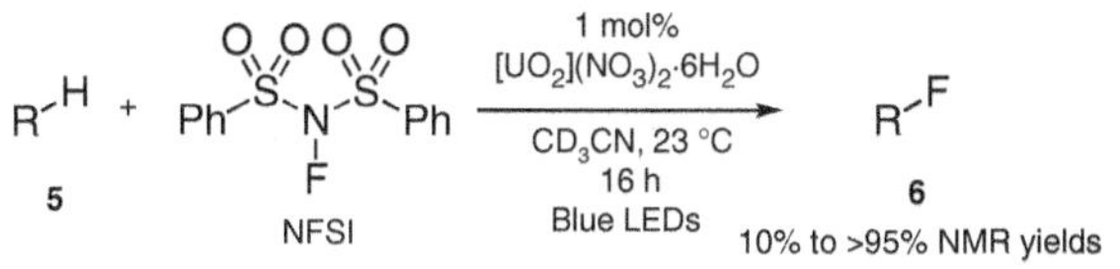

Scheme 7.6 Uranyl salt-catalyzed aliphatic C–H fluorination.

The groups of Sammis, Paquin, Ye, and MacMillan paid attention to a carboxylic acid group [11], which is well known to be oxidized by highly oxidizing photocatalysts [12]. As an example, the work reported by the group of MacMillan is introduced below (Scheme 7.7) [11b]. They used the Ir complex, $[Ir\{dF(CF_3)ppy\}_2(dtbbpy)]PF_6$ ($dF(CF_3)ppy = 2$-(2,4-difluorophenyl)-5-(trifluoromethyl)pyridine, dtbbpy = 4,4′-di-*tert*-butyl-2,2′-bipyridine), as the highly oxidizing photocatalyst. At the initial stage, the excited Ir photocatalyst (****Ir***) undergoes SET to a sacrificial quantity of Selectfluor. Then, the generated strongly oxidizing Ir species (***Ir^+***) undergoes SET from aliphatic carboxylic acid **7** to yield the ground state catalyst (***Ir***) and the carboxyl radical, which undergoes spontaneous extrusion of CO_2 to provide the alkyl radical (˙R). The formed radical ˙R reacts with Selectfluor to forge the C—F bond with concomitant formation of the corresponding *N*-centered radical cation, which serves as an electron acceptor instead of Selectfluor in the subsequent photoredox cycles. The present method can be applied to selective production of fluoroalkanes **6** from various aliphatic carboxylic acids **7**.

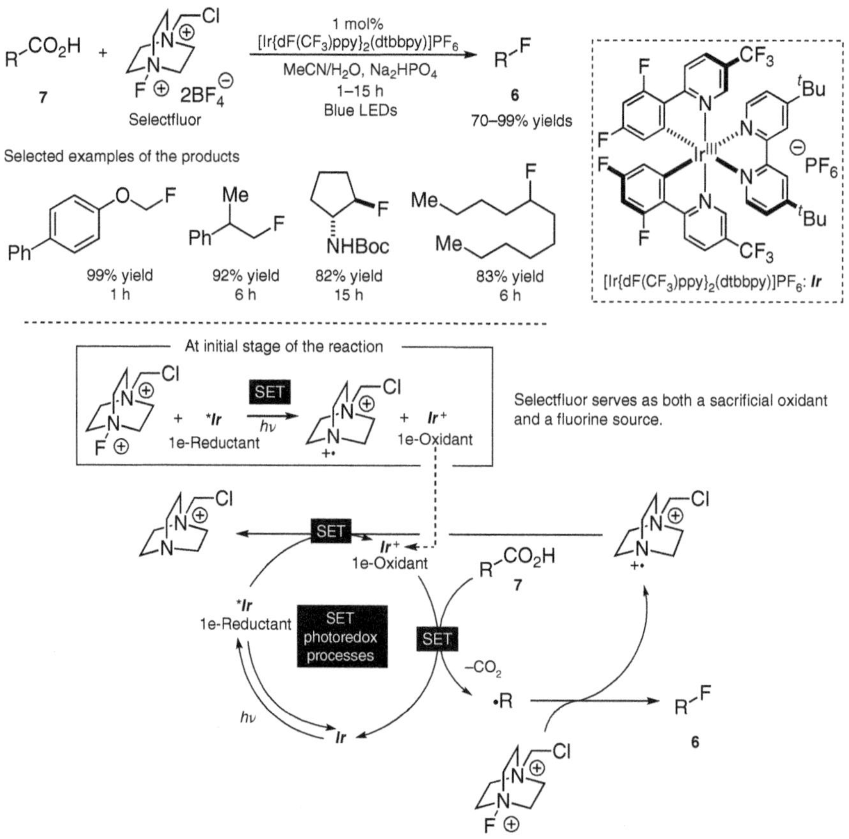

Scheme 7.7 Decarboxylative fluorination of aliphatic carboxylic acids.

In 2014, the group of Lectka disclosed a photocatalytic strategy for alkane fluorination employing UV light ($h\nu = 302$ nm) and 1,2,4,5-tetracyanobenzene (TCB) as a photocatalyst, which overcame the problem of the high oxidation potential of alkanes through the photoredox process, i.e. the reductive quenching cycle (Scheme 7.8) [13]. TCB excited by UV irradiation (*[TCB]) induces 1e-oxidation of alkane **5**. The resultant radical cation ([R–H]$^{\cdot+}$) transfers the highly acidic proton to the solvent to form alkyl radical ($\cdot$R). The generated radical species $\cdot$R reacts with SOMOphile, Selectfluor, to produce fluoroalkane **6** and N-centered cationic radical species, which oxidizes the reduced TCB ([TCB]$^{\cdot-}$). The reactions of alkanes proceed in a fairly selective manner.

Well-designed heteroatom-centered radicals induce selective fragmentation or intramolecular HAT to result in the generation of carbon-centered radicals at the desired positions [14]. In 2018, the Leonori group developed a site-selective remote fluorination via photocatalytically generated iminyl radicals (Scheme 7.9) [14c]. Carboxylic acid-substituted oxime (**8**, **9**) undergoes 1e-oxidation by the excited highly oxidizing organic photocatalyst, Mes–Acr$^+$–Me catalyst (10-methyl-9-(2,4,6-trimethylphenyl)acridinium salt), to generate the iminyl radical

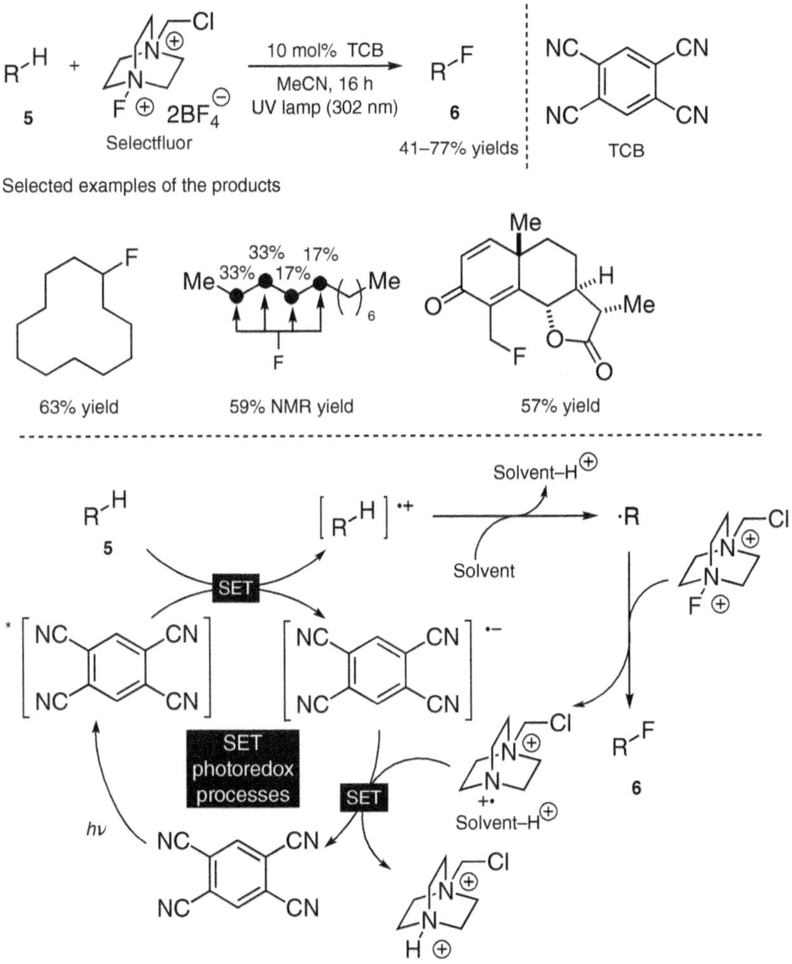

Scheme 7.8 TCB-catalyzed direct aliphatic fluorination.

after CO_2 and acetone extrusion, which induces ring opening via β-scission to form the carbon-centered radical or is translated to the nucleophilic γ-carbon radical through 1,5-HAT. A polarity-matched S_H2 reaction with Selectfluor produces fluorinated nitrile **10** or γ-fluorinated ketones **11** in a selective manner.

Fluorination of arenes is also an important transformation because aryl fluorides are frequently found in active pharmaceuticals and agrochemicals. In addition, introduction of ^{18}F atom is strongly associated with PET imaging as mentioned above. In 2019, Nicewicz and coworkers developed a direct conversion of an arene C—H bond into the C—^{18}F bond by the highly oxidizing organic photocatalyst, Mes–Acr$^+$–Ph (Scheme 7.10) [15]. They previously reported on selective functionalization of arenes through nucleophilic attack of amines and cyanide on the cationic radical intermediate generated by photocatalytic 1e-oxidation [16]. Thus, the concept was extended to F nucleophiles, i.e. when

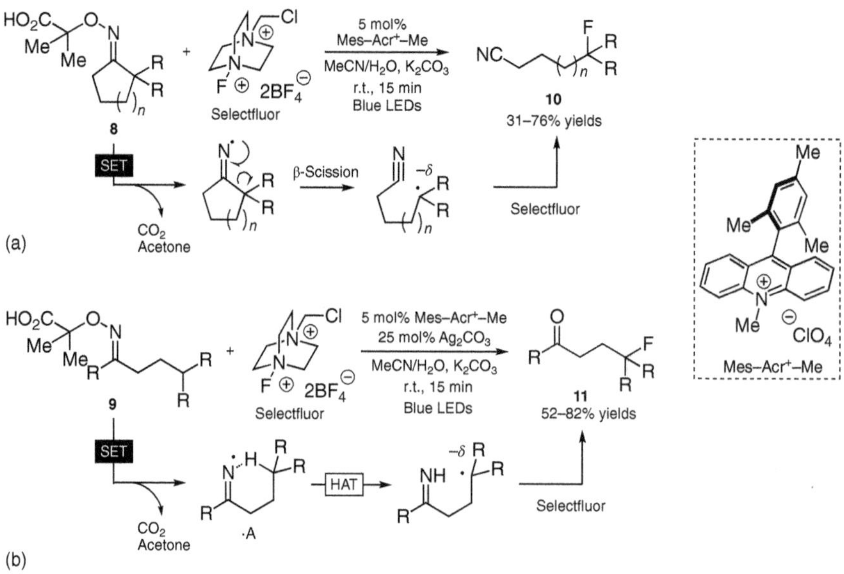

(a)

(b)

Scheme 7.9 Remote C–H fluorination via iminyl radical intermediates.

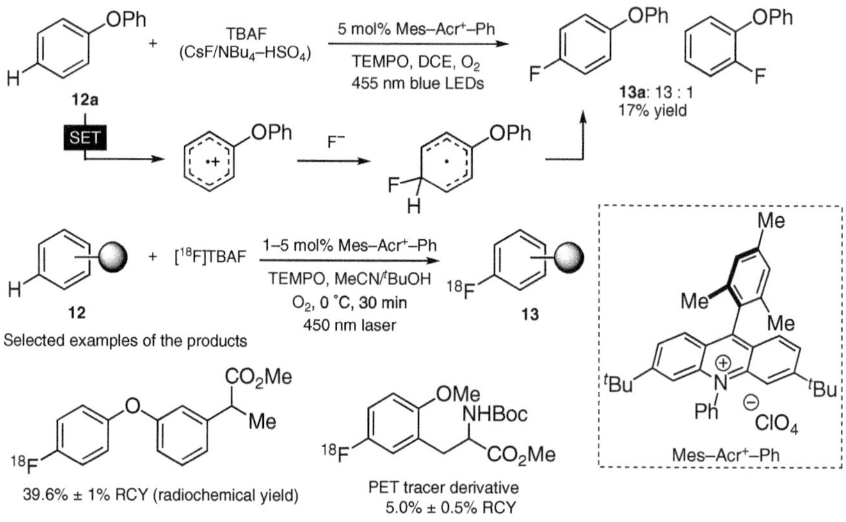

Selected examples of the products

39.6% ± 1% RCY (radiochemical yield)

PET tracer derivative
5.0% ± 0.5% RCY

Scheme 7.10 Direct conversion of arene C–H bond into C–^{18}F bond via cationic aryl radical.

an anionic ^{19}F nucleophile composed of CsF and tetrabutylammonium bisulfate (TBA-HSO$_4$) was reacted with diphenyl ether **12a**, 17% of fluorinated arenes **13a** were obtained as a mixture of *p*- and *o*-isomers (13 : 1). To improve the reaction efficiency, 450 nm laser device was used as the light source, and radiofluorination was carried out with [^{18}F]tetrabutylammonium fluoride (TBAF) prepared from proton bombardment of H$_2$^{18}O and subsequently the product was eluted with tetrabutylammonium bicarbonate. The system could be applied to radiofluorination of various arenes.

In 2020, the group of Ritter reported on a two-step procedure forming arene C—F bonds with excellent site selectivity (Scheme 7.11) [17]. They previously developed site-selective and late-stage aromatic C–H thianthrenation [18], which is followed by fluorination by metallaphotoredox catalysis. They merged Ir photocatalysis by [Ir{dF(CF$_3$)ppy}$_2$(dtbbpy)]PF$_6$ with copper catalysis by [Cu(MeCN)$_4$]BF$_4$. Use of a catalytic amount of copper salt (20 mol%) was viable, but to reduce the amounts of side products, an excess amount of copper salt was used for preparative synthesis. The excited Ir catalyst (***Ir***) undergoes SET from thianthrene, which is formed from 1e-reduction of **14** or **15** at the initial stage, to give the ***Ir*$^-$** species, which is suitable for reduction of arylthianthrenium salt (**14** or **15**). Oxidative ligation of the resultant aryl radical (˙Ar) toward a Cu^{II} species, which is formed from 1e-oxidation of the Cu^I salt by the radical cation of thianthrene, and subsequent addition of fluoride anion affords the Ar–Cu^{III}–F species, which undergoes reductive elimination to produce aryl fluoride **16**. C—H bonds in complex drug-like small molecules were successfully transformed into the C—F bonds at late stages.

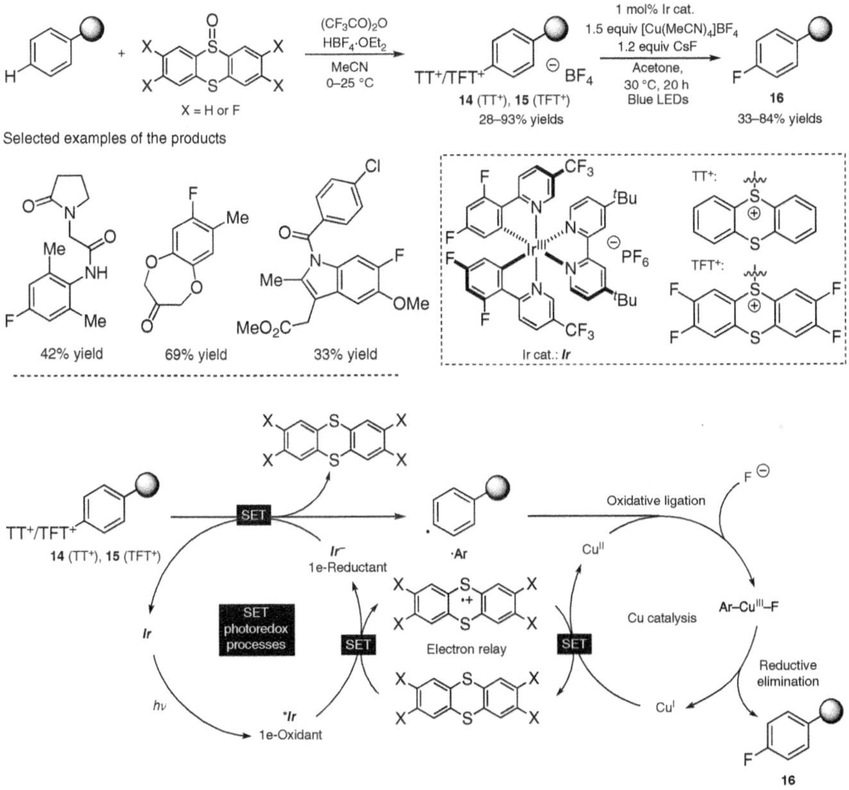

Selected examples of the products

42% yield   69% yield   33% yield

Scheme 7.11 Site-selective late-stage fluorination of aromatic C–H bonds by dual photo/copper catalysis.

7.3 Trifluoromethylation

Photoredox catalysis becomes a useful strategy for the generation of trifluoromethyl radical from various electron-rich or -deficient CF_3 sources. Thus, photocatalytic radical trifluoromethylation of unsaturated C—C bonds such as alkenes, alkynes, and allenes has been developed so far (Scheme 7.12). Highly stereocontrolled reactions are realized by well-designed photocatalytic systems. Recently, several fine reviews on photocatalytic trifluoromethylation of unsaturated bonds have been already published [19]. Thus, in this section, we will focus on site-selective trifluoromethylation of aromatic and aliphatic compounds.

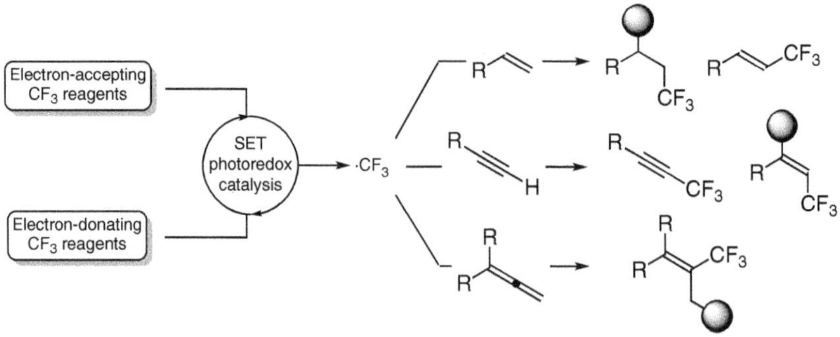

Scheme 7.12 Photoredox-catalyzed trifluoromethylation of unsaturated C—C bonds.

7.3.1 Trifluoromethylation of Aromatic Compounds

The trifluoromethyl motif in aromatic systems is frequently found in pharmacophores. Thus, development of simple and direct trifluoromethylation of unactivated arenes and heteroarenes is an important topic in drug discovery. In 2011, the MacMillan group showed that photoredox catalysis by [Ru(phen)$_3$]Cl$_2$ (phen = phenanthroline) and Ir(Fppy)$_3$ (Fppy = 2-(2,4-difluorophenyl)pyridine) is amenable to the direct trifluoromethylation of arenes under mild reaction conditions using CF_3SO_2Cl **17** as a CF_3 radical source (Scheme 7.13) [20]. A possible reaction mechanism is depicted in Scheme 7.13. The photoexcited metal catalyst (*M) undergoes SET to **17** to give ˙CF_3 and the strong oxidant (M^+). The resultant ˙CF_3 reacts with arenes **12** to afford the corresponding cyclohexadienyl radical, which is oxidized by the oxidant M^+ to give the cyclohexadienyl cation and regenerate M. Deprotonation of the cationic intermediate produces the trifluoromethylated arene **18**. The scope with respect to the substrate with electron-rich and -deficient heteroarenes and unactivated arenes is broad. MacMillan's work was followed by those with various combinations of trifluoromethylating reagents and photocatalysts [21]. Their methods brought excellent outcome in terms of direct conversion of unactivated C(sp^2)—H bonds to C(sp^2)—CF_3 bonds but some substrates suffered from low site selectivity.

Incorporation of a CF_3 group to the specific position should be feasible by using arenes bearing activating groups despite the requirements of prefunctionalization.

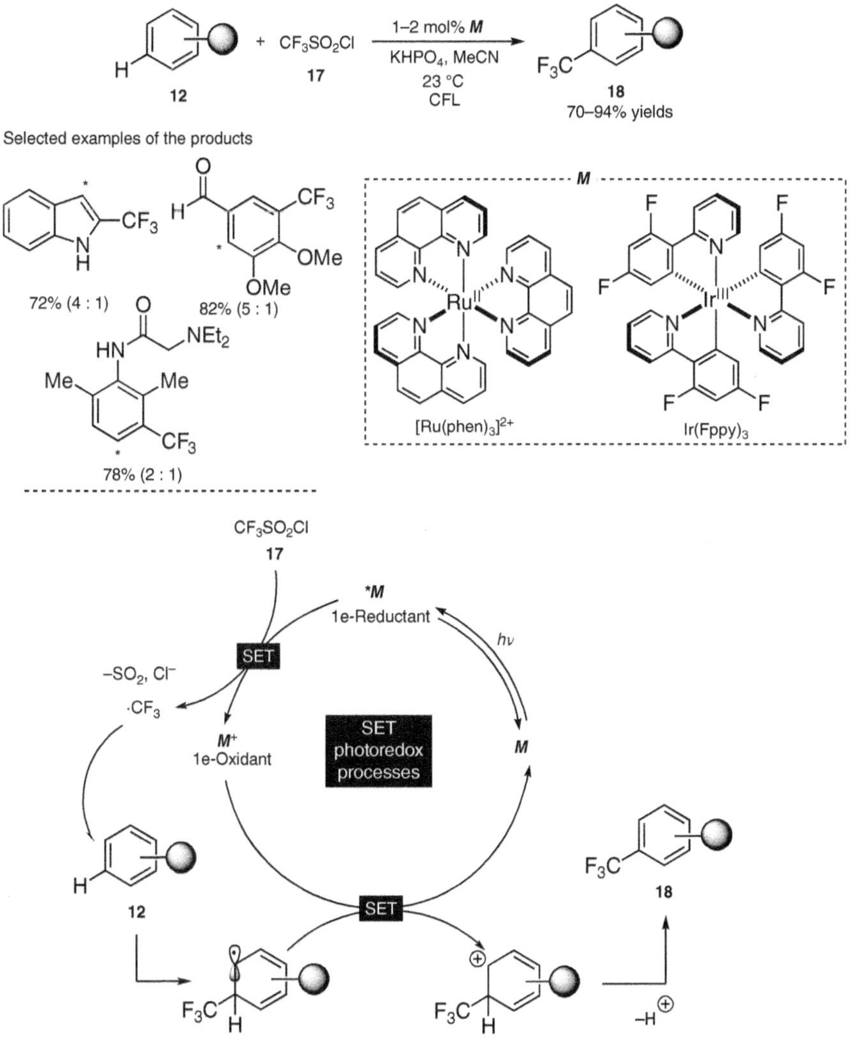

Scheme 7.13 Photocatalytic direct trifluoromethylation of aryl C—H bonds.

The copper-catalyzed trifluoromethylation of boronic acids with nucleophilic and electrophilic trifluoromethylating reagents has been well studied for the past few years [22]. In most of the reactions, Cu–CF$_3$ species are considered to be key active species generated via nucleophilic or electrophilic transfer of the CF$_3$ group to the Cu center. The Sanford group showed that an alternative radical pathway by the action of photoredox catalyst, [Ru(bpy)$_3$]Cl$_2$, leads to trifluoromethylation of areneboronic acids **19** via formation of the Cu–CF$_3$ species (Scheme 7.14) [23].

A plausible mechanism based on dual catalysis process starts with SET from a Cu^I species to *[Ru(bpy)$_3$]$^{2+}$, accompanying the formation of a Cu^{II} species and the strongly reducing [Ru(bpy)$_3$]$^+$. The Ru reductant induces the second SET event to CF$_3$I **20** to provide ·CF$_3$ and [Ru(bpy)$_3$]$^{2+}$ in the ground state. The resultant ·CF$_3$

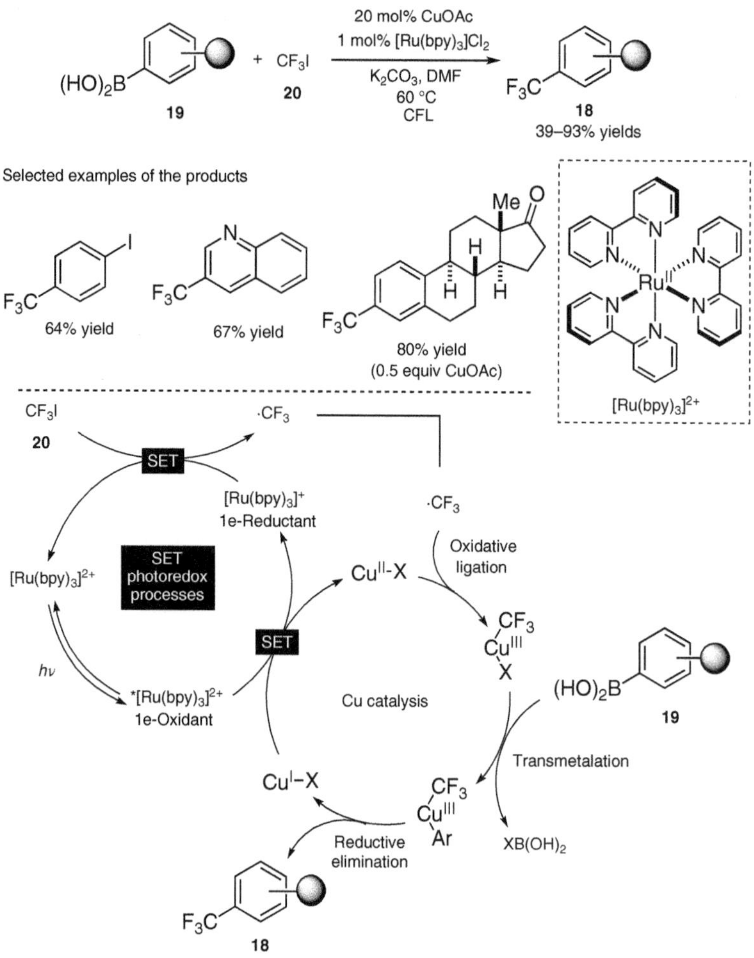

Scheme 7.14 Trifluoromethylation of boronic acids based on dual catalysis.

reacts with the Cu^{II} species to generate a Cu^{III}–CF$_3$ active species via oxidative ligation. Subsequent base-promoted transmetalation between the Cu^{III} species and the areneboronic acid **19** affords a CF$_3$–Cu^{III}–Ar intermediate, which undergoes aryl—CF$_3$ bond-forming reductive elimination to produce trifluoromethylated arene **18** and the Cu^{I} catalyst.

In 2018, the MacMillan group developed a dual copper/photoredox catalytic system of a different type (Scheme 7.15) [24]. They focused on trifluoromethylation of aryl bromides **21** with sulfonium-based electrophilic CF$_3$ reagent **22** through a metallaphotoredox-catalyzed cross-electrophile coupling protocol [25].

Utilization of the highly oxidizing Ir photocatalysts, [Ir(dFFppy)$_2$(4,4′-dCF$_3$bpy)]PF$_6$ or [Ir(dFMeppy)$_2$(4,4′-dCF$_3$bpy)]PF$_6$ (dFFppy = 2-(2,4-difluorophenyl)-5-fluoropyridine, dFMeppy = 2-(2,4-difluorophenyl)-5-methylpyridine, 4,4′-dCF$_3$bpy = 4, 4′-ditrifluoromethyl-2,2′-bipyridine), for 1e-oxidation of tris(trimethylsilyl)silanol

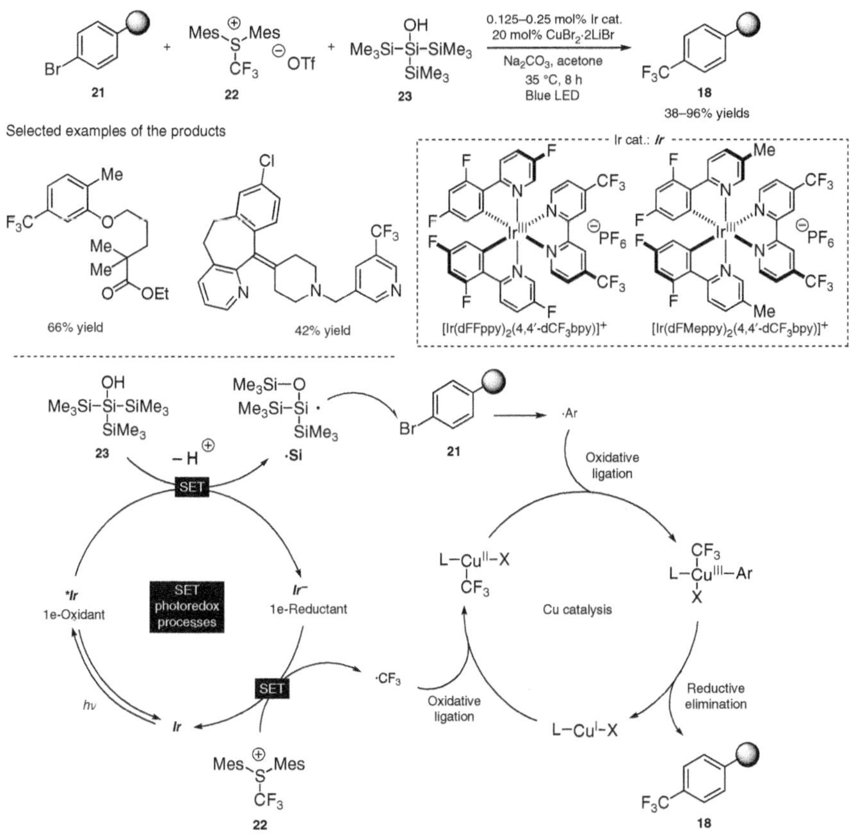

Scheme 7.15 Metallaphotoredox catalysis: trifluoromethylation of bromoarenes.

23 is a key to generation of aryl radical through atom transfer from **21** by silyl radical (˙Si). This species would make the formation of an Ar–Cu^{III}–CF$_3$ species easy compared to the oxidative addition of **21** to a Cu^I species. The photoexcited **Ir*, a strong oxidant, induces SET from tris(trimethylsilyl)silanol **23** accompanying the formation of the reduced *Ir⁻*. Oxidation and deprotonation of supersilanol cause rapid Brook rearrangement-type silyl migration to yield the silyl radical (˙Si), which abstracts the bromine atom to form aryl radical (˙Ar). The *Ir⁻* species concurrently reduces electrophilic CF$_3$ source **22** to generate the CF$_3$ radical (˙CF$_3$). Then, the Cu^I species, which is generated through reduction of the Cu^{II} catalyst precursor by the photocatalyst, undergoes oxidative ligation of ˙CF$_3$ to afford a Cu^{II}–CF$_3$ active species. The aryl radical (˙Ar) is quickly trapped by the Cu^{II} species to yield the key Aryl–Cu^{III}–CF$_3$ adduct, reductive elimination from which provides the trifluoromethylated arene product **18**. A broad range of electronically diverse bromoarenes including heteroaryl bromides were found to give good to excellent yields. High compatibility with various functional groups was also observed.

In 2019, the Ritter group showed that the abovementioned two-step reaction system (Scheme 7.11) could be applied to trifluoromethylation, i.e. selective C–H thianthrenation followed by Cu-mediated photocatalytic trifluoromethylation of arenes (Scheme 7.16) [26]. The combination of $[Ru(bpy)_3](PF_6)_2$ and the "Cu–CF$_3$" reagent prepared by mixing CuSCN, CsF, and TMSCF$_3$ turned out to be the best. The protocol enables site-selective trifluoromethylation of various arenes with broad functional group tolerance. In addition, gram-scale synthesis is viable.

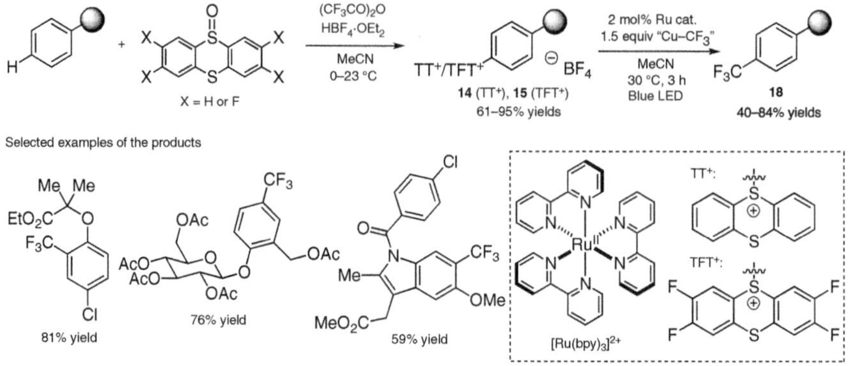

Scheme 7.16 Site-selective late-stage trifluoromethylation of aromatic C–H bonds.

7.3.2 Trifluoromethylative Substitution of Alkyl Bromides

To forge C(sp^3)—CF$_3$ bonds, reaction of the CF$_3$ radical with olefins is one of the reliable methods as shown in Scheme 7.12. On the other hand, in 2019, the MacMillan group applied the metallaphotoredox-catalyzed cross-electrophile coupling protocol mentioned above to the reaction of alkyl bromides **24** with the electrophilic trifluoromethylating reagent **22** (Scheme 7.17) [27]. The corresponding trifluoromethylated alkanes **25** were obtained via site-specific substitution of the bromine atom. In addition, in the present reaction system, the organic photoredox catalyst, 4CZIPN (1,2,3,5-tetrakis(carbazole-9-yl)-4,6-dicyanobenzene), also serves as an efficient photocatalyst instead of the Ir photocatalyst.

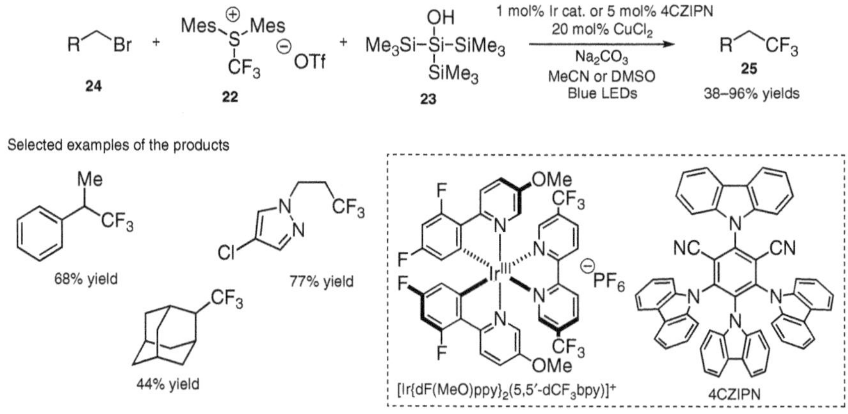

Scheme 7.17 Metallaphotoredox catalysis: trifluoromethylation of bromoalkanes.

7.4 Summary and Outlook

Photocatalytic strategies for selective generation of organic radical species are connected to fluorination and trifluoromethylation. A fine design of the photocatalytic system achieves site-selective construction of $C(sp^3)/C(aryl)$—F/CF_3 bonds. Significant advances have been made by merger of photocatalysis and metal catalysis, which would lead to asymmetric fluorination and trifluoromethylation in the near future. In addition, extension to di- and mono-fluoromethylation still remains underdeveloped. Elaborate design of photocatalytic systems should lead to further successful stereoselective fluorination and various fluoroalkylations.

References

1 Stephenson, C.R.J., Yoon, T.P., and MacMillan, D.W.C. (eds.) (2018). *Visible Light Photocatalysis in Organic Chemistry*. Weinheim, Germany: Wiley-VCH.

2 Capaldo, L. and Ravelli, D. (2017). *Eur. J. Org. Chem.* 2017: 2056–2071.

3 (a) Buckingham, F. and Gouverneur, V. (2016). *Chem. Sci.* 7: 1645–1652. (b) Preshlock, S., Tredwell, M., and Gouverneur, V. (2016). *Chem. Rev.* 116: 719–766.

4 Xia, J.B., Zhu, C., and Chen, C. (2013). *J. Am. Chem. Soc.* 135: 17494–17500.

5 Cantillo, D., de Frutos, O., Rincon, J.A. et al. (2014). *J. Org. Chem.* 79: 8486–8490.

6 Xia, J.B., Zhu, C., and Chen, C. (2014). *Chem. Commun.* 50: 11701–11704.

7 Kee, C.W., Chin, K.F., Wong, M.W. et al. (2014). *Chem. Commun.* 50: 8211–8214.

8 (a) Halperin, S.D., Fan, H., Chang, S. et al. (2014). *Angew. Chem. Int. Ed.* 53: 4690–4693. (b) Halperin, S.D., Kwon, D., Holmes, M. et al. (2015). *Org. Lett.* 17: 5200–5203. (c) Nodwell, M.B., Bagai, A., Halperin, S.D. et al. (2015). *Chem. Commun.* 51: 11783–11786.

9 Teare, H., Robins, E.G., Arstad, E. et al. (2007). *Chem. Commun.*: 2330–2332.

10 West, J.G., Bedell, T.A., and Sorensen, E.J. (2016). *Angew. Chem. Int. Ed.* 55: 8923–8927.

11 (a) Rueda-Becerril, M., Mahe, O., Drouin, M. et al. (2014). *J. Am. Chem. Soc.* 136: 2637–2641. (b) Ventre, S., Petronijevic, F.R., and MacMillan, D.W.C. (2015). *J. Am. Chem. Soc.* 137: 5654–5657. (c) Wu, X., Meng, C., Yuan, X. et al. (2015). *Chem. Commun.* 51: 11864–11867.

12 (a) Xuan, J., Zhang, Z.G., and Xiao, W.J. (2015). *Angew. Chem. Int. Ed.* 54: 15632–15641. (b) Huang, H., Jia, K., and Chen, Y. (2016). *ACS Catal.* 6: 4983–4988. (c) Jin, Y. and Fu, H. (2017). *Asian J. Org. Chem.* 6: 368–385.

13 (a) Bloom, S., Knippel, J.L., and Lectka, T. (2014). *Chem. Sci.* 5: 1175–1178. (b) Bloom, S., McCann, M., and Lectka, T. (2014). *Org. Lett.* 16: 6338–6341.

14 (a) Bume, D.D., Pitts, C.R., Ghorbani, F. et al. (2017). *Chem. Sci.* 8: 6918–6923. (b) Pitts, C.R., Bume, D.D., Harry, S.A. et al. (2017). *J. Am. Chem. Soc.* 139: 2208–2211. (c) Dauncey, E.M., Morcillo, S.P., Douglas, J.J. et al. (2018). *Angew. Chem. Int. Ed.* 57: 744–748. (d) Morcillo, S.P., Dauncey, E.M., Kim, J.H. et al. (2018). *Angew. Chem. Int. Ed.* 57: 12945–12949.

15 Chen, W., Huang, Z., Tay, N.E.S. et al. (2019). *Science* 364: 1170–1174.

16 (a) Romero, N.A., Margrey, K.A., Tay, N.E. et al. (2015). *Science* 349: 1326–1330.
(b) McManus, J.B. and Nicewicz, D.A. (2017). *J. Am. Chem. Soc.* 139: 2880–2883.

17 Li, J., Chen, J., Sang, R. et al. (2020). *Nat. Chem.* 12: 56–62.

18 Berger, F., Plutschack, M.B., Riegger, J. et al. (2019). *Nature* 567: 223–228.

19 (a) Barata-Vallejo, S., Bonesi, S.M., and Postigo, A. (2015). *Org. Biomol. Chem.*
13: 11153–11183. (b) Koike, T. and Akita, M. (2016). *Acc. Chem. Res.* 49:
1937–1945. (c) Tomita, R., Koike, T., and Akita, M. (2017). *Chem. Commun.* 53:
4681–4684. (d) Barata-Vallejo, S., Cooke, M.V., and Postigo, A. (2018). *ACS Catal.*
8: 7287–7307. (e) Han, S., Kim, H., and Oh, E. (2018). *Synthesis* 50: 3346–3358.
(f) Koike, T. and Akita, M. (2018). *Chem* 4: 409–437.

20 Nagib, D.A. and MacMillan, D.W.C. (2011). *Nature* 480: 224–228.

21 (a) Iqbal, N., Choi, S., Ko, E. et al. (2012). *Tetrahedron Lett.* 53: 2005–2008.
(b) Beatty, J.W., Douglas, J.J., Cole, K.P. et al. (2015). *Nat. Commun.* 6: 7919.
(c) Beatty, J.W., Douglas, J.J., Miller, R. et al. (2016). *Chem* 1: 456–472. (d)
Su, Y., Kuijpers, K.P., Konig, N. et al. (2016). *Chem. Eur. J.* 22: 12295–12300.
(e) Corsico, S., Fagnoni, M., and Ravelli, D. (2017). *Photochem. Photobiol. Sci.*
16: 1375–1380. (f) Noël, T., Alcazar, J., Abdiaj, I. et al. (2017). *Synthesis* 49:
4978–4985. (g) Ouyang, Y., Xu, X.H., and Qing, F.L. (2018). *Angew. Chem. Int.
Ed.* 57: 6926–6929. (h) Yang, B., Yu, D., Xu, X.-H. et al. (2018). *ACS Catal.* 8:
2839–2843.

22 (a) Chu, L. and Qing, F.-L. (2010). *Org. Lett.* 12: 5060–5063. (b) Morimoto, H.,
Tsubogo, T., Litvinas, N.D. et al. (2011). *Angew. Chem. Int. Ed.* 50: 3793–3798.
(c) Senecal, T.D., Parsons, A.T., and Buchwald, S.L. (2011). *J. Org. Chem.* 76:
1174–1176. (d) Xu, J., Luo, D.F., Xiao, B. et al. (2011). *Chem. Commun.* 47:
4300–4302. (e) Zhang, C.P., Wang, Z.L., Chen, Q.Y. et al. (2011). *Angew. Chem.
Int. Ed.* 50: 1896–1900. (f) Khan, B.A., Buba, A.E., and Goossen, L.J. (2012).
Chem. Eur. J. 18: 1577–1581.

23 Ye, Y. and Sanford, M.S. (2012). *J. Am. Chem. Soc.* 134: 9034–9037.

24 Le, C., Chen, T.Q., Liang, T. et al. (2018). *Science* 360: 1010–1014.

25 Zhang, P., Le, C.C., and MacMillan, D.W.C. (2016). *J. Am. Chem. Soc.* 138:
8084–8087.

26 Ye, F., Berger, F., Jia, H. et al. (2019). *Angew. Chem. Int. Ed.* 58: 14615–14619.

27 Kornfilt, D.J.P. and MacMillan, D.W.C. (2019). *J. Am. Chem. Soc.* 141: 6853–6858.

8

Asymmetric Fluorination Reactions

Edward Miller and F. Dean Toste

University of California, Berkeley, Department of Chemistry, Berkeley, CA 94720, USA

8.1 Introduction

Fluorinated organic molecules play an important role in pharmaceutical and agrochemical applications, primarily due to the many desirable pharmacological properties fluorine confers onto organic molecules. These effects include increased lipophilicity, tuning of acidity and basicity of surrounding functional groups, blocking oxidizable positions, as well as enhancing specific interactions with proteins [1]. Although 35% of agrochemicals and 20% of pharmaceuticals contain fluorine, fewer than 10 fluorinated natural products have been isolated to date [2, 3]. As a result of the lack of chiral fluorinated building blocks in biology, as well as the immense benefit that fluorination confers on the pharmacological properties of drug molecules, there has been considerable demand to develop stereoselective fluorination methods.

Asymmetric fluorination has historically been a challenging problem due to the inherent difficulty associated with working with fluorine. Traditional electrophilic fluorine sources, such as F_2, $CsSO_3OF$, and metal hypofluorites, are highly reactive, rendering selective fluorination or catalyst-controlled reactions challenging. The use of nucleophilic fluoride sources features its own set of challenges. Specifically, metal fluorides exhibit poor solubility and high basicity in traditional nonpolar solvents, and hydrogen fluoride is a gaseous and highly toxic acid. Despite these challenges, many strategies for selective installation of fluorine have been developed following the introduction of convenient fluorine sources.

In this chapter, we present an overview of the major developments in modern asymmetric fluorination since 2000. To improve readability, the studies highlighted here will be organized by electrophilic and nucleophilic fluorination reagents. While this chapter is primarily focused on catalytic, enantioselective fluorination reactions, select early stoichiometric electrophilic fluorination methodologies will be briefly discussed, as these served as an inspiration for many methods that followed. Emphasis will not be placed on enantioselective fluoroalkylation reactions, or enantioselective functionalization of organofluorine compounds that

Organofluorine Chemistry: Synthesis, Modeling, and Applications,
First Edition. Edited by Kálmán J. Szabó and Nicklas Selander.
© 2021 WILEY-VCH GmbH. Published 2021 by WILEY-VCH GmbH.

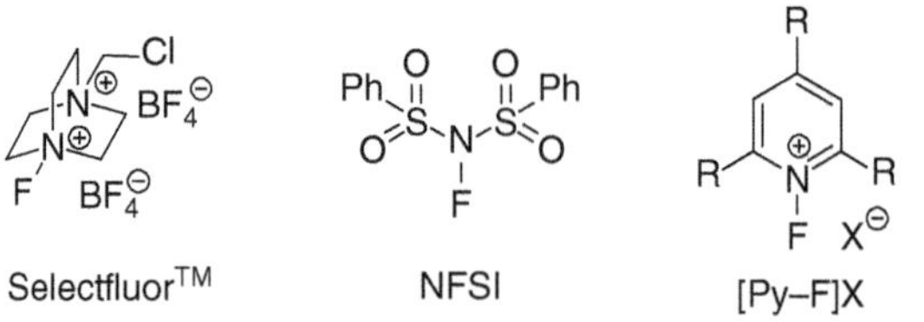

Figure 8.1 Electrophilic fluorinating reagents developed.

Selectfluor™   NFSI   [Py–F]X

generate a C–F-containing stereocenter. Following the discussion of organocatalytic electrophilic processes, metal-catalyzed reactions will be highlighted and finally, a summary of nucleophilic fluorination protocols will be outlined.

8.2 Electrophilic Fluorination

In the late twentieth century, Selectfluor [4], N-fluorobenzenesulfonimide (NFSI) [5], and N-fluoropyridinium salts (Figure 8.1) [6] became available as convenient and bench-stable sources of electrophilic fluorine. The development of these electrophilic reagents allowed for the selective functionalization of prochiral nucleophiles under mild conditions. Initial attempts to utilize these reagents in asymmetric reactions were realized using stoichiometric sources of chiral information. While these studies represent some of the first examples of asymmetric fluorination, the utility of these methods was limited by the inability to render the chiral source catalytic. Nevertheless, these examples served as a foundation from which many of the contemporary catalytic examples would draw inspiration.

8.2.1 Stoichiometric Asymmetric Fluorination

8.2.1.1 Chiral Auxiliary
Initial strategies for enantioselective fluorination chemistry involved the use of chiral auxiliaries to control stereoselectivity. In 1993, Davis et al. demonstrated

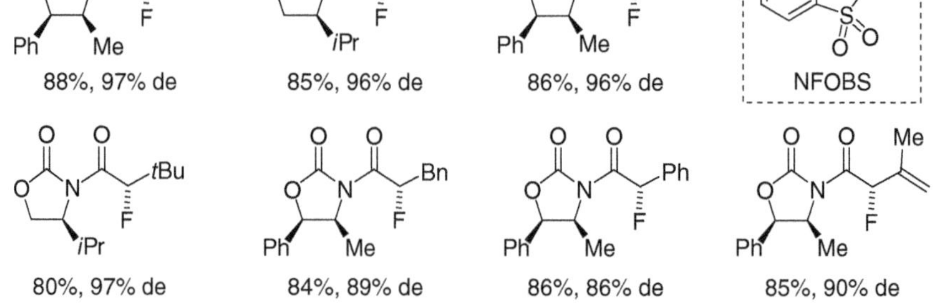

Scheme 8.1 Utilization of oxazolidinones as chiral auxiliaries for asymmetric fluorination

this concept with the use of oxazolidinones to synthesize α-fluoro carbonyl stereo-centers (Scheme 8.1) [7]. The fluorination of oxazolidinones proceeded with high diastereoselectivity (>86% de), where a simple stereochemical model rationalizes that selectivity is controlled by the preferential approach of the fluorinating agent from the less sterically hindered side of the oxazolidinone fragment. The chiral auxiliary was removed with LiBH₄ to access α-fluorohydrins in high yields and enantioselectivities. It was later demonstrated that the α-fluoro oxazolidinones could directly be converted to Weinreb amides, which were used for synthesizing a variety of fluorine-containing ketones with high enantiomeric excess [8]. A number of subsequent studies generated similar products under milder and more user-friendly conditions, either by using an enamide as the substrate [9] or by employing TiCl₄ as a Lewis acid [10].

8.2.1.2 Chiral Reagents

An alternate stoichiometric approach, developed by Differding and Lang (Scheme 8.2) [11], was achieved with camphor derived *N*-fluorosultams as chiral fluorinating reagents. These reagents were found to fluorinate a handful of sodium enolates in low to moderate yields and enantioselectivities. Subsequent reports focused on designing chiral non-camphor-based fluorinating agents to tune enantioselectivity [12–14]; however, many of these reagents featured laborious and multistep syntheses that ultimately limited their synthetic utility.

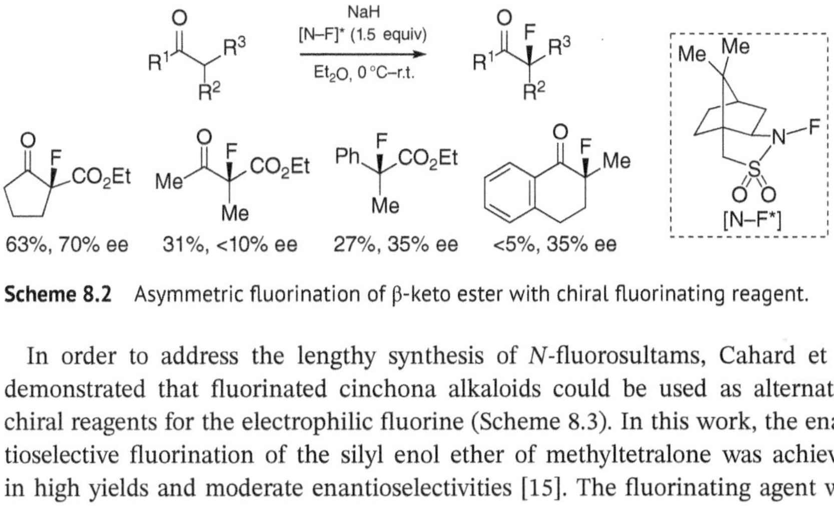

Scheme 8.2 Asymmetric fluorination of β-keto ester with chiral fluorinating reagent.

In order to address the lengthy synthesis of *N*-fluorosultams, Cahard et al. demonstrated that fluorinated cinchona alkaloids could be used as alternative chiral reagents for the electrophilic fluorine (Scheme 8.3). In this work, the enantioselective fluorination of the silyl enol ether of methyltetralone was achieved in high yields and moderate enantioselectivities [15]. The fluorinating agent was readily prepared *in situ* by reacting the parent alkaloid with either NFSI or Select-fluor. The use of unmodified cinchona alkaloids has produced a variety of synthetic methods [16], including the synthesis of both enantiomers of fluoro-thalidomide from one enantiomer of DHQ [17].

Subsequent reports focused on leveraging known synthetic manipulations of cinchona alkaloids to adjust the chiral information surrounding the electrophilic fluorine, which can have a marked effect on enantioselectivity. In 2001, Cahard and coworkers demonstrated that modification of the cinchona alkaloid scaffold allowed for improvement of the yield and enantioselectivity of the fluorination of

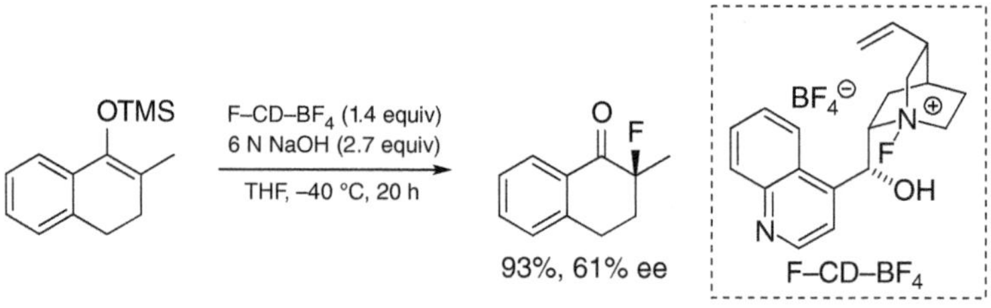

Scheme 8.3 *N*-Fluoro cinchonidinium-mediated asymmetric fluorination of a silyl enol ether.

α-cyano phthalimide. A survey of 11 reagents produced the fluorination product in yields spanning 48–91% and ees of 36–94% (Scheme 8.4) [18].

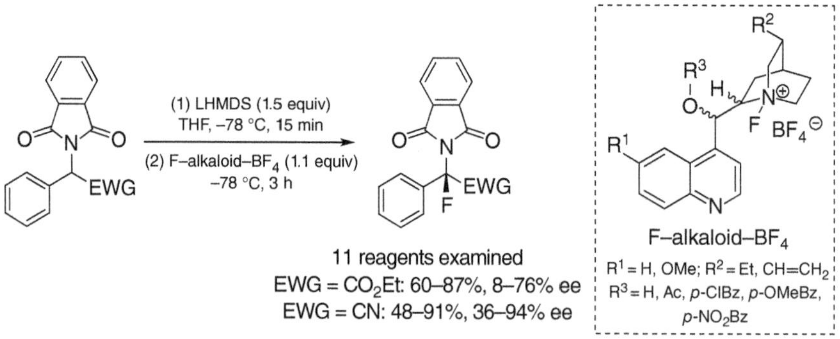

Scheme 8.4 Reagent-controlled asymmetric fluorination of α-fluoro-α-phenylglycine derivatives.

8.2.2 Catalytic Electrophilic Fluorination

8.2.2.1 Organocatalytic Fluorination

Following the development of fluorination methods with stoichiometric *N*-fluorocinchona alkaloid reagents, there was increased interest in the development of catalytic variants of these reactions. As these reagents are formed *in situ* by fluorination of the parent alkaloid, preliminary organocatalytic strategies focused on utilizing the cinchona alkaloid as a fluorine shuttle between Selectfluor and the substrate. Following the development of such methods that utilize tertiary amine catalysis, other organocatalytic strategies were explored. While many of these catalytic strategies initially relied on the cinchona alkaloid framework, many novel, nonnatural product derived catalyst systems were subsequently developed.

8.2.2.1.1 Tertiary Amine Catalysis

With the successful use of cinchona alkaloids as stoichiometric fluorinating agents, the next challenge was rendering these reactions catalytic. Results from stoichiometric variants revealed the modularity of the cinchona alkaloid framework for asymmetric fluorination, allowing for the potential of catalyst control over selectivity. One key problem that led to low enantioselectivities under catalytic conditions was the uncatalyzed background reaction between the electrophilic fluorinating reagent and enolates, which led to racemic fluorination. In 2006, Shibata and coworkers demonstrated that the uncatalyzed background reaction could be mitigated by using a less reactive enol ester substrate and a solvent (dichloromethane) in which Selectfluor has limited solubility (Scheme 8.5) [19]. In this example, the tertiary amine catalyst is fluorinated *in situ* by Selectfluor or NFSI, forming a soluble chiral fluorinating reagent. After enantioselective fluorination of the substrate, the tertiary amine catalyst is regenerated, allowing for the transfer of another electrophilic fluorine.

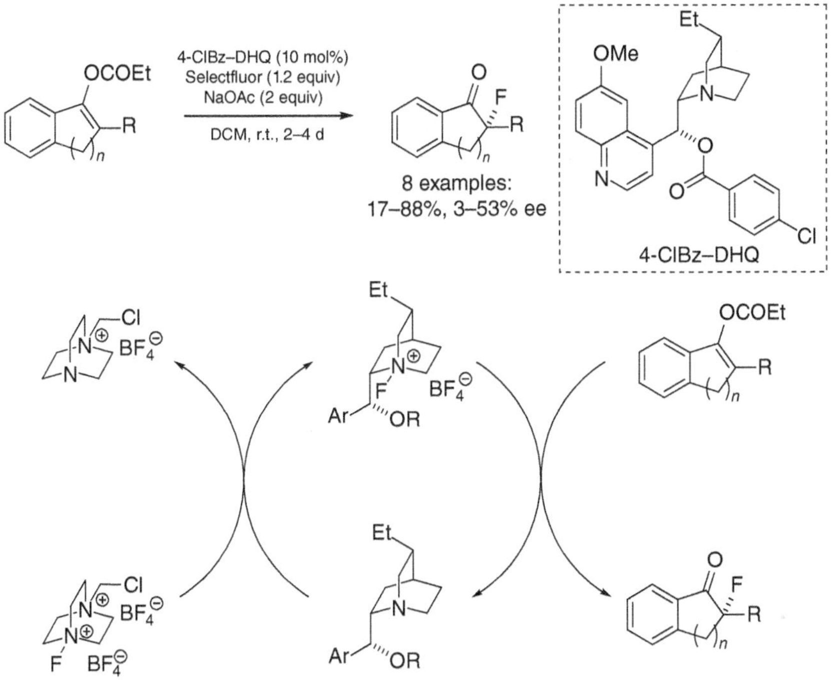

Scheme 8.5 Enantioselective fluorination of enol esters with cinchona alkaloid as tertiary amine catalyst with proposed mechanism for catalytic cycle.

In 2008, Shibata et al. published a report that focused on improved catalytic reactivity utilizing a bis-cinchona alkaloid catalyst. This catalyst affected the enantioselective fluorination of allyl silanes, silyl enol ethers, as well as oxindoles in high yields and enantioselectivities (Scheme 8.6) [20]. A similar electrophilic fluorine transfer mechanism was proposed where the quinuclidine core of the bis-cinchona alkaloid acted as an electrophilic fluorine transfer agent. This strategy was further applied to a variety of different substrate classes, such as isoxazolinones [21] and pyrazolones [22].

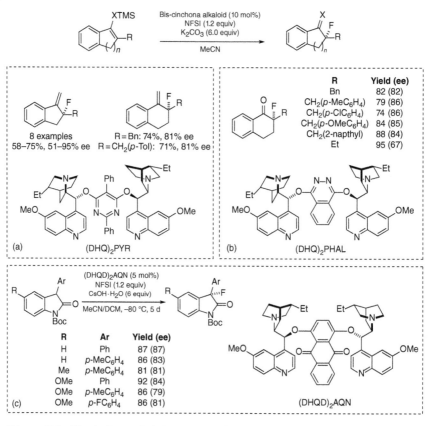

Scheme 8.6 Bis-cinchona alkaloid-catalyzed fluorination of (a) allyl silanes, (b) silyl enol ethers, and (c) and oxindoles.

The tertiary amine catalyst manifold was also applied to the fluorination of non-enolate-based substrates featuring less nucleophilic functional groups. Tu and coworkers demonstrated that (DHQ)$_2$PYR-catalyzed an enantioselective fluorinative semi-pinacol rearrangement to form β-fluoro carbonyls, which are typically difficult to access by other fluorination methods (Scheme 8.7) [23]. Additionally, the Gouverneur group displayed that (DHQ)$_2$PYR effectively catalyzed the enantioselective fluorocyclization of a number of tryptophol and tryptamine derivatives in moderate to high yields, with similarly high enantioselectivities (Scheme 8.8) [24].

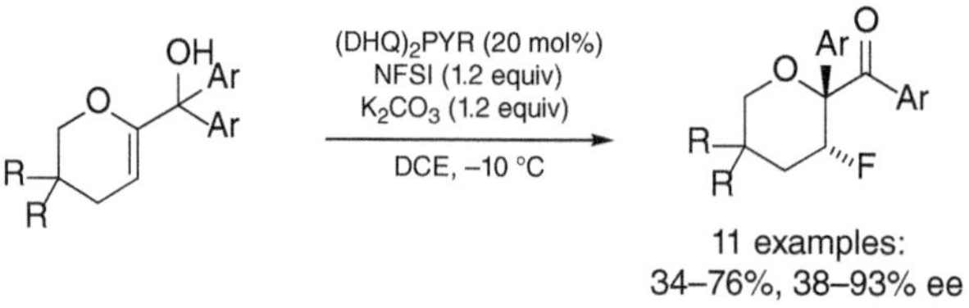

Scheme 8.7 Bis-cinchona alkaloid-catalyzed enantioselective semi-pinacol fluorination.

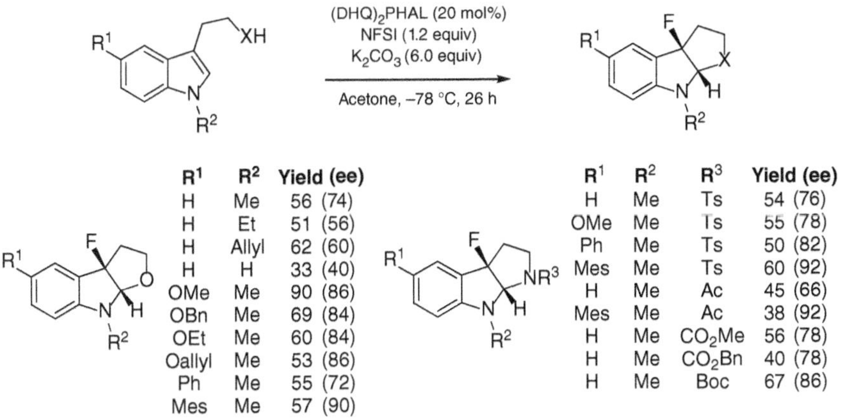

R¹	R²	Yield (ee)
H	Me	56 (74)
H	Et	51 (56)
H	Allyl	62 (60)
H	H	33 (40)
OMe	Me	90 (86)
OBn	Me	69 (84)
OEt	Me	60 (84)
Oallyl	Me	53 (86)
Ph	Me	55 (72)
Mes	Me	57 (90)

R¹	R²	R³	Yield (ee)
H	Me	Ts	54 (76)
OMe	Me	Ts	55 (78)
Ph	Me	Ts	50 (82)
Mes	Me	Ts	60 (92)
H	Me	Ac	45 (66)
Mes	Me	Ac	38 (92)
H	Me	CO₂Me	56 (78)
H	Me	CO₂Bn	40 (78)
H	Me	Boc	67 (86)

Scheme 8.8 Bis-cinchona alkaloid-catalyzed enantioselective fluorocyclization of tryptophols and tryptamines.

8.2.2.1.2 *Chiral Cation Phase Transfer*

Phase transfer catalysis with chiral cations is an important strategy for the electrophilic functionalization of anionic substrates. In 2002, the Kim group employed this strategy for electrophilic fluorination, where benzylated cinchonine catalyzed the enantioselective fluorination of β-keto esters (Scheme 8.9) [25]. It was identified that the benzyl and propargyl fragments were crucial to the enantioselectivity of the reaction, where other substitution patterns led to diminished yield and enantioselectivity. A handful of β-keto esters were fluorinated in good yields and moderate enantioselectivities. In this paradigm, the alkylated cinchona alkaloid cation acts as a phase transfer catalyst, to solubilize the inorganic base into the nonpolar organic solvent. After deprotonation of the substrate, the chiral cation is now ion-paired with the anionic substrate, where fluorination occurs enantioselectively due to the selective blocking of one face of the nucleophile by the chiral cation. This strategy was also applied to asymmetric fluorination of α-cyano esters [26].

Following these earlier examples, the compatibility of various chiral cation phase transfer catalysts to enantioselective fluorination was explored. In 2010, Maruoka and coworkers showed that quaternary ammonium catalyst, based on the binapthyl scaffold, catalyzed the enantioselective fluorination of a variety of benzo and non-benzofused cyclic β-keto esters in high yields and enantioselectivities (Scheme 8.10) [27]. A similar phosphonium salt was reported by Ma and coworkers to catalyze the enantioselective fluorination of lactones, albeit with low to moderate enantioselectivities [28].

In 2014, Waser and coworker reported the use of urea-containing ammonium phase transfer catalysts for the enantioselective fluorination of β-keto esters (Scheme 8.11) [29]. The bifunctional catalysts were proposed to serve two roles; the traditional phase transfer activation of the substrate, as well as activation of the NFSI via hydrogen bonding with the urea moiety. As a result, a variety of substrates were fluorinated in high yield and enantioselectivities. A polyethylene glycol (PEG)

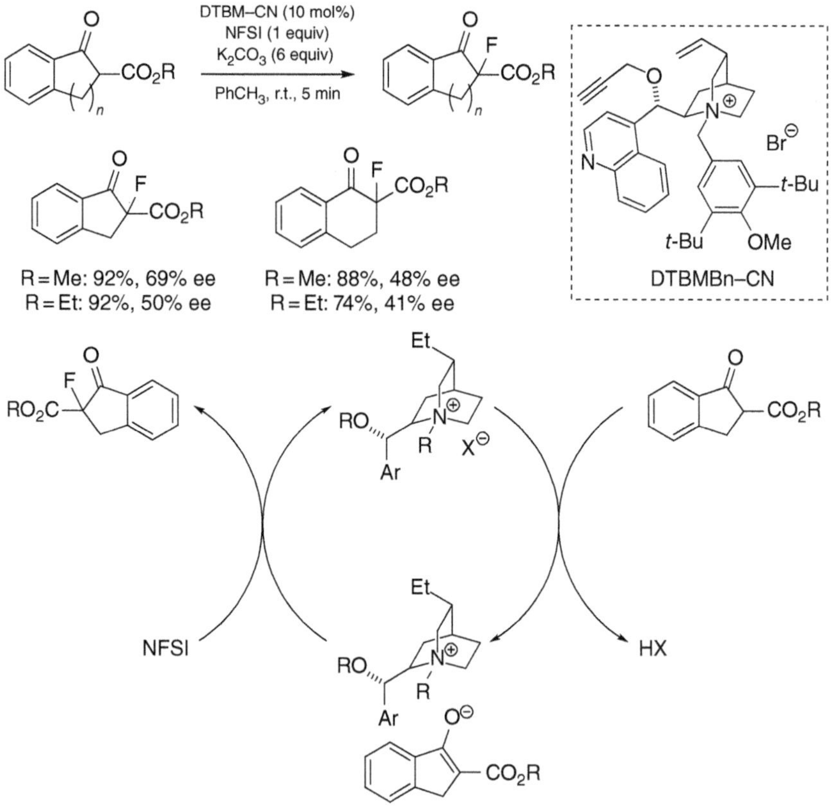

Scheme 8.9 Chiral cation-catalyzed fluorination of β-keto esters with proposed catalytic cycle.

bound version of the catalyst was later synthesized to facilitate re-isolation by precipitation with ether [30].

8.2.2.1.3 Enamine Catalysis

Covalent activation of carbonyl compounds is a mechanistic staple of many enantioselective transformations. In this paradigm, a chiral amine condenses with a carbonyl to form a chiral enamine, which both enhances nucleophilicity and restricts the approach of electrophiles from one enantioface of the enamine π-bond preferentially. Initial attempts to amend this strategy to electrophilic fluorination were conducted by Enders and Hüttl in 2005 with the asymmetric fluorination of cyclohexanone (Scheme 8.12) [31]. A variety of proline derivatives were explored; but yields and enantioselectivities were low in most cases.

This strategy proved more readily amenable to aldehydes, as first demonstrated by both Jørgensen [32] and MacMillan [33] in 2005 (Scheme 8.13a,b). Jørgensen and coworkers utilized a diarylprolinol silyl ether (Hayashi–Jørgensen catalyst), while MacMillan used an imidazolidinone catalyst to achieve high yield and enantioselectivities (>90% ee for both methods). The two methods differ mainly

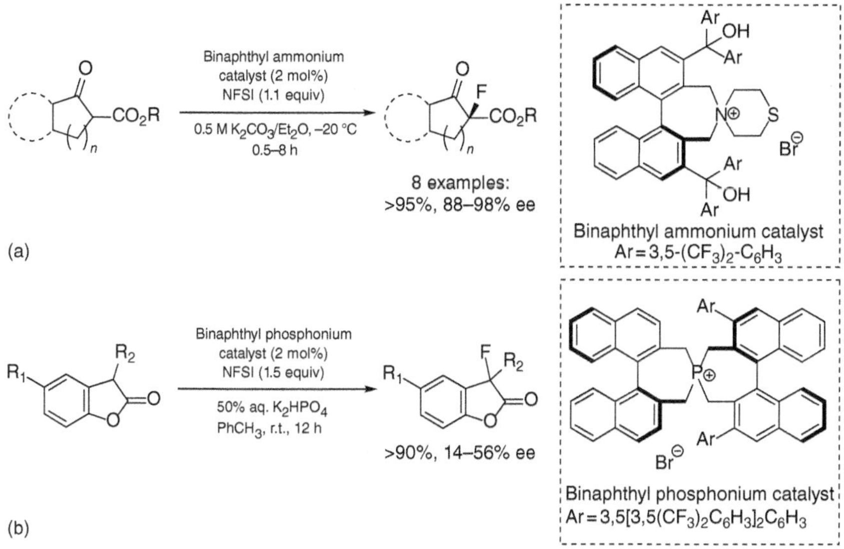

(a)

(b)

Scheme 8.10 Enantioselective fluorination of enolates using tetra-alkyl ammonium (a) and phosphonium (b) phase transfer catalysts.

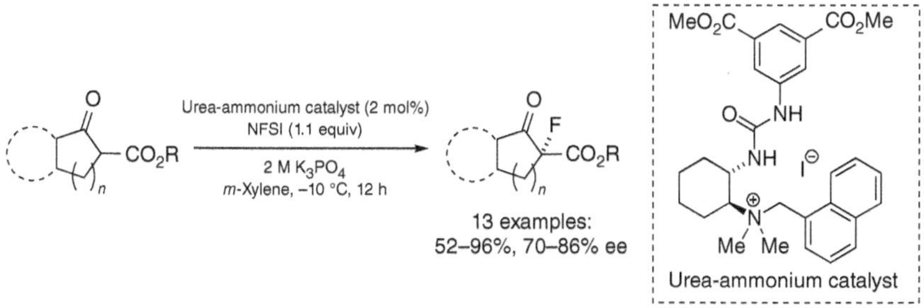

Scheme 8.11 Bifunctional urea–ammonium phase transfer catalyst for enantioselective fluorination of β-keto esters.

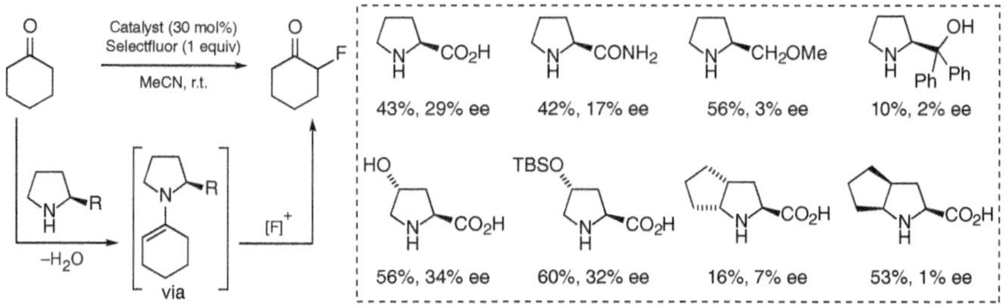

Scheme 8.12 Amine-catalyzed activation of cyclohexanone for electrophilic fluorination.

in stereochemical outcome for the α-fluoro aldehyde, due to the difference in alkene geometry of the enamide. As shown in Scheme 8.13c, the steric influence of the dimethyl-methylene in the MacMillan catalyst favors a configuration with the chiral information on the opposite face of the olefin, as compared to the Hayashi–Jørgensen catalyst. Currently, branching at the α-position is an unsolved problem for this strategy.

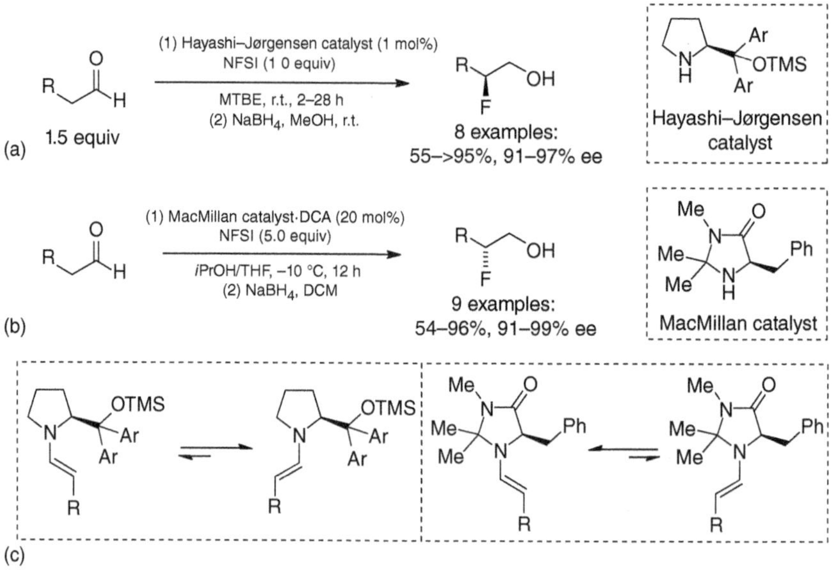

Scheme 8.13 Amine-catalyzed enantioselective fluorination of aldehydes with (a) Hayashi–Jørgensen catalyst and (b) MacMillan imidazolidinone catalyst. (c) Speciation of enamines for different catalyst systems.

Fluorination utilizing the imidazolidinone catalyst was shown to be amenable to ^{18}F fluorination, for positron emission tomography (PET) applications, by the Gouverneur laboratory in 2015 (Scheme 8.14) [34]. In this report, radiochemical conversions of up to 62% were observed, with enantioselectivities above 90% ee. The stereochemical rationale is presumably identical to the MacMillan example shown above.

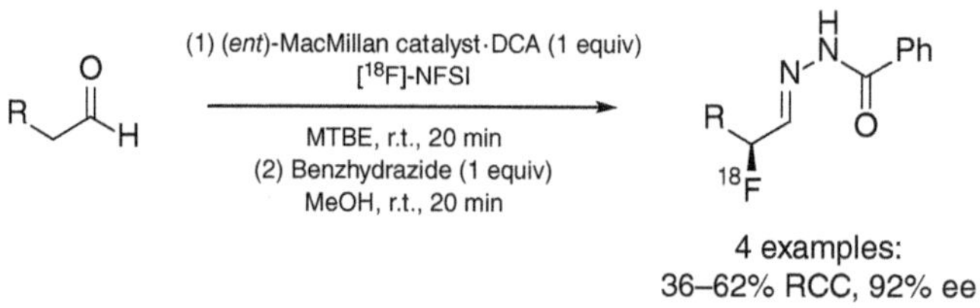

Scheme 8.14 ^{18}F fluorination of aldehydes utilizing enamine catalysis for PET applications.

In addition to the previously displayed compatibility of the stereogenic α-fluoroaldehyde motif toward reduction and condensation with amines, these products were also converted to a propargyl fluoride without loss of stereochemistry by *in situ* alkynylation with the Ohira–Bestmann reagent (Scheme 8.15a) [35]. The propargyl fluorides underwent copper-catalyzed click reactions to form the corresponding triazoles, Sonogashira coupling, as well as other alkynylation reactions in high yields without loss of stereochemistry. Allylic fluorides were also accessed by Wittig olefination of the α-fluoro aldehyde. In 2009, Lindsley and coworker demonstrated that the α-fluoro aldehydes underwent reductive amination to generate β-fluoroamines in high yield without loss of stereochemistry (Scheme 8.15b) [36].

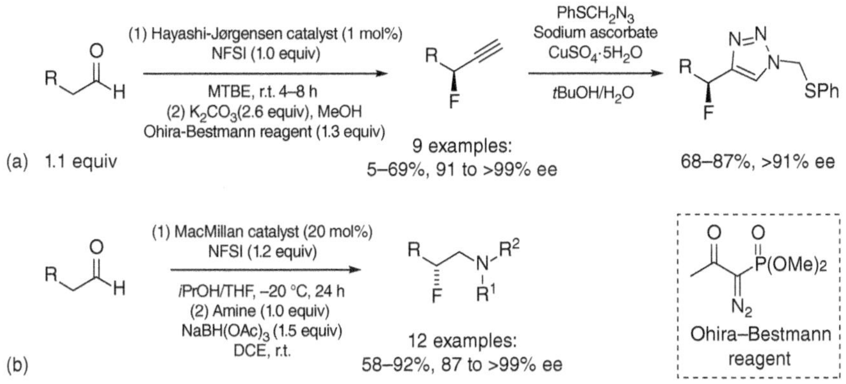

Scheme 8.15 *In situ* alkynylation (a) and reductive amination (b) of stereogenic α-fluoro-aldehydes.

In 2011, a highly enantioselective variant of cyclic ketones was disclosed by the MacMillan laboratory, where an extensive screening of chiral amines revealed that an aminated cinchona alkaloid (NH$_2$–DHCN) provided the fluorination products in high yield and enantioselectivities (Scheme 8.16) [37]. A handful of densely functionalized polycyclic compounds were subjected to fluorination conditions, providing the product in high yield and diastereomeric ratios. The high reactivity and selectivity of the aminated catalyst was proposed to be a result of it serving two mechanistic roles, as an activator of the substrate and as an intramolecular fluorinating agent after fluorine transfer to the tertiary amine. A subsequent computational study revealed that the preferred conformation of the seven-membered ring transition state associated with this intramolecular fluorination is responsible for the observed selectivities [38].

8.2.2.1.4 Chiral Lewis Base Catalysis

Alongside chiral amines, chiral Lewis base catalysts have also been shown to be useful in covalent activation of carbonyl-containing substrates. In 2008, Leckta and coworkers demonstrated that acyl chlorides were activated by cinchona alkaloid derivatives in conjunction with a transition metal Lewis acid (Scheme 8.17) [39]. The

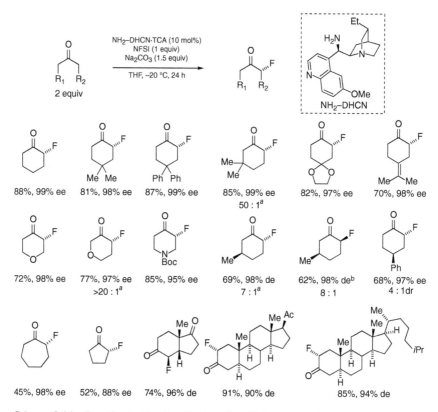

Scheme 8.16 Enantioselective fluorination of cyclic ketones regioselectivity[a] (*ent*)-NH₂–DHCN[b] used.

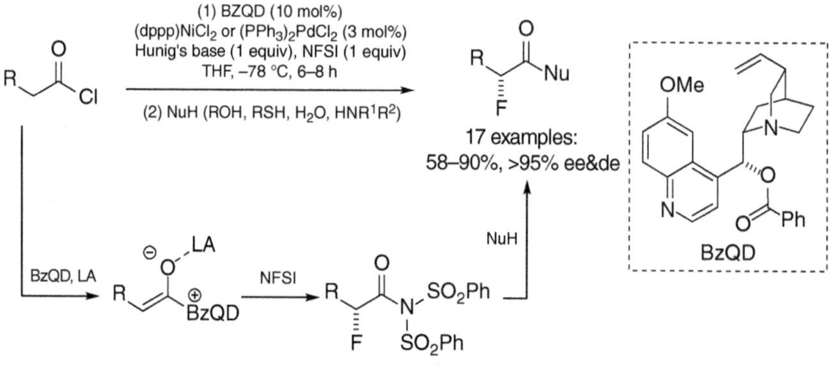

Scheme 8.17 Enantioselective α-fluorination of acid chlorides.

resulting chiral zwitterionic enolate was fluorinated in high yield and selectivity, and then hydrolyzed by the latent nucleophile in solution, enabling catalyst turnover. A *trans*-acylation was conducted, allowing for a divergent approach for the synthesis of fluoro amides and esters enantioselectively. In a closely related follow-up publication, the group showed that this methodology can be applied to natural product functionalization [40].

In 2014, the Fu group demonstrated similar reactivity with a ferrocene-based chiral 4-dimethylamino pyridine (DMAP*) derivative (Scheme 8.18) [41]. In this work, the chiral DMAP* condenses with the ketene to form a zwitterionic enolate, which is preferentially fluorinated in high yield and enantioselectivity from the less sterically hindered face. The catalyst is then regenerated by nucleophilic attack of perfluoro phenoxide. The resulting α-fluoro perfluoro phenol ester product was then converted to a variety of α-fluoro carbonyl-containing products, such as amides, esters, alcohols, and carboxylic acids. Evidence for this mechanism came with stoichiometric experiments, where α-fluoro acylated DMAP* was isolated as a single diastereomer and hydrolyzed to form the desired product.

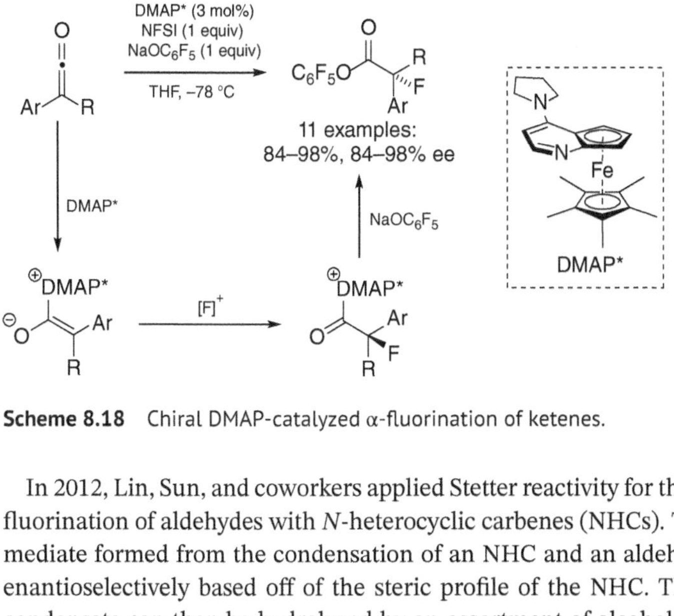

Scheme 8.18 Chiral DMAP-catalyzed α-fluorination of ketenes.

In 2012, Lin, Sun, and coworkers applied Stetter reactivity for the enantioselective fluorination of aldehydes with *N*-heterocyclic carbenes (NHCs). The Breslow intermediate formed from the condensation of an NHC and an aldehyde is fluorinated enantioselectively based off of the steric profile of the NHC. The NHC-substrate condensate can then be hydrolyzed by an assortment of alcohols, allowing for the synthesis of a variety of α-fluoro esters in high enantioselectivities (Scheme 8.19) [42]. In successive reports, similar reactivity was achieved with unactivated aldehydes [43, 44] and cinnamaldehydes [45].

8.2.2.1.5 Chiral Anion Phase Transfer

While chiral cation phase transfer catalysis had proved incredibly useful in the field of asymmetric catalysis, the substrate scope was typically limited to activated nucleophiles or those with low pKas, due to the mode of activation. In order to solve this problem, the Toste group developed a chiral anion phase transfer

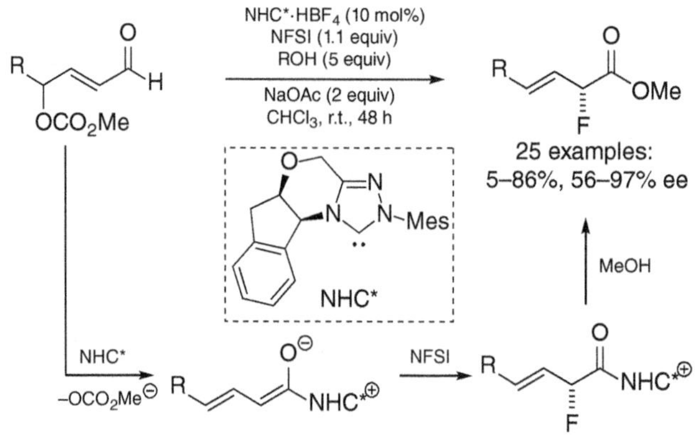

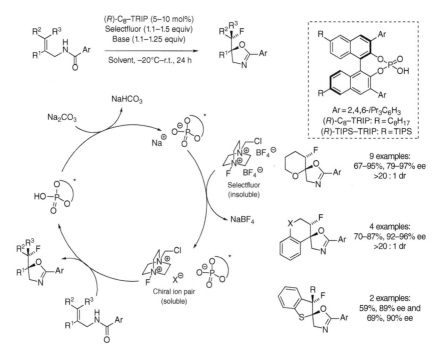

Scheme 8.19 Chiral NHC-catalyzed synthesis of α-fluoro-β,γ-unsaturated esters via Stetter-type reactivity.

(CAPT) approach in 2011, where a number of alkene-containing substrates such as dihydropyran, dihydronaphthalene, and benzothiophene-containing amides, underwent fluorination with high yields and enantioselectivities (Scheme 8.20) [46]. It was proposed that the deprotonated chiral phosphoric acid (CPA) catalyst undergoes salt metathesis with the insoluble Selectfluor to form a soluble chiral ion pair, which fluorinated the substrate enantioselectively. This chiral ion pair with the cationic electrophilic reagent allows non-anionic and relatively unactivated

Scheme 8.20 CAPT-catalyzed fluorocyclization of olefins

substrates to be compatible with this strategy. Mechanistically, the phosphate anion serves the dual purpose of both activating the substrate via hydrogen bonding and ion-pairing with the reagent.

Since this initial report, the CAPT strategy has proved to be a useful method for the construction of fluorinated heterocycles of different substrate classes. In 2017 and 2018, the You group expanded this strategy to the fluorocyclization of tryptophols [47] and tryptamines [48], respectively, utilizing similar reaction conditions (Scheme 8.21).

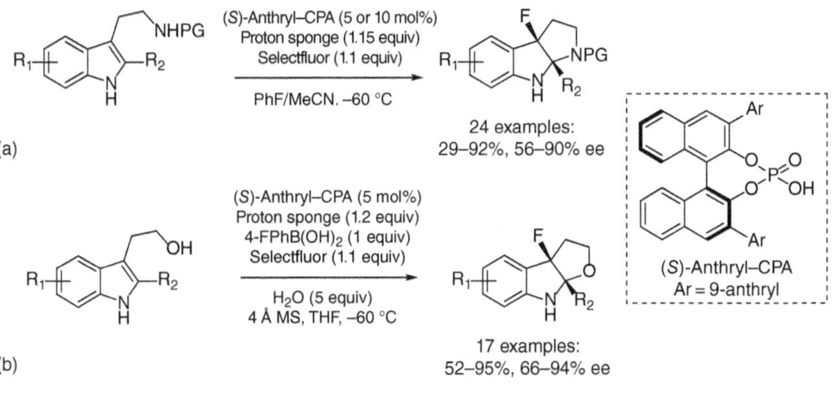

Scheme 8.21 Chiral anion phase transfer-catalyzed fluorocyclization of (a) tryptophols and (b) tryptamines.

In addition to the ability to synthesize fluorinated heterocycles through enantioselective fluorocyclization, CAPT has been used to perform a variety of stereoselective acyclic fluorination reactions. In 2012, the Toste group demonstrated this by expanding the scope of reactions to α-fluorinations, first to enamides [49] followed closely by phenols [50]. For the phenolic substrates, dearomative α-fluorination was observed in high yield and enantioselectivity (Scheme 8.22). Interestingly, for less sterically encumbered substrates, a [4+2] dimerization following fluorination was observed, leading to fluorinated tricyclic compounds with dense functionality. A fluoro-dehydroxy analog of (−)-grandifloracin was synthesized using this methodology.

In 2013 and 2014, the Toste group demonstrated the application of this methodology to allylic fluorinations, with both amide [51] and *in situ*-generated boronic acid-monoester directing groups [52] (Scheme 8.23a,b). In the latter case, condensation of the alcohol with an achiral boronic acid proved to be crucial to both the yield and enantioselectivity of the reaction. By judicious choice of boronic acid, the product was isolated in high yield and enantioselectivity.

In a follow-up study, systematic alteration of the catalyst and boronic acid structure in conjunction with multivariate parameterization and traditional mechanistic investigation revealed that the observed selectivity differences were the result of multiple potential mechanistic paradigms being operative. By studying this effect, the authors developed a system where one chiral catalyst can yield either enantiomer

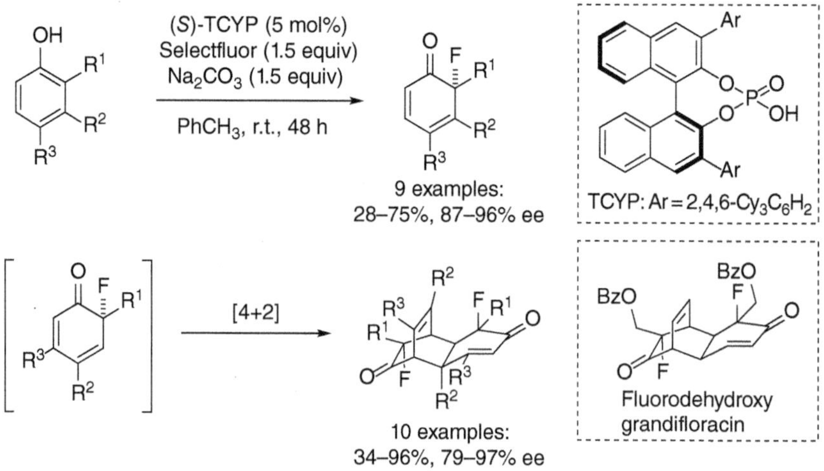

Scheme 8.22 Enantioselective α-fluorination of phenols enabled by CAPT and subsequent [4+2] cycloaddition for sterically unhindered substrates.

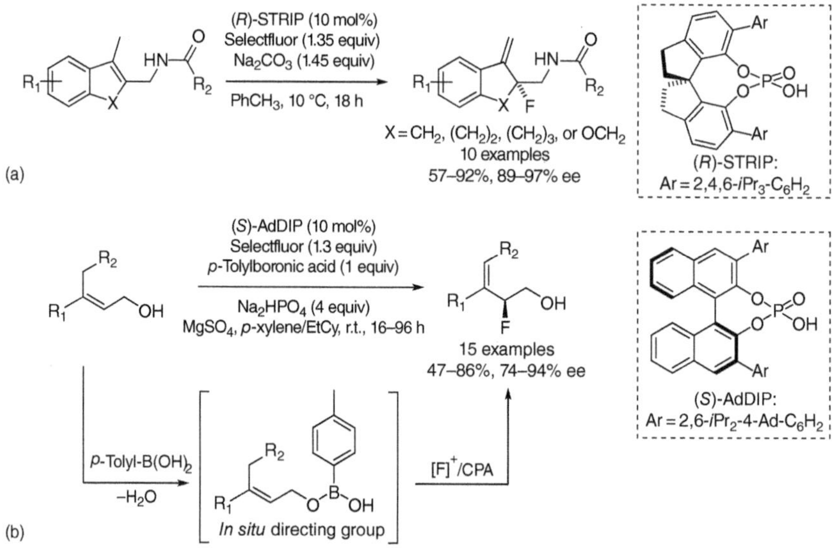

Scheme 8.23 Chiral anion phase transfer-catalyzed allylic fluorination with amides (a) and boronic esters (b) as directing groups.

of the product depending on the identity of the achiral boronic acid (Scheme 8.24) [53]. A similar effect was observed in the homoallylic version of the reaction, where multivariate analysis revealed that a potential T-shaped arene–arene interaction was responsible for the enantioselectivity observed [54].

In 2014, the Alexakis group applied the CAPT approach to a fluorinative aza-semi-pinacol rearrangement to synthesize spirocyclic β-fluoro imines in high yield, diastereoselectivity, and enantioselectivity (Scheme 8.25) [55]. The

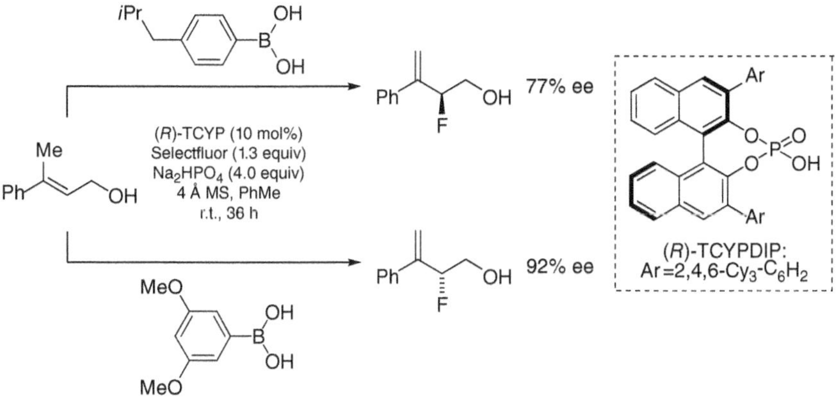

Scheme 8.24 Enantiodivergent fluorination of allylic alcohols with different boronic acid directing groups.

methodology was later expanded to strained alcohols, providing β-fluoro spirocyclic butanones, pentanones, and hexanones in similarly high yield, diastereoselectivity, and enantioselectivity [56]. These cyclic ketones underwent Baeyer–Villiger oxidation, as well as reduction via Red-Al, in high yield without loss of stereochemistry.

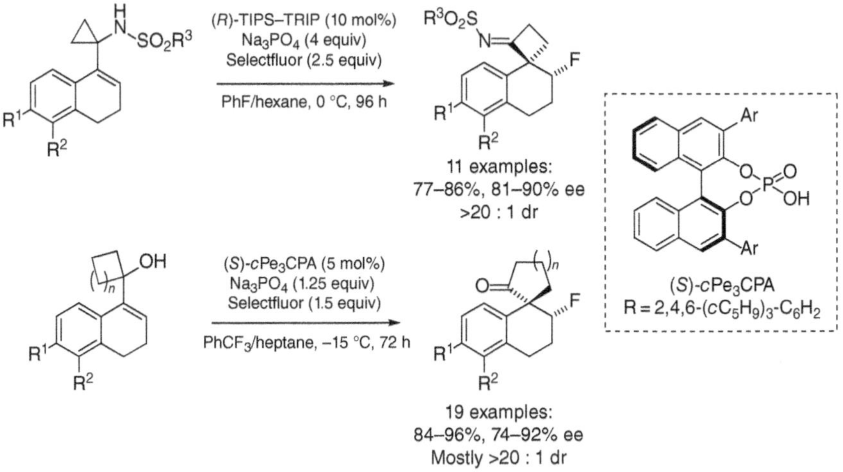

Scheme 8.25 Chiral anion phase transfer-catalyzed fluorinative semi-pinacol rearrangements.

In 2014, the Toste group published a cooperative amine–CPA catalyst system for the α-fluorination of ketones (Scheme 8.26) [57]. Under the reaction conditions, the chiral amine was proposed to condense with the ketone to form an enamine. The enamine N—H bond then acts as a hydrogen bond donor to the catalyst and aids in directing the fluorination event from one face. During the optimization process, the identity of the amine was found to be critical to the reaction, where matched and mismatched effects based off of the chirality of amine–CPA pair were observed in the enantioselectivity of the reaction.

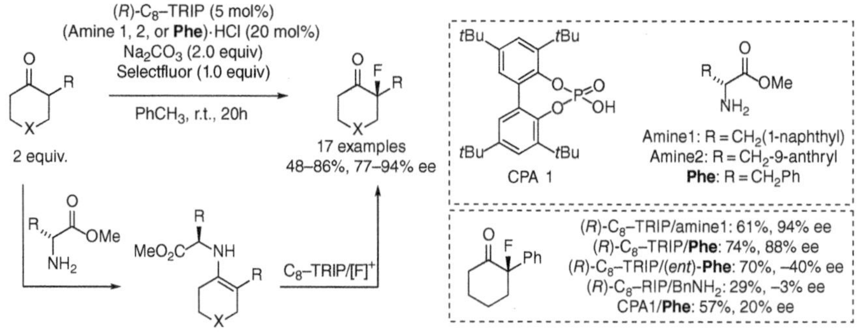

Scheme 8.26 Enamine–CAPT dual catalysis for enantioselective α-fluorination of ketones.

Aside from the single axially CPA-based catalysts, other catalyst frameworks have been developed to achieve CAPT reactivity. In 2012, the Toste group demonstrated that a double axially CPA could be used for the CAPT-catalyzed oxy-fluorination of enamides in high yields, diastereoselectivity, and enantioselectivity (Scheme 8.27) [58]. An intramolecular variant of the transformation was later disclosed for the synthesis of dihydroquinazolones in high yield and enantioselectivity [59].

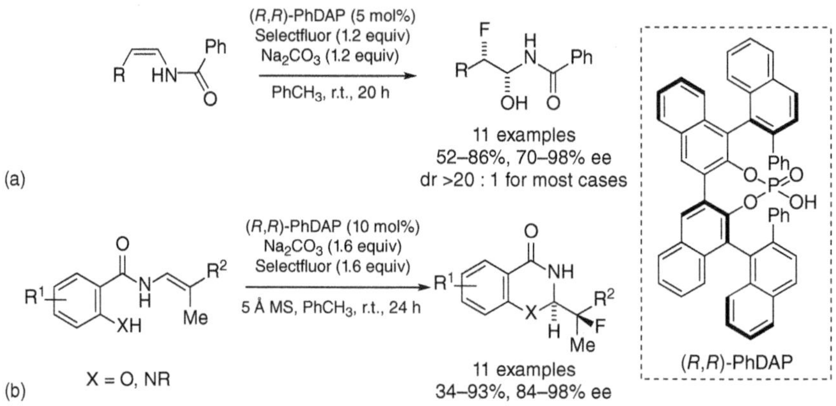

Scheme 8.27 (a) Intermolecular and (b) intramolecular fluoro-functionalization of enamides.

Chiral anion phase transfer catalysts are not limited to the BINOL-derived phosphoric acids. In 2015, the Hamashima group disclosed an enantioselective fluoro-lactonization that proceeds with an alternative chiral anion catalyst (Scheme 8.28a) [60]. This catalyst contains both a carboxylic acid and an alcohol functional group, which are proposed to mimic the dual-purpose nature of the CPA catalyst for both ion-pairing and acting as a hydrogen bond donor. As a probe for their mechanistic proposal, a catalyst featuring a methylated ether at the alcoholic position was synthesized and tested under their reaction conditions. The catalyst produced the racemic product in a diminished 28% yield. Thus, it was posited that this hydrogen bonding interaction was key to catalyst efficacy. In a closely related follow-up report, Hamashima and coworkers displayed the synthetic utility of carboxylic acid-based phase transfer catalyst, by synthesizing a tethered version of the previously reported catalyst (Scheme 8.28b) [61]. This catalyst was hypothesized to undergo double deprotonation, allowing for better cooperation with the doubly cationic Selectfluor. Utilizing this catalyst, the authors affected an enantioselective 6-*endo* fluorocyclization of styrenal amides with high yields and enantioselectivity.

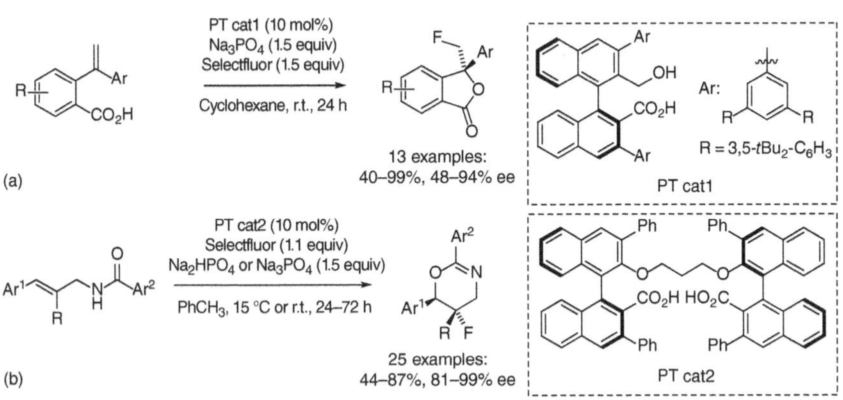

Scheme 8.28 Chiral carboxylic acid phase transfer catalysts for the enantioselective fluorocyclizations of (a) benzoic acids and (b) amides.

8.2.2.2 Transition Metal-Catalyzed Fluorinations

Transition metal-catalyzed enantioselective reactions have historically played an important role in the field of asymmetric fluorination. Initial pioneering examples focused on fluorination of enolates that can engage in two-point binding to a Lewis acidic metal, where chiral information installed on the ligands can direct the facial selectivity between the enolate and the electrophilic fluorine source. A variety of metals and ligand platforms have shown that this is a general strategy for enantioselective fluorination of enolates. In recent years, there have been an increasing number of examples wherein the transition metal is directly involved in the C—F bond forming step via a C–F reductive elimination. These examples typically require transition metals in high oxidation states, which significantly limits the metal and ligand platforms that can be utilized.

8.2.2.2.1 Fluorination of Metal Enolates or Equivalent

As previously mentioned, initial reports focused on substrates that displayed two-point binding with a transition metal catalyst to generate rigid cyclic chiral nucleophiles (Scheme 8.29). As a result, β-keto esters have been the substrate of choice for many early examples of this strategy in developing metal–ligand systems for affecting these transformations. In these systems, the coordination of both carbonyls to one metal reduces the pKa of the substrate significantly, allowing it to be deprotonated with a variety of bases. Upon deprotonation, the substrate–catalyst pair forms a rigid system that is selectively fluorinated from one face, as determined by the chiral ligand. After fluorination, the product is displaced by the starting material, turning over the catalyst.

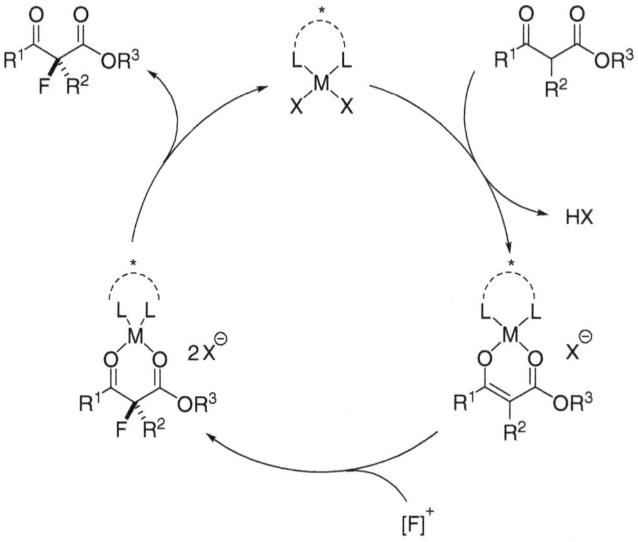

Scheme 8.29 General mechanism for metal-catalyzed fluorination of β-keto esters.

Ti–Taddol Complexes In 2000, Hintermann and Togni published the first metal-catalyzed enantioselective fluorination reaction. The TiCl$_2$(TADDOLato) catalyst effectively fluorinated β-keto esters at room temperature with high enantioselectivity (Scheme 8.30) [62]. An ensuing computational study revealed the possibility of single-electron transfer and radical intermediates in the fluorination mechanism [63]. In 2012, Togni published an expanded substrate scope for the enantioselective α-fluorination of activated carbonyl compounds. In addition to a variety of acyclic and cyclic β-keto esters, thioesters, amides, and diketones were examined [64].

BINAP- and SEGPHOS-Based Ligands Bisphosphine-based ligands comprise a large fraction of metal-catalyzed enantioselective fluorination reactions, most commonly ligated with palladium as the Lewis acid. The first example of this catalyst platform was reported by Sodeoka and coworkers in 2002, where a BINAP–Pd complex acted as a Lewis acid to activate β-ketoesters toward electrophilic fluorination (Scheme 8.31) [65]. A handful of compounds were fluorinated in high yields (49–92%) and high enantioselectivities (83–94% ee). The reaction was performed on gram scale with no observable loss in reactivity or enantioselectivity. From the enantioenriched fluorinated β-keto esters, the authors accessed β-hydroxy and β-amino esters via reduction and Mitsunobu amination. In a subsequent

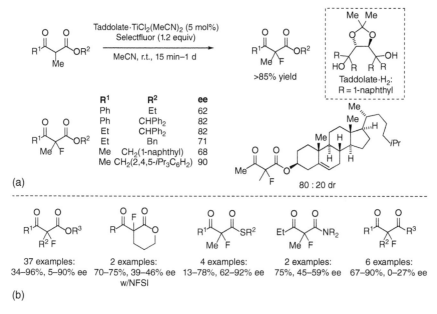

(a)

(b)

Scheme 8.30 Ti–Taddol-catalyzed electrophilic fluorination of β-keto esters (a) and expanded substrate scope (b).

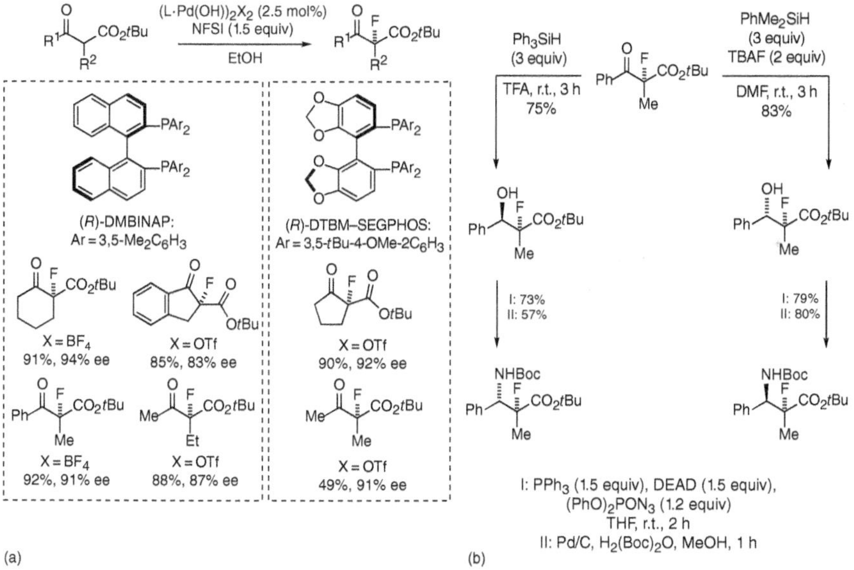

(a)

(b)

Scheme 8.31 (a) Pd-catalyzed fluorination of β-keto esters with BINAP and SEGPHOS ligands. (b) Derivatization of α-fluoro-β-keto esters.

report, Sodeoka and coworkers immobilized the Pd–BINAP catalyst in an ionic liquid, which allowed for catalyst reuse for 11 cycles without loss of reactivity or enantioselectivity [66].

Utilizing this synthetic platform, many substrate classes were fluorinated with high yield and enantioselectivity. Sodeoka and coworkers first expanded the methodology to β-keto phosphonates by utilizing a similar catalyst architecture and conditions (Scheme 8.32a) [67]. This was soon followed by a report from Kim and coworker where α-cyano esters were fluorinated in high yield and enantioselectivity [68]. Since these initial reports, the scope of substrates has expanded to α-cyanophosphonates [69, 70], α-chloro-β-ketophosphonates [71], α-cyanosulfones [72], β-ketoamides [73], and α-ketoesters [74]. The synthetic utility of the Pd–BINAP system for asymmetric fluorination was showcased in a scalable synthesis to a spleen tyrosine kinase (Syk) inhibitor, in which an intermediate was fluorinated with NFSI to produce 44 kg (68%) of product as a single diastereomer (Scheme 8.32b) [75].

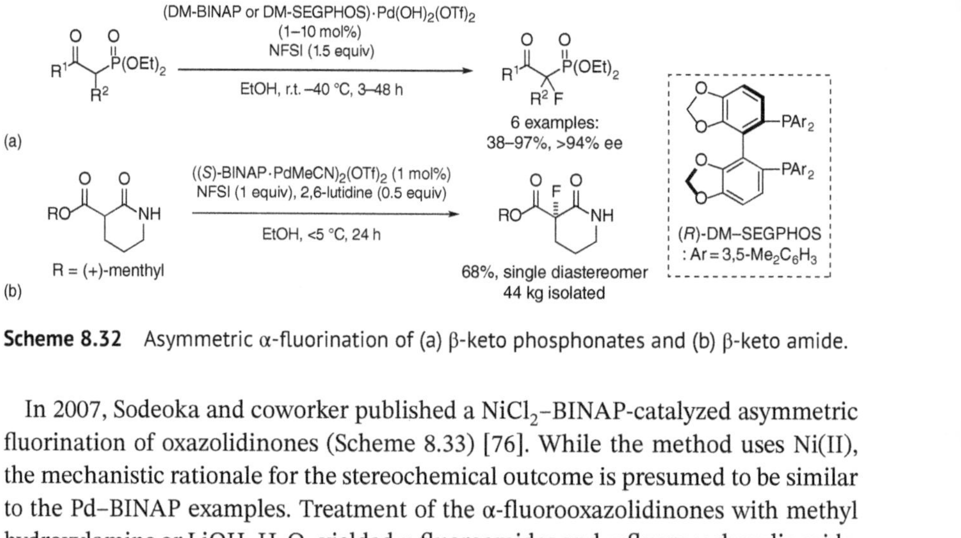

Scheme 8.32 Asymmetric α-fluorination of (a) β-keto phosphonates and (b) β-keto amide.

In 2007, Sodeoka and coworker published a $NiCl_2$–BINAP-catalyzed asymmetric fluorination of oxazolidinones (Scheme 8.33) [76]. While the method uses Ni(II), the mechanistic rationale for the stereochemical outcome is presumed to be similar to the Pd–BINAP examples. Treatment of the α-fluorooxazolidinones with methyl hydroxylamine or $LiOH–H_2O_2$ yielded α-fluoroamides and α-fluorocarboxylic acids, respectively. This method was valuable, as it provided a solution for the incompatibility of monoesters in asymmetric fluorination.

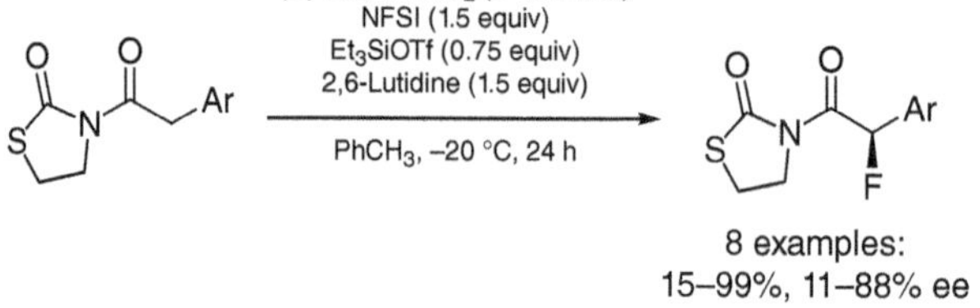

Scheme 8.33 Enantioselective fluorination of oxazolidinones by Ni–BINAP catalyst system.

Oxazoline-Based Ligands Oxazolines, a privileged ligand class for methodologies involving enantioselective transition metal catalysis, were first used for asymmetric fluorination in 2004 by Cahard and coworker. In this example, a Cu(II)–BOX complex catalyzed the fluorination of β-keto esters in high yield and moderate to high enantioselectivities, with as low as 1 mol% catalyst loading (Scheme 8.34a) [77]. Similar to previously published metal-catalyzed examples, enantioinduction was presumed to be a result of the bidentate binding of the substrate to the catalyst, which influences the face available to the electrophile. In a closely related publication, Shibata, Toru, and coworker demonstrated that the enantiomer produced in the reaction can be switched by changing from Cu(OTf)$_2$ to Ni(ClO$_4$)$_2$·6H$_2$O (Scheme 8.34b) [78]. The source of enantiodivergent fluorination was rationalized by alternate binding geometries between the two metals, where Cu(II) and Ni(II) were proposed to be distorted square planar and square pyramidal, respectively.

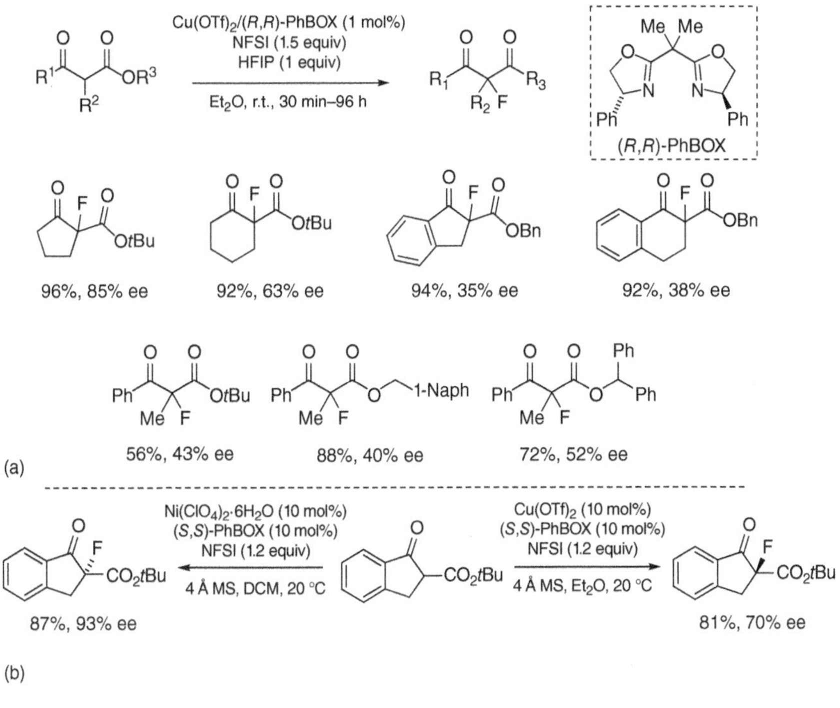

Scheme 8.34 (a) Fluorination of β-keto esters in the presence of a Cu–BOX catalyst. (b) Metal-dependent divergence in enantioselectivity.

In a follow-up publication, Shibata, Toru, and coworker improved the enantioselectivities of the β-keto ester substrate class, as well as expanded the scope of the reaction by utilizing (R,R)-DBFOX-Ph, an oxazoline-containing ligand based on the dibenzofuran core. A variety of cyclic and acyclic β-keto esters as well as oxindoles were fluorinated in high yield (>70%) and high enantioselectivities (>83% ee, Scheme 8.35) [79]. The Xu group adapted this process to flow [80] and ball mill [81] synthetic protocols by using similar tridentate ligand frameworks. PyBox ligands, in conjunction with Lewis acidic lanthanide metal salts, also catalyzed the fluorination of similar β-keto ester substrates in high yield and enantioselectivity [82].

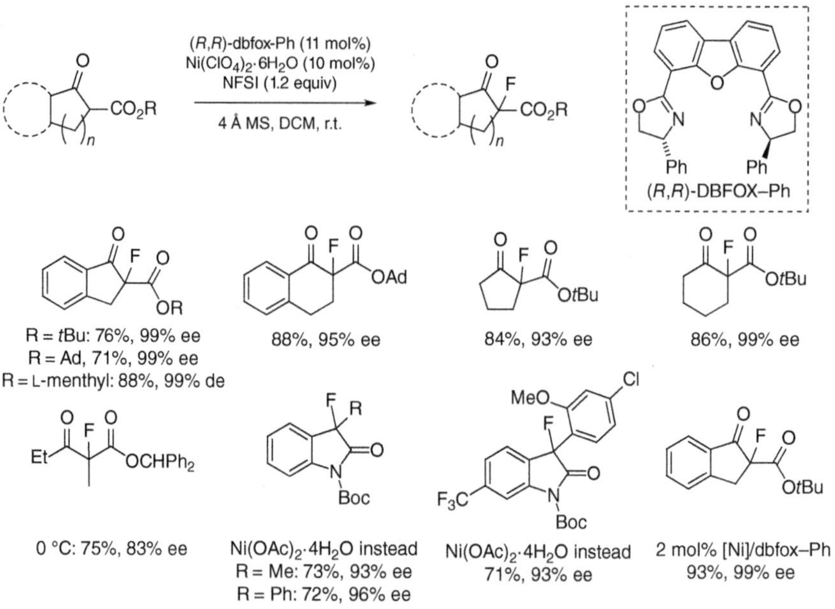

Scheme 8.35 Ni–DBFOX-catalyzed enantioselective fluorination of various substrate classes.

In 2008, Shibata, Toru, and coworker used the DBFOX ligand platform to affect the enantioselective fluorination of malonates in a desymmetrization-like transformation. A variety of acyclic malonates were fluorinated in 81–95% yield and ees >90% (Scheme 8.36) [83]. In 2009, Shibata and coworkers reported the enantioselective fluorination of thiazolidinones with a Ni(ClO₄)₂–DBFOX catalyst system. A variety of benzylic and allylic substituents were examined, wherein generally high yields and enantioselectivities were observed [84].

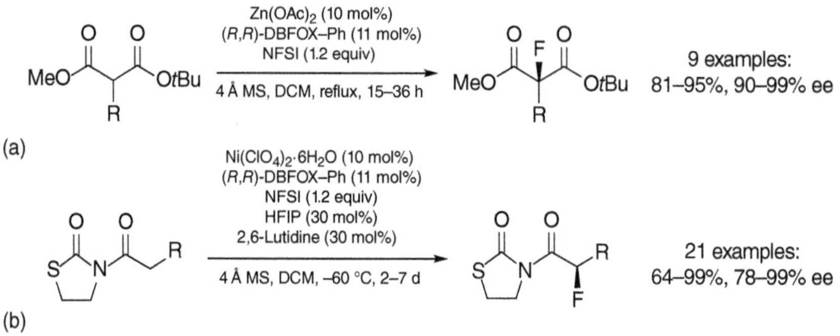

Scheme 8.36 Ni–DBFOX-catalyzed α-fluorination of (a) malonates and (b) thiazolidinones.

Miscellaneous Metal-Catalyzed Reactions A novel strategy for asymmetric fluorination of β-keto esters was demonstrated in 2007 by Shibata, Toru, and coworkers. In this example, an achiral Cu(II)–dmbipy complex intercalated in salmon testes DNA (st-DNA) provided a chiral pocket based off of the chirality of the helix. A variety of substrates were fluorinated in 32–96% yield and with ees of 3–74% (Scheme 8.37) [85]. This represents one of the few examples where supramolecular chiral information is transferred in a fluorination reaction.

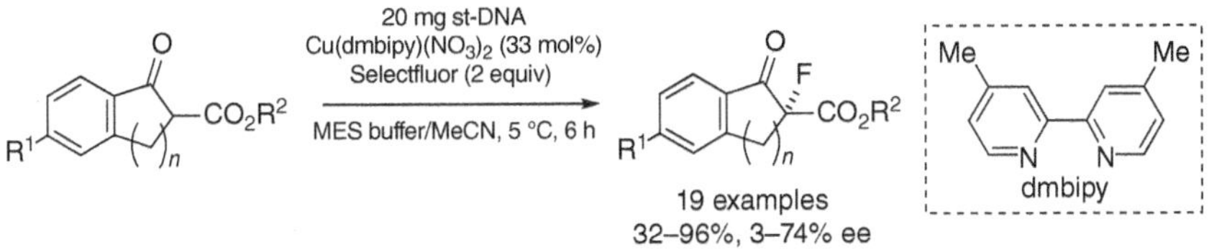

Scheme 8.37 Cu–dmbipy intercalated in st-DNA-catalyzed fluorination of β-keto esters.

In 2012, Shi and coworkers published the use of a bis-NHC ligated palladium complex that catalyzed the enantioselective fluorination of oxindoles in high yield (>80%) and with moderate enantioselectivity (0–59% ee, Scheme 8.38) [86]. Owing to the incompatibility of NHC ligands with electrophilic fluorinating agents, there are not many other examples of similar catalyst systems for enantioselective fluorination.

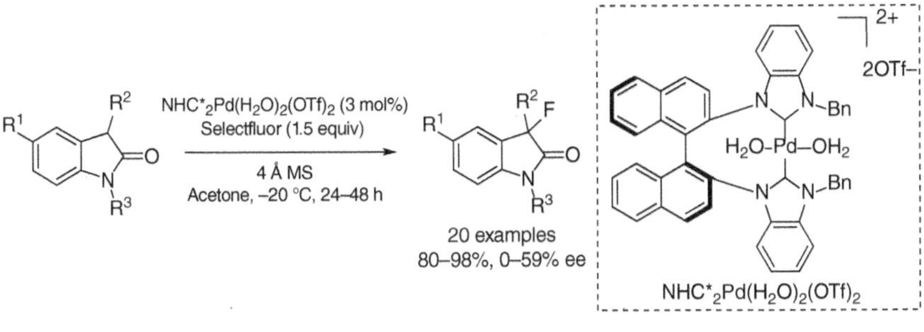

Scheme 8.38 Bis-NHC ligated Pd(II) complex for α-fluorination of oxindoles.

Another ligand class with relatively few examples for electrophilic fluorination is the tetradentate ligand frameworks. The first example of ligands of this type applied to asymmetric fluorination was disclosed by Togni and coworkers. They used a PNNP ruthenium complex to perform an enantioselective fluorination of β-keto esters (Scheme 8.39) [87]. A wide variety of ring sizes, ring positions, lactams, and acyclic esters were examined. While β-keto esters provided the product in high yields (34–96%) and high enantioselectivities (mostly 53–93% ee), lower enantioselectivities (8–11% ee) were observed with β-keto amides. Cobalt–salen complexes [88] and binaphthyl derived salen ligands [89] also catalyzed similar fluorination reactions.

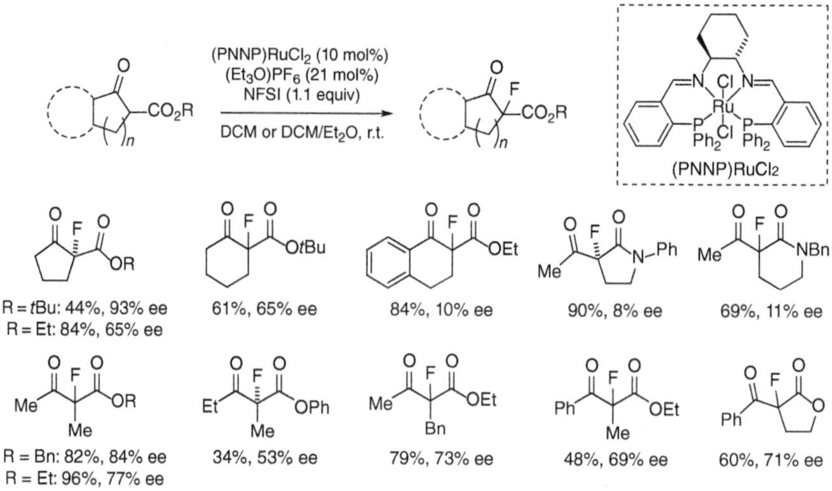

Scheme 8.39 Ru–PNNP-catalyzed α-fluorination of β-keto esters.

While BINAP derived ligands comprise the majority of phosphine-based ligands for enantioselective fluorination, similar reactivity with alternate phosphine-based ligand platforms had been studied. In 2012, Leeuwen and coworkers reported the synthesis of a novel ligand class, SPANphos, which features a wide bite angle bisphosphine motif. In this report, a Pd–SPANphos complex catalyzed the enantioselective fluorination of α-cyano esters in high yields and enantioselectivities (Scheme 8.40) [90].

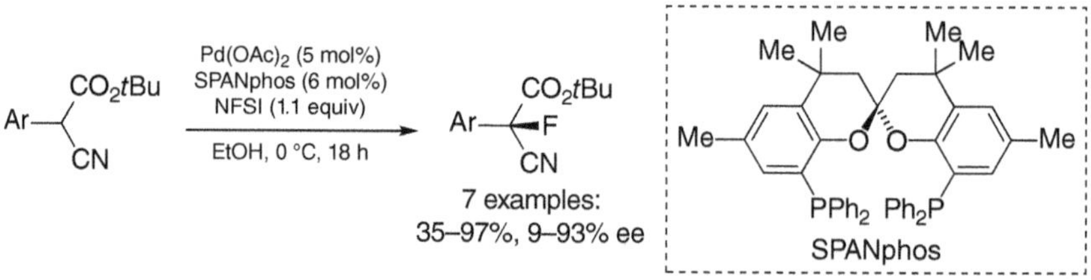

Scheme 8.40 Pd–SPANphos-catalyzed enantioselective fluorination of α-cyano esters.

In 2006, Inanaga and coworkers disclosed the use of a unique scandium perfluorophosphate complex, which catalyzed the asymmetric fluorination of β-keto esters in high yields and moderately high enantioselectivities as well (Scheme 8.41a) [91]. In addition to the atypical catalyst, this report also featured the use of N-fluoropyridinium salts as the electrophilic fluorine source. Another rare use of a Lewis acid catalyst was presented by the Xu group in 2016, where a chiral at iridium complex catalyzed the asymmetric fluorination of acyl imidazoles in high yield and enantioselectivity [92] (Scheme 8.41b).

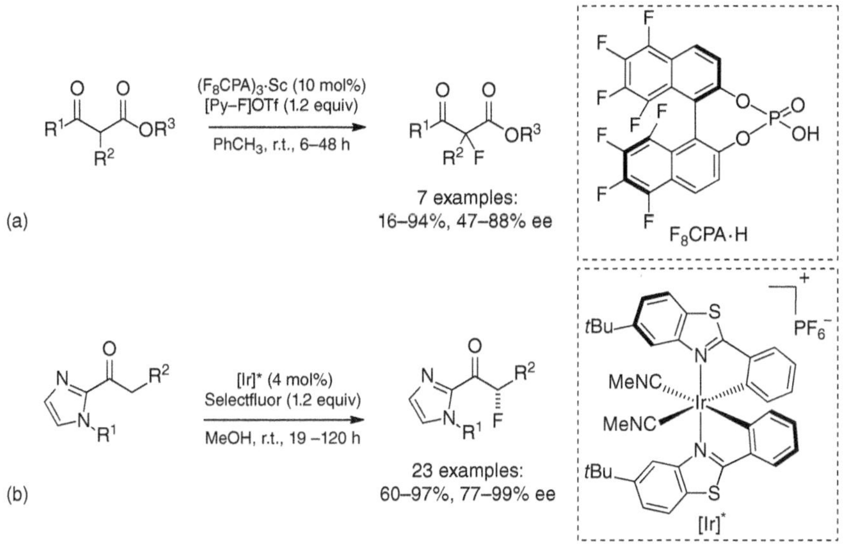

(a) (F$_8$CPA)$_3$·Sc (10 mol%) [Py–F]OTf (1.2 equiv) PhCH$_3$, r.t., 6–48 h

7 examples: 16–94%, 47–88% ee

F$_8$CPA·H

(b) [Ir]* (4 mol%) Selectfluor (1.2 equiv) MeOH, r.t., 19 –120 h

23 examples: 60–97%, 77–99% ee

[Ir]*

Scheme 8.41 (a) Scandium–CPA-catalyzed fluorination of β-keto esters and (b) chiral at iridium complex-catalyzed α-fluorination of acyl imidazoles.

8.2.2.2.2 Fluorination of Metal–Carbon Bonds Without Enolates

In the late 2000s, catalytic C(sp^2)—F bond forming reactions were being studied, facilitated by C–F reductive elimination from high valent metals. These developments resulted in a growing interest in applying this mechanistic paradigm to C(sp^3)—F bond formation, especially in an enantioselective manner. Compared to other metal-catalyzed electrophilic fluorination methods, there are relatively few examples of metal-catalyzed asymmetric fluorination reactions wherein a transition metal undergoes C—F bond reductive elimination. This class of reactions is mainly populated by group 10 elements. The first example was published by the Gagné laboratory in 2013 with an enantioselective fluorinative polyene cyclization-catalyzed by an electrophilic platinum(II) complex (Scheme 8.42) [93]. A range of polyene substrates underwent fluorocyclization utilizing this methodology. Similar to other reports published by the Gagné laboratory, the proposed mechanism begins with a polyene cyclization initiated by a chiral platinum complex, which results in a polycyclic platinum organometallic species. The Pt(II) organometallic complex is then oxidized by XeF$_2$ to a Pt(IV) fluoride. This intermediate undergoes inner-sphere C—F bond reductive elimination to provide the product, as evidenced by the stereoretentive fluorination of the Pt—C bond.

Following this, the Toste group developed a palladium-catalyzed three component fluoroarylation of olefins. A variety of amide-containing styrenes were fluorinated in high yields and high enantioselectivities (Scheme 8.43) [94]. The proposed reaction mechanism is initiated by an enantiodetermining migratory insertion of a Pd–Ar complex into the olefin to arrive at a chiral organometallic palladacycle, where coordination of an intramolecular amide prohibits β-hydride elimination. Oxidation of

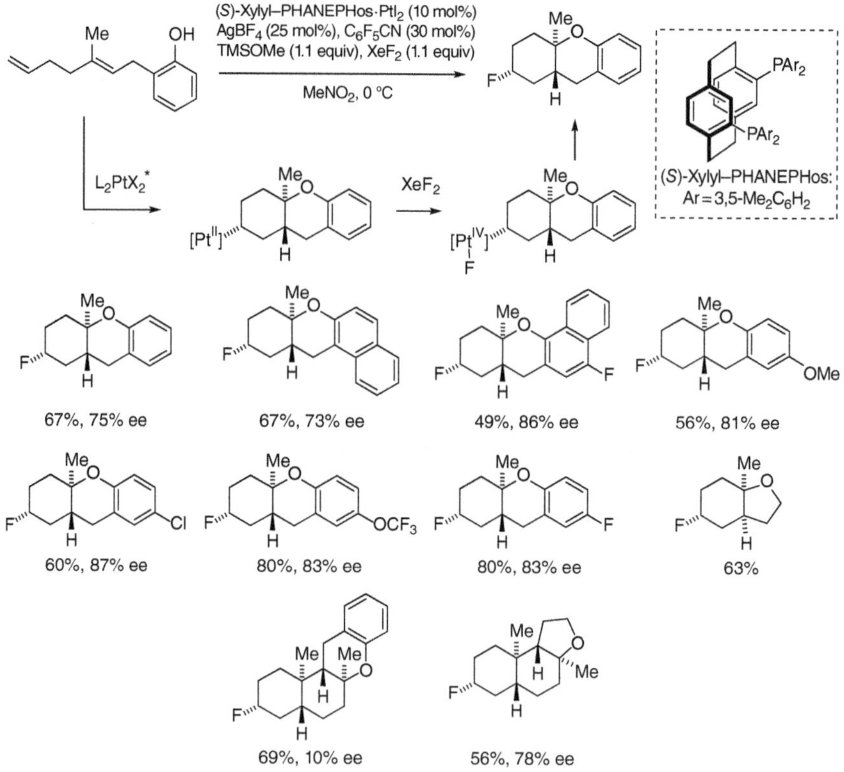

Scheme 8.42 Enantioselective fluorinative polyene cyclization catalyzed by a chiral platinum complex.

Pd(II) complex to Pd(IV) by Selectfluor is followed by reductive elimination to form the C—F bond stereocenter.

In a subsequent report, the Toste group demonstrated the enantioselective 1,1-fluoroarylation of allylic amines (Scheme 8.44a) [95]. Similar to the example described above, this reaction is proposed to initiate with migratory insertion of an aryl–palladium complex into the alkene. The resulting complex undergoes β-hydride elimination and reinsertion with the palladium at the benzylic position. Oxidation to a Pd(IV) fluoride by Selectfluor is followed by C–F reductive elimination. A similar enantioselective 1,1-fluoro arylation reaction was applied to acrylates and acrylamides (Scheme 8.44b) [96]. In 2017, the approach was applied to the enantioselective 1,3-fluoroarylation of chromenes, where product selectivity is rationalized based on a palladium chain walking mechanism, followed by oxidation and C–F reductive elimination [97] (Scheme 8.44c).

In 2018, the Yu group developed an enantioselective C(sp^3)–H fluorination strategy catalyzed by palladium (Scheme 8.45) [98]. In this example, a 2-alkyl substituted benzaldehyde condenses with a chiral amine, which directs a C–H metalation event diastereoselectively. The resulting Pd(II) complex is oxidized with [2,4,6-Me$_3$Py-F]BF$_4$, followed by inner-sphere reductive elimination. A

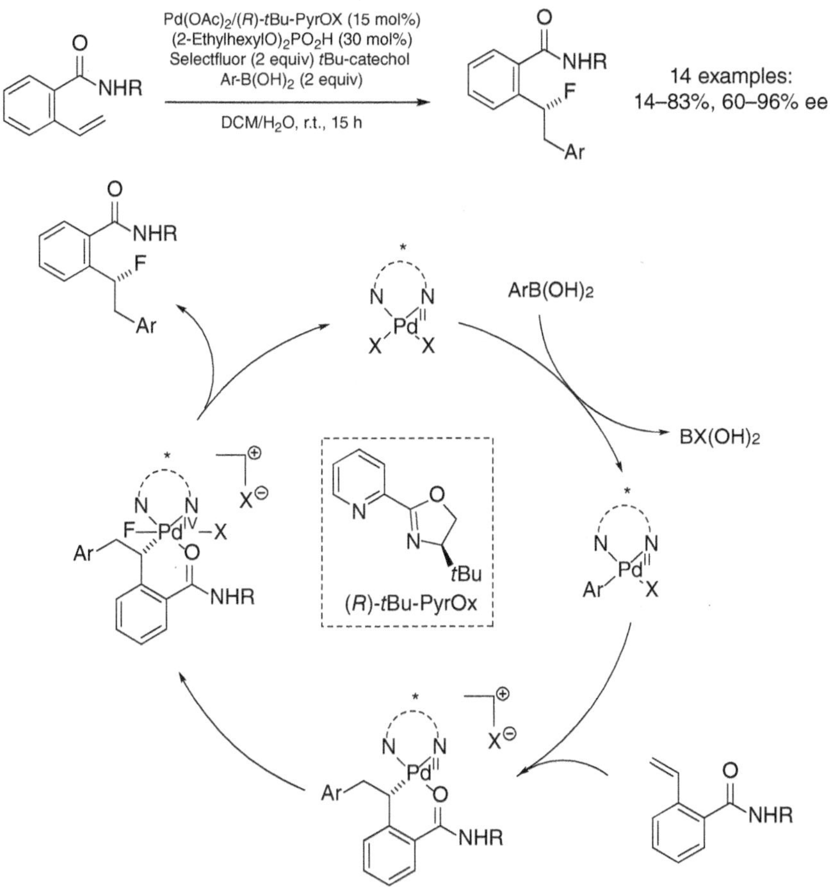

Scheme 8.43 Palladium-catalyzed enantioselective 1,2-fluoro-arylation of styrenes.

variety of 2-alkylbenzaldehydes were fluorinated in moderate yields and excellent enantioselectivity.

8.3 Nucleophilic Fluorination

Fluorination using nucleophilic fluoride sources provides many unique bene-fits over electrophilic methodologies. In addition to avoiding strong chemical oxidants, the use of HF or MF as the source of fluoride renders these reactions atom economical. Additionally, amending these types of reactions to ^{18}F labeling for PET purposes is much simpler, as ^{18}F fluoride is easier to generate and work with than electrophilic ^{18}F sources. Unfortunately, in contrast to electrophilic fluorination, methodologies that utilize nucleophilic fluoride sources are much less common. One main reason for this is the poor solubility of fluoride salts in typical solvents that are useful for catalysis. Additionally, the nucleophilicity of the fluoride ion is significantly diminished in solvents with hydrogen bond donors due to a strong solvation shell. These restrictions make the development of methods

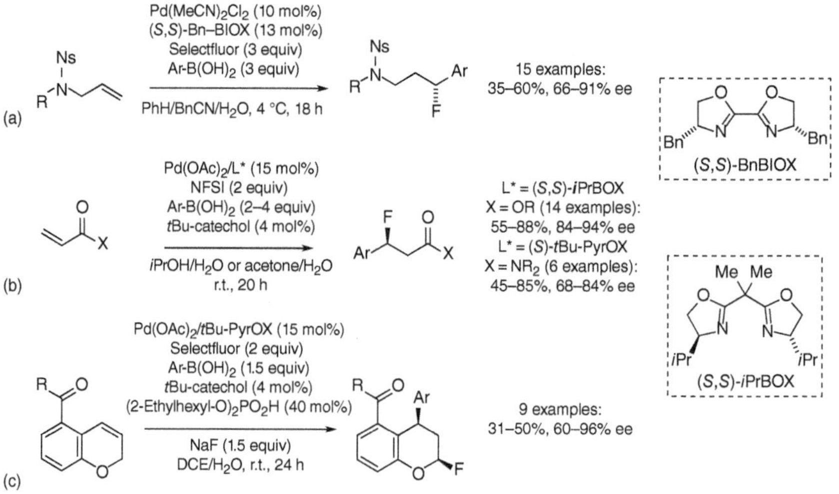

Scheme 8.44 Palladium-catalyzed enantioselective fluoro-arylation of (a) allylic amines, (b) acrylates–acrylamides, and (c) chromenes.

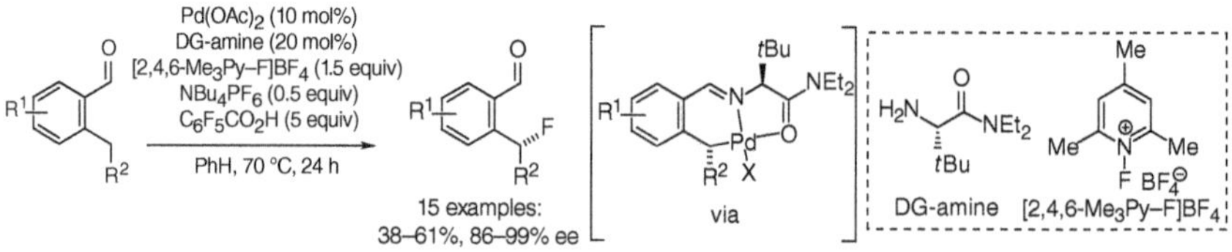

Scheme 8.45 Palladium-catalyzed enantioselective C–H fluorination with *in situ* chiral directing group strategy.

that install fluorine from fluoride difficult. Nevertheless, there are strategies that can incorporate fluoride into an organic structure enantioselectively.

8.3.1 Metal-Catalyzed Nucleophilic Fluorination

8.3.1.1 Ring Opening of Strained Ring Systems

The incorporation of fluoride into organic molecules has been a longstanding challenge for transition metal catalysis due to the strength of M—F bonds. The first reports had to offset this large energetic cost by rendering the overall desired transformation highly exergonic. As a result, these initial methodologies for asymmetric fluorination using fluoride involved the ring opening of strained heterocycles with Lewis acidic metals. The first demonstration of this strategy was by Haufe and coworker in 2000, where treatment of *meso*-epoxides with a chromium salen complex produced the ring-opened fluorohydrin in moderate enantioselectivities (>55% ee, Scheme 8.46) [99]. Unfortunately, high Lewis acid loadings of >50% were required as decreasing this to 10% led to sharp declines in enantioselectivities (<50% ee), presumably due to uncatalyzed background fluorination. In a following study, the authors investigated the effect of other fluorinating agents on enantioselectivity, and used the methodology to synthesize both enantiomers of fluoro lasiodiplodin [100].

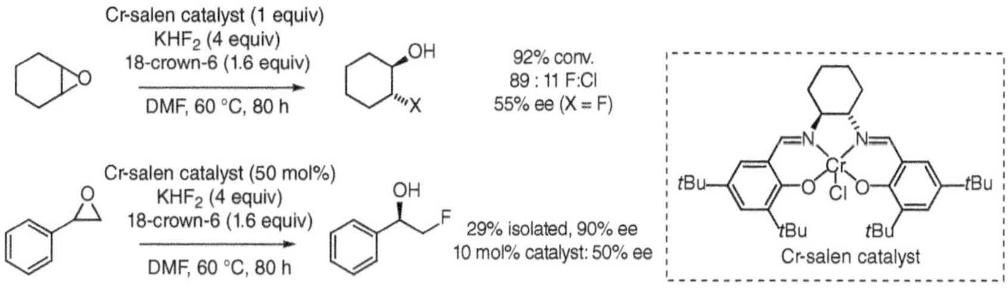

Scheme 8.46 Chromium (I)–salen-catalyzed ring opening fluorination of cyclohexene oxide and styrene oxide.

In 2010, the Doyle group rendered the reaction catalytic in cobalt salen complex with the use of benzoyl fluoride and hexafluoroisopropanol (HFIP) as a slow-release source for HF in the reaction (Scheme 8.47) [101]. This strategy presumably lowers the concentration of free HF in solution, limiting the problem of uncatalyzed racemic ring opening. A variety of *meso*-epoxides underwent ring opening in high yields and enantioselectivities. In a follow-up study, kinetic analysis and mechanistic evidence implicated a dual role for the Co–salen complex as both a Lewis acid activator and forming the active nucleophilic fluorination reagent [102]. The resting state of the catalyst was proposed to exist as a fluoride or hydrogen bifluoride bridged dimer, which the amine cocatalyst served to dissociate, to promote catalysis.

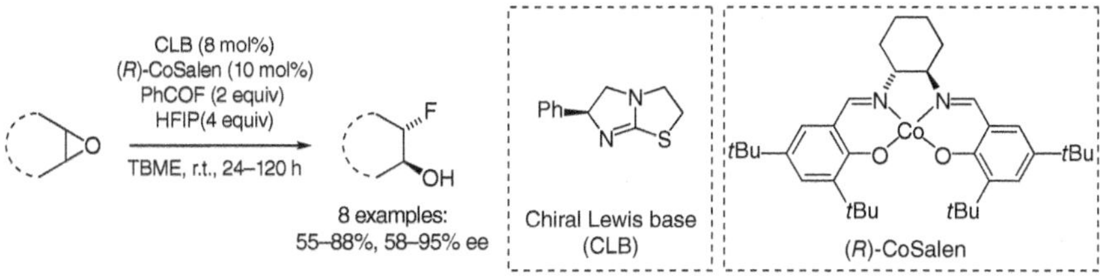

Scheme 8.47 CoSalen–chiral base cooperative catalyst system for asymmetric ring opening fluorination of epoxides.

This methodology was extended to the enantioselective ring opening fluorination of aziridines in 2013, typically proceeding with high yield and enantioselectivity to the β-fluoro amide (Scheme 8.48) [103]. It was found that the 2-picolinamide protecting group played a crucial role in the reaction, as benzamide and 4-picolinamide provided only the racemic product and no detectable product respectively. It was rationalized that Lewis acid coordination by titanium served to activate the aziridine toward nucleophilic attack.

The strategy extended beyond three-membered ring strained heterocycles. In 2012, the Lautens group published an enantioselective ring opening fluorination of an oxabicyclic alkene (Scheme 8.49) [104]. In this report, Et$_3$N·3HF was used as the fluoride source and Rh(I)–Josiphos system catalyzed the transformation.

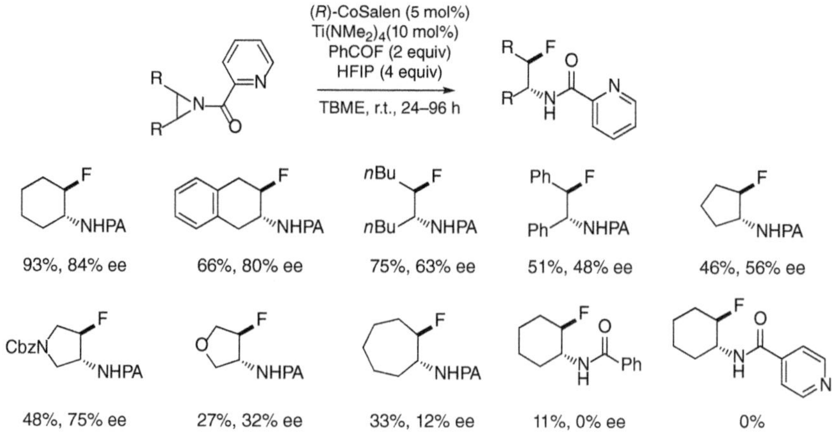

(R)-CoSalen (5 mol%)
Ti(NMe₂)₄(10 mol%)
PhCOF (2 equiv)
HFIP (4 equiv)

TBME, r.t., 24–96 h

93%, 84% ee   66%, 80% ee   75%, 63% ee   51%, 48% ee   46%, 56% ee

48%, 75% ee   27%, 32% ee   33%, 12% ee   11%, 0% ee   0%

Scheme 8.48 CoSalen–Ti(IV)-cocatalyzed asymmetric ring opening fluorination of aziridines.

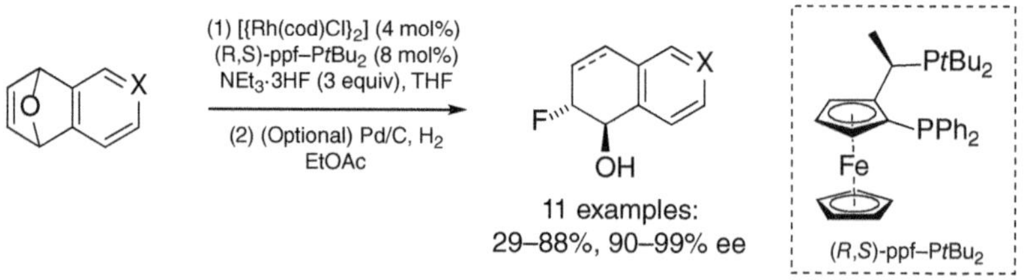

(1) [{Rh(cod)Cl}₂] (4 mol%)
(R,S)-ppf–PtBu₂ (8 mol%)
NEt₃·3HF (3 equiv), THF

(2) (Optional) Pd/C, H₂
EtOAc

11 examples:
29–88%, 90–99% ee

(R,S)-ppf–PtBu₂

Scheme 8.49 Rh–Josiphos-catalyzed asymmetric fluorination of oxabicyclic alkenes.

8.3.1.2 Allylic Functionalization

In 2006, Hintermann, Togni, and coworkers explored allylic substitution for asymmetric fluorination, where evidence from stoichiometric organometallic experiments revealed that many kinetic and thermodynamic problems exist with the transformation [105]. In 2010, Doyle and coworker realized an enantiose-lective palladium-catalyzed allylic substitution of racemic allyl chlorides with silver fluoride. A variety of cyclic allylic chlorides were fluorinated in high yields (56–85%) and high enantioselectivities (85–95% ee, Scheme 8.50a) [106]. Initial mechanistic investigation of the reductive elimination of π-allyl Pd(II) complex revealed the importance of silver as well as the chloride leaving group for reactivity, implying that AgCl formation provided the thermodynamic driving force for the reaction.

The Doyle group expanded this methodology to acyclic allylic chlorides using similar reaction conditions (Scheme 8.50b) [107]. A variety of branched acyclic allylic fluorides were synthesized in high yields (50–88%) and a range of enantiose-lectivities (0–97% ee). While linear allylic chlorides were fluorinated in good yields (66–84%), the enantioselectivities observed were not particularly high (21–71% ee). A mechanistic analysis of the reaction implicated that the C—F bond forming step occurs by nucleophilic attack of neutral Pd–allyl fluoride onto a cationic Pd–allyl complex [108].

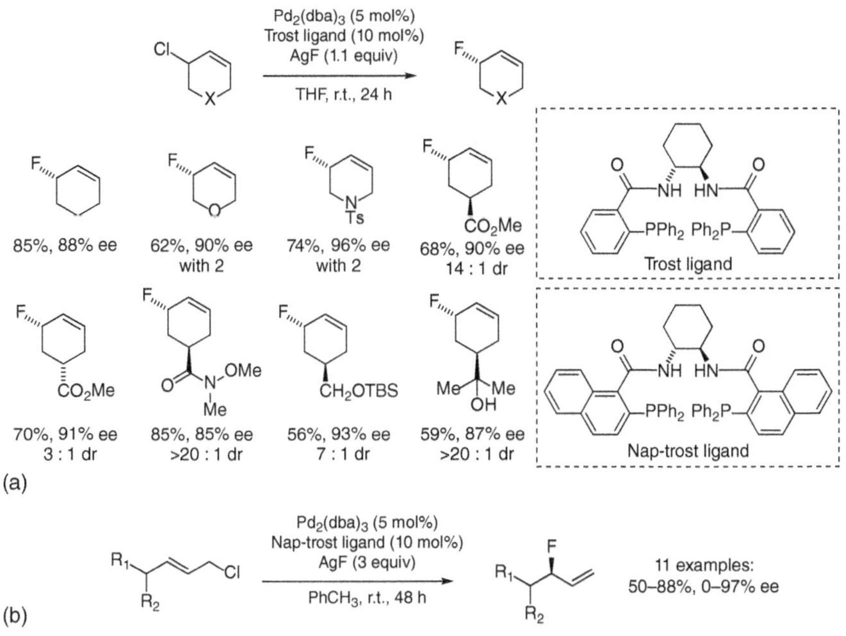

Scheme 8.50 Palladium-catalyzed enantioselective fluorination of (a) cyclic and (b) acyclic allylic chlorides.

In 2015, the Nguyen group published an asymmetric fluorination reaction of racemic secondary allylic trichloroacetimidates catalyzed by a chiral–iridium diene complex. In this report, a number of allylic trichloroacetimidates with varying substitution patterns were fluorinated in high yield and enantioselectivity (Scheme 8.51) [109]. A mechanistic study revealed the importance of trichloroacetimidate leaving group and that the reaction is best described as a dynamic kinetic asymmetric transformation, where π-allyl isomerization allows both enantiomers of starting material to form the final product [110].

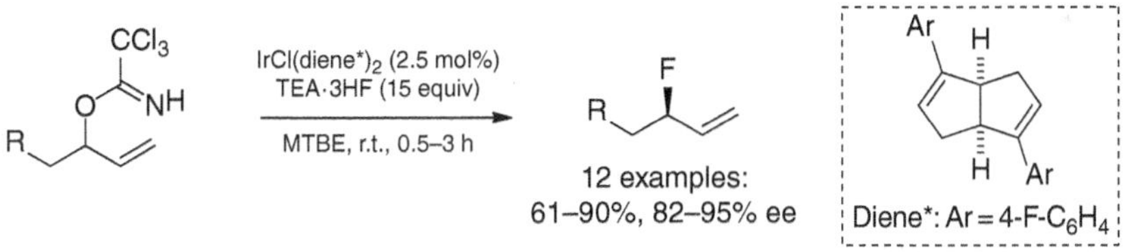

Scheme 8.51 Iridium-catalyzed fluorination of allylic trichloroacetimidates.

8.3.2 Organocatalytic Nucleophilic Fluorination

In contrast to organocatalytic fluorination with electrophilic reagents, there are very few examples of using fluoride and organocatalysis to confer enantioselectivity in a reaction. This is primarily due to the limited ways of activating fluoride with an organocatalyst. One of the first methods to do so was disclosed by Shibata and coworkers in 2014 with the use of hypervalent iodine chemistry (Scheme 8.52) [111].

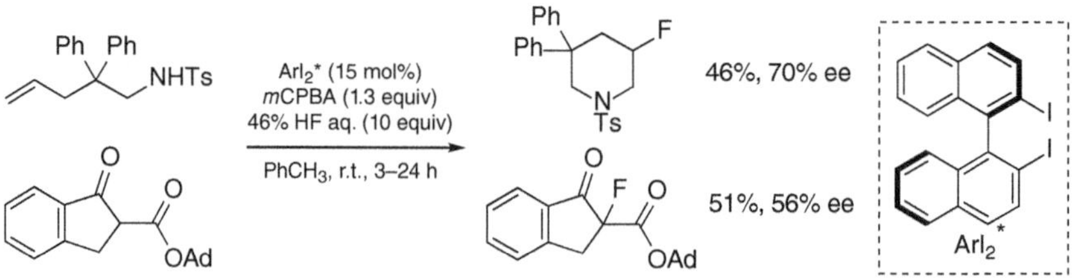

Scheme 8.52 Chiral iodane-catalyzed fluorination.

In 2016, the Jacobsen group published an iodine(III)-catalyzed enantioselective fluorocyclization of styrenal benzoates to fluoro-lactones (Scheme 8.53a) [112]. The proposed mechanism begins with oxidation of the chiral aryl iodide to iodosyl arene via *m*CPBA oxidation. After reaction with HF, the resulting chiral difluoro hypervalent iodine(III) reagent undergoes an *anti*-fluoriodination with the alkene to yield an alkyl iodine(III) species. The iodine(III) undergoes intramolecular S_N2-type substitution by the pendant ester, reducing the catalyst back to the ArI species, and yielding the product after loss of fluoromethane. Evidence for this mechanism is based off the exclusive isolation of the *syn*-diastereomer of the product. This methodology was later extended to the synthesis of aziridines via an asymmetric intramolecular fluoroamination reaction catalyzed by chiral iodine(III) reagents (Scheme 8.53b) [113]. A variety of miscellaneous substrate classes were also demonstrated to be competent under the reaction conditions.

In 2018 the Gouverneur laboratory achieved the asymmetric fluorination of β-bromo thioethers with CsF as the fluoride source, by utilizing a chiral urea catalyst (Scheme 8.54) [114]. A computational investigation of the mechanism suggested that autoionization of the thioether to form an episulfonium occurs first, where the bromide interacts with the catalyst via hydrogen bonding. The bromide then undergoes a salt metathesis with CsF, and the hydrogen bound fluoride is then delivered enantioselectively, opening the episulfonium and forming the product. A variety of racemic β-bromo thioethers were fluorinated in good yield and enantioselectivity. In a follow-up study, this strategy was applied to β-chloro amines via a similar approach [115].

8.4 Summary and Conclusions

Since the year 2000, a number of key strategies for the enantioselective incorporation of fluorine into organic molecules have been developed. Many of these strategies focused on developing methods centered around the use of stable electrophilic fluorine sources in conjunction with transition metal and organic catalysts. These strategies typically feature easily activated substrates with low pKas to render the reactions relatively mild. A challenge that has recently begun to be addressed is the application

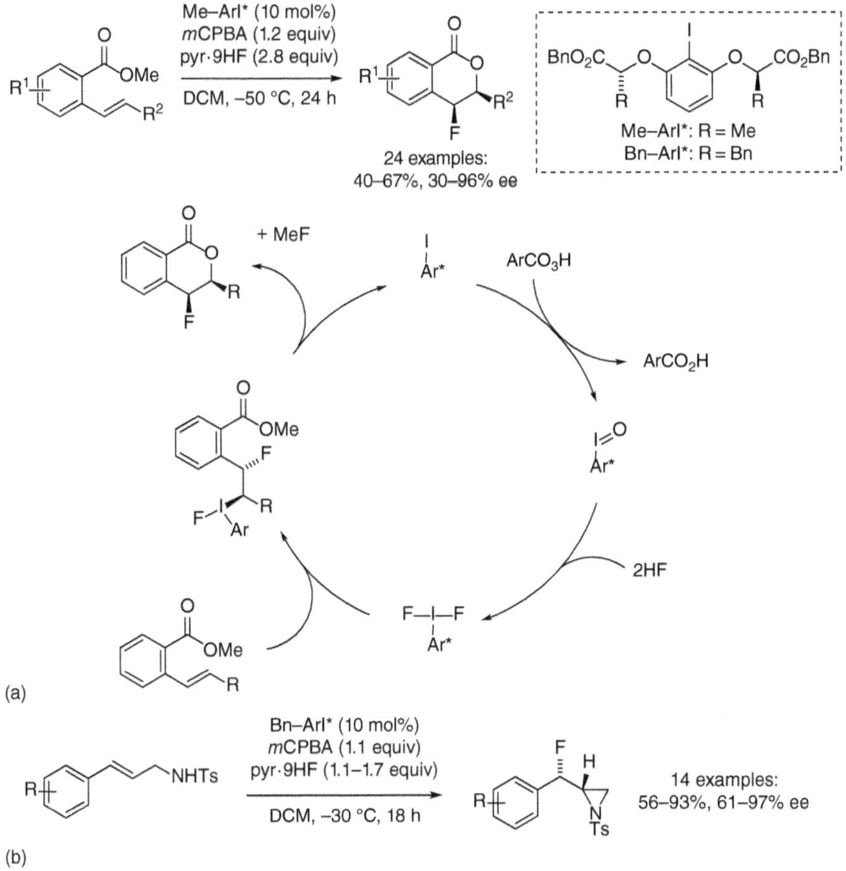

(a)

(b)

Scheme 8.53 Chiral aryl iodide-catalyzed enantioselective fluorocyclization of (a) styrenal benzoates and (b) allylic amines with representative proposed mechanism.

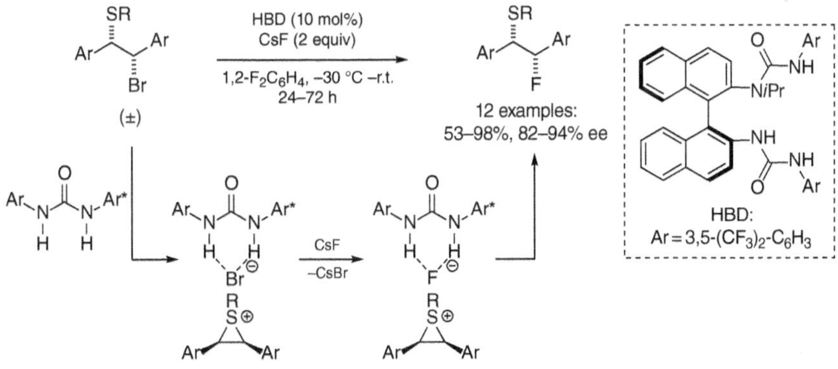

Scheme 8.54 Chiral urea-catalyzed phase transfer fluorination of β-bromo thioethers.

of enantioselective electrophilic fluorination to relatively unactivated substrates via CAPT, enamine, or transition metal catalysis. Additionally, while offering its own set of challenges, nucleophilic fluorination strategies have begun to be developed, providing distinct advantages over electrophilic fluorination, such as atom economy, avoiding harsh oxidants, and amenability to ^{18}F.

References

1 Gillis, E.P., Eastman, K.J., Hill, M.D. et al. (2015). *J. Med. Chem.* **58**: 8315–8359.

2 Chan, K.K.J. and O'Hagan, D. (2012). The rare fluorinated natural products and biotechnological prospects for fluorine enzymology. In: *Methods of Enzymology*, vol. 516 (ed. D.A. Hopwood), 219–235. Oxford, UK: Elsevier Inc.

3 Deng, H., O'Hagan, D., and Schaffrath, C. (2004). *Nat. Prod. Rep.* **21**: 773–784.

4 Banks, R.E., Murtagh, V., An, I. et al. (2007). 1-(Chloromethyl)-4-fluoro-1,4-diazoniabicyclo[2.2.2]octane bis(tetrafluoroborate). In: *Encyclopedia of Reagents for Organic Synthesis (EROS)*, 1–7. Wiley.

5 Differding, E., Poss, A.J., Cahard, D. et al. (2019). *N*-Fluoro-*N*-[phenylsulfonyl]-benzenesulfonamide. In: *Encyclopedia of Reagents for Organic Synthesis (EROS)*, 1–17. Wiley.

6 Zgonnik, V., Mazières, M.-R., and Plaquevent, J.-C. (2010). 1-Fluoro-2,4,6-trimethylpyridinium tetrafluoroborate. In: *Encyclopedia of Reagents for Organic Synthesis (EROS)*, 1–4. Wiley.

7 Davis, F.A., Zhou, P., and Murphy, C.K. (1993). *Tetrahedron Lett.* **34**: 3971–3974.

8 Davis, F.A. and Kasu, P.V.N. (1998). *Tetrahedron Lett.* **39**: 6135–6138.

9 Xu, Y.-S., Tang, Y., Feng, H.-J. et al. (2015). *Org. Lett.* **17**: 572–575.

10 Alvarado, J., Herrmann, A.T., and Zakarian, A. (2014). *J. Org. Chem.* **79**: 6206–6220.

11 Differding, E. and Lang, R.W. (1988). *Tetrahedron Lett.* **29**: 6087–6090.

12 Takeuchi, Y., Suzuki, T., Satoh, A. et al. (1999). *J. Org. Chem.* **64**: 5708–5711.

13 Liu, Z., Shibata, N., and Takeuchi, Y. (2000). *J. Org. Chem.* **65**: 7583–7587.

14 Shibata, N., Liu, Z., and Takeuchi, Y. (2000). *Chem. Pharm. Bull.* **48**: 1954–1958.

15 Cahard, D., Audouard, C., Plaquevent, J.-C. et al. (2000). *Org. Lett.* **2**: 3699–3701.

16 Wang, M., Wang, B.M., Shi, L. et al. (2005). *Chem. Commun.*: 5580–5582.

17 Takeshi, Y., Suzuki, Y., Ito, E. et al. (2011). *Org. Lett.* **13**: 470–473.

18 Mohar, B., Baudoux, J., Plaquevent, J.-C. et al. (2001). *Angew. Chem. Int. Ed.* **40**: 4214–4216.

19 Fukuzumi, T., Shibata, N., Sugiura, M. et al. (2006). *J. Fluorine Chem.* **127**: 548–551.

20 Ishimaru, T., Shibata, N., Horikawa, T. et al. (2008). *Angew. Chem. Int. Ed.* **47**: 4157–4161.

21 Zhang, H., Wang, B., Cui, L. et al. (2015). *Eur. J. Org. Chem.*: 2143–2147.

22 Bao, X., Wei, S., Zou, L. et al. (2016). *Tetrahedron: Asymmetry* 27: 436–441.

23 Chen, Z.-M., Yang, B.-M., Chen, Z.-H. et al. (2012). *Chem. Eur. J.* 18: 12950–12954.

24 Lozano, O., Blessley, G., Martinez Del Campo, T. et al. (2011). *Angew. Chem. Int. Ed.* 50: 8105–8109.

25 Kim, D.Y. and Park, E.J. (2002). *Org. Lett.* 4: 545–547.

26 Park, E.J., Kim, M.H., and Kim, Y.K. (2004). *J. Org. Chem.* 69: 6897–6899.

27 Wang, X., Lan, Q., Shirakawa, S. et al. (2010). *Chem. Commun.* 46: 321–323.

28 Zhu, C.-L., Fu, X.-Y., Wei, A.-J. et al. (2013). *J. Fluorine Chem.* 150: 60–66.

29 Novacek, J. and Waser, M. (2014). *Eur. J. Org. Chem.*: 802–809.

30 Zhang, B., Shi, L., Guo, R. et al. (2016). *Arkivoc*: 363–370.

31 Enders, D. and Hüttl, M.R.M. (2005). *Synlett*: 991–993.

32 Marigo, M., Fielenbach, D., Braunton, A. et al. (2005). *Angew. Chem. Int. Ed.* 44: 3703–3706.

33 Beeson, T.D. and MacMillan, D.W.C. (2005). *J. Am. Chem. Soc.* 127: 8826–8828.

34 Buckingham, F., Kirjavainen, A.K., Forsback, S. et al. (2015). *Angew. Chem. Int. Ed.* 54: 13366–13369.

35 Jiang, H., Falcicchio, A., Jensen, K.L., et. al., *J. Am. Chem. Soc.* 2009, 131, 7153–7157

36 Fadeyi, O.O. and Lindsley, C.W. (2009). *Org. Lett.* 11: 943–946.

37 Kwiatkowski, P., Beeson, T.D., Conrad, J.C. et al. (2011). *J. Am. Chem. Soc.* 133: 1738–1741.

38 Lam, Y.-H. and Houk, K.N. (2014). *J. Am. Chem. Soc.* 136: 9556–9559.

39 Paull, D.H., Scerba, M.T., Alden-Danforth, E. et al. (2008). *J. Am. Chem. Soc.* 130: 17260–17261.

40 Erb, J., Alden-Danforth, E., Kopf, N. et al. (2010). *J. Org. Chem.* 75: 969–971.

41 Lee, S.Y., Neufeind, S., and Fu, G.C. (2014). *J. Am. Chem. Soc.* 136: 8899–8902.

42 Zhao, Y.M., Cheung, M.S., Lin, Z. et al. (2012). *Angew. Chem. Int. Ed.* 51: 10359–10363.

43 Li, F., Wu, Z., and Wang, J. (2015). *Angew. Chem. Int. Ed.* 54: 656–659.

44 Dong, X., Yang, W., Hu, W. et al. (2015). *Angew. Chem. Int. Ed.* 54: 660–663.

45 Wang, L., Jiang, X., Chen, J. et al. (2019). *Angew. Chem. Int. Ed.* 58: 7410–7414.

46 Rauniyar, V., Lackner, A.D., Hamilton, G.L. et al. (2011). *Science* 334: 1681–1684.

47 Liang, X.-W., Cai, Y., and You, S.-L. (2018). *Chin. J. Chem.* 36: 925–928.

48 Liang, X.-W., Liu, C., Zhang, W. et al. (2017). *Chem. Commun.* 53: 5531–5534.

49 Phipps, R.J., Hiramatsu, K., and Toste, F.D. (2012). *J. Am. Chem. Soc.* 134: 8376–8379.

50 Phipps, R.J. and Toste, F.D. (2013). *J. Am. Chem. Soc.* 135: 1268–1271.

51 Wu, J., Wang, Y.-M., Drljevic, A. et al. (2013). *Proc. Natl. Acad. Sci. U.S.A.* 110: 13729–13733.

52 Zi, W., Wang, Y.-M., and Toste, F.D. (2014). *J. Am. Chem. Soc.* 136: 12864–12867.

53 Neel, A.J., Milo, A., Sigman, M.S. et al. (2016). *J. Am. Chem. Soc.* 138: 3863–3875.

54 Coelho, J.A.S., Matsumoto, A., Orlandi, M. et al. (2018). *Chem. Sci.* 9: 7153–7158.

55 Romanov-Michailidis, F., Pupier, M., Besnard, C. et al. (2014). *Org. Lett.* 16: 4988–4991.

56 Romanov-Michailidis, F., Romanova-Michaelides, M., Pupier, M. et al. (2015). *Chem. Eur. J.* 21: 5561–5583.

57 Yang, X., Phipps, R.J., and Toste, F.D. (2014). *J. Am. Chem. Soc.* 136: 5225–5228.

58 Honjo, T., Phipps, R.J., Rauniyar, V. et al. (2012). *Angew. Chem. Int. Ed.* 51: 9684–9688.

59 Hiramatsu, K., Honjo, T., Rauniyar, V. et al. (2016). *ACS Catal.* 6: 151–154.

60 Egami, H., Asada, J., Sato, K. et al. (2015). *J. Am. Chem. Soc.* 137: 10132–10135.

61 Egami, H., Niwa, T., Sato, H. et al. (2018). *J. Am. Chem. Soc.* 140: 2785–2788.

62 Hintermann, L. and Togni, A. (2000). *Angew. Chem. Int. Ed.* 39: 4359–4362.

63 Piana, S., Devillers, I., Togni, A. et al. (2002). *Angew. Chem. Int. Ed.* 41: 979–982.

64 Bertogg, A., Hintermann, L., Huber, D.P. et al. (2012). *Helv. Chim. Acta* 95: 353–403.

65 Hamashima, Y., Yagi, K., Takano, H. et al. (2002). *J. Am. Chem. Soc.* 124: 14530–14531.

66 Hamashima, Y., Takano, H., Hotta, D. et al. (2003). *Org. Lett.* 5: 3225–3228.

67 Hamashima, Y., Suzuki, T., Shimura, Y. et al. (2005). *Tetrahedron Lett.* 46: 1447–1450.

68 Kim, H.R. and Kim, D.Y. (2005). *Tetrahedron Lett.* 46: 3115–3117.

69 Moriya, K.-I., Hamashima, Y., and Sodeoka, M. (2007). *Synlett*: 1139–1142.

70 Kang, Y.K., Cho, M.J., Kim, S.M. et al. (2007). *Synlett*: 1135–1138.

71 Woo, S.B., Suh, C.W., Koh, K.O. et al. (2013). *Tetrahedron Lett.* 54 (26): 3359–3362.

72 Kang, Y.K. and Kim, D.Y. (2011). *Tetrahedron Lett.* 52: 2356–2358.

73 Kwon, S.J. and Kim, D.Y. (2015). *J. Fluorine Chem.* 180: 201–207.

74 Suzuki, S., Kitamura, Y., Lectard, S. et al. (2012). *Angew. Chem. Int. Ed.* 51: 4581–4585.

75 Curtis, N.R., Davies, S.H., Gray, M. et al. (2015). *Org. Process Res. Dev.* 19: 865–871.

76 Suzuki, T., Hamashima, Y., and Sodeoka, M. (2007). *Angew. Chem. Int. Ed.* 46: 5435–5439.

77 Ma, J.-A. and Cahard, D. (2004). *Tetrahedron: Asymmetry* 15: 1007–1011.

78 Shibata, N., Ishimaru, T., Nagai, T. et al. (2004). *Synlett*: 1703–1706.

79 Shibata, N., Kohno, J., Takai, K. et al. (2005). *Angew. Chem. Int. Ed.* 44: 4204–4207.

80 Wang, B., Wang, Y., Jiang, Y. et al. (2018). *Org. Biomol. Chem.* 16: 7702–7710.

81 Wang, Y., Wang, H., Jiang, Y. et al. (2017). *Green Chem.* 19: 1674–1677.

82 Granados, A., Sarró, P., and Vallribera, A. (2019). *Molecules* 24: 1141.

83 Reddy, D.S., Shibata, N., Nagai, J. et al. (2008). *Angew. Chem. Int. Ed.* 47: 164–168.

84 Reddy, D.S., Shibata, N., Horikawa, T. et al. (2009). *Chem. Asian J.* 4: 1411–1415.

85 Shibata, N., Yasui, H., Nakamura, S. et al. (2007). *Synlett*: 1153–1157.

86 Zhang, R., Wang, D., Xu, Q. et al. (2012). *Chin. J. Chem.* 30: 1295–1304.

87 Althaus, M., Becker, C., Togni, A. et al. (2007). *Organometallics* 26: 5902–5911.

88 Kawatsura, M., Hayashi, S., Komatsu, Y. et al. (2010). *Chem. Lett.* 39: 466–467.

89 Zheng, L.S., Wei, Y.-L., Jiang, K.-Z. et al. (2014). *Adv. Synth. Catal.* 356: 3769–3776.

90 Jacquet, O., Clément, N.D., Blanco, C. et al. (2012). *Eur. J. Org. Chem.*: 4844–4852.

91 Suzuki, S., Furuno, H., Yokoyama, Y. et al. (2006). *Tetrahedron: Asymmetry* 17: 504–507.

92 Xu, G.-Q., Liang, H., Fang, J. et al. (2016). *Chem. Asian J.* 11: 3355–3358.

93 Cochrane, N.A., Nguyen, H., and Gagne, M.R. (2013). *J. Am. Chem. Soc.* 135: 628–631.

94 Talbot, E.P.A., Fernandes, T.A., McKenna, J.M. et al. (2014). *J. Am. Chem. Soc.* 136: 4101–4104.

95 He, Y., Yang, Z., Thornbury, R.T. et al. (2015). *J. Am. Chem. Soc.* 137: 12207–12210.

96 Miró, J., del Pozo, C., Toste, F.D. et al. (2016). *Angew. Chem. Int. Ed.* 55: 9045–9049.

97 Thornbury, R.T., Saini, V., Fernandes, T.A. et al. (2017). *Chem. Sci.* 8: 2890–2897.

98 Park, H., Verma, P., Hong, K. et al. (2018). *Nat. Chem.* 10: 755–762.

99 Bruns, S. and Haufe, G. (2000). *J. Fluorine Chem.* 104: 247–254.

100 Haufe, G., Bruns, S., and Runge, M. (2001). *J. Fluorine Chem.* 112: 55–61.

101 Kalow, J.A. and Doyle, A.G. (2010). *J. Am. Chem. Soc.* 132: 3268–3269.

102 Kalow, J.A. and Doyle, A.G. (2011). *J. Am. Chem. Soc.* 133: 16001–16012.

103 Kalow, J.A. and Doyle, A.G. (2013). *Tetrahedron* 69: 5702–5709.

104 Zhu, J., Tsui, G.C., and Lautens, M. (2012). *Angew. Chem. Int. Ed.* 51: 12353–12356.

105 Hintermann, L., Läng, F., Maire, P. et al. (2006). *Eur. J. Inorg. Chem.*: 1397–1412.

106 Katcher, M.H. and Doyle, A.G. (2010). *J. Am. Chem. Soc.* 132: 17402–17404.

107 Katcher, M.H., Sha, A., and Doyle, A.G. (2011). *J. Am. Chem. Soc.* 133: 15902–15905.

108 Katcher, M.H., Norrby, P.-O., and Doyle, A.G. (2014). *Organometallics* 33: 2121–2133.

109 Zhang, Q., Stockdale, D.P., Mixdorf, J.C. et al. (2015). *J. Am. Chem. Soc.* 137: 11912–11915.

110 Sorlin, A.M., Mixdorf, J.C., Rotella, M.E. et al. (2019). *J. Am. Chem. Soc.* 141: 14843–14852.

111 Suzuki, S., Kamo, T., Fukushi, K. et al. (2014). *Chem. Sci.* 5: 2754–2760.

112 Woerly, E.M., Banik, S.M., and Jacobsen, E.N. (2016). *J. Am. Chem. Soc.* 138: 13858–13861.

113 Mennie, K.M., Banik, S.M., Reichert, E.C. et al. (2018). *J. Am. Chem. Soc.* 140: 4797–4802.

114 Pupo, G., Ibba, F., Ascough, D.M.H. et al. (2018). *Science* 360: 638–642.

115 Pupo, G., Vicini, A.C., Ascough, D.M.H. et al. (2019). *J. Am. Chem. Soc.* 141: 2878–2883.

9

The Self-Disproportionation of Enantiomers (SDE): Fluorine as an SDE-Phoric Substituent

Jianlin Han[1], Santos Fustero[2], Hiroki Moriwaki[3], Alicja Wzorek[4], Vadim A. Soloshonok[5,6] and Karel D. Klika[7]

[1] Nanjing Forestry University, College of Chemical Engineering, No.159 Longpan Road, Nanjing, Jiangsu 210037, China
[2] Universidad de Valencia, Departamento de Química Orgánica, Calle Dr. Moliner, 50, Burjassot, Valencia E-46100, Spain
[3] Hamari Chemicals Ltd., 1-4-29 Kunijima, Higashi-Yodogawa-ku, Osaka 533-0024, Japan
[4] Jan Kochanowski University in Kielce, Institute of Chemistry, Faculty of Natural Sciences, Uniwersytecka 7, Kielce 25-406, Poland
[5] University of the Basque Country (UPV/EHU), Department of Organic Chemistry I, Faculty of Chemistry, Paseo Manuel Lardizábal 3, San Sebastián E-20018, Spain
[6] IKERBASQUE, Basque Foundation for Science, Alameda Urquijo 36-5, Plaza Bizkaia, Bilbao E-48011, Spain
[7] German Cancer Research Center (DKFZ), Molecular Structure Analysis, NMR Spectroscopy Analysis Unit, Im Neuenheimer Feld 280, Heidelberg D-69120, Germany

9.1 Introduction

The self-disproportionation of enantiomers (SDE) [1–4] is now a well-established descriptor for a phenomenon that, apart from recrystallization, is nevertheless not widely appreciated in all its guises. The acronym SDE refers to the disproportionation of enantiomers across resulting fractions whenever a physicochemical process is applied to a scalemate (i.e. a non-racemic, enantioenriched sample). Processes for which the SDE phenomenon has been observed include recrystallization, sublimation, distillation, chromatography, and the application of a force field. The underlying basis of the SDE is the association of molecules [5–7] and the differentiation between homo- and heterochiral associates with diastereomeric-type physicochemical properties. Molecules containing particular functional groups, structural elements, or other moieties can have a propensity for exhibiting the SDE phenomenon, especially to a high degree, thereby leading to the descriptor SDE-phoric [8] for such entities, and fluorine is a prime example of an SDE-phoric substituent.

This chapter is not a review of all possible examples of fluorine-containing compounds that have exhibited SDE manifestations; rather, selected cases are presented to highlight the role that fluorine plays as an SDE-phoric structural element in the processes giving rise to the SDE, i.e. the propensity of fluorine-containing compounds to exhibit SDE more commonly and with an enhanced magnitude in comparison to other compounds generally, and their non-fluorine-containing counterparts in particular, and to provide insight into the processes and interactions that are responsible for the SDE phenomenon. The role that fluorine plays is central in many extreme cases of the SDE, particularly those involving sublimation and distillation. Fluorine has many special properties, and the ones that seem to have the most impact on the SDE are considered. Since organic chemists are well versed in SDE via recrystallization (SDEvR), this aspect of the SDE will not be covered. Crystallization though, could be a genuine source for homochirality, and Blackmond [9–11] has provided persuasive arguments for this case. The necessity for workers to routinely perform SDE tests to ensure the accuracy of the enantiomeric excess (ee) of samples that are likely to be scalemic, e.g. those obtained from catalytic asymmetric transpositions, is emphasized. Although the SDE represents a danger in such work, it also represents unconventional means of enantiopurification. The SDE phenomenon has been reviewed previously, generally [12, 13], on specific processes, e.g. chromatography [14–17] and sublimation [18], and for SDE-phoric structural types such as sulfoxides [19, 20], amides [19, 21], amino acids and their derivatives [22], and fluorine-containing compounds [3, 4, 23, 24].

The propensity of molecules to associate in solution or to be organized in the solid state permeates other aspects of their properties and behavior and is not limited to the SDE phenomenon, for example, effects are also observed in spectroscopy. Clearly, for measurements of crystalline material, spectra can differ depending on whether enantiopure or racemic crystals are examined, and this has been noted in the case of IR, Raman, solid-state NMR, etc. as well as obviously in X-ray crystallography and powder diffraction. Indeed, it has long been known that optical rotation measurements are concentration dependent due to the molecular association – the well-known *Horeau effect* [25, 26]. These cases will not be examined here but one case that has particular relevance to SDE via chromatography (SDEvC) is the phenomenon of self-induced diastereomeric anisochronism (SIDA) in solution-state NMR – another relatively unappreciated phenomenon – and will therefore be included. The consequences of intermolecular association for chiral molecules in scalemic samples are neatly summarized in Figure 9.1.

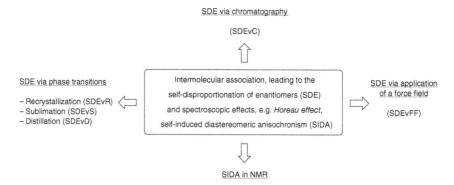

Figure 9.1 The consequences of intermolecular association for chiral molecules in scalemic samples where the diastereomeric-type properties of such associations can entirely determine the SDE or otherwise contribute significantly to it.

9.2   General Concepts and the Role of Fluorine in the Manifestation of the SDE

In addition to the terms SDE and SDE-phoric, since the SDE occurs via numerous processes, the terminology follows the form SDEvR when the process occurs via recrystallization, and likewise an S refers to SDE via sublimation (SDEvS), a C refers to SDEvC, a D refers to SDE via distillation (SDEvD), while FF refers to SDE via force field (SDEvFF) as depicted in Figure 9.1. Other important definitions include the SDE magnitude, defined [27] as

$$\text{SDE magnitude } (\Delta\text{ee}) = \text{ee}_{\text{fraction with the highest ee}} - \text{ee}_{\text{fraction with the lowest ee}}$$

The range of ees (the SDE range) over which the SDE appreciably occurs [with respect to a nominated minimum level of enantiopurity of the fraction(s) containing the enantiopure material and/or SDE yield (vide infra) and/or Δee] [13]:

$$\text{SDE range } (R_{\text{ee}}) = \text{ee}_{\text{sample with highest ee exhibiting SDE}} - \text{ee}_{\text{sample with lowest ee exhibiting SDE}}$$

It is worth noting that there is no theoretical limit to the SDE and it can be observed even for scalemates of very low ee, even as low as 0.05% ee. The reason for defining a range is simply for consideration of reasonable practical applications and also as a means to further quantify the magnitude of the SDE for comparative purposes among compounds or conditions.

The amount of enantiopure material that can be yielded by an SDE process is dependent on the ee of the starting material and in theory the maximum amount of enantiopure material obtainable by SDEvC or SDEvFF is effectively very close to the amount of excess enantiomer present, i.e. the ee [13]:

$$\text{Maximum theoretical yield for SDEvC/FF } (Y_{\text{max,SvC/FF}}) \approx \text{ee}$$

Thus the practical SDE yield as a percentage can be expressed as the isolated amount of the enantiopure material (with respect to a nominated minimum enantiopurity) divided by Y_{max} (converted to mass) and multiplied by 100 [13]:

$$\text{SDE yield } (Y_{\text{SDE}}) = \text{amount of the enantiopure material}/(Y_{\text{max}} \text{ as mass}) \times 100$$

The quintessential basis for the SDE is the preference for homo- or heterochiral states (associates) and the different physicochemical properties of these associates, akin to diastereomers, to each other and to the single (free or unassociated) molecules. Depending on the particular process concerned, the associates do not need to extend infinitely as in crystal lattices and they may form exclusively either dimeric associates or oligomers, either of fixed or variable order in the latter case. Importantly, it is not necessarily a case of exclusively one over the other (homo- vs. heterochiral) and often a dynamic equilibrium is in effect for some SDE processes, usually also involving single molecules (Figure 9.2). In can immediately be seen, though, that while the *R,R* and *S,S* homochiral associates remain enantiomeric with respect to each other, the *R,R* and *S,S* homochiral associates in comparison with the *S,R* heterochiral associate have a formally diastereomeric relationship and therefore the former pair are separable from the latter under achiral conditions.

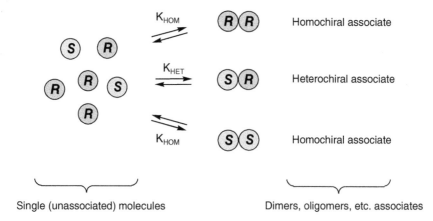

Figure 9.2 The dynamic equilibria of a chiral compound that forms homo- and heterochiral associates will yield different distributions of the enantiomers between the various associates depending on their concentrations and the values of K_{HOM} and K_{HET}. The formation of temporal associates during processes such as chromatography, and possibly also during distillation and sublimation, may involve either dimeric or higher order oligomeric associates of variable size.

The formation of temporal associates during processes such as chromatography, and possibly also during distillation and sublimation, may involve either dimeric or higher order oligomeric associates of variable size. Though it is often difficult to discern the precise nature of the associates with respect to either or both structure and size (and where precisely these associates may form or where their presence has a determinate effect for multi-environment processes), it is the unequal distribution of the enantiomers between the various associates – single molecules and homo- and heterochiral associates – that give rise to the SDE in conjunction with either thermodynamic and/or kinetic factors. Ultimately, it is the unequal contents of the enantiomers in a scalemate that affect the result due to the equilibrium constants involved, yielding different distributions among the associates. Since (1) single molecules can obviously be separable from associates, (2) homo- and heterochiral associates of the same order (dimeric, trimeric, etc.) can also be separated from each other as they are diastereomeric and thereby have different physicochemical properties, and (3) oligomeric homo- and heterochiral associates of differing average order can also be separated from each other as they are of differing average molecular weights, the SDE can arise from either one or a combination of these processes. Systems where the material is present only as non-interacting single molecules cannot give rise to the SDE in an achiral environment. Finally, it is important to point out that the SDE phenomenon is not chirogenic, i.e. if the fractions resulting from the applied physicochemical process were all recombined, the same mass and ee for the sample would be conserved.

Fluorine is an element possessing many unique and peculiar properties [3], and as a constituent of a molecule it imparts many unique attributes to the molecule. One elemental property and three attributes that fluorine imparts to a molecule upon its inclusion are of particular relevancy for the SDE phenomenon and are the reasons why fluorine has such an influential effect on the SDE. First is fluorine's high electronegativity, leading to the enhancement of attractive intermolecular interactions, particularly hydrogen bonds when possible or dipole–dipole interactions, and thus the formation of associates (or a greater extent of associate formation). On the other hand, there can also be strong repulsive electrostatic interactions between molecules containing $-CF_3$ or $-COCF_3$ moieties due to the resulting high electron densities [28–31]. Repulsive electrostatic interactions can accentuate immensely the differences in energy between the homo- and heterochiral associates, thereby affecting the SDE, and $-CF_3$ groups especially have been found to be SDE-phoric groups. The inclusion of fluorine into a molecule can also lead to increased solubility for the resulting compound in low-polarity solvents, thereby allowing the use of more apolar eluents in SDEvC, resulting, in turn, in a greater Δee due to reduced competing analyte–solvent interactions based on either hydrogen bonds or dipole–dipole interactions. Finally, the inclusion of fluorine in a molecule can impart considerable volatility to the resulting compound, which is pivotal for processes involving the gaseous phase, viz. distillation and sublimation and it is telling that the two clear-cut cases of SDEvD and many of the examples of SDEvS involve compounds containing fluorine. Interestingly though, thus far there have been no reports of SDE via GC or of chiral-state dependencies in gaseous-phase spectroscopies such as MS for fluorine-containing compounds.

Overall, due to the unique and peculiar properties of fluorine, the presence of fluorine or fluorinated substituents in the structure of a chiral compound can have a profound effect on the magnitude of the SDE. The SDE is always present to some degree for all scalemic samples, it should be noted, and fluorine simply lends itself to amplifying the SDE by accentuating either attractive or repulsive intermolecular interactions.

9.3 The SDE Phenomenon

9.3.1 SDE via Distillation

A most remarkable result is that of SDEvD. Thus far, only two clear-cut examples of SDEvD have been reported, viz. for *N*-trifluoroacetyl valine methyl ester (**1**) [32] and isopropyl (3,3,3-trifluoro)lactate (**2**) [33] (Figure 9.3), and it is telling that both examples involve fluorine-containing molecules. While the SDEvD of **1** was not studied in detail, the more meticulous account for the SDEvD of **2** revealed differences in bps between the racemate and scalemates with the largest difference being a remarkable 50 °C for scalemates from 50% to 70% ee. Interestingly, the distillate of **2** could either be enantioenriched or -depleted, indicating the SDEvD process to be quite complex. For example, while a sample of 74.1% ee yielded a distillate of 81.7% ee, in contrast, a sample of 40.2% ee yielded a distillate of 33.2% ee. In analogy to recrystallization, a eutectic point (ep) can be declared based on the bps and the consequent change in the distillate ee (at c. 60% ee for **2**).

Strong intermolecular interactions must be present to explain the behavior of **2**, at least in the liquid state. Indeed, hydrogen bonding was found to be present by

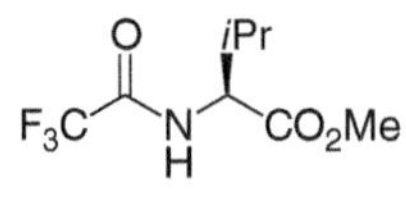

F₃C—C(=O)—N(H)—CH(iPr)—CO₂Me

1

F₃C—C(OH)—CO₂iPr

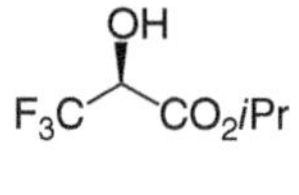

2

Figure 9.3 Structures of *N*-trifluoroacetyl valine methyl ester (**1**) [32] and isopropyl (3,3,3-trifluoro)lactate (**2**) [33]. Sources: Koppenhoefer and Trettin [32], Katagiri et al. [33].

IR [33] and also by low-angle X-ray diffraction [34, 35]. The strength of the hydrogen bonding and the extent of the hydrogen bonding network or repulsive properties between the molecules, in particular between $-CF_3$ groups [36, 37], are clearly accentuated by the $-CF_3$ group as the non-fluorine-containing analog of **2** failed to exhibit SDEvD [33]. However, if there was just an overwhelming bias toward either homo- or heterochiral association, then it would be expected that at all times, the distillate would be either enantioenriched or -depleted. Since this is not the case, either one or other of the associations is strongly preferred but with sufficient dependence on concentration, or the difference between the two types of association is not that dramatic, or the interactions are more complex than simple dimeric association (e.g. long homochiral chains as evidenced in the solid state [34, 38] and perhaps also in the liquid state [35]), and possibly also the process is compounded by gas-phase intermolecular interactions that have been postulated but are unproved [33, 34]. IR spectra of low (17%) and high (75%) ee samples of **2** seem to indicate that homochiral associations are indeed strongly favored [33]. The strong preference for homochiral interactions was shown not only in the solid state [34, 38] whereby it is difficult to even obtain racemic crystals, but also by size-exclusion chromatography (SEC) [38] (vide infra).

So far, SDEvD has defied definitive explanation, and causes have been ascribed to both kinetics [32] and thermodynamics [16, 35], though both are likely to contribute. Moreover, SDEvD may only be possible if molecules form at least homo- and/or heterochiral dimeric associates in the gaseous phase to effect differentiation, though SIDA was not observed for **2** in the gas phase (Klika and Alvarez, unpublished results). SDEvD is a truly astonishing phenomenon, and though Δee is sizeable for **2**, SDEvD is unlikely to be a method put into common practice for enantiopurification purposes. SDEvD is also unlikely to be a cause for concern in terms of unintentionally altering the ee of a sample given that SDEvD is likely to be only a very rare event. Not only is volatility a necessity for SDEvD – something that fluorine is able to impart to a molecule upon its inclusion – but it is also likely that strong intermolecular interactions are required, again, an attribute that fluorine can also impart to a molecule upon its inclusion since it is notable that the non-fluorine-containing analog of **2** did not exhibit SDEvD.

9.3.2 SDE via Sublimation

Sublimation bears many similarities to crystallization (as the compound is in a fluid phase and forms crystals directly from the fluid phase), but there are also some stark differences since two solid phases are involved – the starting material and the sublimed material in the solid → gas → solid phase transitions. Sublimation is actually a more complex process compared with recrystallization in practice as it

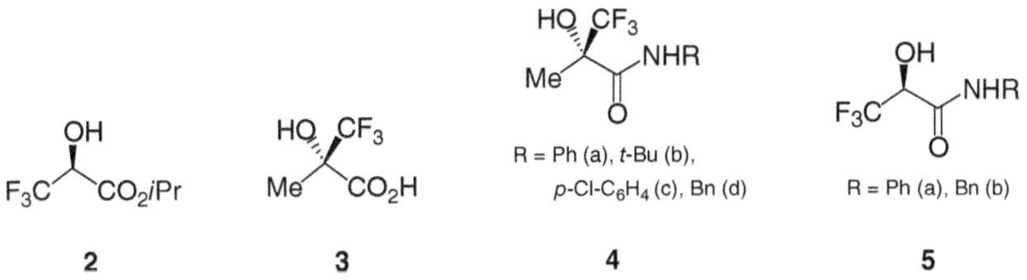

Figure 9.4 Structures of compounds **2** [42], **3** [43], **4** [44], and **5** [45] that have exhibited SDEvS. Sources: Yasumoto et al. [42, 44, 45], Soloshonok et al. [43].

cannot normally be conducted under equilibrium conditions for practical purification purposes and is thus conducted under primarily kinetic conditions [10]. The makeup of the sample to be sublimed is crucial in terms of whether a mechanical mixture consisting of either R and S enantiopure crystals or enantiopure and racemic crystals is used regarding the progress of the sublimation in terms of enantiopurification. Some of the seemingly irreconcilable differences in the literature were simply down to this [10]. One of the working rationales for the SDEvS, with the initial step being a solid → gas phase transition, is that due to the differences in the crystallographic lattice structures of racemic and enantiopure crystals, sublimation rates differ between racemic and enantiopure crystals, similar to the differences seen for other phase transitions such as melting and dissolution. Like crystallization, Blackmond and others [10, 39–41] have also proposed that sublimation, too, could be a possible origin for homochirality. Since for sublimation there is an obvious need for compound volatility, it is unsurprising that there is a disproportionate representation of fluorine-containing compounds exhibiting SDEvS.

Depicted in Figure 9.4 are some fluorine-containing compounds **2** [42], **3** [43], **4** [44], and **5** [45] that have exhibited SDEvS, while listed in Table 9.1 are the sublimation results conducted on these compounds. What is immediately apparent is how simple and efficient the process can be for these compounds. As a matter of fact, for some examinations, sublimation apparatuses were not even required due to sufficient volatilities of the compounds (and also the need not to collect the sublimate). Theoretical investigations [46, 47] and practical examinations [48] of the SDEvS of **3** have been conducted.

Self-disproportionation of enantiomers via sublimation can obviously be a very viable means to effect enantiopurification of a scalemate – if the compound is sufficiently volatile and thermally stable of course. In fact, fluorine could deliberately be judiciously incorporated into a molecule to enhance the volatility of the molecule, and this tactic has been successfully demonstrated using (hexafluoro)pivalic acid as a sublimation enabling tag [49] for the enantiopurification of an amine by SDEvS under ambient conditions followed by removal of the (hexafluoro)pivaloyl group to regenerate the amine. Indeed, SDEvS can be so proficient that there is even a danger of altering the ee of a scalemate during rotary evaporation [50, 51], an everyday occurrence in organic laboratories of course. The problems extend to the storage of such samples whereby the homogeneity of the sample can be compromised within the container [52]. The dangers are thus clear and omnipresent and workers

Table 9.1 SDEvS results for compounds and **2** [42], **3** [43], **4** [44], and **5** [45].

Compound	Conditions	Starting ee (%)	Change in sublimate or ee of sublimate[a]	Remainder ee (%)	References
(S)-**2**	Ambient, open air, 0.92 h	21	Enantiodepleted	>99.9	[42]
(S)-**2**	Ambient, open air, 1 h	37	Enantiodepleted	>99.9	[42]
(S)-**2**	Ambient, open air, 0.75 h	59	Enantiodepleted	>99.9	[42]
(S)-**2**	Ambient, open air, 0.25 h	79	Enantiodepleted	>99.9	[42]
(S)-**3**	60 °C, 3 h	76	48%	80	[43]
(S)-**3**	Ambient, open air, 57 h	80	Racemic	>99.9	[43]
(S)-**4a**	80 °C, 56 h	88	Enantiodepleted	97	[44]
(S)-**4b**	rt	70	Enantiodepleted	98	[44]
(S)-**4b**	rt, 27 h	82	Enantiodepleted	>99	[44]
(S)-**4c**	80 °C, 23 h	77	Enantiodepleted	>99	[44]
(S)-**4d**	60 °C	79	Enantiodepleted	73	[44]
(S)-**5a**	80 °C, 12 h	87	Enantiodepleted	96	[45]
(S)-**5a**	80 °C, 15.5 h	67	Enantioenriched	18	[45]
(S)-**5b**	50 °C, 131 h	67	Enantioenriched	27	[45]

a) Enantioenrichment/-depletion assumed if the sublimate was not collected.
Source: Yasumoto et al. [42, 44, 45], Soloshonok et al. [43].

should take due care when dealing with scalemic samples of volatile, crystalline compounds.

9.3.3 SDE via Chromatography

Self-disproportionation of enantiomers via chromatography is a very complex process. During chromatographic development, there are various environments to which the analytes are exposed such as the interstitial voids with an environment akin to the solution state, within the stationary-phase pores where the polarity of the environment is opposite to the eluent, and the environment of the stationary-phase adsorbent surface, which can drastically alter the associate preference [53]. Each of these environments can affect the position of the equilibrium and the associate preference and thus influence the progress of the chromatography. So sensitive is SDEvC to any altered position of the equilibrium and, particularly, the associate preference that even a seemingly inconsequential change in the eluent [54–56] or stationary phase [8, 57] can lead to a reversal of the elution order between the excess enantiomer and the racemic portions. The SDEvC phenomenon has been the subject of theoretical analyses [5, 58–62] and the key points of these treatments are as follows: (1) complete separation of the excess enantiomer and racemic portions is not possible; (2) while the second-eluting component will, theoretically, always be contaminated with the first-eluting component irrespective of how much the chromatography is extended and how late the fractions are collected, the content of the first-eluting component can nevertheless drop to negligible levels in practice; and (3) it is possible, however, to obtain initial fractions containing the first-eluting component completely free of the second-eluting component [13].

Self-disproportionation of enantiomers via chromatography has been observed in many forms of chromatography, e.g. HPLC, MPLC, flash, gravity-driven column, and SEC, with numerous examples having been reported thus far [3, 4, 12–17, 19–24], again with fluorine-containing compounds prominent among the reports. Notably, some of the largest Δees, SDE yields, and best separations of the enantiomeric and racemic portions have been obtained for fluorine-containing compounds. In Sections 9.3.3.1–9.3.3.3, selected examples of fluorine-containing compounds that have exhibited SDEvC have been allotted to categories depending on whether they possess a –CF$_3$, a C$_q$–F$_{1/2}$, or a –COCF$_3$ moiety. Where quantification of the SDEvC has been reported, selected values have been tabulated.

9.3.3.1 SDEvC for Compounds Containing a –CF$_3$ Moiety

Structures of compounds containing a –CF$_3$ moiety that have exhibited SDEvC are depicted in Figure 9.5 with selected results presented in Table 9.2.

It is interesting to note that while isopropyl (3,3,3-trifluoro)lactate (**2**) only exhibited [38] a very low Δee by normal-phase column chromatography, due to the excessively strong preference for homochiral associates, it exhibited an extremely high Δee by SEC. It is also notable that neither of the non-fluorinated analogs of **6a** and **6c** exhibited SDEvC when tested [2], thus clearly revealing the definitive role a –CF$_3$ moiety plays in promoting SDEvC. What is most remarkable, though, is the ability of SDEvC to furnish enantiopure fractions when the conditions are optimal, and even to provide high SDE yields.

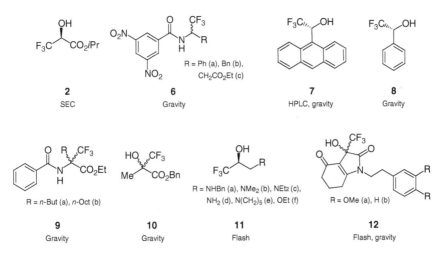

Figure 9.5 Structures of CF$_3$-containing compounds **2** [38], **6** [2, 63], **7** [2, 64], **8** [2], **9** [2], **10** [2], **11** [65], and **12** [66] that have exhibited SDEvC by achiral chromatography. Source: Aceña et al. [38], Soloshonok [2], Soloshonok and Berbasov [63], Sorochinsky et al. [65], Ogawa et al. [66], Matusch and Coors [64].

Table 9.2 Selected SDEvC results for CF_3-containing compounds **2** [38], **6** [2, 63], **11** [65], and **12** [66] examined by chromatography using silica gel.[a]

Compound	Conditions; eluent	Starting ee (%)	First fraction ee (%)	Last fractiond ee (%) (Y_{SDE}, %)	References
(*S*)-**2**	SEC, polystyrene gel; $CHCl_3$	75	99.9	~Racemic	[38]
(*S*)-**6a**	Gravity; *n*-hex/EtOAc (5 : 1)	66.6	59.9	78.8	[2]
(*S*)-**6b**	Gravity; *n*-hex/EtOAc (5 : 1)	66.6	58.4	99.9	[2]
(*S*)-**6c**	Gravity; *n*-hex/EtOAc (5 : 1)	66.6	8.1	>99.9	[2]
(*R*)-**6c**	Gravity; *n*-hex/EtOAc (5 : 1)	66.6	8.1	>99.9	[63]
(*R*)-**6c**	Gravity; *n*-hex/EtOAc (5 : 1)	95.0	88.3	>99.9	[63]
(*S*)-**11a**	Flash; *c*-hex/C_6H_6/*t*-Bu_2O (1 : 1 : 0.1)	75	51	>99 (35)	[65]
(*S*)-**11b**	Flash; *c*-hex/C_6H_6/*t*-Bu_2O (1 : 1 : 0.1)	75	31	>99 (39)	[65]
(*S*)-**11c**	Flash; *c*-hex/C_6H_6/*t*-Bu_2O (1 : 1 : 0.1)	75	33	>99 (41)	[65]
(*S*)-**11d**	Flash; *c*-hex/C_6H_6/*t*-Bu_2O (1 : 1 : 0.1)	75	28	>99 (44)	[65]
(*S*)-**11e**	Flash; *c*-hex/C_6H_6/*t*-Bu_2O (1 : 1 : 0.1)	75	30	>99 (40)	[65]
(*S*)-**11f**	Flash; *c*-hex/C_6H_6/*t*-Bu_2O (1 : 1 : 0.1)	75	21	>99 (53)	[65]
12a	Gravity; Et_2O	52.1	99.9	33.0	[66]
12a	Flash; Et_2O	52.0	99.9	28.3	[66]
12b	Flash; Et_2O	37.5	82.3	26.1	[66]

a) Unless stated otherwise.
Source: Aceña et al. [38], Soloshonok [2], Soloshonok and Berbasov [63], Sorochinsky et al. [65], Ogawa et al. [66].

9.3.3.2 SDEvC for Compounds Containing a C_q–$F_{1/2}$ Moiety

Structures of compounds containing a C_q–$F_{1/2}$ moiety that have exhibited SDEvC are depicted in Figure 9.6 with selected results presented in Table 9.3.

It is worth noting that for this set of compounds, the Δees, while still significant, are much smaller generally in comparison to the CF_3-containing compounds. Interestingly, the 3′-fluorinated analog **13a** of thalidomide (**13b**) exhibited a reversed order of elution with the racemic portion eluting first in comparison with thalidomide (**13b**) [67]. In fact, thalidomide (**13b**) performed better than its 3′-fluorinated analog **13a**. The incorporation of fluorine into a molecule likely alters the relative stabilities of the homo- and heterochiral associates, perhaps so much so that the association preference can even be inverted, thus potentially leading to a reversed order of elution. However, in the case of **13a** and **13b**, the incorporation of fluorine into the 3′ position of thalidomide (**13b**) considerably alters the conformation

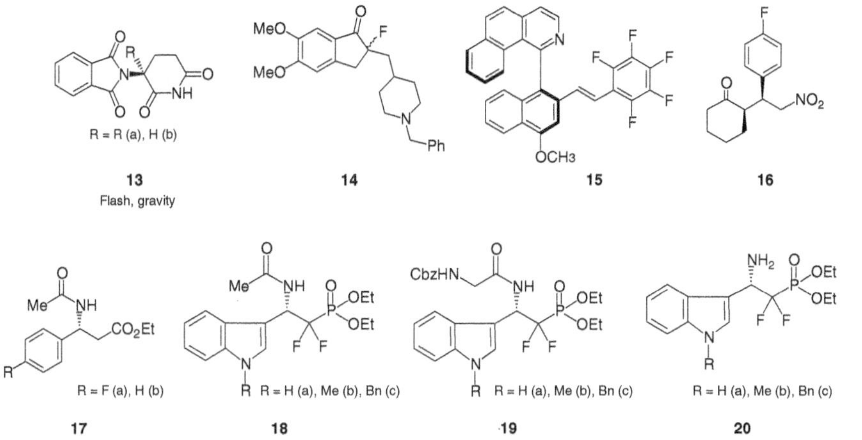

Figure 9.6 Structures of $C_q-F_{1/2}$-containing compounds **13** [67], **14** [68], **15** [69], **16** [70], **17** [57], **18** [56], **19** [56], and **20** [56] that have exhibited SDEvC by gravity-driven achiral column chromatography (unless otherwise indicated). Sources: Maeno et al. [67, 68], Zheng and You [69], Reyes-Rangel et al. [70], Kwiatkowska et al. [56], Wzorek et al. [57].

of the molecule [67] and hence a valid comparison cannot be made in this case regarding solely the electronic effects of fluorine on the stability of the associates. The starting ees for the SDE using column and flash chromatography of thalidomide (**13b**) are comparable and provide an evaluation of the column and flash chromatographic methods, and reveal the positive effect of improved chromatographic efficiency for the SDEvC [71] for the latter method with Δees of 53.5% and 65%, respectively. Compound **17b** also provided excellent results, though it was comparable to its non-fluorinated analog **17a** [57]. In this case, the fluorine is located far from the expected site of intermolecular interaction and thus it is unsurprising that there is little difference between the two compounds. More examples of SDEvC for C_q–F-containing compounds are presented further on in Section 9.5 as examples of how the SDEvC has the potential to hamper correct reporting of ees if workers are unaware of the phenomenon.

9.3.3.3 SDEvC for Compounds Containing a –COCF₃ Moiety

Structures of compounds containing a $-COCF_3$ moiety that have exhibited SDEvC are depicted in Figure 9.7 with selected results presented in Table 9.4.

Also for $COCF_3$-containing compounds, the Δees, while still significant, are generally much smaller in comparison to the CF_3-containing compounds. This is unsurprising as the electronegativity of fluorine can reduce the hydrogen bonding capability of the carbonyl oxygen, with hydrogen bonding as one of the main intermolecular interactions giving rise to the SDE. There are, however, some surprising results. For instance, the series of **22** [27, 54] using column chromatography is most interesting, since while the non-fluorinated N-acetyl 1-phenylethylamine (**22b**) exhibits a strong SDEvC with a Δee of 54.5%, the three perfluorinated R derivatives – trifluoromethyl (**22a**), pentafluoroethyl (**22d**), and heptafluoropropyl

Table 9.3 Selected SDEvC results for $C_q - F_{1/2}$-containing compounds **13** [67], **14** [68], **15** [69], **16** [70], **17** [57], **18** [56], **19** [56], and **20** [56] examined by gravity-driven column chromatography[a] using silica gel.[a]

Compound	Conditions; eluent	Starting ee (%)	First fraction ee (%) (Y_{SDE}, %)	Last fraction ee (%)	References
(R)-**13a**	n-Hex/EtOAc[b)c)]	34.2	30.0	70.6	[67]
(R)-**13a**	Flash[b)], n-hex/EtOAc[c)]	27.3	9.4	50.4	[67]
(R)-**13a**	Aluminum oxide; n-hex/EtOAc[c)]	37.2	27.4	38.6	[67]
(R)-**13a**	Mesoporous silica; n-hex/EtOAc[c)]	34.2	32.2	56.1	[67]
(S)-**13b**	n-Hex/EtOAc[b)c)]	35.5	80.9	27.4	[67]
(S)-**13b**	Flash; n-hex/EtOAc[c)]	36.2	80.0	15.0	[67]
(S)-**13b**	Aluminum oxide; n-hex/EtOAc[c)]	31.1	34.5	21.6	[67]
(S)-**13b**	Mesoporous silica; n-hex/EtOAc[c)]	35.5	86.9	20.7	[67]
14	n-Hex/EtOAc (1 : 4)	44.0	72.0	29.0	[68]
(aS)-**15**	Petroleum ether/EtOAc (10 : 1)	77	99	4	[69]
(1R,2S)-**16**	n-Hex/EtOAc (4 : 1)	82	86	80	[70]
(1R,2S)-**16**	CH₂Cl₂	82	84	82	[70]
(R)-**17a**	c-Hex/MTBE (1 : 2)	79.8	>99.9 (37.1)	36.6	[57]
(R)-**17b**	c-Hex/MTBE (1 : 2)	69.4	>99.9 (21.7)	33.2	[57]
(S)-**18b**	c-Hex/EtOAc (1 : 8)	64.2	96.0	41.4	[56]
(S)-**19b**	MTBE	65.4	86.9	52.4	[56]
(S) **20b**	c Hex/EtOAc (1 : 8)	66.2	81.8	59.2	[56]

a) Unless stated otherwise.
b) Silica gel wetted with 10% wt/wt with water.
c) Eluent ratio not given.
Source: Maeno et al. [67, 68], Zheng and You [69], Reyes-Rangel et al. [70], Kwiatkowska et al. [56], Wzorek et al. [57].

R¹ = R² = iPr, R3 = c-hex (a);
R¹ = Me, R² = H, R³ = iPr (b)

21
Gravity

R = CF₃ (a), Me (b), CF₂CF₂CF₃ (c),
CF₂CF₃ (d), CH₂F (e), CHF₂ (f)

22
MPLC, gravity

R = CF₃ (a), Me (b), CH₂F (c), CHF₂ (d)

23
Gravity

Figure 9.7 Structures of COCF₃-containing compounds **21** [72], **22** [27, 54, 55], and **23** [73] that have exhibited SDEvC by achiral chromatography. Source: Charles and Gil-Av [72], Wzorek et al. [27, 54], Nakamura et al. [55], Hosaka et al. [73].

Table 9.4 Selected SDEvC results for $COCF_3$-containing compounds **21** [72], **22** [27, 54], and **23** [73] examined by gravity-driven column chromatography using silica gel.

Compound	Conditions; eluent	Starting ee (%)	First fraction ee (%) (Y_{SDE}, %)	Last fraction ee (%)	References
(S,S)-**21a**	n-Hex/EtOAc (19 : 1)[a]	67.0	88.0	12.5	[72]
(S)-**21b**	n-Hex/EtOAc (9 : 1)[a]	70.7	86.5	25.0	[72]
(R)-**22a**	c-Hex/MTBE (10 : 1)	63.2	46.6	67.6	[54]
(R)-**22b**	c-Hex/MTBE (1 : 2)	70.0	99.9 (3.6)	45.4	[27]
(R)-**22c**	c-Hex/MTBE (20 : 1)	65.0	61.0	72.0	[54]
(R)-**22d**	c-Hex/MTBE (20 : 1)	68.2	45.2	95.2	[54]
(R)-**22e**	c-Hex/MTBE (2 : 1)	62.4	67.2	57.4	[54]
(R)-**22f**	c-Hex/MTBE (4 : 1)	64.9	68.6	60.6	[54]
23a	n-Hex/MTBE (10 : 1)	73.0	71.0	74.0	[73]
23b	n-Hex/MTBE (1 : 2)	75.2	92.0	68.4	[73]
23c	n-Hex/MTBE (5 : 2)	65.6	67.0	58.4	[73]
23d	n-Hex/MTBE (4 : 1)	70.5	74.6	64.6	[73]

a) Silica gel wetted with 6% wt/wt with water.
Source: Charles and Gil-Av [72], Wzorek et al. [27, 54], Hosaka et al. [73].

(**22c**) – all show a reversed order of elution with Δees of −21.0%, −50.0%, and −11.0%, respectively. The monofluoro (**22e**) and difluoro (**22f**) analogs of **22b** retain the same order of elution as **22b**, but with reduced Δees of 9.8% and 8.0%, respectively, and are thus intermediate along the changeover to **22a** and reveal a clear, continuous trend. A very similar trend was also observed [73] for the series **23** (Δees of 23.6%, 8.6%, 10.0%, and −3.1% for CH_3 to CF_3, respectively). The same reduction in Δee was also observed [55] for **22a** in comparison with **22b** when using MPLC, 23% and 71%, respectively, though the elution order did not reverse under the applied conditions where n-hex/EtOAc was used as the eluent. Thus, when fluorine is present in a molecule, it not only has the ability to greatly accentuate SDE effects, it can also mitigate any SDE effects present to a high degree in the protiated analog by creating opposing influences, even nullifying the effects altogether in some cases. In this example, the dramatic change between **22a** and **22b**, a reversal in elution order nonetheless, is duly reflected in the NMR by SIDA (vide infra).

As can be seen from the aforementioned results, the SDEvC can be an effective means of enantiopurification when the conditions are right – typically an R_f of 0.2 using a solvent system that will not interfere with the likely intermolecular interaction is considered a good starting point [2, 8, 27, 54, 56, 57, 63, 73–76], e.g. protic solvents and carbonyl solvents such as acetone are not recommended for intermolecular interactions based on hydrogen bonding or dipole–dipole interactions, respectively.

9.4 The SIDA Phenomenon

Another chiral phenomenon very much related to SDE is SIDA in solution-state NMR. The basis of the SIDA phenomenon is that with association, molecules in solution can be present either as single molecules (SM), in homochiral associates (HOM), or in heterochiral associates (HET). For a dynamic equilibrium that is fast on the NMR timescale, the resulting observed chemical shift (δ_{obs}) of a nucleus is the population-weighted (based on mole fraction, χ) average of the δ values of these three states [6, 77]:

$$\delta_{obs} = \chi_{SM} \cdot \delta_{SM} + \chi_{HOM} \cdot \delta_{HOM} + \chi_{HET} \cdot \delta_{HET}$$

The result is that the δ values of an enantiopure sample will not necessarily match those of a racemic sample even under identical conditions and the δ values can be very concentration, solvent, and temperature dependent [77]. Moreover, a scalemic sample can potentially have distinct NMR signals for some nuclei for the two enantiomers present since the mole fractions of each enantiomer within each state will not necessarily be the same. When distinct signals are present, integration of the signals provides the ee of the sample directly without any chiral induction such as the addition of a chiral selector (CS). In other words, the enantiopurity (% ee) of a sample can be determined simply by routine, achiral NMR without recourse to such techniques as chiral shift reagents or HPLC using chiral stationary phases. The SIDA phenomenon has recently been well reviewed and explained [77]. The phenomenon is illustrated schematically in Figure 9.8 for a hypothetical scalemic sample.

Here again, fluorine is able to play a central role in the manifestation of the phenomenon by enhancing the formation of intermolecular interactions (notably via hydrogen bonds) or by accentuating the difference in stability between the homo- and heterochiral associates (hence accentuating the difference in the δ values). Surprisingly though, there are only four literature reports of SIDA involving fluorine-containing compounds. In the case of the pharmaceutical intermediate **24** [78] (Figure 9.9), the fluorine may not be playing a key role in the process given its distal position from the likely points of intermolecular interaction. A key role for fluorine in the process, however, is more likely in the 2'-amino analog of the cruciferous phytoalexin spirobrassinin **25** where signal separation was particularly good in both the ^{1}H and ^{13}C NMR spectra for several signals [79]. Interestingly, in the case of 1-phenyl-2,2,2-trifluoroethanol (**8**) [6], signal separation was not observed though some signals were noted to be shifted in a comparison between the enantiopure and racemic solutions. A series of scalemic solutions at the same concentrations as the enantiopure and racemic solutions revealed a progression from the δ for the enantiopure solution to the δ for the racemic solution for these signals. This behavior of migrating signals without splitting across an enantiomeric titration [6, 80] has been termed atypical SIDA (aSIDA) [6].

Other examples of SIDA involving fluorine-containing compounds include the synthetic intermediate **26** [81] and the drug candidate **27** [81] (Figure 9.10) where the concentration dependency of the δ values for the methyl signals was found to be

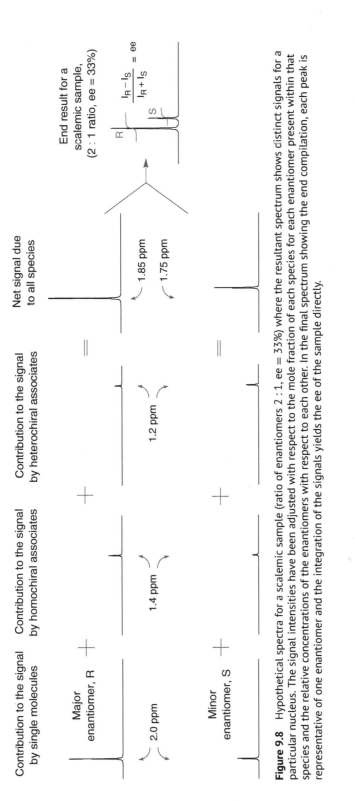

Figure 9.8 Hypothetical spectra for a scalemic sample (ratio of enantiomers 2 : 1, ee = 33%) where the resultant spectrum shows distinct signals for a particular nucleus. The signal intensities have been adjusted with respect to the mole fraction of each species for each enantiomer present within that species and the relative concentrations of the enantiomers with respect to each other. In the final spectrum showing the end compilation, each peak is representative of one enantiomer and the integration of the signals yields the ee of the sample directly.

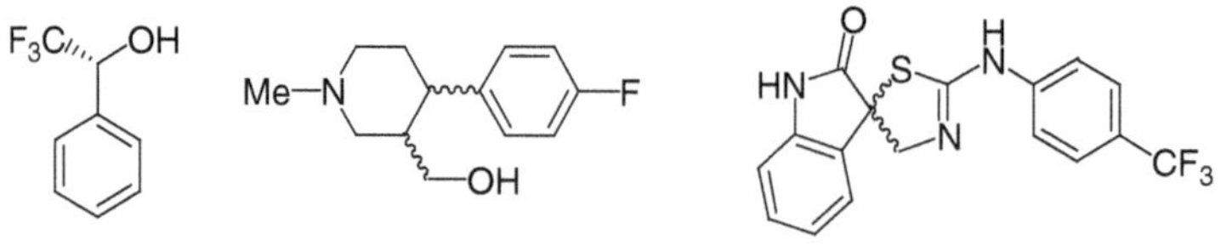

Figure 9.9 Structures of 1-phenyl-2,2,2-trifluoroethanol (**8**) [6], **24** [78], and **25** [79]. Source: Nieminen et al. [6], Navrátilová and Potáček [78], Budovská et al. [79].

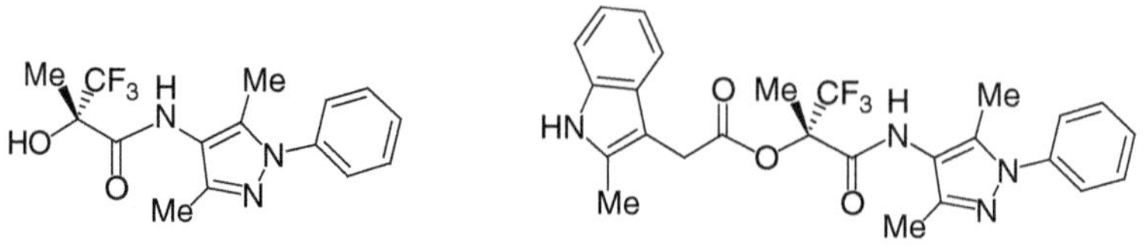

Figure 9.10 Structures of the synthetic intermediate **26** [81] and the drug candidate **27** [81]. Source: Baumann et al. [81].

particularly strong, leading to confusion among the workers regarding the veracity and structure of the enantiomer pairs that had been synthesized.

As stated above, when fluorine is present in a molecule, it not only has the ability to greatly accentuate the effects of these chiral phenomena (both SDE and SIDA), it can also mitigate any effects present to a high degree in the protiated analog by creating opposing influences, even nullifying the effects altogether in some cases. For example, while *N*-acetyl 1-phenylethylamine (**22b**) showed strong SIDA effects for the H-2 methyl signal, the effect was much reduced in its trifluoro analog **22a** (Figure 9.11) (Klika, Alvarez, Wzorek, and Soloshonok, unpublished results). The reduction in the separation of the signals could arise from either a significant shift in the equilibria back toward single molecules, a change (most likely a reduction) in the preference between the homo- and heterochiral associates, and/or relative changes in the δ values of the various solution-state species.

The relevance of SIDA to the SDE is that it can be a practical and accessible test for possible susceptibility to a high magnitude of the SDE, particularly for SDEvC [6], as well as providing an indication of the preferred solution-state association by means of an enantiomeric titration [6, 80] using NMR or by other NMR methods such as diffusion, T_1, or T_2 measurements, for example. Unfortunately, in addition to providing a readily accessible means to measure the ee supremely easily, SIDA can also cause considerable problems with regard to confusion, misidentification, and incorrect evaluation of purity if workers are unaware of the phenomenon.

9.5 Conclusions and Recommendations

It is apparent that the SDE represents a danger regarding alteration of the ee of a sample for any work involving scalemates if due care is not undertaken, particularly

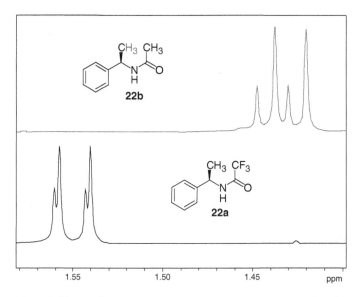

Figure 9.11 The ^{1}H NMR spectra of *N*-acetyl 1-phenylethylamine (**22b**, top trace) and its trifluoro analog **22a** (bottom trace) for the region containing the H-2 methyl signal. Note the strongly reduced SIDA effect for **22a**. The solutions are of the same analyte concentration in CDCl$_3$. Source: Based on Klika, Alvarez, Wzorek, and Soloshonok, unpublished results.

for compounds that tend to show a propensity for exhibiting the SDE, such as fluorine-containing compounds as has been amply demonstrated here. This is especially true for compounds containing a –CF$_3$ group as it is a particularly effective SDE-phoric group and due caution should be applied when dealing with scalemic samples of molecules containing this moiety. Incongruously, while all workers would be aghast if the ee of a product following recrystallization was reported as representing the direct result of an asymmetric transformation – acutely aware that such a process could well result in alteration of the ee – many workers are seemingly oblivious to the possibility that alteration of the ee of a sample can also occur as a result of routine gravity-driven column chromatography if due care is not taken to avoid fractionation of the eluting product material. It cannot be emphasized enough that the dangers of alteration of the ee are present whenever any physicochemical process is applied to a sample and the material of interest is fractionated. Thus, routine testing for occurrences of the SDE [82] must be considered an obligatory practice to follow when undertaking the purification of a sample if measurement of the ee is to be made afterwards to correctly assess the true outcome of an asymmetric transformation. Ideally, measurement of the ee should be made prior to any application of a physicochemical process for purification purposes, especially if fractionation of the portion containing the compound in question is made. Fortunately, some groups are now adopting the practice of routine testing for occurrences of the SDE [83]. For example, compounds **28a** [84], **28b** [85], **28c** [86], **29** [86], **30** [86], and **31** [87] (Figure 9.12) – where the SDE might not be expected to be so strong due to the presence of only a single fluorine atom – were

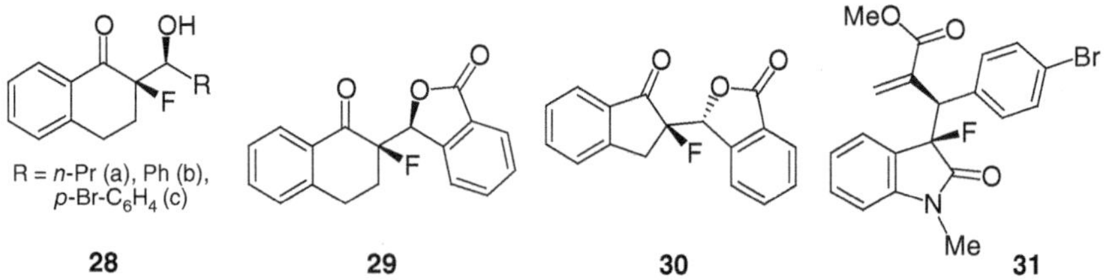

R = *n*-Pr (a), Ph (b),
** *p*-Br-C₆H₄ (c)**

28   **29**   **30**   **31**

Figure 9.12 Structures of compounds **28** [84–86], **29** [86], **30** [86], and **31** [87]. Sources: Zhang et al. [84, 86], Xie et al. [85], Zhu et al. [87].

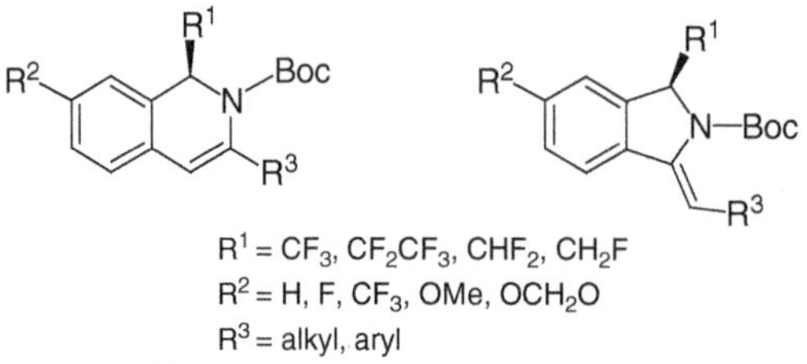

R¹ = CF₃, CF₂CF₃, CHF₂, CH₂F
R² = H, F, CF₃, OMe, OCH₂O
R³ = alkyl, aryl

32   **33**

Figure 9.13 Structures of compounds **32** [88] and **33** [88]. Source: Reprinted with permission from Fustero et al. [88]. Copright 2013, American Chemical Society.

all tested for the SDE both via sublimation and chromatography. For compounds **28–30**, the SDEvS tests proved negative while for **31** a small (Δee = 2%) effect only was found. For the SDEvC tests, only for **28a** was a negative result found while for compounds **28b,c–31** Δees of up to 11% were found, thereby clearly showing the worth of conducting SDE tests. However, it seems that the practice of routine testing for occurrences of the SDE is yet to be widely adopted. Interestingly, the SDE can cause confusion for workers, either mistaking the change in ee as a racemization of the sample or simply not understanding the SDE process, i.e. chirality is neither created nor lost during the process [13]. For example, the change in ee for the compounds **32** and **33** (Figure 9.13) was attributed [88] to "parcial (sic) racemization of compounds **32** and **33** was observed probably due to an enantiomer self-disproportionation effect on achiral silica gel", despite both compounds **32** and **33** appearing to be configurationally stable.

Looking for SIDA effects can also be a practical and accessible test for the possible potential for SDEvC by demonstrating the occurrence of intermolecular association in solution while more specific testing for SDE via column chromatography can be accomplished by TLC examinations. Indeed, it seems that not enough use has been made of such tactics and it is highly recommended for workers to do so. On the other hand, the SDE, particularly SDEvC and SDEvS, represents an additional, unconventional means of enantiopurification to add to the available repertoire of enantiopurification methods, particularly on a macro scale. Indeed, as stated earlier, fluorine could deliberately be judiciously incorporated into a molecule to effect enantiopurification by the SDE, and this tactic has been successfully demonstrated using (hexafluoro)pivalic acid as a sublimation enabling tag [49] for the enantiopurification of an amine by sublimation under ambient conditions. Similarly, the concept of pseudo-SDE (CS-assisted SDE resolution of racemates) has also been successfully demonstrated [89, 90]. The process mimics true SDEvC but uses a CS to effect enantiodifferentiation instead of the excess enantiomer by having a closely eluting, structurally similar CS to the substrate. In addition, pseudo-SDE can provide an indication of the elution order between dimeric associates and monomers as well as the preference between homo- and heterochiral associates.

In summary, due to the unique and peculiar properties of fluorine, the presence of fluorine or fluorinated substituents in the structure of a chiral compound can have profound effects on the magnitude of the SDE by accentuating intermolecular interactions or by altering the homo-/heterochiral associate preference. Though fluorine can greatly accentuate the SDE, it can also, conversely, diminish the SDE when it is already present to a high degree in the protiated analog, to the point of nullifying the SDE altogether in some cases or even reversing the effect (e.g. elution order reversal in SDEvC). The SDE is always present to some degree for all scalemic samples, it should be noted, and fluorine simply lends itself to amplifying the SDE, by enhancing either attractive or repulsive intermolecular interactions. The potential consequences of ignoring the SDE phenomenon are significant errors in reporting the stereochemical outcome of catalytic asymmetric transformations in terms of the enantiomeric composition of the catalysts and products as well as peculiarities in the properties of chiral pharmaceuticals. Since there are a rapidly growing number of fluorine-containing drugs on the pharmaceutical market [91–95], the necessity for regulations with a clear understanding of the SDE properties for the safe handling, production, processing, storage, and medical applications of fluorine-containing drugs is paramount. While the many unique properties that fluorine can impart to a molecule upon its incorporation are widely appreciated by many workers, the prevalence of fluorine-containing compounds among the set of compounds that exhibit extreme SDE behavior is much less appreciated. But how wondrous it is that enantiopurification of a sample can be effected by simple, routine distillation, a truly remarkable result and a direct consequence of the special properties of fluorine!

References

1 Klika, K.D. and Soloshonok, V.A. (2014). Terminology related to the phenomenon "self-disproportionation of enantiomers" (SDE). *Helv. Chim. Acta* 97: 1583–1589.

2 Soloshonok, V.A. (2006). Remarkable amplification of the self-disproportionation of enantiomers on achiral-phase chromatography columns. *Angew. Chem. Int. Ed.* 45: 766–769.

3 Sorochinsky, A.E., Aceña, J.L., and Soloshonok, V.A. (2013). Self-disproportionation of enantiomers of chiral, non-racemic fluoroorganic compounds: role of fluorine as enabling element. *Synthesis* 45: 141–152.

4 Soloshonok, V.A. and Berbasov, D.O. (2006). Self-disproportionation of enantiomers on achiral phase chromatography. One more example of fluorine's magic powers. *Chim. Oggi – Chem. Today* 24: 44–47.

5 Schurig, V. (2009). Elaborate treatment of retention in chemoselective chromatography – The retention increment approach and non-linear effects. *J. Chromatogr. A* 1216: 1723–1736.

6 Nieminen, V., Murzin, D.Yu., and Klika, K.D. (2009). NMR and molecular modeling of the dimeric self-association of the enantiomers of 1,1′-bi-2-naphthol and 1-phenyl-2,2,2-trifluoroethanol in the solution state and their relevance to enantiomer self-disproportionation on achiral-phase chromatography (ESDAC). *Org. Biomol. Chem.* 7: 537–542.

7 Klika, K.D., Budovská, M., and Kutschy, P. (2010). NMR spectral enantioresolution of spirobrassinin and 1-methoxyspirobrassinin enantiomers using (*S*)-(−)-ethyl lactate and modeling of spirobrassinin self-association for rationalization of its self-induced diastereomeric anisochromism (SIDA) and enantiomer self-disproportionation on achiral-phase chromatography (ESDAC) phenomena. *J. Fluorine Chem.* 131: 467–476.

8 Wzorek, A., Sato, A., Drabowicz, J., and Soloshonok, V.A. (2016). Self-disproportionation of enantiomers via achiral gravity-driven column chromatography: a case study of *N*-acyl-α-phenylethylamines. *J. Chromatogr. A* 1467: 270–278.

9 Klussmann, M., White, A.J.P., Armstrong, A., and Blackmond, D.G. (2006). Rationalization and prediction of solution enantiomeric excess in ternary phase systems. *Angew. Chem. Int. Ed.* 45: 7985–7989.

10 Klussmann, M. and Blackmond, D.G. (2007). Spoilt for choice: assessing phase behavior models for the evolution of homochirality. *Chem. Commun.*: 3990–3996.

11 Blackmond, D.G. and Klussmann, M. (2007). Investigating the evolution of biomolecular homochirality. *AIChE J.* 53: 1–8.

12 Sorochinsky, A.E. and Soloshonok, V.A. (2013). Self-disproportionation of enantiomers of enantiomerically enriched compounds in differentiation of enantiomers II. In: *Topics in Current Chemistry*, vol. 341 (ed. V. Schurig), 301–340. Springer-Verlag GmbH.

13 Han, J., Kitagawa, O., Wzorek, A. et al. (2018). The self-disproportionation of enantiomers (SDE): a menace or an opportunity? *Chem. Sci.* 9: 1718–1739.

14 Soloshonok, V.A., Roussel, Ch., Kitagawa, O., and Sorochinsky, A.E. (2012). Self-disproportionation of enantiomers via achiral chromatography: a warning and an extra dimension in optical purifications. *Chem. Soc. Rev.* 41: 4180–4188.

15 Martens, J. and Bhushan, R. (1992). Resolution of enantiomers with achiral phase chromatography. *J. Liq. Chromatogr. Related Technol.* 15: 1–27.

16 Martens, J. and Bhushan, R. (2014). Purification of enantiomeric mixtures in enantioselective synthesis: overlooked errors and scientific basis of separation in achiral environment. *Helv. Chim. Acta* 97: 161–187.

17 Martens, J. and Bhushan, R. (2016). Enantioseparations in achiral environments and chromatographic systems. *Isr. J. Chem.* 56: 990–1009.

18 Han, J., Nelson, D.J., Sorochinsky, A.E., and Soloshonok, V.A. (2011). Self-disproportionation of enantiomers via sublimation; new and truly green dimension in optical purification. *Curr. Org. Synth.* 8: 310–317.

19 Drabowicz, J., Jasiak, A., Wzorek, A. et al. (2017). Self-disproportionation of enantiomers (SDE) of chiral sulfoxides, amides and thioamides via achiral chromatography. *Arkivoc* 2017: 557–578.

20 Han, J., Soloshonok, V.A., Klika, K.D. et al. (2018). Chiral sulfoxides: advances in asymmetric synthesis and problems with the accurate determination of the stereochemical outcome. *Chem. Soc. Rev.* 47: 1307–1350.

21 Wzorek, A., Sato, A., Drabowicz, J., and Soloshonok, V.A. (2016). Self-disproportionation of enantiomers (SDE) of chiral nonracemic amides via achiral chromatography. *Isr. J. Chem.* 56: 977–989.

22 Han, J., Wzorek, A., Kwiatkowska, M. et al. (2019). The self-disproportionation of enantiomers (SDE) of amino acids and their derivatives. *Amino Acids* 51: 865–889.

23 Han, J., Wzorek, A., Klika, K.D., and Soloshonok, V.A. (2018). Fluorine-containing pharmaceuticals and the phenomenon of the self-disproportionation of enantiomers, Ch. 10. In: *Late-Stage Fluorination of Bioactive Molecules and Biologically-Relevant Substrates* (ed. A. Postigo). Amsterdam: Elsevier.

24 Han, J., Wzorek, A., Klika, K.D., and Soloshonok, V.A. (2020). The role of fluorine in the self-disproportionation of enantiomers (SDE) phenomenon of scalemic samples of fluoroorganics, Ch. 6. In: *Frontiers of Organofluorine Chemistry* (ed. I. Ojima). London: World Scientific Publishing Co.

25 Horeau, A. and Guetté, J.P. (1974). Interactions diastereoisomeres d'antipodes en phase liquid. *Tetrahedron* 30: 1923–1931.

26 Horeau, A. (1969). Interactions d'enantiomeres en solution; influence sur le pouvoir rotatoire: purete optique et purete enantiomerique. *Tetrahedron Lett.* 10: 3121–3124.

27 Wzorek, A., Sato, A., Drabowicz, J. et al. (2015). Enantiomeric enrichments via the self-disproportionation of enantiomers (SDE) by achiral, gravity-driven column chromatography: a case study using *N*-(1-phenylethyl)acetamide for optimizing the enantiomerically pure yield and magnitude of the SDE. *Helv. Chim. Acta* 98: 1147–1159.

28 Soloshonok, V.A., Kirilenko, A.G., Fokina, N.A. et al. (1994). Chemo-enzymatic approach to the synthesis of each of the four isomers of α-alkyl-β-fluoroalkyl-substituted β-amino acids. *Tetrahedron: Asymmetry* 5: 1225–1228.

29 Soloshonok, V.A. and Ono, T. (1996). The effect of substituents on the feasibility of azomethine–azomethine isomerization: new synthetic opportunities for biomimetic transamination. *Tetrahedron* 52: 14701–14712.

30 Bravo, P., Farina, A., Kukhar, V.P. et al. (1997). Stereoselective additions of α-lithiated alkyl *p*-tolylsulfoxides to *N*-PMP(fluoroalkyl)aldimines. An efficient approach to enantiomerically pure fluoro amino compounds. *J. Org. Chem.* 62: 3424–3425.

31 Soloshonok, V.A., Ono, T., and Soloshonok, I.V. (1997). Enantioselective biomimetic transamination of β-keto carboxylic acid derivatives. an efficient asymmetric synthesis of β-(fluoroalkyl) β-amino acids. *J. Org. Chem.* 62: 7538–7539.

32 Koppenhoefer, B. and Trettin, U. (1989). Is it possible to affect the enantiomeric composition by a simple distillation process? *Fresenius' Z. Anal. Chem.* 333: 750.

33 Katagiri, T., Yoda, C., Furuhashi, K. et al. (1996). Separation of an enantiomorph and its racemate by distillation: strong chiral recognizing ability of trifluorolactates. *Chem. Lett.* 25: 115–116.

34 Katagiri, T. and Uneyama, K. (2001). Chiral recognition by multicenter single proton hydrogen bonding of trifluorolactates. *Chem. Lett.* 30: 1330–1331.

35 Katagiri, T., Takahashi, S., Tsuboi, A. et al. (2010). Discrimination of enantiomeric excess of optically active trifluorolactate by distillation: evidence for a multi-center hydrogen bonding network in the liquid state. *J. Fluorine Chem.* 131: 517–520.

36 Soloshonok, V.A., Cai, C., Hruby, V.J., and Van Meervelt, L. (1999). Asymmetric synthesis of novel highly sterically constrained (2S,3S)-3-methyl-3-trifluoro-methyl- and (2S,3S,3R)-3-trifluoromethyl-4-methylpyroglutamic acids. *Tetrahedron* 55: 12045–12058.

37 Soloshonok, V.A., Avilov, D.V., and Kukhar, V.P. (1996). Asymmetric aldol reactions of trifluoromethyl ketones with a chiral Ni(II) complex of glycine: stereo-controlling effect of the trifluoromethyl group. *Tetrahedron* 52: 12433–12442.

38 Aceña, J.L., Sorochinsky, A.E., Katagiri, T., and Soloshonok, V.A. (2013). Unconventional preparation of racemic crystals of isopropyl 3,3,3-trifluoro-2-hydroxypropanoate and their unusual crystallographic structure: the ultimate preference for homochiral intermolecular interactions. *Chem. Commun.* 49: 373–375.

39 Perry, R.H., Wu, C., Nefliu, M., and Cooks, G. (2007). Serine sublimes with spontaneous chiral amplification. *Chem. Commun.*: 1071–1073.

40 Fletcher, S.P., Jagt, R.B.C., and Feringa, B.L. (2007). An astrophysically-relevant mechanism for amino acid enantiomer enrichment. *Chem. Commun.*: 2578–2580.

41 Cintas, P. (2008). Sublime arguments: rethinking the generation of homochirality under prebiotic conditions. *Angew. Chem. Int. Ed.* 47: 2918–2920.

42 Yasumoto, M., Ueki, H., Ono, T. et al. (2010). Self-disproportionation of enantiomers of isopropyl 3,3,3-(trifluoro)lactate via sublimation: sublimation rates vs. enantiomeric composition. *J. Fluorine Chem.* 131: 535–539.

43 Soloshonok, V.A., Ueki, H., Yasumoto, M. et al. (2007). Phenomenon of optical self-purification of chiral non-racemic compounds. *J. Am. Chem. Soc.* 129: 12112–12113.

44 Yasumoto, M., Ueki, H., and Soloshonok, V.A. (2010). Self-disproportionation of enantiomers of α-trifluoromethyl lactic acid amides via sublimation. *J. Fluorine Chem.* 131: 540–544.

45 Yasumoto, M., Ueki, H., and Soloshonok, V.A. (2010). Self-disproportionation of enantiomers of 3,3,3-trifluorolactic acid amides via sublimation. *J. Fluorine Chem.* 131: 266–269.

46 Tsuzuki, S., Orita, H., Ueki, H., and Soloshonok, V.A. (2010). First principle lattice energy calculations for enantiopure and racemic crystals of α-(trifluoromethyl)lactic acid: is self-disproportionation of enantiomers controlled by thermodynamic stability of crystals? *J. Fluorine Chem.* 131: 461–466.

47 Tonner, R., Soloshonok, V.A., and Schwerdtfeger, P. (2011). Theoretical investigations into the enantiomeric and racemic forms of α-(trifluoromethyl)lactic acid. *Phys. Chem. Chem. Phys.* 13: 811–817.

48 Albrecht, M., Soloshonok, V.A., Schrader, L. et al. (2010). Chirality-dependent sublimation of α-(trifluoromethyl)-lactic acid: relative vapor pressures of racemic, eutectic, and enantiomerically pure forms, and vibrational spectroscopy of isolated (*S*,*S*) and (*S*,*R*) dimers. *J. Fluorine Chem.* 131: 495–504.

49 Ueki, H., Yasumoto, M., and Soloshonok, V.A. (2010). Rational application of self-disproportionation of enantiomers via sublimation – a novel methodological dimension for enantiomeric purifications. *Tetrahedron: Asymmetry* 21: 1396–1400.

50 Doucet, H., Fernandez, E., Layzell, T.P., and Brown, J.M. (1999). The scope of catalytic asymmetric hydroboration/oxidation with rhodium complexes of 1,1′-(2-diarylphosphino-1-naphthyl)isoquinolines. *Chem. Eur. J.* 5: 1320–1330.

51 Abás, S., Arróniz, C., Molins, E., and Escolano, C. (2018). Access to the enantiopure pyrrolobenzodiazepine (PBD) dilactam nucleus via self-disproportionation of enantiomers. *Tetrahedron* 74: 867–871.

52 Carman, R.M. and Klika, K.D. (1992). Partially racemic compounds as brushtail possum urinary metabolites. *Aust. J. Chem.* 45: 651–657.

53 Dutta, S. and Gellman, A.J. (2017). Enantiomer surface chemistry: conglomerate versus racemate formation on surfaces. *Chem. Soc. Rev.* 46: 7787–7839.

54 Wzorek, A., Kamizela, A., Sato, A., and Soloshonok, V.A. (2017). Self-disproportionation of enantiomers (SDE) via achiral gravity-driven column chromatography of *N*-fluoroacyl-1-phenylethylamines. *J. Fluorine Chem.* 196: 37–43.

55 Nakamura, T., Tateishi, K., Tsukagoshi, S. et al. (2012). Self-disproportionation of enantiomers of non-racemic chiral amine derivatives through achiral chromatography. *Tetrahedron* 68: 4013–4017.

56 Kwiatkowska, M., Marcinkowska, M., Wzorek, A. et al. (2019). The self-disproportionation of enantiomers (SDE) via column chromatography of β-amino-α,α-difluorophosphonic acid derivatives. *Amino Acids* 51: 1377–1385.

57 Wzorek, A., Sato, A., Drabowicz, J. et al. (2016). Remarkable magnitude of the self-disproportionation of enantiomers (SDE) via achiral chromatography: application to the practical-scale enantiopurification of β-amino acid esters. *Amino Acids* 48: 605–613.

58 Nicoud, R.-M., Jaubert, J.-N., Rupprecht, I., and Kinkel, J. (1996). Enantiomeric enrichment of non-racemic mixtures of binaphthol with non-chiral packings. *Chirality* 8: 234–243.

59 Jung, M. and Schurig, V. (1992). Computer simulation of three scenarios for the separation of non-racemic mixtures by chromatography on achiral stationary phases. *J. Chromatogr. A* 605: 161–166.

60 Gil-Av, E. and Schurig, V. (1994). Resolution of non-racemic mixtures in achiral chromatographic systems: a model for the enantioselective effects observed. *J. Chromatogr. A* 666: 519–525.

61 Baciocchi, R., Zenoni, G., Mazzotti, M., and Morbidelli, M. (2002). Separation of binaphthol enantiomers through achiral chromatography. *J. Chromatogr. A* 944: 225–240.

62 Baciocchi, R., Mazzotti, M., and Morbidelli, M. (2004). General model for the achiral chromatography of enantiomers forming dimers: application to binaphthol. *J. Chromatogr. A* 1024: 15–20.

63 Soloshonok, V.A. and Berbasov, D.O. (2006). Self-disproportionation of enantiomers of (*R*)-ethyl 3-(3,5-dinitrobenzamido)-4,4,4-trifluorobutanoate on achiral silica gel stationary phase. *J. Fluorine Chem.* 127: 597–603.

64 Matusch, R. and Coors, C. (1989). Chromatographic separation of the excess enantiomer under achiral conditions. *Angew. Chem. Int. Ed.* 28: 626–627.

65 Sorochinsky, A.E., Katagiri, T., Ono, T. et al. (2013). Optical purifications via self-disproportionation of enantiomers by achiral chromatography: case study of a series of α-CF$_3$-containing secondary alcohols. *Chirality* 25: 365–368.

66 Ogawa, S., Nishimine, T., Tokunaga, E. et al. (2010). Self-disproportionation of enantiomers of heterocyclic compounds having a tertiary trifluoromethyl alcohol center on chromatography with a non-chiral system. *J. Fluorine Chem.* 131: 521–524.

67 Maeno, M., Tokunaga, E., Yamamoto, T. et al. (2015). Self-disproportionation of enantiomers of thalidomide and its fluorinated analogue via gravity-driven achiral chromatography: mechanistic rationale and implications. *Chem. Sci.* 6: 1043–1048.

68 Maeno, M., Kondo, H., Tokunaga, E., and Shibata, N. (2016). Synthesis of fluorinated donepezil by palladium-catalyzed decarboxylative allylation of α-fluoro-β-keto ester with tri-substituted heterocyclic alkene and the self-disproportionation of its enantiomers. *RSC Adv.* 6: 85058–85062.

69 Zheng, J. and You, S.-L. (2014). Construction of axial chirality by rhodium-catalyzed asymmetric dehydrogenative heck coupling of biaryl compounds with alkenes. *Angew. Chem. Int. Ed.* 53: 13244–13247.

70 Reyes-Rangel, G., Vargas-Caporali, J., and Juaristi, E. (2017). Asymmetric Michael addition reaction organocatalyzed by stereoisomeric pyrrolidine sulfinamides under neat conditions. A brief study of self-disproportionation of enantiomers. *Tetrahedron* 73: 4707–4718.

71 Han, J., Wzorek, A., Soloshonok, V.A., and Klika, K.D. (2019). The self-disproportionation of enantiomers (SDE): the effect of scaling down, potential problems versus prospective applications, possible new occurrences, and unrealized opportunities? *Electrophoresis* 40: 1869–1880.

72 Charles, R. and Gil-Av, E. (1984). Self-amplification of optical activity by chromatography on an achiral adsorbent. *J. Chromatogr. A* 298: 516–520.

73 Hosaka, T., Imai, T., Wzorek, A. et al. (2019). The self-disproportionation of enantiomers (SDE) of α-amino acid derivatives: facets of steric and electronic properties. *Amino Acids* 51: 283–294.

74 Wzorek, A., Klika, K.D., Drabowicz, J. et al. (2014). The self-disproportionation of the enantiomers (SDE) of methyl *n*-pentyl sulfoxide via achiral, gravity-driven column chromatography: a case study. *Org. Biomol. Chem.* 12: 4738–4746.

75 Suzuki, Y., Han, J., Kitagawa, O. et al. (2015). A comprehensive examination of the self-disproportionation of enantiomers (SDE) of chiral amides via achiral, laboratory-routine, gravity-driven column chromatography. *RSC Adv.* 5: 2988–2993.

76 Terada, S., Hirai, M., Honzawa, A. et al. (2017). Possible case of halogen bond-driven self-disproportionation of enantiomers (SDE) via achiral chromatography. *Chem. Eur. J.* 23: 14631–14638.

77 Szakács, Z., Sánta, Z., Lomoschitz, A., and Szántay, C. Jr., (2018). Self-induced recognition of enantiomers (SIRE) and its application in chiral NMR analysis. *Trends Anal. Chem.* 109: 180–197.

78 Navrátilová, H. and Potáček, M. (2001). Enantiomer enrichment of (+)-(3R,4S)-4-(4-fluorophenyl)-3-hydroxymethyl-1-methylpiperidine by crystallization. *Enantiomer* 6: 333–337.

79 Budovská, M., Tischlerová, V., Mojžiš, J. et al. (2017). 2′-Aminoanalogues of the cruciferous phytoalexins spirobrassinin, 1-methoxyspirobrassinin and 1-methoxyspirobrassinol methyl ether: synthesis and anticancer properties. *Tetrahedron* 73: 6356–6371.

80 Klika, K.D., Budovská, M., and Kutschy, P. (2010). Enantiodifferentiation of phytoalexin spirobrassinin derivatives using the chiral solvating agent (R)-(+)-1,1′-bi-2-naphthol in conjunction with molecular modeling. *Tetrahedron: Asymmetry* 21: 647–658.

81 Baumann, A., Wzorek, A., Soloshonok, V.A. et al. (2020). Potentially mistaking enantiomers for different compounds due to the self-induced diastereomeric anisochronism (SIDA) phenomenon. *Symmetry* 12(7): 1106.

82 Soloshonok, V.A., Wzorek, A., and Klika, K.D. (2017). A question of policy: should tests for the self-disproportionation of enantiomers (SDE) be mandatory for reports involving scalemates? *Tetrahedron: Asymmetry* 28: 1430–1434.

83 Walęcka-Kurczyk, A., Walczak, K., Kuźnik, A. et al. (2020). The synthesis of α-aminophosphonates via enantioselective organocatalytic reaction of 1-(N-acylamino)alkylphosphonium salts with dimethyl phosphite. *Molecules* 25: Art. No. 405.

84 Zhang, L., Xie, C., Dai, Y. et al. (2016). Catalytic asymmetric detrifluoroacetylative aldol reactions of aliphatic aldehydes for construction of C–F quaternary stereogenic centers. *J. Fluorine Chem.* 184: 28–35.

85 Xie, C., Wu, L., Han, J. et al. (2015). Assembly of fluorinated quaternary stereogenic centers through catalytic enantioselective detrifluoroacetylative aldol reactions. *Angew. Chem. Int. Ed.* 54: 6019–6023.

86 Zhang, L., Zhang, W., Sha, W. et al. (2017). Detrifluoroacetylative generation and chemistry of fluorine containing tertiary enolates. *J. Fluorine Chem.* 189: 2–9.

87 Zhu, Y., Mao, Y., Mei, H. et al. (2018). Palladium-catalyzed asymmetric allylic alkylations of colby proenolates with MBH carbonates: enantioselective access to quaternary C–F oxindoles. *Chem. Eur. J.* 24: 8994–8998.

88 Fustero, S., Ibáñez, I., Barrio, P. et al. (2013). Gold-catalyzed intramolecular hydroamination of o-alkynylbenzyl carbamates: a route to chiral fluorinated isoindoline and isoquinoline derivatives. *Org. Lett.* 15: 832–835.

89 Tateishi, K., Tsukagoshi, S., Nakamura, T. et al. (2013). Chiral initiator-induces self-disproportionation of enantiomers via achiral chromatography: application to enantiomer separation of racemate. *Tetrahedron Lett.* 54: 5220–5223.

90 Goto, M., Tateishi, K., Ebine, K. et al. (2016). Chiral additive induced self-disproportionation of enantiomers under MPLC conditions: preparation of enantiomerically pure samples of 1-(aryl)ethylamines from racemates. *Tetrahedron: Asymmetry* 27: 317–321.

91 Mei, H., Han, J., Klika, K.D. et al. (2020). Applications of fluorine-containing amino acids for drug design. *Eur. J. Med. Chem.* 186: Art. No. 111826.

92 Mei, H., Han, J., Fustero, S. et al. (2019). Fluorine-containing drugs approved by the FDA in 2018. *Chem. Eur. J.* 25: 11797–11819.

93 Wang, J., Sánchez-Roselló, M., Aceña, J.L. et al. (2014). Fluorine in pharmaceutical industry: fluorine-containing drugs introduced to the market in the last decade (2001–2011). *Chem. Rev.* 114: 2432–2506.

94 Zhou, Y., Wang, J., Gu, Z. et al. (2016). Next generation of fluorine-containing pharmaceuticals, compounds currently in phase II–III clinical trials of major pharmaceutical companies: new structural trends and therapeutic areas. *Chem. Rev.* 116: 422–518.

95 Izawa, K., Aceña, J.L., Wang, J. et al. (2016). Small-molecule for ebola virus (EBOV) disease treatment. *Eur. J. Org. Chem.* 2016: 8–16.

10

DFT Modeling of Catalytic Fluorination Reactions: Mechanisms, Reactivities, and Selectivities

Yueqian Sang, Biying Zhou, Meng-Meng Zheng, Xiao-Song Xue, and Jin-Pei Cheng

Nankai University, State Key Laboratory of Elemento-Organic Chemistry, College of Chemistry, Tianjin 300071, China

10.1 Introduction

The increasing importance of organofluorine compounds in the fields of medicinal, agricultural, and materials science has spurred enormous efforts in the development of new catalytic methods for their synthesis [1, 2]. The catalytic synthesis of fluorinated compounds, however, is far from trivial and represents a challenging research area in organic synthetic chemistry because the rules that govern the synthesis of non-fluorinated analogs cannot be transposed to fluorinated compounds [3]. Consequently, most catalytic methods for fluorine incorporation have been discovered through serendipity or trial-and-error practices, rather than by rational design. In this regard, a molecular-level understanding of these processes is beneficial for further advances in the field.

With the exponential growth of computing power and the rapid development of more accurate methods and efficient algorithms, computational chemistry has become an invaluable tool for the deduction and exploration of the origins of catalysis and selectivity [4–8]. Quantum mechanical modeling methodologies, particularly the density functional theory (DFT) methods, can now be applied to the real chemical systems that are studied by experimentalists, allowing one to gain insights into the mechanisms of catalytic processes where the active catalytic species or intermediates have been challenging or impractical to study via experimental means.

Over the past years, much of the mechanistic understanding of catalytic fluorination reactions has arisen from computational investigations or studies combining both experimental and computational explorations. Computations have also led to the establishment of cation and radical donor ability scales of electrophilic fluorinating and fluoroalkylating reagents as well as the detailed structure–reactivity relationships [9–16]. These results have served chemists in the rational design, optimization, and prediction of novel reagents and new reactions. For the sake of space, in this chapter, we chose not to include computational studies on thermodynamic

Organofluorine Chemistry: Synthesis, Modeling, and Applications,
First Edition. Edited by Kálmán J. Szabó and Nicklas Selander.

parameters of fluorinating reagents [9]. Instead, we focus on recent computational mechanistic studies of selected catalytic fluorination reactions, highlighting significant new concepts and insights obtained from these studies. We will also refrain from describing the details of various computational methods, which have already appeared in some excellent and comprehensive reviews [17–20].

10.2 DFT Modeling of Transition Metal-Catalyzed Fluorination Reactions

10.2.1 Ti-Catalyzed Fluorination Reaction

The first transition metal-catalyzed enantioselective fluorination was achieved by the Togni group in 2000 [21]. The α-fluorination of β-ketoesters with Selectfluor in the presence of 5 mol% of TiCl$_2$(TADDOLato) (**1** or **2**) catalyst afforded the desired products in high yields with good enantioselectivity (Figure 10.1a). In collaboration with the Rothlisberger group, they subsequently carried out a combined QM/MM computational and experimental study to gain deeper mechanistic understanding [23]. It was assumed that the β-ketoester coordinates with the catalyst as an enolate

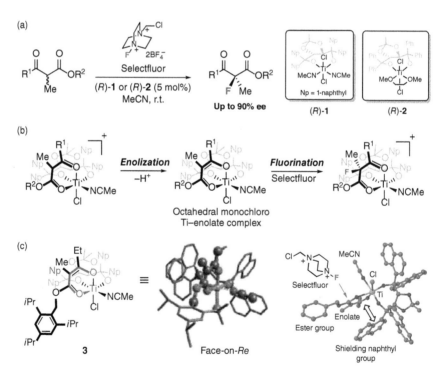

Figure 10.1 (a) TiCl$_2$(TADDOLato)-catalyzed enantioselective α-fluorination of β-ketoesters. (b) Proposed mechanism of enolization and fluorination. (c) Structure of the most stable Ti(enolato) complex and the proposed stereocontrol model [22]. Source: Piana et al. [23]. Reprinted with permission of John Wiley & Sons.

to give an octahedral monochloro Ti–enolate complex, which is fluorinated by Selectfluor (Figure 10.1b). Calculations revealed that the most stable isomer of the Ti–enolate complex (**3**) has the chloride ligand in the axial position (Figure 10.1c). The enolate fragment and the face-on oriented naphthyl group are almost perfectly parallel. As a result, the *Re*-face of the enolate in **3** is completely shielded and an electrophilic attack of the fluorinating agent can only occur from the opposite side. This stereocontrol model correctly predicts the observed absolute configuration of the product. Both computational and experimental results supported a single-electron transfer (SET) mechanism for the fluorination step.

10.2.2 Mn-Catalyzed Fluorination Reactions

In 2012, the Groves group developed an efficient aliphatic C—H bond fluorination reaction that employed a manganese tetramesitylporphyrin (Mn(TMP)Cl) (**4**) as the catalyst and silver fluoride/tetrabutylammonium fluoride trihydrate (AgF/TBAF·3H$_2$O) as the fluoride source (Figure 10.2a) [24]. Simple hydrocarbons, substituted cyclic molecules, terpenoids, and steroid derivatives were selectively fluorinated in good yields. The proposed catalytic cycle (Figure 10.2b) is composed of four steps, including (1) oxidation of the resting Mn(TMP)F catalyst to afford a reactive O=Mn^V(TMP)F species, (2) hydrogen atom abstraction from the substrate by O=Mn^V(TMP)F to produce a C-centered radical and a HO–Mn^{IV}(TMP)–F rebound intermediate, (3) an OH/F ligand exchange in HO–Mn^{IV}(TMP)–F to form *trans*-Mn^{IV}(TMP)F$_2$, and (4) fluorine atom transfer from the *trans*-Mn^{IV}(TMP)F$_2$ to

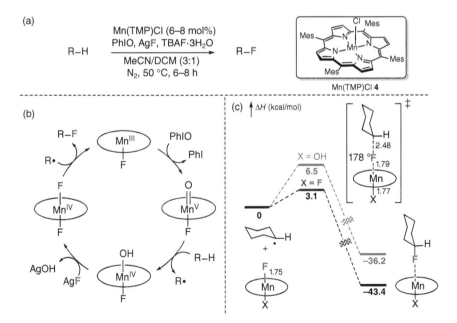

Figure 10.2 (a) Manganese porphyrin-catalyzed C—H bond fluorination. (b) Proposed catalytic cycle. (c) PCM-B3LYP/6-311++G(d,p)+LACVP**(Mn)//B3LYP/6-31G(d,p)+LACVP**(Mn) energy landscape for the fluorine atom transfer process.

the incipient substrate radical to furnish the fluorinated product. DFT calculations showed that the fluorine atom transfer from *trans*-Mn^{IV}F$_2$ to a cyclohexyl radical is quite facile with an activation barrier as low as 3.0 kcal/mol (Figure 10.2c). The barrier of fluorine transfer for its analogous HO–Mn^{IV}–F was found to be ~3 kcal/mol higher. The anticipation that *trans*-Mn^{IV}(TMP)F$_2$ would be an excellent radical fluorinating agent was verified by both their experiments and Zhang et al.'s computations [25].

With the manganese porphyrin catalyst Mn(TMP)Cl **4**, the Groves group also developed the first catalytic decarboxylative fluorination reaction with a less basic nucleophilic fluoride source, triethylamine trihydrofluoride (Et$_3$N·3HF) (Scheme 10.1) [26]. The proposed mechanism is shown in Figure 10.3a. Two likely pathways were postulated for the activation of the carboxylic acid. Pathway A involves the *in situ* formation of an iodine(III) carboxylate ester that oxidizes the manganese(III) porphyrin to a fluoromanganese(IV) intermediate with concomitant decarboxylation. Pathway B proceeds through a direct hydrogen atom abstraction from the hydroxyl group of the carboxylic acid by an oxomanganese(V) porphyrin intermediate. Mechanistic experiments and DFT calculations support Pathway A. Calculations showed that the *in situ*-formed iodine(III) carboxylate ester first forms a stable adduct with the manganese(III) porphyrin, which then undergoes a facile dissociation at the iodine center with a barrier of 18.2 kcal/mol (Figure 10.3b). In the dissociation transition state (TS), significant elongations of the I—O and I—F bonds were observed with a concurrent contraction of the Mn—F bond, consistent with a dissociation of the carboxyl radical from iodine with concomitant fluorine atom transfer to Mn^{III} to afford *trans*-Mn^{IV}F$_2$.

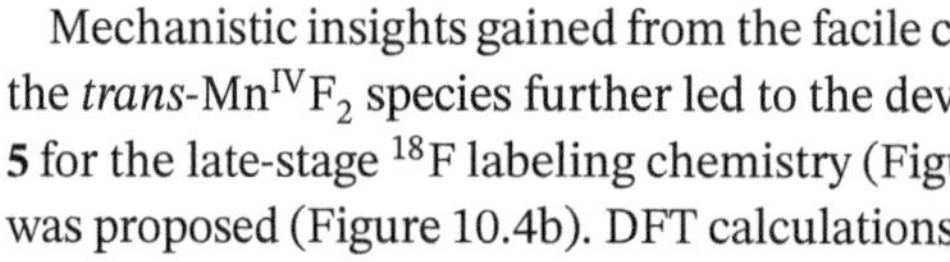

$$\text{R–COOH} \quad \xrightarrow[\text{DCE, 45 °C, 45 min –1.5 h}]{\substack{\text{Mn(TMP)Cl (2.5 mol\%)} \\ \text{PhIO, Et}_3\text{N·3HF}}} \quad \text{R–F}$$

Scheme 10.1 Mn(TMP)Cl-catalyzed decarboxylative fluorination of carboxylic acids. Source: Huang et al. [26].

Mechanistic insights gained from the facile capture of substrate carbon radicals by the *trans*-Mn^{IV}F$_2$ species further led to the development of a Mn(salen)OTs catalyst **5** for the late-stage ^{18}F labeling chemistry (Figure 10.4a) [27]. A similar mechanism was proposed (Figure 10.4b). DFT calculations supported the fluorine transfer reactivity of *trans*-Mn^{IV}(salen)(OH)F complex. The computed activation barrier of fluorine transfer from the *trans*-Mn^{IV}(salen)(OH)F complex to the benzyl radical is only 9.6 kcal/mol, and the overall fluorine transfer process is exergonic by −38.0 kcal/mol (Figure 10.4c).

10.2.3 Fe-Catalyzed Fluorination Reactions

In 2016, Cook and coworkers developed a mild, amide-directed fluorination of benzylic, allylic, and unactivated C—H bonds catalyzed by low-cost Fe(OTf)$_2$ (Scheme 10.2a) [28]. The reaction proceeds in high yield with broad functional

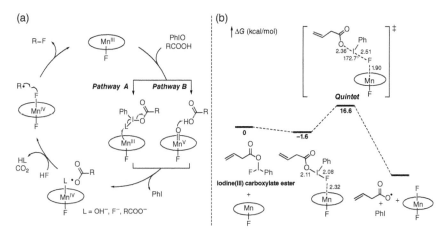

Figure 10.3 (a) Proposed catalytic cycle. (b) CPCM-M06/6-311++G(d,p)+SDD(Mn,I)// M06/6-31G(d)+SDD(Mn,I) free energy profile for the formation of carboxyl radical through the interaction of an iodine(III) carboxylate complex and a manganese(III) porphyrin.

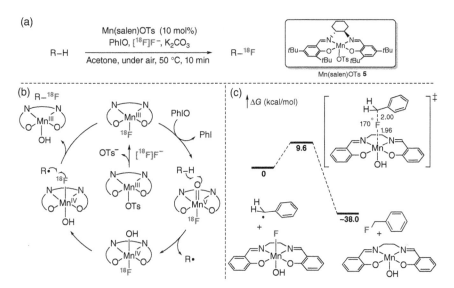

Figure 10.4 (a) Mn(salen)OTs-catalyzed ^{18}F labeling of benzylic C—H bonds. (b) Proposed catalytic cycle. (c) CPCM-B3LYP/6-311++G(2df,2p)+SDD(Mn)//B3LYP/6-31G(d)+SDD(Mn) free energy profile for the fluorine transfer step.

group tolerance. Crossover experiments and DFT calculations suggest that an organometallic pathway should be more feasible than a free radical mechanism for this reaction. As shown in Figure 10.5, the reaction begins with a rate-determining N—F bond cleavage by Fe(OTf)$_2$, followed by a facile 1,5-hydrogen atom transfer to generate the benzylic radical. From the benzylic radical, fluorine atom abstraction could occur from either the starting fluoroamide (a free radical mechanism) or the in situ-formed [Fe^{III}]-F (an organometallic pathway). The organometallic pathway,

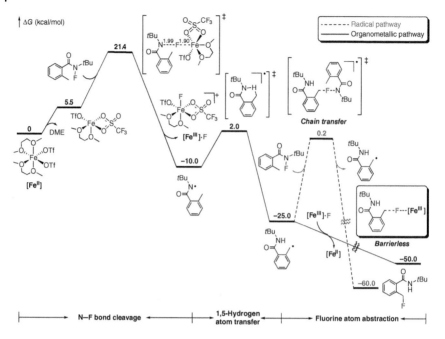

Figure 10.5 (U)M06/cc-pVTZ(-f)+LACV3P**(Fe)//(U)M06/6-31G(d,p)-LACVP**(Fe) free energy profile for Fe-catalyzed C−H bond fluorination.

where the metal fluoride [Fe^{III}]-F donates the fluorine atom back to the resulting benzylic radical to furnish the fluorinated product, was calculated to be a barrierless process [28]. In contrast, the free radical atom transfer pathway has a barrier as high as 25.2 kcal/mol. Guided by the computational mechanistic insights, the same group has recently extended this approach to a site-selective fluorination of remote C(sp^3)—H bond of adamantoyl-based fluoroamides (Scheme 10.2b) [29].

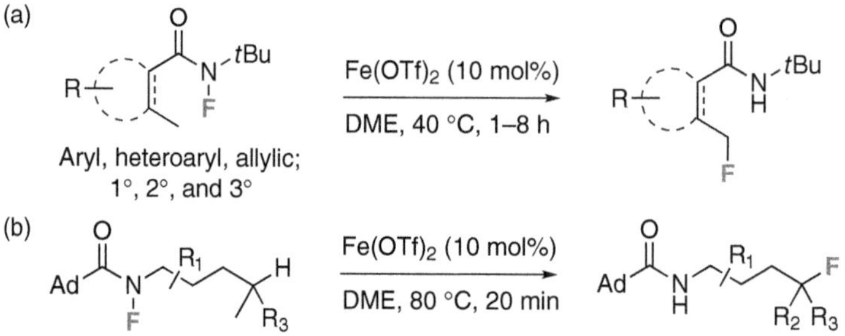

Scheme 10.2 Fe(OTf)$_2$-catalyzed C−H bond fluorination directed by (a) fluoroamide and (b) adamantoyl-based fluoroamide. Source: Groendyke et al. [28], Pinter et al. [29].

10.2.4 Rh-Catalyzed Fluorination Reactions

Recently, the Szabó group has developed several highly efficient methodologies to introduce F, CF$_3$, and SCF$_3$ groups to diazocarbonyl compounds with a dirhodium catalyst $Rh_2(OAc)_4$ [30–32]. In collaboration with the Himo group, they carried out a detailed computational study on the mechanisms of the Rh-catalyzed geminal oxyfluorination and oxytrifluoromethylation of diazocarbonyl compounds with fluoro-benziodoxole **6** and Togni reagent **7** (Scheme 10.3) [33]. The computed mechanism of the Rh-catalyzed oxyfluorination is shown in Figure 10.6, which involves the following five steps: (1) N_2 dissociation to form a Rh–carbene intermediate, (2) alcohol insertion and proton transfer to give a stable Rh–enol intermediate, (3) concerted proton transfer/electrophilic addition of the hypervalent iodine reagent to enol, (4) isomerization of the hypervalent iodine **6**, and (5) ligand coupling to afford the final product. The dirhodium catalyst was found to lower the barrier for the N_2 dissociation by stabilizing the carbene intermediate and to facilitate the isomerization of **6** by coordinating to the fluorine atom center. The N_2 dissociation is the rate-determining step with a computed free energy barrier of 22.6 kcal/mol. Rh-catalyzed geminal oxytrifluoromethylation follows a similar mechanism. The main difference lies in the isomerization step: for reagent **6** the fluorine atom can bind to Rh to facilitate the isomerization, whereas this is not possible for reagent **7** with a trifluoromethyl group. Consequently, the isomerization step becomes rate determining for oxytrifluoromethylation.

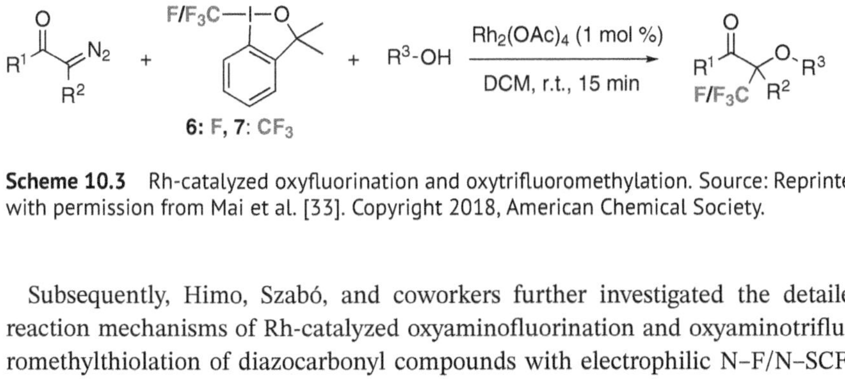

Scheme 10.3 Rh-catalyzed oxyfluorination and oxytrifluoromethylation. Source: Reprinted with permission from Mai et al. [33]. Copyright 2018, American Chemical Society.

Subsequently, Himo, Szabó, and coworkers further investigated the detailed reaction mechanisms of Rh-catalyzed oxyaminofluorination and oxyaminotrifluoromethylthiolation of diazocarbonyl compounds with electrophilic N–F/N–SCF$_3$-based reagents (N-fluorobenzenesulfonimide (NFSI, **8**) and N-trifluoromethylthiodibenzenesulfonimide (**9**), Scheme 10.4) [34]. Calculations showed that the two reactions follow the same reaction mechanism (Figure 10.7). The catalytic cycle begins with N_2 dissociation to give a Rh–carbene intermediate, followed by a nucleophilic attack of tetrahydrofuran on the carbene. Then, a Rh coordination change leads to the Rh–enolate intermediate, providing the crucial alkene for subsequent electrophilic attack by N–F/N–SCF$_3$ reagents. Electrophilic attack via an S_N2-like transition state introduces the F or the SCF$_3$ moiety. The barrier for trifluoromethylthiolation is lower than that for fluorination due to the lower energy of σ^* N–SCF$_3$ than that of σ^* N–F molecular orbital. Finally, the nucleophilic attack of the remaining amino groups furnishes the final products. Similar to the Rh-catalyzed oxyfluorination of diazoketones with hypervalent fluoroiodine [33], the N_2 dissociation was identified as the rate-determining step for Rh-catalyzed oxyaminofluorination and oxyaminotrifluoromethylthiolation.

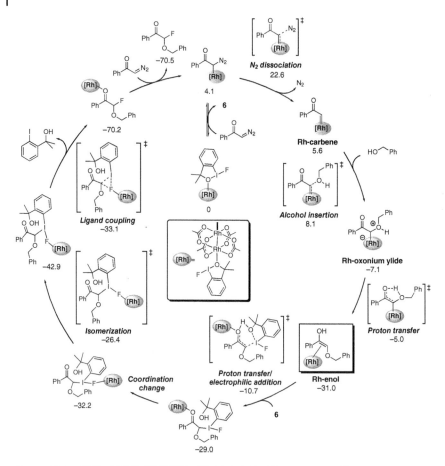

Figure 10.6 CPCM-B3LYP-D3/6-311+G(2d,2p)+LANL2DZdp(I)+LANL2DZ(Rh)//
B3LYP/6-31G(d,p)+LANL2DZdp(I)+LANL2DZ(Rh) computed mechanism for Rh-catalyzed
oxyfluorination of diazoketone. Energies are in kcal/mol. Source: Mai et al. [33]. Reprinted
with permission of American Chemical Society.

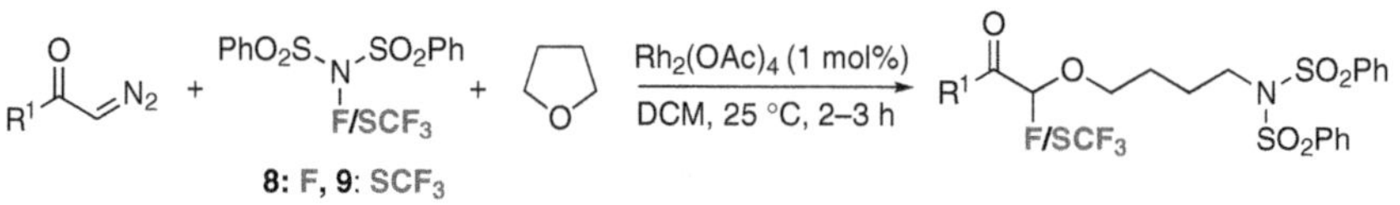

Scheme 10.4 Rh-catalyzed oxyaminofluorination and oxyaminotrifluoromethylthiolation.
Source: Reprinted with permission from Mai et al. [34]. Copyright 2018, American Chemical
Society.

Most recently, Huang, Yu, and coworkers developed a Rh-catalyzed cascade
arylfluorination reaction of α-diazoketoesters with arylboronic acids and **NFSI**
for facile synthesis of α-aryl-α-fluoroketoesters (Scheme 10.5) [35]. Based on their
experimental and computational results, a plausible catalytic cycle was proposed
(Figure 10.8), including transmetalation, diazo coupling reaction, migratory
carbene insertion, outer-sphere fluorination, and regeneration of the catalyst.

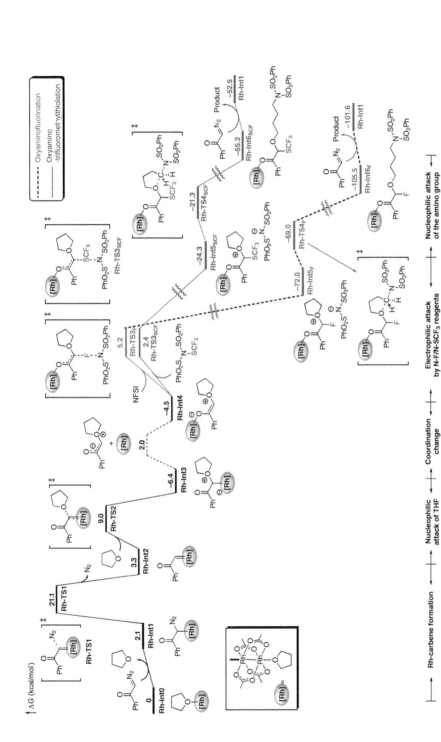

Figure 10.7 CPCM-B3LYP-D3/6-311+G(2d,2p)+LANL2DZ(Rh)//B3LYP/6-31G(d,p)+LANL2DZ(Rh) free energy profile for Rh-catalyzed oxyaminofluorination (dashed line) and oxyaminotrifluoromethylthiolation (solid line). Source: Mai et al. [34]. Reprinted with permission of American Chemical Society.

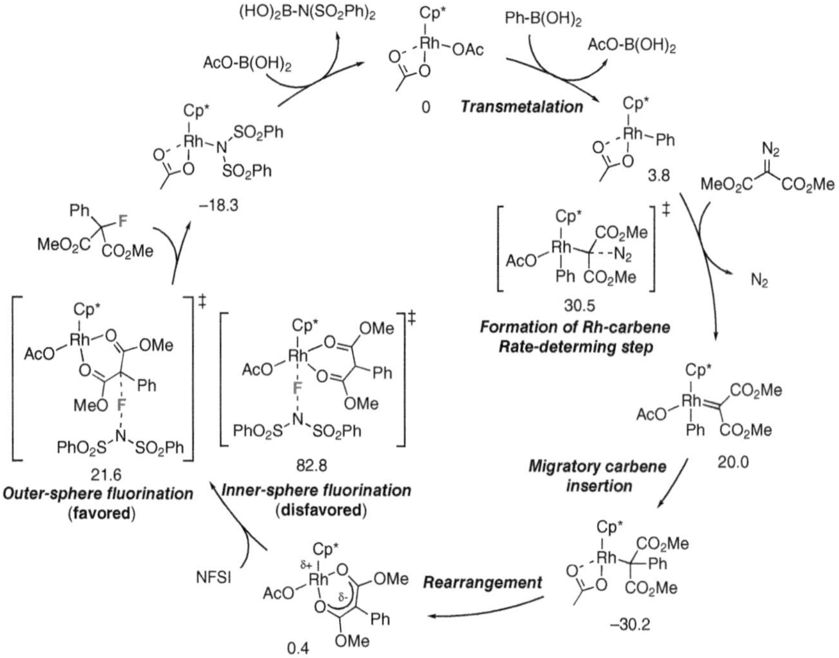

Figure 10.8 Proposed catalytic cycle for Rh-catalyzed arylfluorination of α-diazoketoesters based on DFT free energies (in kcal/mol).

Compared with the outer-sphere fluorination pathway, the inner-sphere one was calculated to be disfavored by more than 60 kcal/mol because the HOMO of diketonato–rhodium(III) complex is mainly contributed from the diketonato carbon. A similar observation was made by Lynam, Slattery, and coworkers [36].

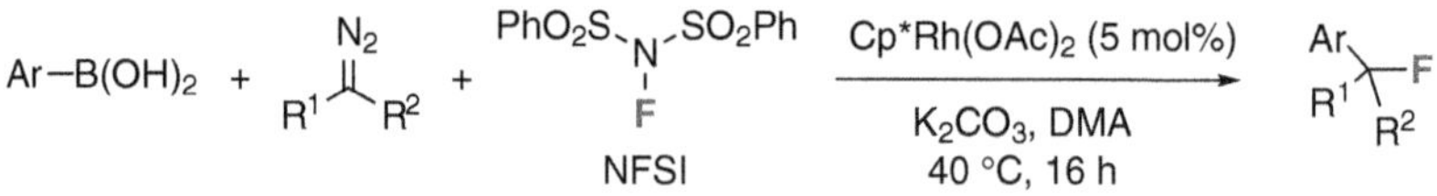

Scheme 10.5 Rh-catalyzed arylfluorination of α-diazoketoesters. Source: Reprinted with permission from Ng et al. [35]. Copyright 2019, Royal Society of Chemistry.

10.2.5 Ir-Catalyzed Fluorination Reactions

Most recently, Gutierrez, Nguyen, and coworkers reported a combined experimental and computational study on the mechanism of the diene-ligated Ir-catalyzed regio- and enantioselective allylic fluorination (Scheme 10.6) [37]. Based on mechanistic experiments and DFT calculations, they proposed a possible mechanism for the fluorination of racemic allylic trichloroacetimidates (Figure 10.9). The C—F bond formation via an outer-sphere nucleophilic fluoride attack on the π-allyl iridium intermediates was identified as the regio- and stereodetermining step. The barriers for nucleophilic attack at the least substituted carbon were calculated to be >8 kcal/mol higher in energy than at the most substituted carbon, consistent with

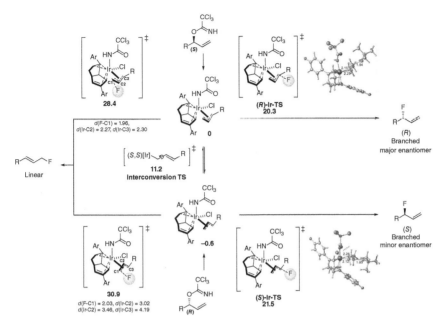

Figure 10.9 CPCM-B3LYP-D3/6-311+G(d,p)+SDD(Ir)//CPCM-M06/6-31G(d,p)-LANL2DZ(Ir) computed reaction pathway and the regio- and stereoselectivity-determining TSs. Energies are in kcal/mol. Source: Sorlin et al. [37]. Reprinted with permission of American Chemical Society.

the experimentally observed high branched-to-linear ratios (b:l > 99 : 1). The high branched-to-linear selectivity was attributed to stronger interactions between the developing $\pi_{C=C}$ moiety and iridium in the branched transition states. For the enantioselectivity of branched product, transition state **(R)-Ir-TS** leading to the experimentally observed major enantiomer was calculated to be 1.2 kcal/mol lower in energy than diastereomeric transition state **(S)-Ir-TS**, in reasonable agreement with the experimental result ($ee_{calc} = 76\%$ vs. $ee_{exp} = 93\%$). The energy difference stems from the unfavorable steric interaction between the allylic and NHCOCl$_3$ moieties.

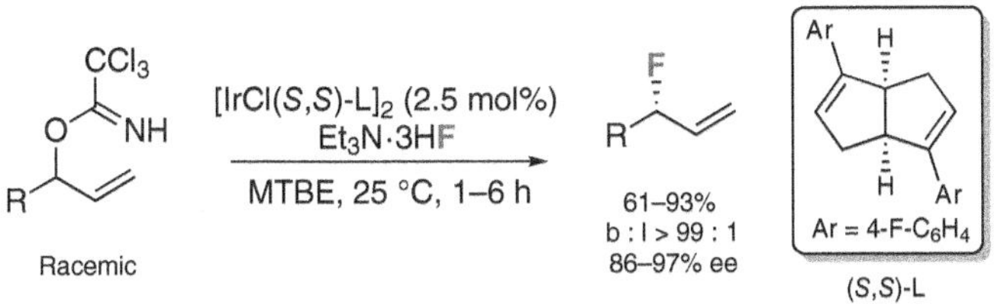

Scheme 10.6 Ir-catalyzed selective fluorination of allylic trichloroacetimidate.

10.2.6 Pd-Catalyzed Fluorination Reactions

10.2.6.1 Pd-Catalyzed Nucleophilic Fluorination

Despite the widespread use of Pd(0)/Pd(II) catalysis for Ar—X bond formation (X = C, N, O, and S), an analogous efficient Ar—F bond-forming process

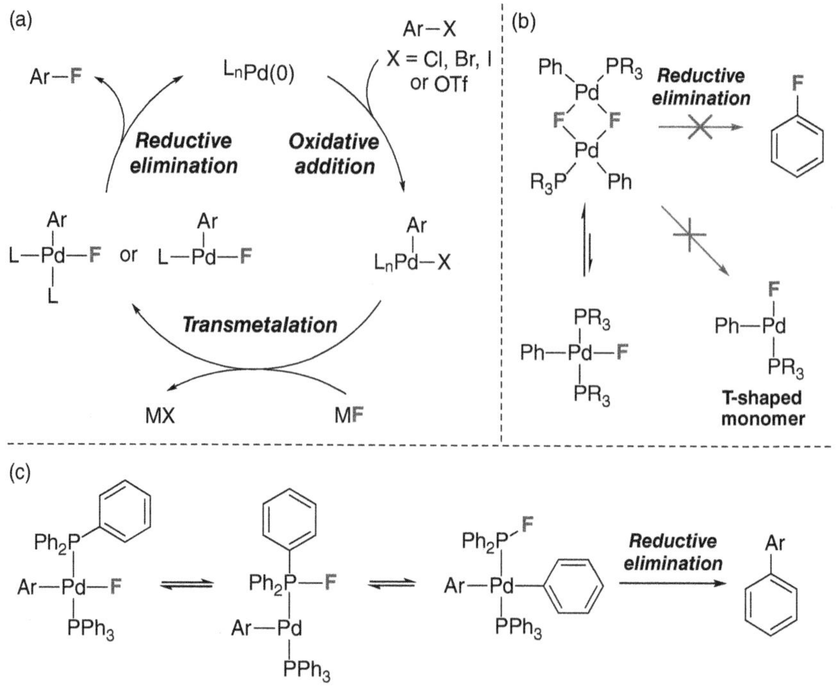

Figure 10.10 (a) Proposed Pd(0)/Pd(II) catalytic cycle for aryl fluorination. (b) Challenges associated with Ar–F cross-coupling. (c) Reaction pathway observed in the thermal decomposition of [L$_2$Pd^{II}(Ar)F] (L = PPh$_3$) complexes.

(Figure 10.10a) was unknown until recently. Grushin's early systematic studies on the [L$_n$M(Ar)F] complexes (M = Pd, Rh, L: ligands) showed that such complexes did not undergo the desired C–F reductive elimination under any conditions investigated (Figure 10.10b) [38]. Instead, undesired reaction pathways involving the supporting ligands and fluoride dominated the chemistry of these complexes (Figure 10.10c). This was ascribed to both the competing, kinetically preferred pathways involving the ligand-based P–F reductive elimination as well as the high barrier to Ar–F reductive elimination.

To elucidate the specific features of the coordination environment of Pd(II) that could enable the Ar–F reductive elimination as a feasible elementary C—F bond-forming reaction in this type of catalytic transformation, Yandulov and Tran conducted a computational and experimental evaluation of the thermodynamic and kinetic feasibility of Ar–F reductive elimination from a series of [L$_n$Pd^{II}(Ar)F] complexes in 2007 [39]. DFT calculations revealed that Ar–F reductive elimination from three-coordinate T-shaped [LPd^{II}(Ar)F] (L = NHC, PR$_3$), irrespective of dimerization is a feasible elementary C—F bond-forming reaction, whereas coordination of a strongly bonding fourth ligand to Pd results in pronounced stabilization of the Pd(II) reactant and increases the activation barrier beyond the practical range (Figure 10.11a). Cui and Saeys made a similar observation in their computational study on a series of phosphine ligands [40]. Yandulov found

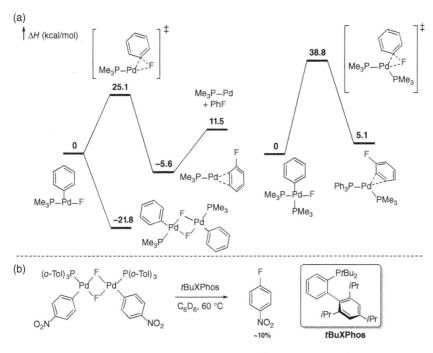

Figure 10.11 (a) Computed reactivity profile of [LPd^{II}(Ph)F] and [L$_2$Pd^{II}(Ph)F] (L = PMe$_3$). (b) First reported aryl C−F bond formation from a fluoride-bridged arylpalladium dimer.

that the aggregation of the reactive three-coordinate [LPd^{II}(Ar)F] into a stable fluoride-bridged dimer is another critical obstacle to Ar–F reductive elimination in practice. Promisingly, they demonstrated that bulky phosphine ligands such as Buchwald's P(C$_6$H$_4$-2-Trip)(t-Bu)$_2$ (**tBuXPhos**) could promote the C–F elimination by preventing the dimerization process (Figure 10.11b). Indeed, the Buchwald group achieved a breakthrough in 2009 by reporting the Pd-catalyzed nucleophilic fluorination of aryl triflates (Scheme 10.7a) [41]. The key to the success of this Pd catalyst system was the use of the sterically demanding biaryl monophosphines such as **tBuBrettPhos** and **HGPhos** (Scheme 10.7a,b).

The Buchwald group subsequently developed a new fluorinated biaryl monophosphine ligand (**AlPhos**) that enabled a room-temperature regioselective Pd-catalyzed fluorination of aryl triflates (Scheme 10.7c) [42]. DFT calculations and crystallographic analysis provided insights into the origin of its higher reactivity. Calculations revealed an attractive C − H···F interaction in the ground-state structure of the [(**HGPhos**)Pd^{II}(Ph)F] complex **10** (Figure 10.12a). Such interaction is, however, absent in [(**AlPhos**)Pd^{II}(Ph)F] complex **11** (Figure 10.12b), which results in a ground-state destabilization relative to **10**, thus making it more active to undergo a C–F reductive elimination. Indeed, the calculated activation barrier of the C–F reductive elimination of **11** is 0.7 kcal/mol lower than that of **10**, coinciding with the observed reactivity enhancement of **AlPhos**.

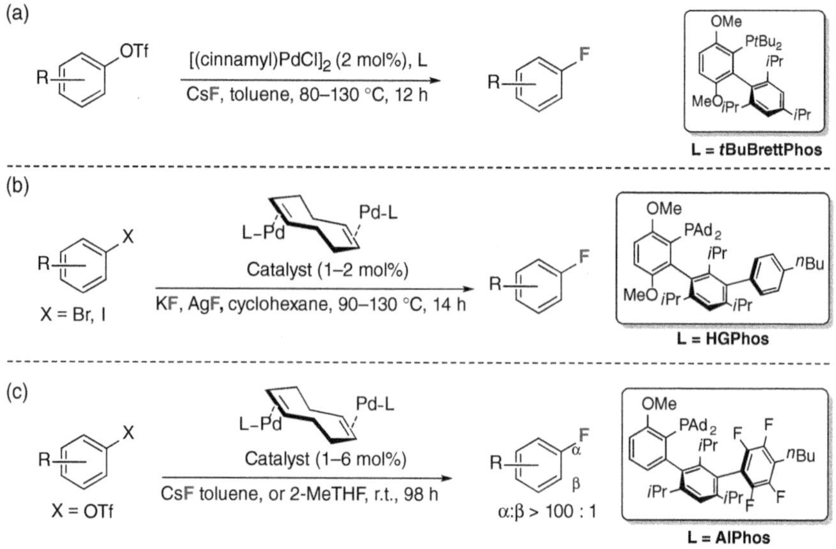

(a)

(b)

(c)

Scheme 10.7 Pd-catalyzed fluorination of aryl halides and triflates enabled by (a) *t*BuBrettPhos, (b) HGPhos, and (c) AlPhos.

Interestingly, the C–F reductive elimination barriers of the 2/3-thienyl-substituted [(**AlPhos**)Pd^{II}(Ar)F] (Ar: thienyl) complexes were calculated to be at least 5 kcal/mol higher in comparison to phenyl-substituted complex and could be lowered substantially by addition of a phenyl group adjacent to the Pd center (Figure 10.13a) [43]. Guided by the insights from DFT calculations, the Buchwald group successfully extended the methodology to the preparation of five-membered heteroaryl fluorides (Figure 10.13b).

Most theoretical and experimental studies of Pd-catalyzed aromatic fluorination mainly focused on the crucial reductive elimination step. Recently, Pliego reported a complete potential energy profile of a model Pd-catalyzed fluorination reaction using an accurate theoretical method mPW2-PLYP [44]. The results suggest that a successful ligand for Pd-catalyzed aromatic fluorination requires not only a monophosphine ligand that avoids dimerization but also a strong repulsion to both phenyl and fluorine bonded to the palladium in the [LPd^{II}(Ph)F] complex.

The Doyle group developed a series of Pd-catalyzed highly regio- and enantioselective fluorinations of allylic chlorides with AgF as the fluoride source (Scheme 10.8) [45, 46]. In 2014, they conducted a joint computational and experimental study on the mechanism of Pd-catalyzed allylic fluorination [47]. The results pointed to a homobimetallic mechanism (Figure 10.14a,b), in which C—F bond formation occurs by an outer-sphere nucleophilic attack of a neutral allylpalladium fluoride **12** on a cationic allylpalladium electrophile **13**. The optimized TSs for the internal and terminal attack of **12** on **13** are quite late and product-like. The calculated $\Delta\Delta G^{\ddagger}$ value of 1.9 kcal/mol for branched/linear product is in excellent agreement with the experimental b:l ratio of 36 : 1 ($\Delta\Delta G^{\ddagger}$ = 2.2 kcal/mol).

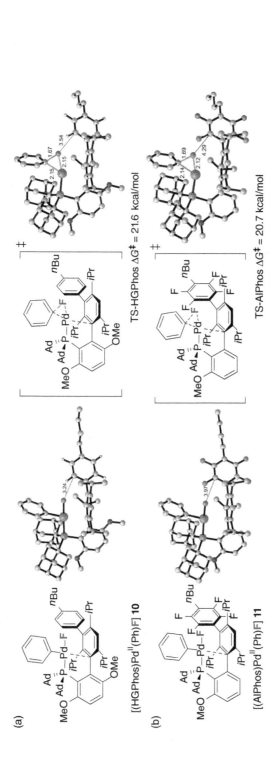

Figure 10.12 SMD-M06/6-311+G(d,p)+SDD(Pd)//B3LYP/6-31G(d)+SDD(Pd) calculated structures of [LPd(Ph)F] complexes and C−F reductive elimination TSs for (a) L = **HGPhos** and (b) L = **AlPhos**.

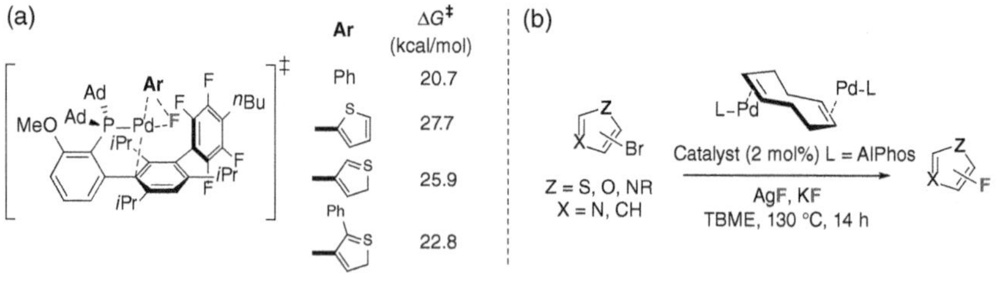

(a)

Ar	$\Delta G^{\ddagger}$ (kcal/mol)
Ph	20.7
(thiophene-2-yl)	27.7
(thiophene-3-yl)	25.9
Ph (5-Ph-thiophene-2-yl)	22.8

(b)

Z = S, O, NR
X = N, CH

Catalyst (2 mol%) L = AlPhos

AgF, KF
TBME, 130 °C, 14 h

Figure 10.13 (a) Substituent effects on C—F reductive elimination from [(**AlPhos**)Pd^{II}(Ar)F] complexes computed at the SMD-M06/6-311+G(d,p)+SDD(Pd)//B3LYP/6-31G(d)+SDD(Pd) level of theory. (b) Pd-catalyzed fluorination of five-membered heteroaryl bromides.

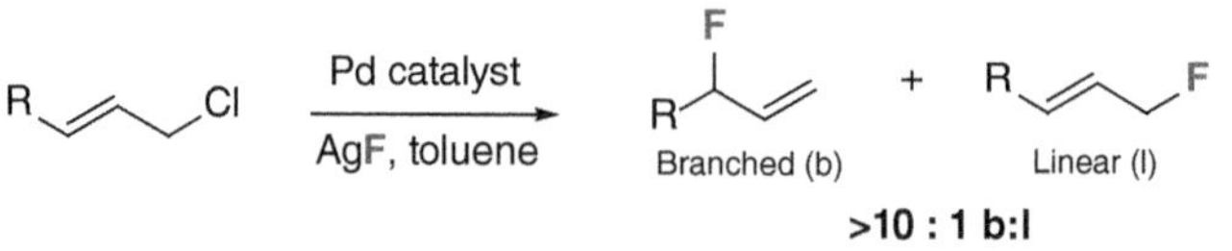

R—CI → Branched (b) + Linear (l)

Pd catalyst

AgF, toluene

>10 : 1 b:l

Scheme 10.8 Pd-catalyzed fluorination of allylic chlorides with AgF.

10.2.6.2 Pd-Catalyzed Electrophilic Fluorination

In 2006, the Sanford group achieved a breakthrough in the field of C—F bond formation that disclosed the first Pd-catalyzed ligand-directed benzyl and aryl C—H bond fluorinations with electrophilic N–F reagents (Figure 10.15a) [48]. This pioneering work marks the beginning of the field of Pd-catalyzed C–H fluorination. The reaction was proposed to proceed through a Pd(II)/Pd(IV) redox cycle (Figure 10.15b), where the C—F bond-forming reductive elimination from Pd(IV)–F serves as a critical step. Experimental and computational studies by Ritter and coworkers [49] and Sanford and coworkers [50, 51] on the mechanism of stoichiometric C–F reductive elimination from Pd(IV)–F complexes supported the feasibility of the proposed Pd(II)/Pd(IV)–F catalytic cycle and encouraged the development of subsequent Pd-catalyzed C–H fluorinations.

In 2015, Fleurat-Lessard, Hierso, and coworkers reported a general protocol for the Pd-catalyzed mono- and difluorination of substituted arylpyrazoles with NFSI (Scheme 10.9) [52]. The palladacycle formed via C–H activation was characterized as the resting state of the Pd(II) catalyst (Figure 10.16). Interestingly, DFT calculations suggest that a previously unseen outer-sphere direct fluorination mechanism is favored over the commonly assumed Pd(II)/Pd(IV) process by ~17 kcal/mol. A key Pd(II) monomer was identified by mass spectrometry, providing support for the unprecedented mechanistic scenario.

In 2016, Sun et al. discovered a remarkable ligand effect on the oxidation of Pd(II) to Pd(IV) in Pd-catalyzed, directing group-controlled C(sp^3)—H bond fluorination reactions [53]. The fluorination occurs in higher yields in the presence of external quinoxaline (Scheme 10.10). DFT calculations revealed that the quinoxaline ligand could significantly stabilize the Pd(II) and Pd(IV) intermediates and lower the barrier for the oxidation of Pd(II) to Pd(IV) (Figure 10.17). Computational insights

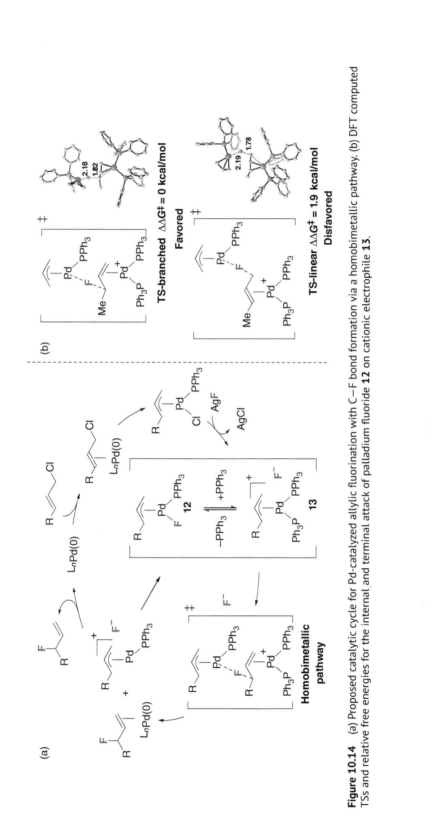

Figure 10.14 (a) Proposed catalytic cycle for Pd-catalyzed allylic fluorination with C—F bond formation via a homobimetallic pathway. (b) DFT computed TSs and relative free energies for the internal and terminal attack of palladium fluoride **12** on cationic electrophile **13**.

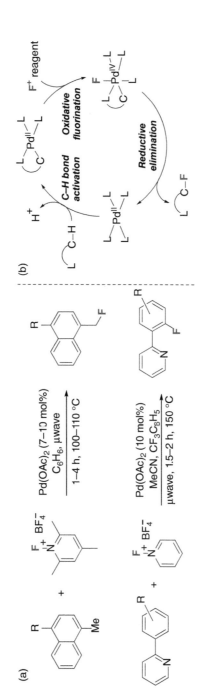

Figure 10.15 (a) Pd-catalyzed C–H fluorination with F⁺ reagents developed by Sanford et al. (b) Proposed catalytic cycle. Source: (a) Modified from Hull et al. [48].

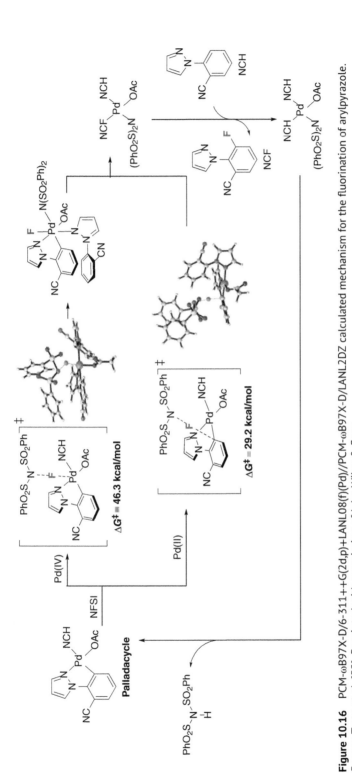

Figure 10.16 PCM-ωB97X-D/6-311++G(2d,p)+LANL08(f)(Pd)//PCM-ωB97X-D/LANL2DZ calculated mechanism for the fluorination of arylpyrazole.
Source: Testa et al. [52]. Reprinted with permission of John Wiley & Sons.

further led to the development of 8-aminoquinoxaline-directed fluorination, wherein the substrate itself serves as the ligand to assist the Pd(II) oxidation.

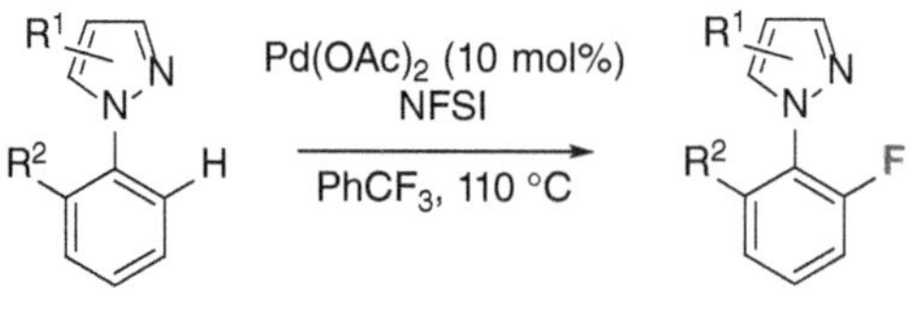

Scheme 10.9 Pd-catalyzed fluorination of substituted arylpyrazoles with NFSI. Source: Reprinted with permission from Testa et al. [52]. Copyright 2015, John Wiley & Sons.

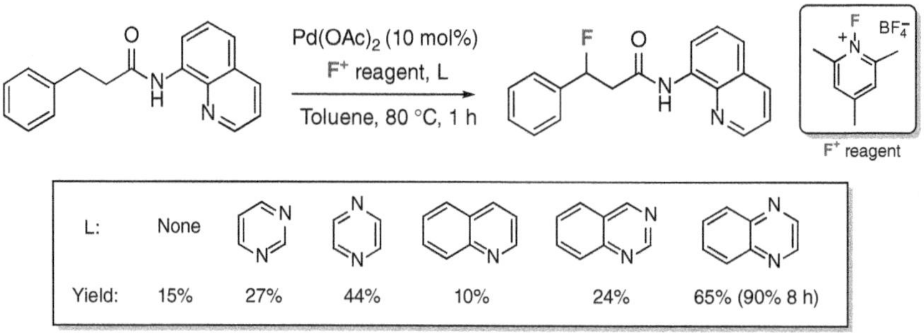

L:	None					
Yield:	15%	27%	44%	10%	24%	65% (90% 8 h)

Scheme 10.10 Pd-catalyzed quinoxaline ligand-accelerated C(sp³)−H bond fluorination.

Most recently, Xu, Lou, and coworkers conducted a detailed mechanism study on the effect of the anion ligand on C—H bond fluorination [54] (Scheme 10.11). DFT calculations revealed that the presence of nitrate/nitrite anion ligand significantly decreased the energy barrier of C—F bond reductive elimination from the Pd(IV) intermediates (Figure 10.18). Guided by computational insights, the authors developed the selective C—H bond fluorination of various azobenzenes.

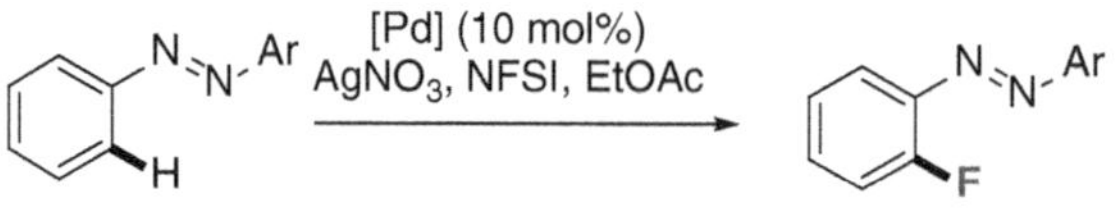

Scheme 10.11 Pd-catalyzed selective C−H bond fluorination of azobenzene.

In 2018, the Ritter group developed an undirected, Pd-catalyzed aromatic C—H bond fluorination with electrophilic N–F reagents (Scheme 10.12) [55]. Computational and experimental mechanistic studies indicate that the reaction involves a highly reactive electrophilic Pd(IV) fluoride (**15**), which is generated catalytically through oxidation of the cationic Pd(II) complex (**14**) by Selectfluor. Complex **15** is able to fluorinate arenes that Selectfluor cannot fluorinate (Figure 10.19). This is because **15** has a higher single-electron reduction potential than Selectfluor, and the fluorination proceeds through a single transition state (**TS**$_{FCET}$) via fluoride-coupled

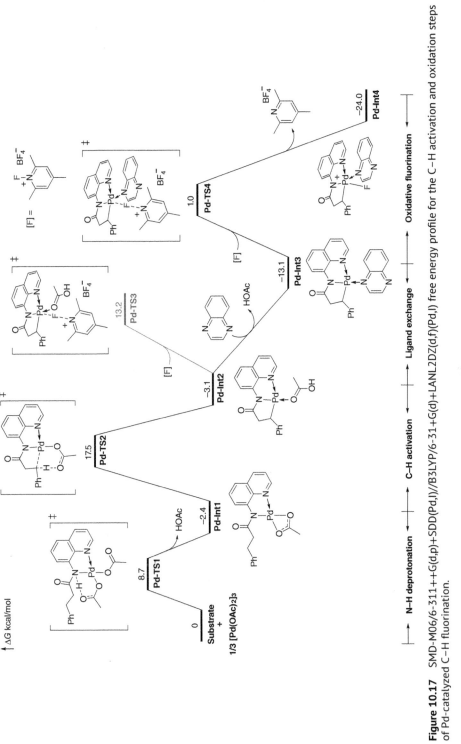

Figure 10.17 SMD-M06/6-311++G(d,p)+SDD(Pd,I)//B3LYP/6-31+G(d)+LANL2DZ(d,f)(Pd,I) free energy profile for the C–H activation and oxidation steps of Pd-catalyzed C–H fluorination.

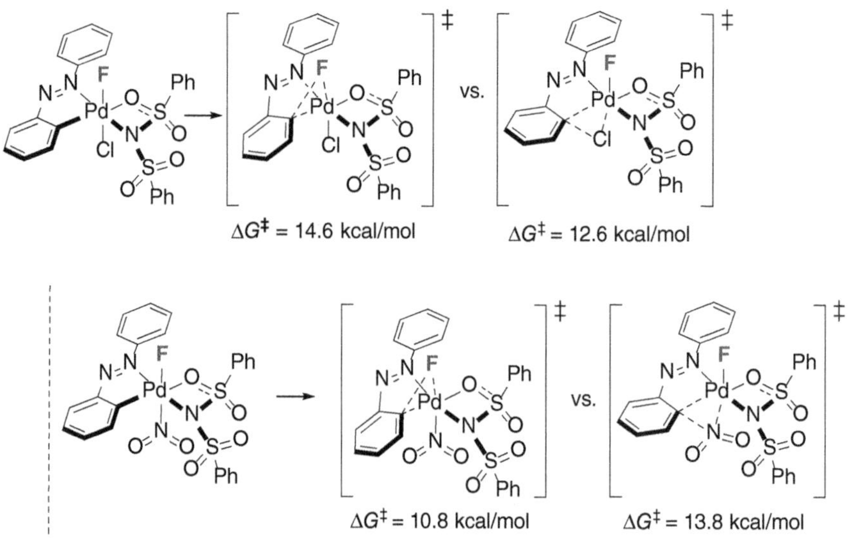

$\Delta G^{\ddagger} = 14.6$ kcal/mol $\Delta G^{\ddagger} = 12.6$ kcal/mol

$\Delta G^{\ddagger} = 10.8$ kcal/mol $\Delta G^{\ddagger} = 13.8$ kcal/mol

Figure 10.18 Computed C–F reductive elimination from Pd(IV) involving Cl and NO$_2$ ligands.

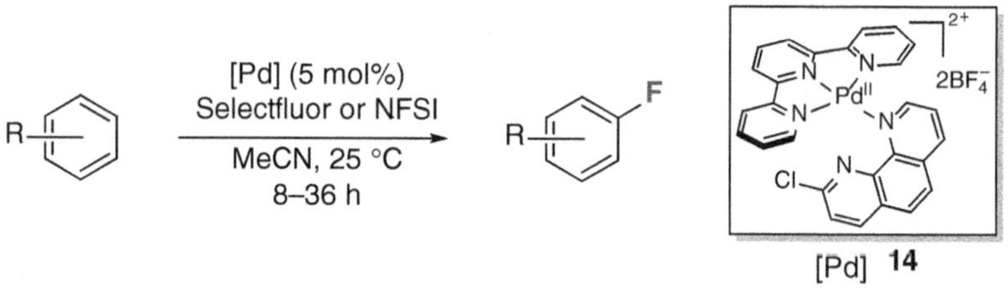

Scheme 10.12 Pd-catalyzed direct, non-chelation-assisted fluorination. Source: Reprinted with permission from Yamamoto et al. [55]. Copyright 2018, Springer Nature.

electron transfer. The unusual mechanism of catalysis enabled by the high-valent Pd(IV) electrophile offers a new avenue for the development of catalytic C–H functionalization reactions.

10.2.7 Cu-Catalyzed Fluorination Reactions

10.2.7.1 Cu-Catalyzed Nucleophilic Fluorination

Along with Pd-catalyzed nucleophilic fluorination reactions, Cu-based catalytic systems have also been developed in recent years for nucleophilic aryl–X fluorination. The feasibility of an oxidative addition/reductive elimination sequence for fluorination of aryl halides through a redox Cu(I)/Cu(III) catalytic cycle was first demonstrated by the Ribas group based on a family of designed macrocyclic substrates (Figure 10.20a) [56]. Experimental and computational results supported the Cu(I)/Cu(III) catalytic cycle involving aryl–X oxidative addition at the Cu(I)

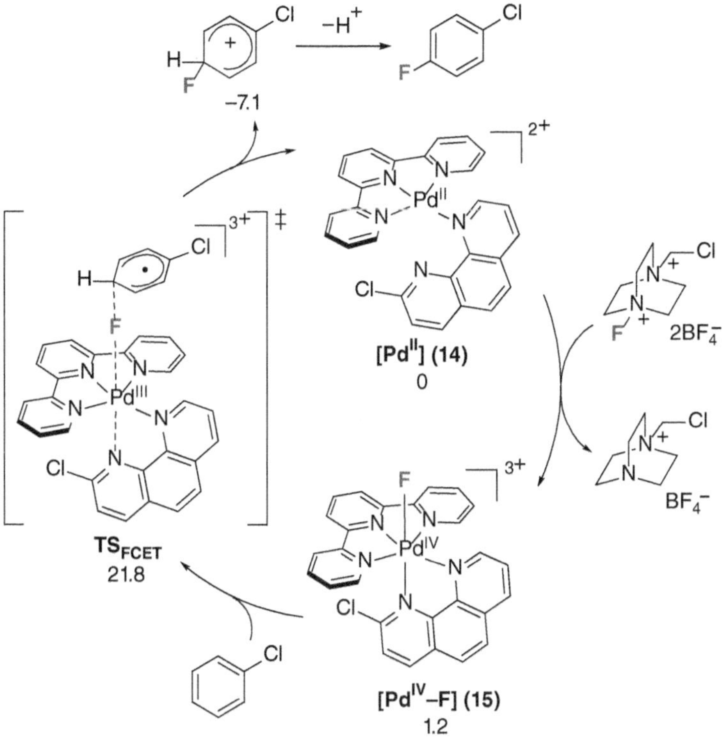

Figure 10.19 Proposed catalytic cycle for the fluorination of chlorobenzene with CPCM-TPSS0-D3-(BJ)/def2-QZVP+def2-SD(Pd)//PBE0-D3-(BJ)/def2-TZVP+def2-SD(Pd) free energy data in kcal/mol.

center, followed by halide exchange and reductive elimination (Figure 10.20b). The computed [Cu^{III}–F] intermediate features a distorted square pyramidal structure, wherein fluoride occupies the axial position *cis* to aryl group (Figure 10.20c). The aryl–F bond-forming reductive elimination at Cu(III) center has a small energy barrier of 16.2 kcal/mol and is exergonic by 4.1 kcal/mol, which is consistent with a fast and irreversible C–F reductive elimination. Intriguingly, aryl–Cl reductive elimination shows a much higher barrier of 26.9 kcal/mol and is endergonic by 24.7 kcal/mol, suggesting the ease of a downhill reversal oxidative addition between Cu(I) and aryl chloride.

In 2013, Sanford and coworkers developed a mild Cu-catalyzed fluorination of unsymmetrical diaryliodonium salts [Mes(Ar)I]$^+$ with KF as the fluoride source (Figure 10.21a) [57]. This protocol preferentially fluorinates the less sterically hindered aryl ligand on iodine(III). The authors conducted a synergistic investigation of experiment and computation into the detailed mechanism of this transformation [58]. The results support a Cu(I)/Cu(III) catalytic cycle (Figure 10.21b). The oxidative aryl transfer from diaryliodonium cation [Mes(Ar)I]$^+$ to the Cu(I) catalyst is the rate-determining step (Figure 10.22). The selectivity for fluorination of Ar vs. Mes was attributed to the steric hindrance of 2,6-methyl groups in Mes.

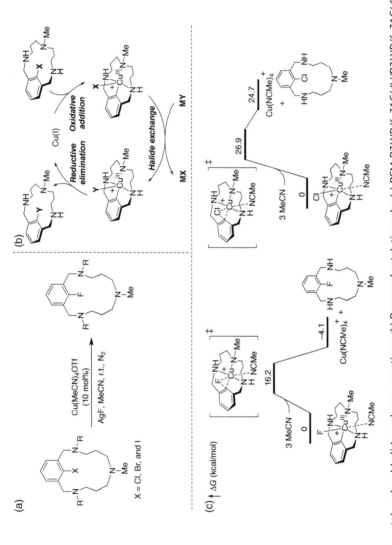

Figure 10.20 (a) Cu-catalyzed aryl halide exchange reactions. (b) Proposed catalytic cycle. (c) PCM-B3LYP/6-31G(d)//B3LYP/6-31G(d) free energy profiles for C–X elimination.

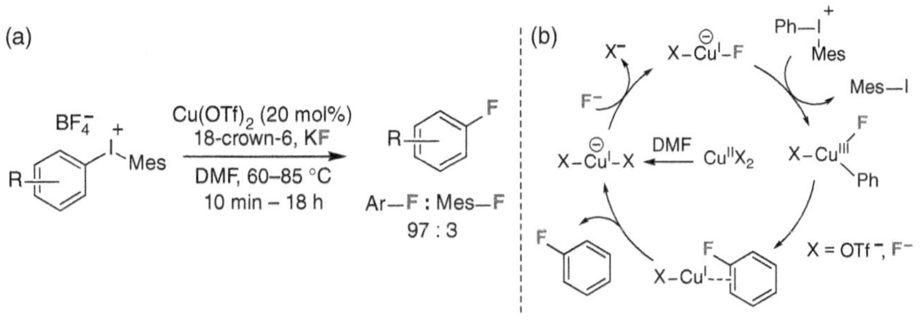

(a)

(b)

Figure 10.21 (a) Cu-catalyzed nucleophilic fluorination of diaryliodonium salts. (b) Proposed catalytic cycle.

10.2.7.2 Cu-Mediated Radical Fluorination

In 2012, the Lectka group developed a catalytic methodology for the monofluorination of aliphatic substrates with a polycomponent catalytic system involving Selectfluor, a putative radical precursor N-hydroxyphthalimide, an anionic phase-transfer catalyst, and a Cu(I) bisimine complex [59]. Two years later, they simplified the protocol and conducted an in-depth mechanistic study of the system they devised (Scheme 10.13) [60]. Based on exhaustive experimental and computational studies, they proposed a radical chain mechanism in which Cu(I) acts as an initiator through an inner-sphere SET reduction of Selectfluor to generate the true catalyst (or chain carrier), doubly cationic radical **TEDA²⁺·** (TEDA, N-(chloromethyl)triethylenediamine, Figure 10.23). The radical dication **TEDA²⁺·** then performs hydrogen atom abstraction on an alkane to form an ammonium salt and an alkyl radical. The resultant alkyl radical abstracts a fluorine atom from Selectfluor to furnish the fluorinated product and regenerate the radical dication to enter the catalytic cycle. Hydrogen atom abstraction occurs via an early transition state and is the rate-determining step, while fluorine atom transfer is a barrierless process. Both processes are highly exothermic. Donahue's ionic curve crossing theory and transition state calculations successfully explained the observed preference for monofluorination. Charge analysis of hydrogen atom abstraction TSs revealed that an electron-withdrawing group such as fluorine on alkyl substrate destabilizes the developing positive charge on the hydrogen atom being transferred, thus advocating for hydrogen atom abstraction of an alkane over a fluoroalkane.

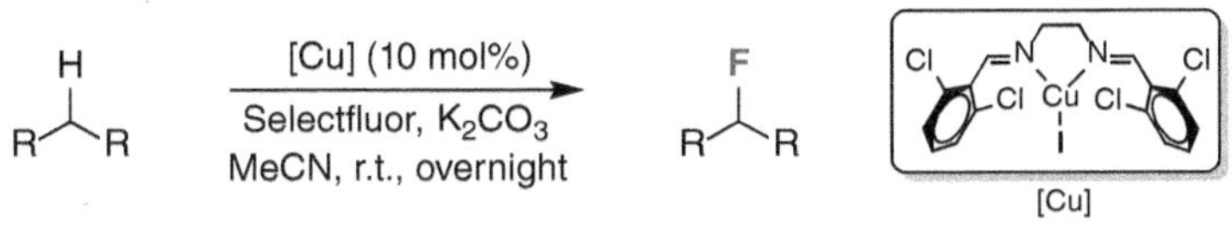

Scheme 10.13 Catalytic monofluorination of C(sp³)−H bonds enabled by Cu(I) and Selectfluor.

In 2014, Zhang et al. reported a Cu-catalyzed highly regioselective and efficient radical aminofluorination of styrenes with NFSI as both fluorine and nitrogen

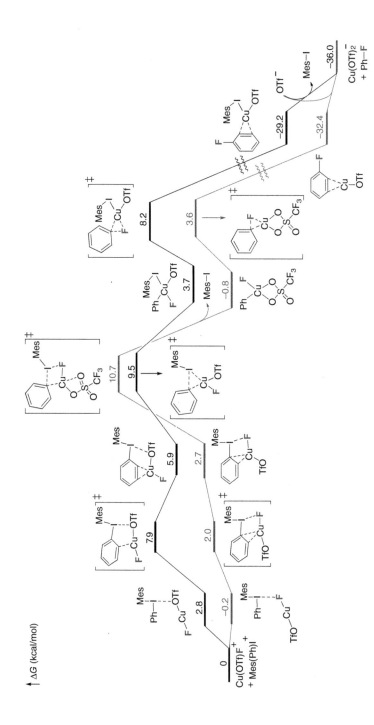

Figure 10.22 IEFPCM-BP86/6-311+G(2d,p)+def2-QZVP(Cu,I) free energy profile for Cu-catalyzed nucleophilic fluorination of diaryliodonium salts.

source (Scheme 10.14) [61]. Mechanistic experiments and DFT calculations supported a radical mechanism (Figure 10.24). That is, the reaction begins with the oxidative addition of Cu(I) to NFSI to give a Cu(III) species (**Int1-Cu^{III}**), which could lead to a Cu(II)-stabilized nitrogen radical (**Int1-Cu^{II}**) through fast equilibration. A concerted aminocupration pathway from **Int1-Cu^{III}** has a barrier as high as 25.2 kcal/mol. In contrast, radical addition of **Int1-Cu^{II}** to styrene, which generates a benzylic radical, exhibits a much lower energy barrier of 10.7 kcal/mol. The benzylic radical then abstracts a fluorine atom from NFSI to produce the aminofluorination product and release the nitrogen radical. The capture of the nitrogen radical by the Cu(II) would regenerate **Int1-Cu^{II}** to continue the radical pathway.

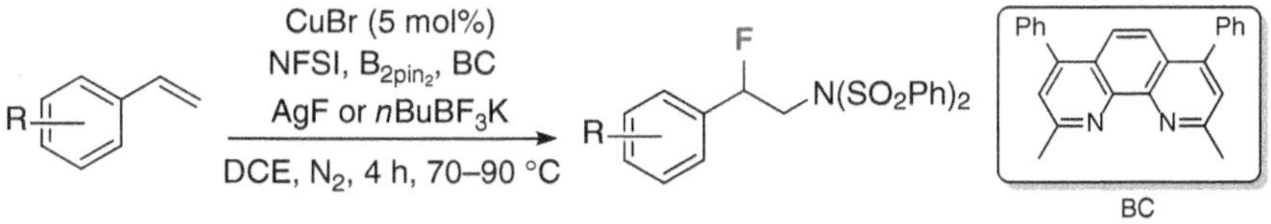

Scheme 10.14 Cu-catalyzed radical aminofluorination of styrenes with NFSI. Source: Modified from Zhang et al. [61].

In 2018, Li, Yu, and coworkers developed a Cu-catalyzed fluorotrifluoromethylation of unactivated alkenes with CsF as the fluoride source and Umemoto reagent as the CF$_3$ source (Scheme 10.15) [62]. A plausible mechanism involving fluorine atom transfer from Cu(II)–F complexes to alkyl radicals was proposed (Figure 10.25a). DFT calculations showed that the fluorine atom abstraction from $[Cu_2(bpy)_2F_2]^{2+}$ by *tert*-butyl radical has an energy barrier of 13.3 kcal/mol (Figure 10.25b). A higher barrier of 16.0 kcal/mol was computed for the isopropyl radical, consistent with the experimental observation that the *tert*-butyl radical is more reactive than the isopropyl radical. Atomic charge analysis revealed that the approach of alkyl radical to the fluoride is accompanied by a partial electron transfer from the radical to the Cu complex, which explains why the *tert*-butyl radical is much easier to be oxidized than the isopropyl radical.

10.2.8 Ag-Catalyzed Fluorination Reactions

In 2015, Liu, Lin, and coworkers reported an experimental and computational study on the Ag-catalyzed aminofluorination of alkynyl-imine substrates with NFSI or

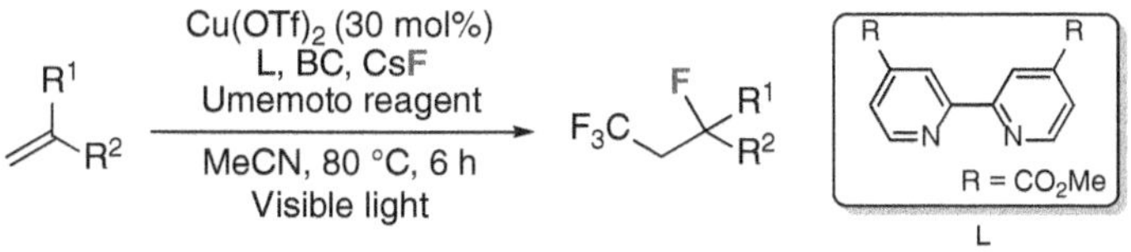

Scheme 10.15 Cu-catalyzed fluorotrifluoromethylation of unactivated alkenes. Source: Reprinted with permission from Liu et al. [62]. Copyright 2018, American Chemical Society.

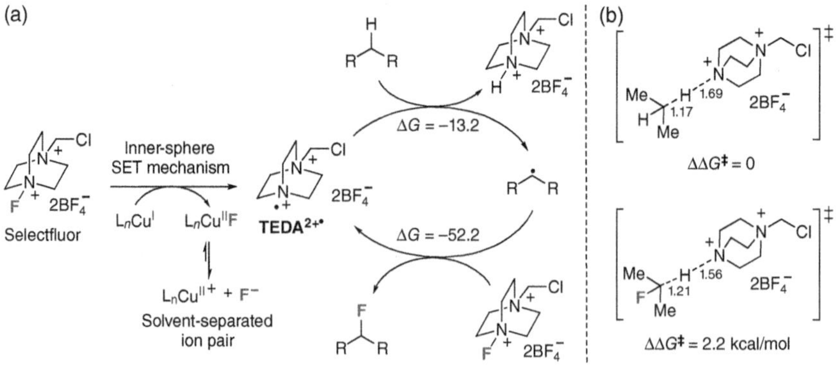

Figure 10.23 (a) Proposed catalytic cycle for monofluorination of C(sp³)—H bond initiated by Cu(I). (b) B3LYP/6-311++G(d,p) calculated hydrogen atom abstraction TSs.

Selectfluor [63] (Scheme 10.16). They successfully isolated and characterized a novel mesoionic carbene silver complex (MIC)$_2$Ag(I). DFT calculations revealed that the mechanism of C—F bond formation depends on the choice of different electrophilic fluorinating reagents. The σ-metathesis process is the favorable pathway over the redox process for the C—F bond formation with NFSI due to its coordination ability with Ag(I) (Figure 10.26a). Alternatively, the redox fluorination process is involved for Selectfluor due to its high oxidative potential (Figure 10.26b). A disilver-assisted σ-metathesis process was proposed for the Ag-catalyzed intramolecular aminofluorination of activated allenes with NFSI [64].

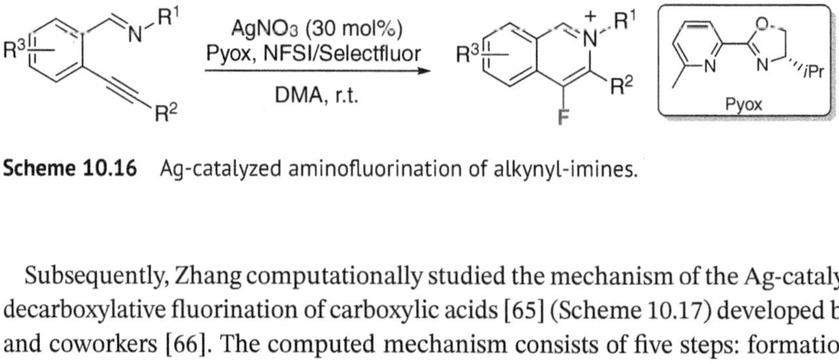

Scheme 10.16 Ag-catalyzed aminofluorination of alkynyl-imines.

Subsequently, Zhang computationally studied the mechanism of the Ag-catalyzed decarboxylative fluorination of carboxylic acids [65] (Scheme 10.17) developed by Li and coworkers [66]. The computed mechanism consists of five steps: formation of Ag(I)-carboxylate, oxidation by Selectfluor, homolytic cleavage of O—Ag(II) bond, loss of CO$_2$, and fluorine abstraction (Figure 10.27). Interestingly, the formation of Ag(II)–F and Ag(III)–F intermediates was calculated to be significantly endergonic ($\Delta G > 40$ kcal/mol). These results argue against the initially proposed mechanism involving the intermediacies of Ag(III)–F and Ag(II)–F, but are consistent with Flowers II and coworkers experimental results [67]. Zhang further studied the mechanism of Ag-catalyzed intramolecular aminofluorination of alkenes [68]. DFT calculations correctly predict the preference for the formation of a 5-*exo* over 6-*endo* cyclization product.

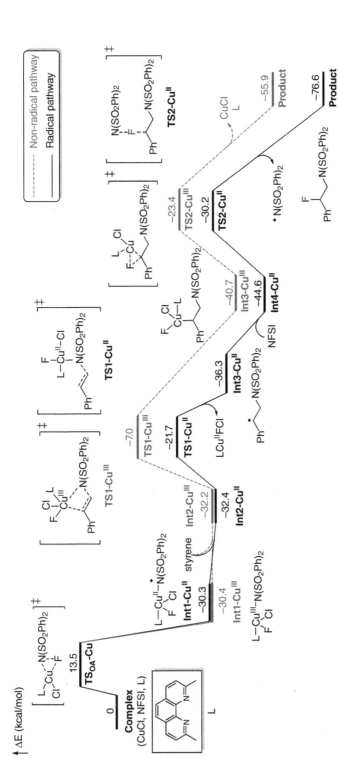

Figure 10.24 B3LYP/6-31G(d) energy profile for Cu-catalyzed aminofluorination of styrene. Source: Zhang et al. [61]. Reprinted with permission of John Wiley & Sons.

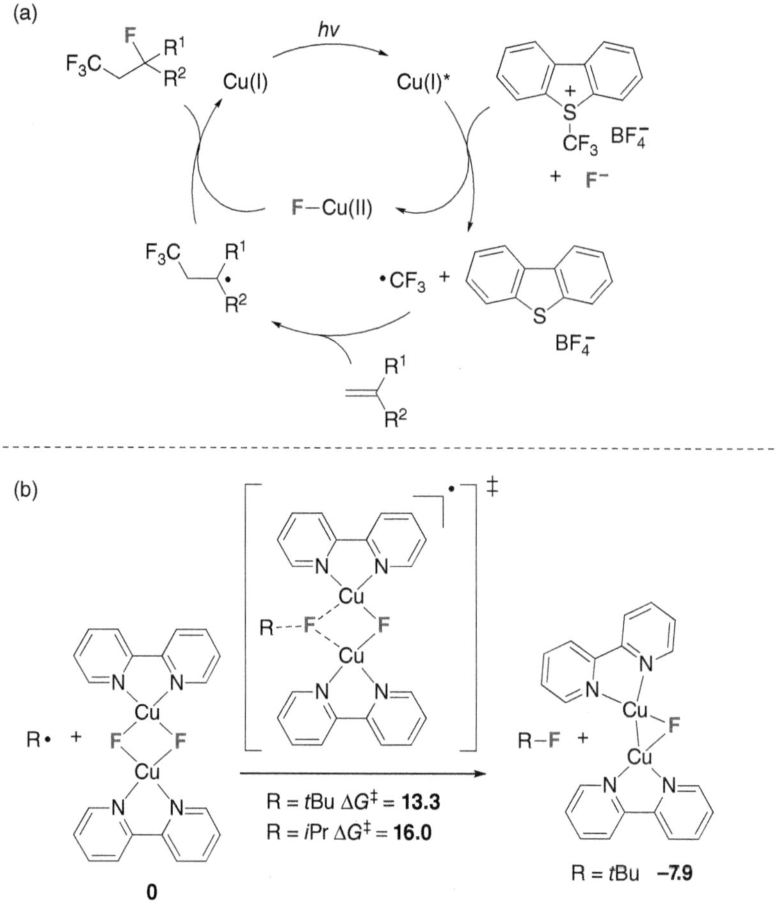

(a)

(b)

R = tBu ΔG‡ = 13.3
R = iPr ΔG‡ = 16.0

R = tBu −7.9

0

Figure 10.25 (a) Proposed catalytic cycle for the fluorotrifluoromethylation of unactivated alkenes. (b) IEFPCM-B3LYP/6-311+G(d)//B3LYP/6-31G(d) free energy profile for fluorine atom transfer. Energies are in kcal/mol.

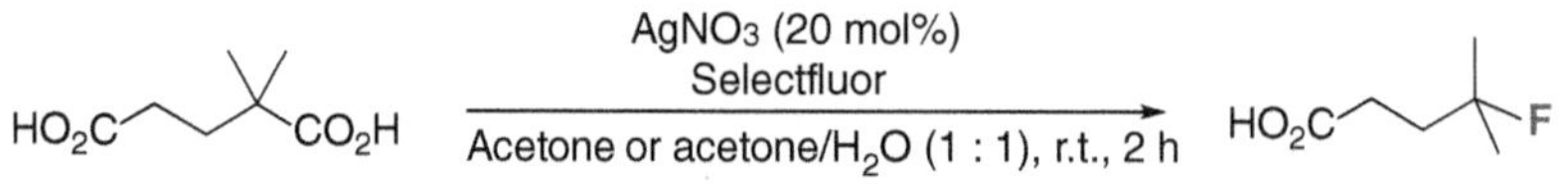

Scheme 10.17 Ag-catalyzed decarboxylative fluorination of carboxylic acid. Source: Reprinted with permission from Yin et al. [66]. Copyright 2012, American Chemical Society.

Recently, Tang, Xu, and coworkers developed a novel Ag-catalyzed anti-Markovnikov hydroxyfluorination reaction of styrenes with Selectfluor (Scheme 10.18) [69]. Preliminary mechanistic experiments and DFT computations supported a radical mechanism (Figure 10.28), which was proposed to start with the coordination of styrene to Ag(H$_2$O)$^+$, followed by oxidation of Selectfluor to give

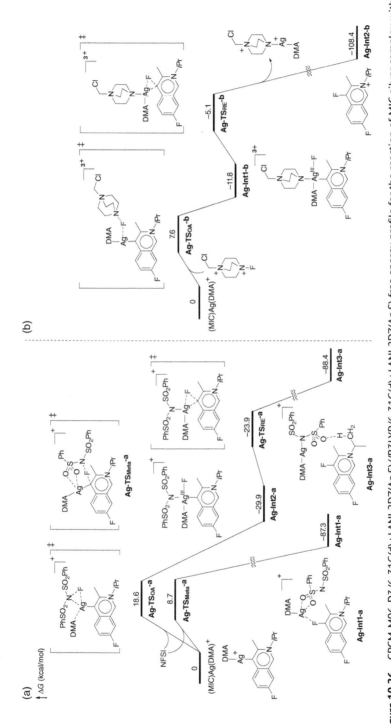

Figure 10.26 CPCM-M06-D3/6-31G(d)+LANL2DZ(Ag,S)//B3LYP/6-31G(d)+LANL2DZ(Ag,S) free energy profile for the reaction of MIC silver complex with (a) NFSI and (b) Selectfluor (the corresponding σ-metathesis pathway could not be located).

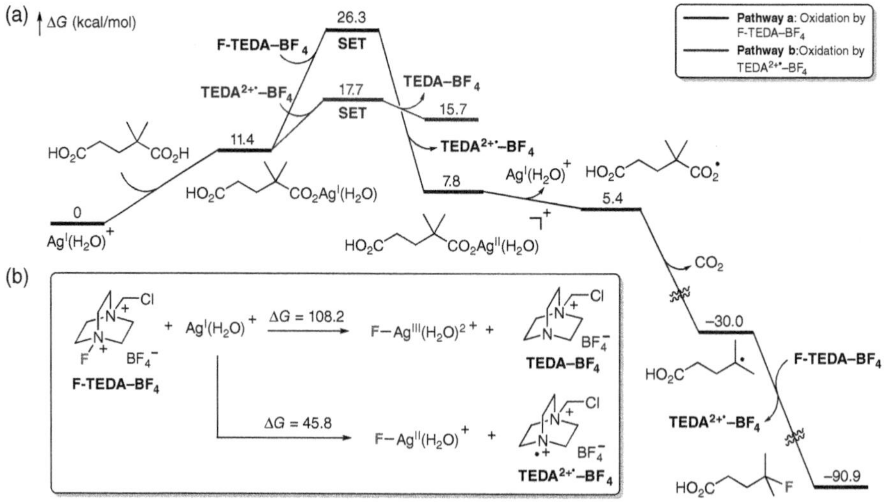

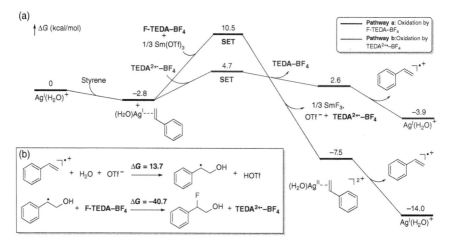

Figure 10.27 CPCM-M06/6-311+G(2df,2p)+SDD(Ag)//B3LYP/6-31G(d)+LANL2DZ+f(Ag) free energy profile for decarboxylative fluorination of carboxylic acid. The SET barriers were derived from outer-sphere Marcus–Hush model and Savéant's model.

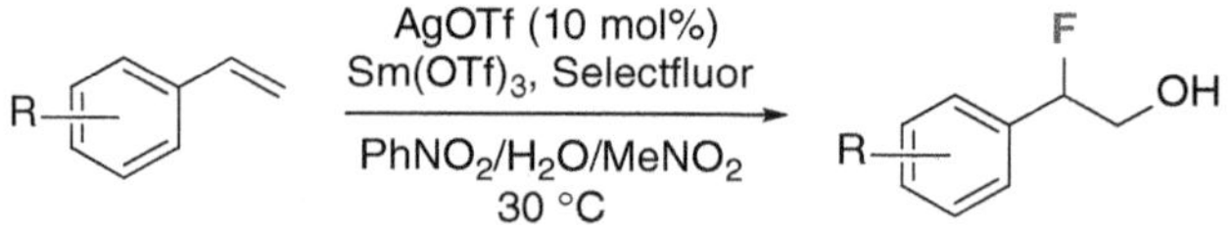

Scheme 10.18 Ag-catalyzed anti-Markovnikov hydroxyfluorination of styrenes with Selectfluor. Source: Li et al. [69]. Reprinted with permission of American Chemical Society.

Figure 10.28 SMD-B3LYP/6-311+G(d,p)+SDD(Ag)+ ECP51MWB(Sc) free energy profile for the hydroxyfluorination of styrene. SET barriers were derived from outer-sphere Marcus–Hush model and Savéant's model.

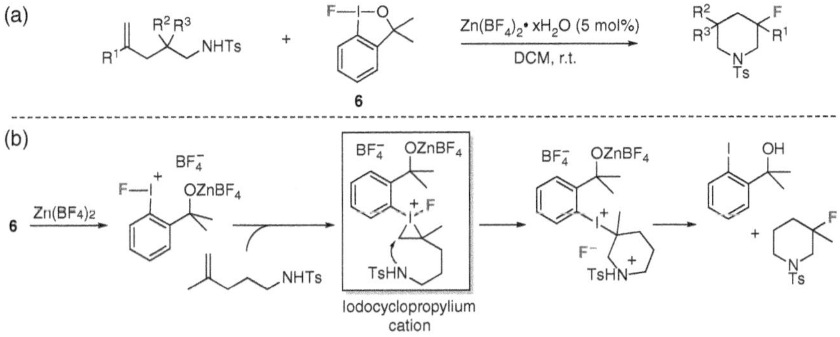

Figure 10.29 (a) Zn-catalyzed aminofluorination of alkenes and (b) proposed mechanism.

Ag(II) species and **TEDA$^{2+\cdot}$–BF$_4$**. Calculations suggested that Sm(OTf)$_3$ promotes the oxidation step by stabilizing the generated fluoride via ligand exchange to give SmF$_3$. The resulting Ag(II) species oxidizes the ligated styrene to form a styrene radical cation and regenerates the catalyst Ag(H$_2$O)$^+$. The styrene radical cation reacts with water to give the benzylic radical, which then abstracts a fluorine atom from Selectfluor to afford the final product. The whole reaction was calculated to be highly exergonic. The oxidation of styrene–Ag(H$_2$O)$^+$ complex by **TEDA$^{2+\cdot}$–BF$_4$** is more facile than by Selectfluor, suggesting that the catalytic cycle should prefer to go along Pathway **b** after launching via Pathway **a**.

10.2.9 Zn-Catalyzed Fluorination Reactions

In 2017, Himo, Szabó, and coworker [70] reported an elaborate computational study of Zn-catalyzed fluorination reaction of alkenes with fluoro-benziodoxole **6** [71] (Figure 10.29a). Following the activation model of the analogous trifluoromethyl-benziodoxole reagent **7**, the initially proposed mechanism involved the addition of Zn-activated reagent **6** to the double bond of the alkene to give an iodocyclopropylium cation (Figure 10.29b). However, the putative iodocyclopropylium intermediate could not be located. Based on DFT calculations, they proposed a plausible metathesis mechanism (Figure 10.30) in which Zn coordinates to the fluorine atom instead of oxygen of the reagent **6**. At nearly the same time, Xue and coworkers reported the same "F-coordination" activation model for reagent **6** in the Ag-mediated geminal difluorination of styrenes [72, 73]. It was unraveled that the Zn-mediated isomerization of **6** leading to an isomerized reagent with lower lying LUMO (a better electrophile) is a vital activation step for the fluorination. Thereafter, the reaction should proceed through a metathesis step to form the C—F bond, and by a nucleophilic substitution to give the final aminofluorination product. The nucleophilic substitution is the rate-determining step with a computed energy barrier of 23.1 kcal/mol. The zinc ion was found to lower the barriers of all steps in this reaction. Their mechanism rationalizes the experimentally observed regioselectivity and can be generalized to Brønsted acid-catalyzed fluorination reactions using reagent **6**.

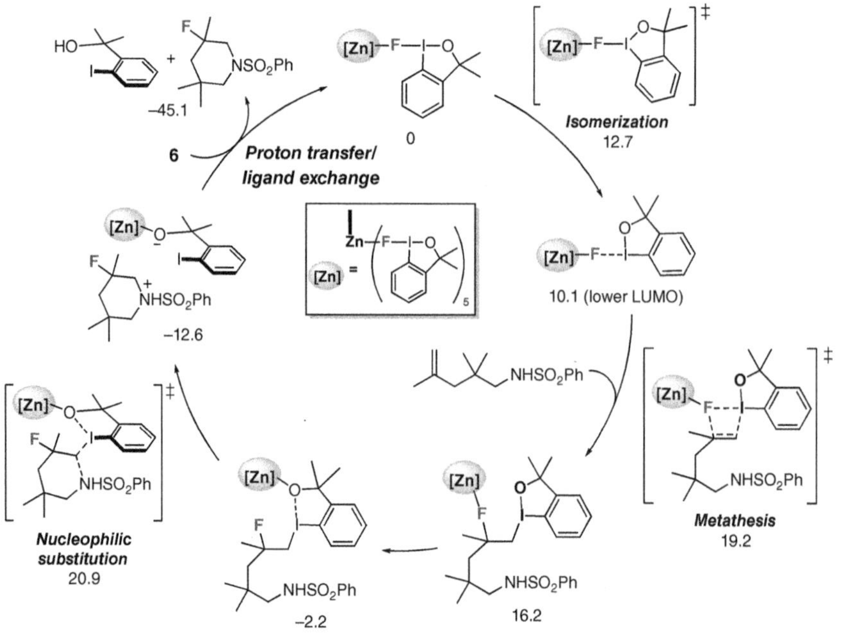

Figure 10.30 SMD-B3LYP-D3-(BJ)/6-311+G(2d,2p)+LANL2DZdp(I)+LANL2DZ(Zn)//B3LYP/6-31G(d,p)+LANL2DZdp(I)+LANL2DZ(Zn) calculated mechanism for Zn-catalyzed aminofluorination of alkenes. Energies are in kcal/mol.

10.3 DFT Modeling of Organocatalytic Fluorination Reactions

10.3.1 Fluorination Reactions Catalyzed by Chiral Amines

10.3.1.1 Chiral Secondary Amines-Catalyzed Fluorination Reactions

The year 2000 saw the rebirth of organocatalysis. Particularly, the rediscovery of the versatile catalytic nature of proline by List et al. [74] and the disclosure of the imidazolidinone-catalyzed highly enantioselective Diels–Alder reaction by MacMillan and coworkers [75] are the most important milestones in the field of asymmetric organocatalysis. The rapid advance in enantioselective enamine catalysis [76–80] was quickly extended to fluorination. In 2005, several groups reported almost simultaneously the chiral secondary amines catalyzed enantioselective fluorination of aldehydes [81–84].

In 2008, Jørgensen and coworkers performed DFT calculations to explain the origin of the enantioselectivity of trimethylsilyl diarylprolinol (**16**) catalyzed α-fluorination of aldehydes (Scheme 10.19) [85]. The stereochemical outcome of these transformations depends on the approach of the electrophile to the enamine intermediate formed by the reaction of **16** and aldehydes. B3LYP/6-31G(d) was used to calculate the relative free energies of the four enamines from propanal and **16**. The results showed that (*E*)-*anti* and (*E*)-*syn* containing an *E*-configuration are more stable than (*Z*)-*anti* and (*Z*)-*syn* having *Z*-configuration (Figure 10.31a).

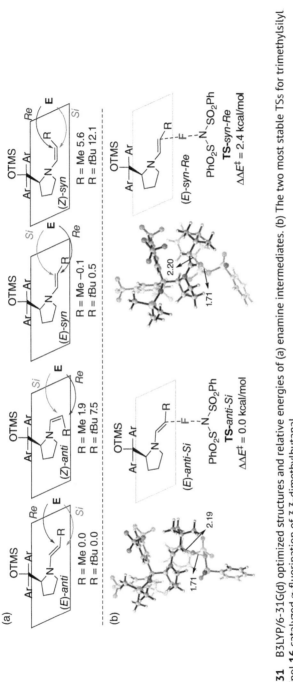

Figure 10.31 B3LYP/6-31G(d) optimized structures and relative energies of (a) enamine intermediates. (b) The two most stable TSs for trimethylsilyl diarylprolinol **16** catalyzed α-fluorination of 3,3-dimethylbutanal.

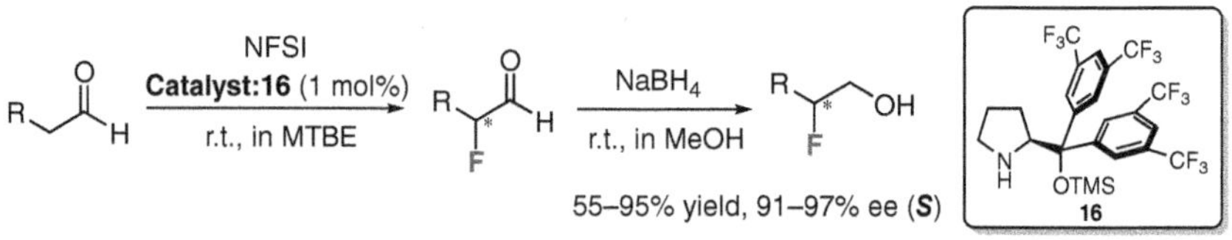

Scheme 10.19 Trimethylsilyl diarylprolinol-catalyzed enantioselective fluorination of aldehydes. Source: Modified from Dinér et al. [85].

Although DFT calculations predicted that both *anti* and *syn* E enamines should be present in the reaction system as a mixture, only (*E*)-*anti* enamine was observed in solution by NMR [86, 87].

Figure 10.31b shows the lowest energy enantiomeric TSs for α-fluorination of 3,3-dimethylbutanal catalyzed by **16**. The **TS-*anti-Si***, where the incoming NFSI attacks the *anti-E* enamine from the less shielded "bottom" (*Si*) face, leads to the major (*S*)-α-fluoroaldehyde observed experimentally. The incoming NFSI attacks on the "bottom" (*Re*) face of the *syn-E*-enamine (**TS-*syn-Re***), giving the minor (*R*)-enantiomer. The **TS-*anti-Si*** is more stable than **TS-*syn-Re*** by 2.4 kcal/mol, corresponding to 96% ee in favor of the (*S*)-product, which agrees well with the level of enantioselectivity observed experimentally (97% ee). The energy difference was attributed to a staggered conformation of the substituents around the forming C—F and the breaking N—F bonds in the major enantiomer, which is in contrast to a more eclipsed arrangement around these bonds found in the minor enantiomer. Calculations also showed that attack on the "top" (*Re* face) of the *anti-E* enamine is disfavored (≥9.5 kcal/mol) due to shielding of the enamine by the bulky substituent in the catalyst. It was thus suggested that the enantioselectivity originates from different populations of **TS-*anti-Si*** and **TS-*syn-Re***, rather than from an approach of the electrophile from the more hindered side of the catalyst.

10.3.1.2 Chiral Primary Amines-Catalyzed Fluorination Reactions

Cinchona alkaloids play an important role in asymmetric fluorination. In 2014, the Houk group [88] computationally studied the origin of selectivity of the cinchona primary amine **17**-catalyzed fluorination of cyclic ketones developed by MacMillan and coworkers (Scheme 10.20) [89], assuming that the reactions proceed by dual activation of the ketone and the fluorine source NFSI (fluorine transfer from NFSI to catalyst). The TSs for the intramolecular N-to-C fluorine transfer to either face of the enamine formed from cyclohexanone and the model catalyst **17a** were located (Figure 10.32). The (*R*)-**TS**$_{chair}$ leading to the major enantiomer is 6.8 kcal/mol lower in energy than (*S*)-**TS**$_{boat,}$ which leads to the minor enantiomer. This calculation corresponded well with the high enantioselectivity observed in experimental studies. The energy difference between (*R*)-**TS**$_{chair}$ and (*S*)-**TS**$_{boat}$ was attributed to (1) the preferred conformation of the seven-membered ring (the chair conformation vs. the boat conformation), and (2) the steric bulk of the C9-quinoline of the organocatalyst. The stereoselectivity model proposed is applicable to more elaborate substrate structures.

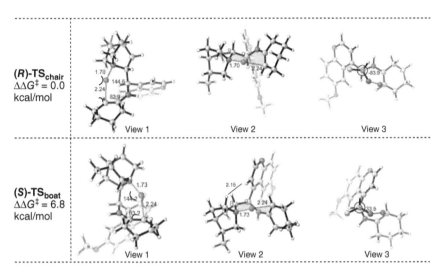

Scheme 10.20 Cinchona alkaloid-primary amine-catalyzed fluorination of cyclohexanone. Source: Modified from Kwiatkowski et al. [89].

(R)-TS$_{chair}$
$\Delta\Delta G^{\ddagger} = 0.0$ kcal/mol

View 1   View 2   View 3

(S)-TS$_{boat}$
$\Delta\Delta G^{\ddagger} = 6.8$ kcal/mol

View 1   View 2   View 3

Figure 10.32 IEFPCM-B3LYP-D3-(BJ)/def2-TZVPP//IEFPCM-B3LYP/6-31G(d) calculated relative free energies of the stereodetermining TSs for **17a**-catalyzed fluorination of cyclohexanone.

Encouraged by Houk's stereoselectivity model [88], Higashi and coworkers developed an efficient synthetic method for proximal α-regioselective fluorination of α-branched enals, providing allylic fluorides in good yields and enantioselectivity (Scheme 10.21) [90]. DFT calculations were conducted to understand the α-regioselectivity and stereoselectivity. The calculated seven-membered fluorine transfer transition states (R)-TS$_{MeSO_3^-}$ and (S)-TS$_{MeSO_3^-}$ are shown in Figure 10.33. The origin of the high α-regioselectivity was ascribed to a much closer distance of the fluorine atom on quinuclidine to the α-position than to the γ-position of the

Scheme 10.21 Cinchona primary amine **18**-catalyzed proximal fluorination of α-branched enals. Source: Reprinted with permission from Arimitsu et al. [90]. Copyright 2017, American Chemical Society.

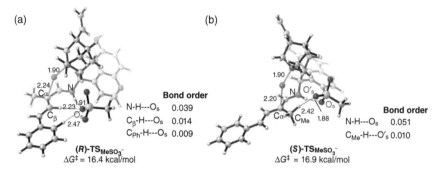

Figure 10.33 SMD-M06-2X/6-31G(d,p) calculated TSs and activation free energies for cinchona primary amine-catalyzed fluorination of α-branched enal.

dienamine. The (R)-$TS_{MeSO_3^-}$ is lower in energy than (S)-$TS_{MeSO_3^-}$ by 0.5 kcal/mol in the presence of $MeSO_3^-$. Further analysis revealed a considerable contribution of the nonclassical $C-H\cdots O$ hydrogen bonds between C_β-H and $MeSO_3^-$ to the stabilization of (R)-$TS_{MeSO_3^-}$, suggesting that a conjugate base of a Brønsted acid could modulate the stereochemistry.

In 2018, the Luo group reported a reagent-controlled enantioselectivity switch in the asymmetric α-fluorination of β-ketocarbonyls catalyzed by a chiral primary amine catalyst **19** [91]. A simple change of the fluorination reagent switched the enantioselectivity: namely, the use of N-fluoropyridinium (**NFPy**) salt led to S-selectivity, while the reaction with NFSI gave R-enantiomer (Scheme 10.22). The authors proposed a dual hydrogen-bonding mode for the NFSI-based R-selective process and an electrostatic model for the **NFPy**-based S-selective process. The computed S-selective (S)-TS_{NFPy} was favored over the R-selective (R)-TS_{NFPy} by

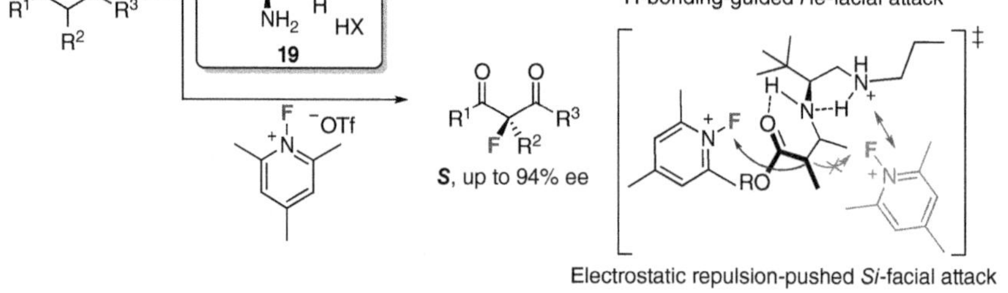

Scheme 10.22 Chiral primary amine **19**-catalyzed asymmetric fluorination of β-ketocarbonyls.

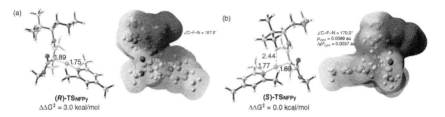

Figure 10.34 Calculated TSs and the electrostatic surface potential for chiral primary amine **19**-catalyzed asymmetric fluorination of β-ketocarbonyl with **NFPy**. Source: You et al. [91]. Licenced under CC BY 3.0.

3.0 kcal/mol, consistent with the experimental observation (Figure 10.34). The energy difference between (S)-TS_{NFPy} and (R)-TS_{NFPy} was mainly attributed to the unfavorable electrostatic repulsion between the ammonium and pyridinium moieties in (R)-TS_{NFPy}.

10.3.2 Tridentate Bis-Urea Catalyzed Fluorination Reactions

In 2018, Gouverneur, Paton, and coworkers achieved a breakthrough in asymmetric nucleophilic fluorination by merging hydrogen bonding with phase-transfer catalysis (Scheme 10.23) [92]. This protocol afforded enantioenriched β-fluorosulfides with a safe fluoride source CsF and a chiral N-alkyl bis-urea catalyst **20**. Molecular dynamics (MD) simulations and DFT calculations were undertaken to gain insights into the mechanism and origins of catalysis and stereoselectivity. Comparison of the computed free energy profile for an achiral urea **21**-catalyzed and -uncatalyzed fluorination revealed that (1) unfavorable halide anion exchange is responsible for the prohibitively high energetic span of the uncatalyzed pathway due to the much higher lattice energy of CsF relative to that of CsBr, and (2) the key catalytic role of the urea is to promote a halide anion exchange through preferential stabilization of fluoride via hydrogen bonding in solution (Figure 10.35). Calculations revealed an energetic preference for the formation of a tridentate hydrogen bonding complex between the chiral bis-urea catalyst and the fluoride ion. This provided an important clue for catalyst optimization and led to the identification of the optimum N-alkyl bis-urea catalyst **20**. Computations predicted (S,S)-product

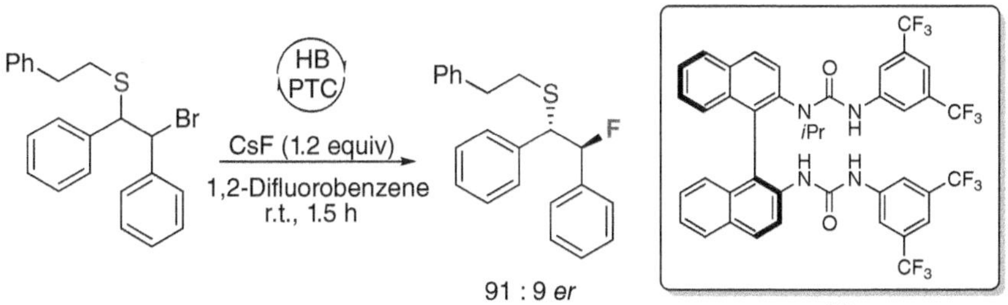

Scheme 10.23 Asymmetric nucleophilic fluorination by hydrogen bonding phase-transfer catalysis. Source: Pupo et al. [92].

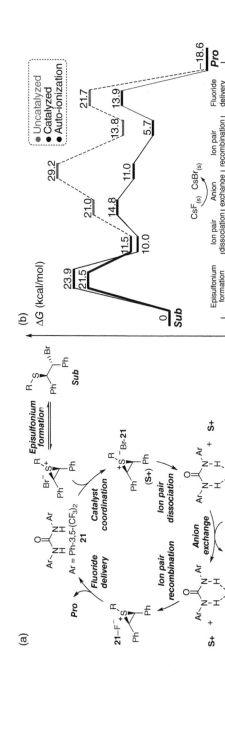

Figure 10.35 COSMO-ωB97X-D3/(ma)-def2-TZVPP//CPCM-M06-2X/def2-SVP(TZVPPD) energy profile for urea-catalyzed fluorination and the uncatalyzed one.

(a)

(b)

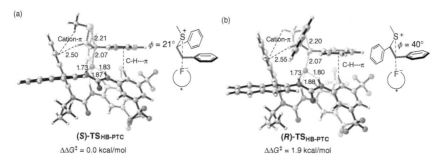

(S)-TS_{HB-PTC}

ΔΔG‡ = 0.0 kcal/mol

(R)-TS_{HB-PTC}

ΔΔG‡ = 1.9 kcal/mol

Figure 10.36 COSMO-ωB97X-D3/(ma)-def2-TZVP//CPCM-M06-2X/def2-SVP(TZVPPD) calculated enantiomeric TSs and their relative free energies for **20**-catalyzed fluorination.

formation in 96.5 : 3.5 er, which is in good agreement with the experimental value (91 : 9 er). Stronger CH$\cdots\pi$ and cation$\cdots\pi$ interactions and the better conjugation between the phenyl ring and the benzylic carbon contribute to the stabilization of the favored transition state (S)-**TS**$_{\text{HB-PTC}}$ (Figure 10.36).

Most recently, Gouverneur, Paton, and coworkers further demonstrated that the hydrogen bonding phase-transfer catalysis could enable access to valuable β-fluoroamines in high yields and enantioselectivities with KF as the fluoride source (Scheme 10.24) [93]. In accordance with their previous observations, MD simulations showed that the N-alkylated catalyst **22** forms a stable tridentate fluoride complex. The DFT-computed selectivity of 95 : 5 er is in excellent agreement with experimental values. The enantioselectivity was attributed to (1) a stronger cation$\cdots\pi$ interaction and more effective conjugation of the phenyl ring with the bonds that are forming and breaking in the (S)-**TS**$_{\text{HB-PTC-22}}$, and (2) destabilizing steric repulsion in the (R)-**TS**$_{\text{HB-PTC-22}}$. The orientation of the N-substituents of aziridinium ion in the most stable enantiomeric TSs is pointing away from the catalytic pocket, which explains the wide substituent tolerance in these positions (Figure 10.37).

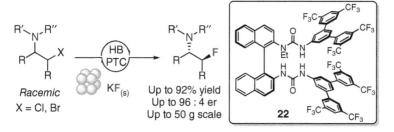

Racemic
X = Cl, Br

KF$_{(s)}$

Up to 92% yield
Up to 96 : 4 er
Up to 50 g scale

Scheme 10.24 Hydrogen bonding phase-transfer catalytic synthesis of enantioenriched β-fluoroamines with KF(s). Source: Pupo et al. [93].

10.3.3 Hypervalent Iodine-Catalyzed Fluorination Reactions

In 2018, Xue, Houk, Jacobsen, and coworkers [94] reported a computational mechanistic study on aryl iodide **23**-catalyzed asymmetric migratory geminal difluorination reaction of β-substituted styrenes developed by the Jacobsen group

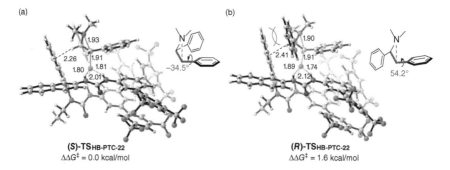

(a)

(S)-TS_{HB-PTC-22}
ΔΔG‡ = 0.0 kcal/mol

(b)

(R)-TS_{HB-PTC-22}
ΔΔG‡ = 1.6 kcal/mol

Figure 10.37 COSMO-ωB97X-D3/(ma)-def2-TZVPP//CPCM-M06-2X/def2-SVP(TZVPPD) calculated lowest energy enantiomeric TSs and their relative free energies for **22**-catalyzed fluorination.

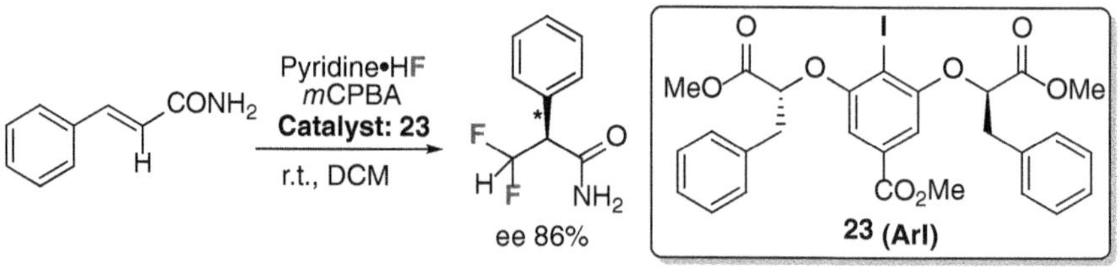

Scheme 10.25 Aryl iodide **23**-catalyzed asymmetric geminal difluorination of cinnamamide.

[95] (Scheme 10.25). The computed mechanism (Figure 10.38) consists of four steps: (1) activation of the *in situ*-formed active catalyst iodoarene difluoride (ArIF$_2$), (2) 1,2-fluoroiodination, (3) bridging phenonium ion formation via S$_N$2 reductive displacement, and (4) regioselective fluoride addition. The 1,2-fluoroiodination was found to be the stereocontrolling step. The computed enantiomeric excess of 95% *S* agrees qualitatively with the enantioselectivity observed experimentally (86% *S* ee) (Figure 10.39a). In the transition structure (*S*)-TS2$_{ArI}$ (leading to major enantiomer), the styrenyl olefin and the C—I bond of catalyst adopt a favorable spiro arrangement, and the phenyl group of the substrate is accommodated in a binding pocket through slipped π·· ·π stacking interaction. Although the transition structure (*R*)-TS2$_{ArI}$ (leading to minor enantiomer) has similar binding of the phenyl, it adopts a less favorable near periplanar arrangement due to the clash between the phenyl group of the substrate and the ester carbonyl group at the arm of catalyst (Figure 10.39b). This stereocontrol model is applicable to other substrates.

In the same year, Falivene, Rueping, and coworkers reported a new chiral iodoarene catalyst **24** for the asymmetric fluorination of carbonyl compounds, affording the products with a quaternary stereocenter in high enantioselectivities (Scheme 10.26) [96]. A combination of computational and experimental studies shed insights into the mechanism and the origin of enantioselectivity. Based on their experiments, they assumed an *in situ* formation and activation of the Ar*–I–F$_2$ by the HF molecule. The favored mechanism involves deprotonation

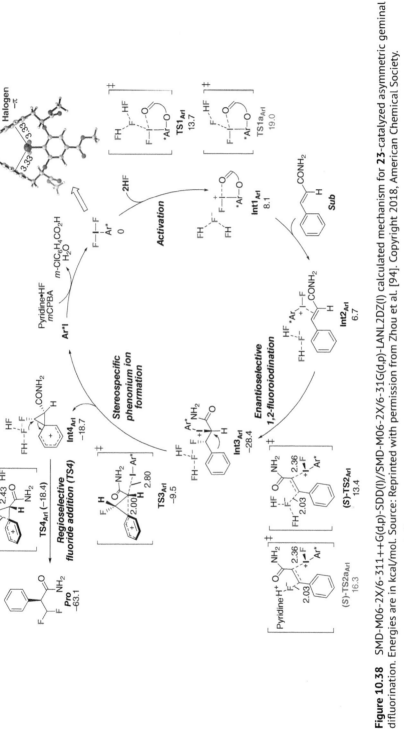

Figure 10.38 SMD-M06-2X/6-311++G(d,p)-SDD(I)//SMD-M06-2X/6-31G(d,p)-LANL2DZ(I) calculated mechanism for **23**-catalyzed asymmetric geminal difluorination. Energies are in kcal/mol. Source: Reprinted with permission from Zhou et al. [94]. Copyright 2018, American Chemical Society.

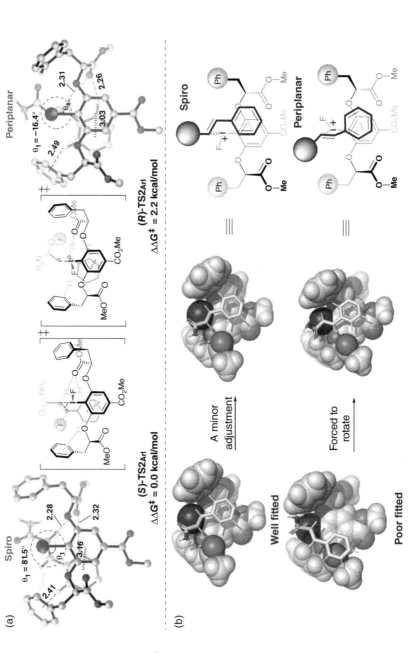

Figure 10.39 (a) Optimized enantiomeric TS geometries and their relative free energies. (b) Space-filling model of *Re-* and *Si*-face coordination of styrenes to the I(III) center of catalyst. Source: Reprinted with permission from Zhou et al. [94]. Copyright 2018, American Chemical Society.

of the substrate with $[Ar^*–I–F]^+/[HF_2]^-$, followed by $[Ar^*–I–F]^+$ migration to form α—C-bonded hypervalent iodine intermediate **Int3**$_{ArI-24}$ (Figure 10.40a). This intermediate undergoes a stepwise Ar^*–I elimination to furnish the product (Figure 10.40a, dark grey pathway). The $[Ar^*–I–F]^+$ migration was identified as the enantiodetermining step. The computed enantiodetermining transition state **(R)-TS2**$_{ArI-24}$ is more stable than **(S)-TS2**$_{ArI-24}$ by 2.3 kcal/mol (Figure 10.40b) in good agreement with the selectivity observed experimentally. The stereoselectivity was attributed to two factors: (1) the destabilizing repulsion that comes from the short distances (<4.0 Å) between carbon atoms of the substrate and the catalyst in the unfavored **(S)-TS2**$_{ArI-24}$, and (2) the stronger stabilizing C – H···O interactions between an oxygen atom of the substrate and C—H bonds of the catalyst in the favored **(R)-TS2**$_{ArI-24}$.

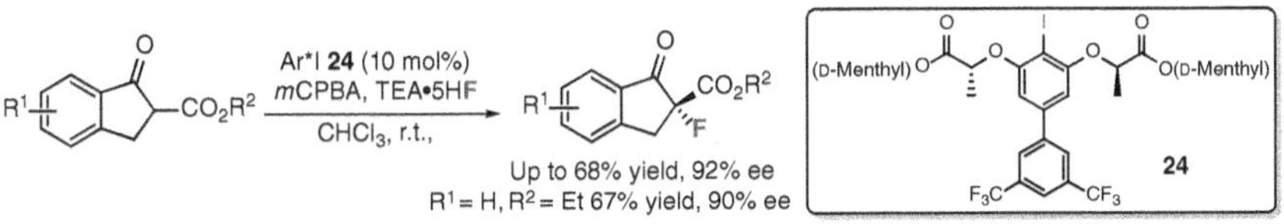

Scheme 10.26 Chiral iodoarene **24**-catalyzed asymmetric fluorination of ketoesters. Source: Pluta et al. [96]. Reprinted with permission of American Chemical Society.

10.3.4 N-Heterocyclic Carbene-Catalyzed Fluorination Reactions

Lin, Sun, and coworkers developed the first N-heterocyclic carbene (NHC)-catalyzed enantioselective fluorination reaction [97]. In the presence of an appropriate combination of a precatalyst **25**, a base, and an additive, the C—F bond formation occurs efficiently at the α-position of enals, affording the enantioenriched β,γ-unsaturated α-fluoroester products (Scheme 10.27). With the aid of DFT calculations, a reaction mechanism was proposed (Figure 10.41). DFT calculations suggested that the C—F bond formation step be both enantioselectivity determining and rate limiting. The barrier for fluorination at the γ-position is higher than that at the α-position by 7.1 kcal/mol, consistent with the observed excellent α-regioselectivity. This preference was attributed to the relatively high electron density at α–C vs. γ–C of the dienolate intermediate A, likely due to the proximate positive charge of the triazolium moiety. It was proposed that the relative stability of the dienolate intermediate A and B is relevant to the observed stereochemical outcome.

In 2017, Wei, Tang, and coworkers [98] conducted a computational study on the reaction mechanism, stereoselectivity, and chemoselectivity of NHC precursor **(26)**-catalyzed oxidative α-fluorination of aliphatic aldehyde (Scheme 10.28) [99].

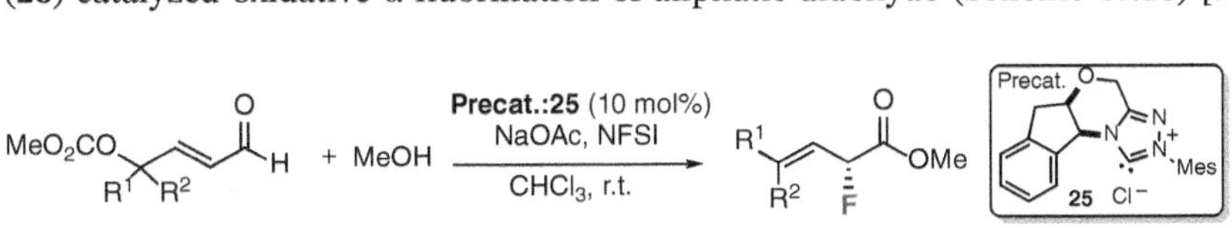

Scheme 10.27 N-heterocyclic carbene-catalyzed enantioselective fluorination reaction.

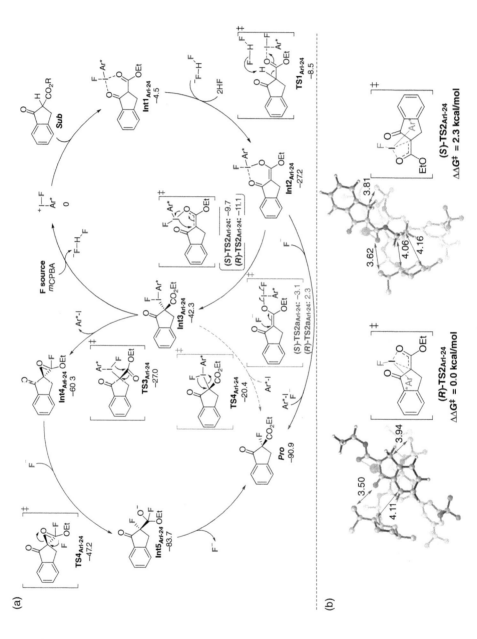

Figure 10.40 PCM-M06/TZVP//PBE-D3/SVP calculated (a) mechanism for chiral iodoarene **24**-catalyzed asymmetric fluorination of ketoesters and (b) enantiomeric TSs.

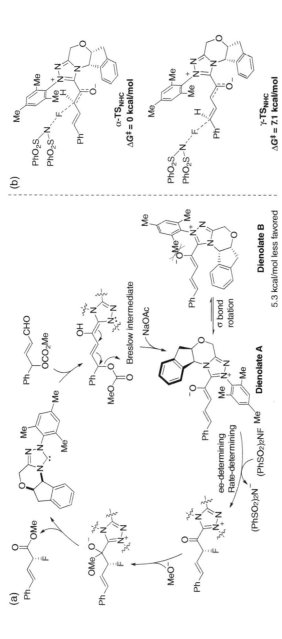

Figure 10.41 (a) Proposed catalytic cycle of *N*-heterocyclic carbene-catalyzed enantioselective fluorination reaction. (b) Computed TSs for fluorination at the α- and γ-positions.

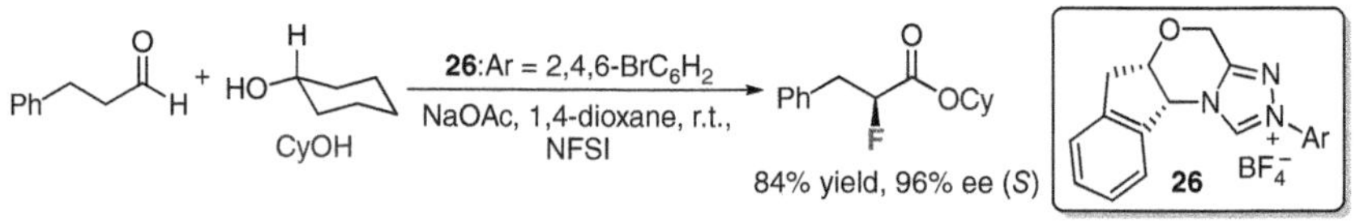

Scheme 10.28 *N*-heterocyclic carbene-catalyzed enantioselective α-fluorination of aliphatic aldehydes. Source: Reprinted with permission from Li et al. [99]. Copyright 2015, John Wiley & Sons.

They found that the most favorable pathway contains eight steps (Figure 10.42), including (1) the addition of NHC to aldehyde, (2) AcOH-assisted formation of the Breslow intermediate, (3) addition of NFSI on the carbonyl carbon of the Breslow intermediate, (4) AcOH-assisted elimination of HF to afford an NHC-bound intermediate, (5) NaOAc-assisted deprotonation to give an enolate intermediate, (6) enantioselective fluorination, (7) the AcO$^-$-assisted esterification, and (8) the dissociation of the NHC catalyst to furnish the monofluorinated ester product. The fluorination was computationally assigned as the stereodetermining step. The energy barriers of the addition of NFSI to the *Si*-face and *Re*-face of the enolate intermediate **Int5$_{NHC}$** are 22.9 and 27.4 kcal/mol (Figure 10.43), respectively. This corresponds to an enantiomeric excess of >99%, which is close to the experimentally observed value of 96% ee. This enantioselectivity was attributed to the more efficient non-covalent interactions (NCIs) (including C−H···O, C−H···π, π···π stacking interactions) in **(S)-TS6$_{NHC}$**. The likely competing reaction pathways producing the by-products of aliphatic ester and difluorinated ester were calculated to be kinetically unfavorable. Global reactivity index analysis showed that NHC promotes the reaction by enhancing the nucleophilicity of the enolate intermediate and the electrophilicity of the monofluorinated intermediate.

10.4 DFT Modeling of Enzymatic Fluorination Reaction

O'Hagan et al. isolated the first native fluorinating enzyme from the bacterium *Streptomyces cattleya* in 2002 [100] and obtained the crystal structures in 2004 [101]. Fluorinase catalyzes the reaction between *S*-adenosyl-L-methionine (SAM)

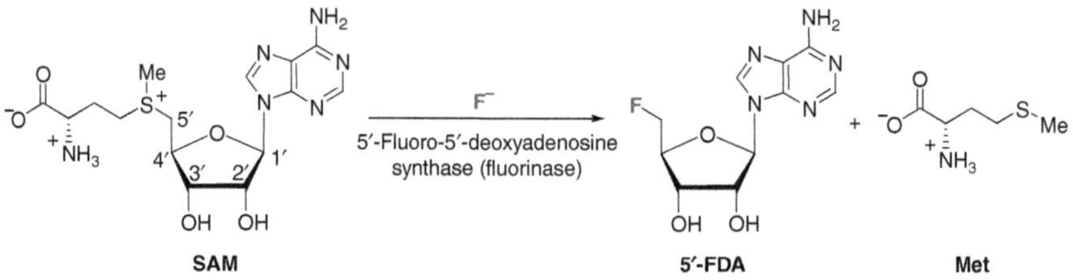

Scheme 10.29 Fluorination of *S*-adenosyl-L-methionine with fluoride to produce 5′-fluoro-5′-deoxyadenosine and L-methionine catalyzed by fluorinase.

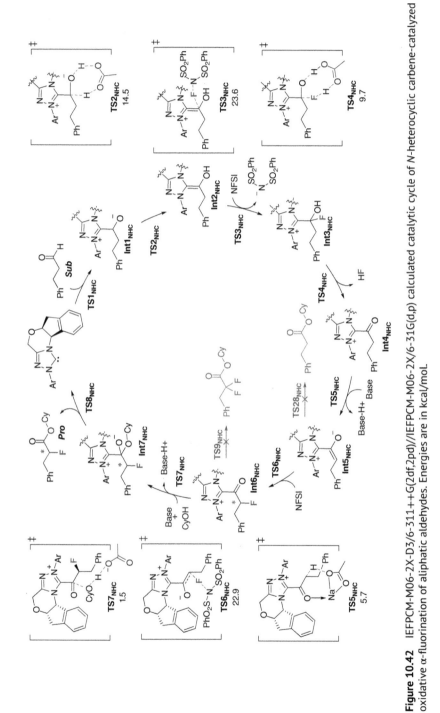

Figure 10.42 IEFPCM-M06-2X-D3/6-311++G(2df,2pd)//IEFPCM-M06-2X/6-31G(d,p) calculated catalytic cycle of *N*-heterocyclic carbene-catalyzed oxidative α-fluorination of aliphatic aldehydes. Energies are in kcal/mol.

(a)

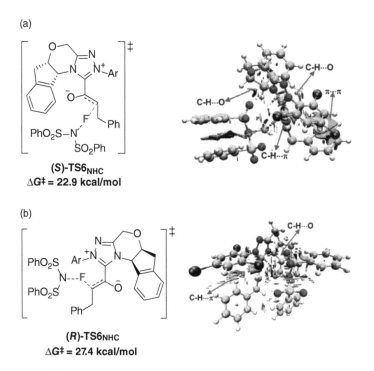

(S)-TS6$_{NHC}$
ΔG‡ = 22.9 kcal/mol

(b)

(R)-TS6$_{NHC}$
ΔG‡ = 27.4 kcal/mol

Figure 10.43 The relative free energies and NCI analysis of the stereocontrolling TSs. Source: Reprinted with permission from Wang et al. [98]. Copyright 2017, Royal Society of Chemistry.

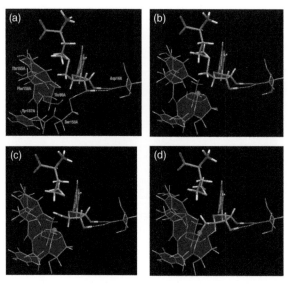

Figure 10.44 QM/MM-optimized structures of (a) the SAM-enzyme complex, (b) the reactant complex, (c) the transition state for nucleophilic attack, and (d) the product complex. Source: Senn et al. [102]. Reprinted with permission of American Chemical Society.

and fluoride to produce 5′-fluoro-5′-deoxyadenosine (5′-FDA) and L-methionine (Scheme 10.29). To gain insights into the mechanism of this unusual enzymatic reaction and to elucidate the role of the enzyme in catalysis, Senn et al. computationally studied the C—F bond-forming step in the fluorinase via DFT and QM/MM calculations in 2005 (Figure 10.44) [102]. For the purpose of comparison, they also explored the intrinsic reactivity patterns of SAM and fluoride by DFT calculations with a polarizable continuum model for water. They found that the C—F bond is formed via an S_N2-type nucleophilic attack of fluoride on SAM, and the computed activation barrier is 22 kcal/mol in solution but only 13 kcal/mol in the enzyme. Accordingly, the fluorinase reduces the energy barrier by 9 kcal/mol, which corresponds to a rate acceleration of more than 6 orders of magnitude at room temperature. The major part of this barrier reduction was ascribed to the structural rearrangements for preparing and placing the reactants in a position suitable for reaction.

10.5 Conclusions

Quantum mechanical methods are indispensable for elucidating the complex catalytic mechanisms and origins of selectivities. In this chapter, we have reviewed recent DFT studies of selected catalytic fluorination reactions. The improved understanding of relevant mechanistic details has facilitated further advancement of the field. We hope that the selected studies presented in this chapter could convincingly showcase the power and opportunities of implementing computational tools to rationalize or even to predict reactivities and the selectivity of catalytic fluorination reactions.

Besides the catalytic fluorination reactions, quantum mechanical methods have been successfully applied to catalytic fluoroalkylation reactions. In particular, the Schoenebeck group has demonstrated perfectly the application of computation and experiment in concert to facilitate the rational development of catalytic fluoroalkylation reactions [103], including trifluoromethylation [104–107], trifluoromethylthiolation [108–111], trifluoromethylselenolation [112, 113], and trifluoromethyltellurolation [114]. The synergistic use of computation and experiment would usually gain more insights than isolated techniques.

We hoped that this review would further encourage the use of computational approaches or the combined use of computation and experiment in mechanistic studies and the development of new catalytic fluorination/fluoroalkylation reactions.

Acknowledgments

Financial support from the National Natural Science Foundation of China (Grants 21772098, 21933004, and 21402099) and the Fundamental Research Funds for the Central Universities is gratefully acknowledged.

References

1 Campbell, M.G. and Ritter, T. (2015). *Chem. Rev.* 115: 612–633.

2 Champagne, P.A., Desroches, J., Hamel, J.-D. et al. (2015). *Chem. Rev.* 115: 9073–9174.

3 Cahard, D. and Bizet, V. (2014). *Chem. Soc. Rev.* 43: 135–147.

4 Cheong, P.H.-Y., Legault, C.Y., Um, J.M. et al. (2011). *Chem. Rev.* 111: 5042–5137.

5 Sperger, T., Sanhueza, I.A., Kalvet, I., and Schoenebeck, F. (2015). *Chem. Rev.* 115: 9532–9586.

6 Ahn, S., Hong, M., Sundararajan, M. et al. (2019). *Chem. Rev.* 119: 6509–6560.

7 Harvey, J.N., Himo, F., Maseras, F., and Perrin, L. (2019). *ACS Catal.* 9: 6803–6813.

8 Vogiatzis, K.D., Polynski, M.V., Kirkland, J.K. et al. (2019). *Chem. Rev.* 119: 2453–2523.

9 Li, M., Xue, X.-S., and Cheng, J.-P. (2020). *Acc. Chem. Res.* 53: 182–197.

10 Li, M., Zheng, H., Xue, X.-S., and Cheng, J.-P. (2018). *Tetrahedron Lett.* 59: 1278–1285.

11 Li, M., Zhou, B., Xue, X.-S., and Cheng, J.-P. (2017). *J. Org. Chem.* 82: 8697–8702.

12 Yang, J.-D., Wang, Y., Xue, X.-S., and Cheng, J.-P. (2017). *J. Org. Chem.* 82: 4129–4135.

13 Li, M., Guo, J., Xue, X.-S., and Cheng, J.-P. (2016). *Org. Lett.* 18: 264–267.

14 Li, M., Wang, Y., Xue, X.-S., and Cheng, J.-P. (2017). *Asian J. Org. Chem.* 6: 235–240.

15 Li, M., Xue, X.-S., Guo, J. et al. (2016). *J. Org. Chem.* 81: 3119–3126.

16 Xue, X.-S., Wang, Y., Li, M., and Cheng, J.-P. (2016). *J. Org. Chem.* 81: 4280–4289.

17 Cohen, A.J., Mori-Sánchez, P., and Yang, W. (2012). *Chem. Rev.* 112: 289–320.

18 Zhao, Y. and Truhlar, D.G. (2008). *Acc. Chem. Res.* 41: 157–167.

19 Neese, F. (2009). *Coord. Chem. Rev.* 253: 526–563.

20 Cramer, C.J. (2002). *Essentials of Computational Chemistry Theories and Models.* New York, NY: Wiley.

21 Hintermann, L. and Togni, A. (2000). *Angew. Chem. Int. Ed.* 39: 4359–4362.

22 Hintermann, L., Perseghini, M., and Togni, A. (2011). *Beilstein J. Org. Chem.* 7: 1421–1435.

23 Piana, S., Devillers, I., Togni, A., and Rothlisberger, U. (2002). *Angew. Chem. Int. Ed.* 41: 979–982.

24 Liu, W., Huang, X., Cheng, M.-J. et al. (2012). *Science* 337: 1322–1325.

25 Du, X., Zhang, H., Lu, Y. et al. (2017). *Comput. Theor. Chem.* 1115: 330–334.

26 Huang, X., Liu, W., Hooker, J.M., and Groves, J.T. (2015). *Angew. Chem. Int. Ed.* 54: 5241–5245.

27 Huang, X., Liu, W., Ren, H. et al. (2014). *J. Am. Chem. Soc.* 136: 6842–6845.

28 Groendyke, B.J., AbuSalim, D.I., and Cook, S.P. (2016). *J. Am. Chem. Soc.* 138: 12771–12774.

29 Pinter, E.N., Bingham, J.E., AbuSalim, D.I., and Cook, S.P. (2020). *Chem. Sci.* 11: 1102–1106.

30 Yuan, W., Eriksson, L., and Szabó, K.J. (2016). *Angew. Chem. Int. Ed.* 55: 8410–8415.

31 Yuan, W. and Szabó, K.J. (2016). *ACS Catal.* 6: 6687–6691.

32 Lübcke, M., Yuan, W., and Szabó, K.J. (2017). *Org. Lett.* 19: 4548–4551.

33 Mai, B.K., Szabó, K.J., and Himo, F. (2018). *ACS Catal.* 8: 4483–4492.

34 Mai, B.K., Szabó, K.J., and Himo, F. (2018). *Org. Lett.* 20: 6646–6649.

35 Ng, F.N., Chan, C.M., Li, J. et al. (2019). *Org. Biomol. Chem.* 17: 1191–1201.

36 Milner, L.M., Pridmore, N.E., Whitwood, A.C. et al. (2015). *J. Am. Chem. Soc.* 137: 10753–10759.

37 Sorlin, A.M., Mixdorf, J.C., Rotella, M.E. et al. (2019). *J. Am. Chem. Soc.* 141: 14843–14852.

38 Grushin, V.V. (2009). *Acc. Chem. Res.* 43: 160–171.

39 Yandulov, D.V. and Tran, N.T. (2007). *J. Am. Chem. Soc.* 129: 1342–1358.

40 Cui, L. and Saeys, M. (2011). *ChemCatChem* 3: 1060–1064.

41 Watson, D.A., Su, M., Teverovskiy, G. et al. (2009). *Science* 325: 1661–1664.

42 Sather, A.C., Lee, H.G., De La Rosa, V.Y. et al. (2015). *J. Am. Chem. Soc.* 137: 13433–13438.

43 Milner, P.J., Yang, Y., and Buchwald, S.L. (2015). *Organometallics* 34: 4775–4780.

44 Pliego, J.R. Jr., (2019). *J. Phys. Chem. A* 123: 9850–9856.

45 Katcher, M.H. and Doyle, A.G. (2010). *J. Am. Chem. Soc.* 132: 17402–17404.

46 Katcher, M.H., Sha, A., and Doyle, A.G. (2011). *J. Am. Chem. Soc.* 133: 15902–15905.

47 Katcher, M.H., Norrby, P.-O., and Doyle, A.G. (2014). *Organometallics* 33: 2121–2133.

48 Hull, K.L., Anani, W.Q., and Sanford, M.S. (2006). *J. Am. Chem. Soc.* 128: 7134–7135.

49 Furuya, T., Benitez, D., Tkatchouk, E. et al. (2010). *J. Am. Chem. Soc.* 132: 3793–3807.

50 Racowski, J.M., Gary, J.B., and Sanford, M.S. (2012). *Angew. Chem. Int. Ed.* 51: 3414–3417.

51 Ball, N.D. and Sanford, M.S. (2009). *J. Am. Chem. Soc.* 131: 3796–3797.

52 Testa, C., Roger, J., Scheib, S. et al. (2015). *Adv. Synth. Catal.* 357: 2913–2923.

53 Sun, H., Zhang, Y., Chen, P. et al. (2016). *Adv. Synth. Catal.* 358: 1946–1957.

54 Mao, Y.-J., Luo, G., Hao, H.-Y. et al. (2019). *Chem. Commun.* 55: 14458–14461.

55 Yamamoto, K., Li, J., Garber, J.A.O. et al. (2018). *Nature* 554: 511–514.

56 Casitas, A., Canta, M., Solà, M. et al. (2011). *J. Am. Chem. Soc.* 133: 19386–19392.

57 Ichiishi, N., Canty, A.J., Yates, B.F., and Sanford, M.S. (2013). *Org. Lett.* 15: 5134–5137.

58 Ichiishi, N., Canty, A.J., Yates, B.F., and Sanford, M.S. (2014). *Organometallics* 33: 5525–5534.

59 Bloom, S., Pitts, C.R., Miller, D.C. et al. (2012). *Angew. Chem. Int. Ed.* 51: 10580–10583.

60 Pitts, C.R., Bloom, S., Woltornist, R. et al. (2014). *J. Am. Chem. Soc.* 136: 9780–9791.

61 Zhang, H., Song, Y., Zhao, J. et al. (2014). *Angew. Chem. Int. Ed.* 53: 11079–11083.

62 Liu, Z., Chen, H., Lv, Y. et al. (2018). *J. Am. Chem. Soc.* 140: 6169–6175.

63 Liu, Q., Yuan, Z., Wang, H.-y. et al. (2015). *ACS Catal.* 5: 6732–6737.

64 Zhang, X. (2017). *Theor. Chem. Acc.* 136: 102–110.

65 Zhang, X. (2016). *Comput. Theor. Chem.* 1082: 11–20.

66 Yin, F., Wang, Z., Li, Z., and Li, C. (2012). *J. Am. Chem. Soc.* 134: 10401–10404.

67 Patel, N.R. and Flowers, R.A. (2015). *J. Org. Chem.* 80: 5834–5841.

68 Zhang, X. (2017). *J. Phys. Org. Chem.* 30: e3655.

69 Li, Y., Jiang, X., Zhao, C. et al. (2017). *ACS Catal.* 7: 1606–1609.

70 Zhang, J., Szabó, K.J., and Himo, F. (2017). *ACS Catal.* 7: 1093–1100.

71 Yuan, W. and Szabó, K.J. (2015). *Angew. Chem. Int. Ed.* 54: 8533–8537.

72 Zhou, B., Yan, T., Xue, X.-S., and Cheng, J.-P. (2016). *Org. Lett.* 18: 6128–6131.

73 Zhou, B., Xue, X.-s., and Cheng, J.-p. (2017). *Tetrahedron Lett.* 58: 1287–1291.

74 List, B., Lerner, R.A., and Barbas, C.F. (2000). *J. Am. Chem. Soc.* 122: 2395–2396.

75 Ahrendt, K.A., Borths, C.J., and MacMillan, D.W.C. (2000). *J. Am. Chem. Soc.* 122: 4243–4244.

76 Yang, X., Wu, T., Phipps, R.J., and Toste, F.D. (2015). *Chem. Rev.* 115: 826–870.

77 Lectard, S., Hamashima, Y., and Sodeoka, M. (2010). *Adv. Synth. Catal.* 352: 2708–2732.

78 Lin, J.-H. and Xiao, J.-C. (2014). *Tetrahedron Lett.* 55: 6147–6155.

79 Zhu, Y., Han, J., Wang, J. et al. (2018). *Chem. Rev.* 118: 3887–3964.

80 Valero, G., Companyó, X., and Rios, R. (2011). *Chem. Eur. J.* 17: 2018–2037.

81 Enders, D. and Hüttl, M.R.M. (2005). *Synlett* 2005: 0991–0993.

82 Steiner, D.D., Mase, N., and Barbas, C.F. III, (2005). *Angew. Chem. Int. Ed.* 44: 3706–3710.

83 Beeson, T.D. and MacMillan, D.W.C. (2005). *J. Am. Chem. Soc.* 127: 8826–8828.

84 Marigo, M., Fielenbach, D., Braunton, A. et al. (2005). *Angew. Chem. Int. Ed.* 44: 3703–3706.

85 Dinér, P., Kjærsgaard, A., Lie, M.A., and Jørgensen, K.A. (2008). *Chem. Eur. J.* 14: 122–127.

86 Schmid, M.B., Zeitler, K., and Gschwind, R.M. (2011). *J. Am. Chem. Soc.* 133: 7065–7074.

87 Schmid, M.B., Zeitler, K., and Gschwind, R.M. (2011). *Chem. Sci.* 2: 1793–1803.

88 Lam, Y.-h. and Houk, K.N. (2014). *J. Am. Chem. Soc.* 136: 9556–9559.

89 Kwiatkowski, P., Beeson, T.D., Conrad, J.C., and MacMillan, D.W.C. (2011). *J. Am. Chem. Soc.* 133: 1738–1741.

90 Arimitsu, S., Yonamine, T., and Higashi, M. (2017). *ACS Catal.* 7: 4736–4740.

91 You, Y.E., Zhang, L., and Luo, S. (2017). *Chem. Sci.* 8: 621–626.

92 Pupo, G., Ibba, F., Ascough, D.M.H. et al. (2018). *Science* 360: 638–642.

93 Pupo, G., Vicini, A.C., Ascough, D.M.H. et al. (2019). *J. Am. Chem. Soc.* 141: 2878–2883.

94 Zhou, B., Haj, M.K., Jacobsen, E.N. et al. (2018). *J. Am. Chem. Soc.* 140: 15206–15218.

95 Banik, S.M., Medley, J.W., and Jacobsen, E.N. (2016). *Science* 353: 51–54.

96 Pluta, R., Krach, P.E., Cavallo, L. et al. (2018). *ACS Catal.* 8: 2582–2588.

97 Zhao, Y.-M., Cheung, M.S., Lin, Z., and Sun, J. (2012). *Angew. Chem. Int. Ed.* 51: 10359–10363.

98 Wang, Y., Qiao, Y., Wei, D., and Tang, M. (2017). *Org. Chem. Front.* 4: 1987–1998.

99 Li, F., Wu, Z., and Wang, J. (2015). *Angew. Chem. Int. Ed.* 54: 656–659.

100 O'Hagan, D., Schaffrath, C., Cobb, S.L. et al. (2002). *Nature* 416: 279–279.

101 Dong, C., Huang, F., Deng, H. et al. (2004). *Nature* 427: 561–565.

102 Senn, H.M., O'Hagan, D., and Thiel, W. (2005). *J. Am. Chem. Soc.* 127: 13643–13655.

103 Sperger, T., Sanhueza, I.A., and Schoenebeck, F. (2016). *Acc. Chem. Res.* 49: 1311–1319.

104 Anstaett, P. and Schoenebeck, F. (2011). *Chem. Eur. J.* 17: 12340–12346.

105 Nielsen, M.C., Bonney, K.J., and Schoenebeck, F. (2014). *Angew. Chem. Int. Ed.* 53: 5903–5906.

106 Pu, M., Sanhueza, I.A., Senol, E., and Schoenebeck, F. (2018). *Angew. Chem. Int. Ed.* 57: 15081–15085.

107 Keaveney, S.T. and Schoenebeck, F. (2018). *Angew. Chem. Int. Ed.* 57: 4073–4077.

108 Yin, G., Kalvet, I., Englert, U., and Schoenebeck, F. (2015). *J. Am. Chem. Soc.* 137: 4164–4172.

109 Yin, G., Kalvet, I., and Schoenebeck, F. (2015). *Angew. Chem. Int. Ed.* 54: 6809–6813.

110 Dürr, A.B., Yin, G., Kalvet, I. et al. (2016). *Chem. Sci.* 7: 1076–1081.

111 Kalvet, I., Guo, Q., Tizzard, G.J., and Schoenebeck, F. (2017). *ACS Catal.* 7: 2126–2132.

112 Aufiero, M., Sperger, T., Tsang, A.S.-K., and Schoenebeck, F. (2015). *Angew. Chem. Int. Ed.* 54: 10322–10326.

113 Dürr, A.B., Fisher, H.C., Kalvet, I. et al. (2017). *Angew. Chem. Int. Ed.* 56: 13431–13435.

114 Sperger, T., Guven, S., and Schoenebeck, F. (2018). *Angew. Chem. Int. Ed.* 57: 16903–16906.

11

Current Trends in the Design of Fluorine-Containing Agrochemicals

Peter Jeschke

Bayer AG, Research & Development, Crop Science, Pest Control, Building 6550, Alfred-Nobel-Str. 50, Monheim am Rhein D-40789, Germany

11.1 Introduction

Today, agriculture is confronted with enormous challenges, from production of suffi-cient high-quality food, feed, fuel, and fiber, to water use and environmental impacts and issues combined with a continually growing world population [1].

Modern agricultural chemistry has to support farmers by providing innovative agrochemicals with more favorable environmental and toxicological profiles, used in changing and applied agriculture [2]. That requires continuously innovative solu-tions for existing and future challenges, such as climate change, development of resistance to target pests, shifting pest spectra, increased regulatory hurdles, renew-able raw materials, or demands resulting from food chain partnerships.

Agrochemical companies with a strong research and development focus will have the opportunity to shape the future of agriculture by delivering cost-efficient and innovative integrated solutions. In this context, the introduction of fluorine atoms into an active ingredient (a.i.) is still an important tool to modulate the properties of novel crop protection products.

11.2 Role of Fluorine in the Design of Modern Agrochemicals

For nearly 40 years halogen-containing a.i.s have had a significant role in the devel-opment of innovative agrochemicals. In the past decade, there has been a remark-able rise in the number of commercial products containing fluorine atoms and/or fluorine-containing substituents (Figure 11.1).

At present, these include, for example, trifluoromethyl (F_3C) and difluoromethyl (F_2HC), or trifluoromethoxy (F_3C-O) and difluoromethoxy (F_2HC-O) groups at aryl or heterocyclic aryl moieties and fragments such as [F_2HC-H_2C]-2,2-difluoroethyl, [F_2HC-SO_2]-difluoromethyl-sulfonyl, [$(F_3C)_2FC$]-heptafluoro-iso-propyl, or [$(F_3C)_2MeOC$]-1-methoxy-heptafluoro-iso-propyl [3].

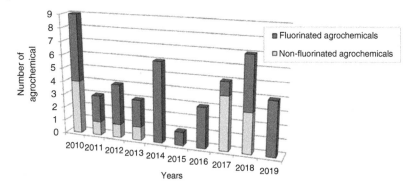

Figure 11.1 Launch of commercial non-fluorinated and fluorinated agrochemicals in the timeframe 2010–2019.

The continuing importance of fluorine atoms and/or fluorine-containing substituents in the design of agrochemicals has often been described and can be attributed to the well-known steric (resemblance to hydrogen, C—F bond length: 1.35 Å, van der Waals radius: 1.47 Å) and electronic (extreme electronegativity) properties, and polarity of the carbon–halogen bond (large dipole moment C–F: $\mu = 1.51$ D) [4]. Owing to the large dipole moment of the C—F bond, fluorine substitution may result in substantial conformation changes through various stereoelectronic interactions [5]. The effect of polyfluorinated isopropyl moieties on molecular conformation observed in the solid state was rationalized by high-level quantum mechanical calculations. The energy profiles along the variation of torsional angle for the fluorinated motifs [(F$_3$C)$_2$FC]-heptafluoro-iso-propyl, [(F$_3$C)$_2$HOC]-1-hydroxy-heptafluoro-iso-propyl, or [(F$_3$C)$_2$MeOC]-1-methoxy-heptafluoro-iso-propyl show a preference for the coplanar arrangement of the phenyl ring and the C—F and C—OH bond with a rotational barrier of 7.74 and 6.62 kcal/mol, respectively [6].

In addition, improving of the metabolic, oxidative, and thermal stability (C—F bond energy: 485 kJ/mol, metabolic stabilization), the pKa value effects (H-bonding, protein binding affinity), and the influence of fluorine on physicochemical properties such as molecular lipophilicity (key parameter for absorption and transport, important for bioavailability) can be underlined. In this context, the impact on metabolic stability of polyfluorinated isopropyl substituents on aromatic rings was demonstrated [7].

Lipophilicity is an important physicochemical parameter quantified by the logP-value; it also determines ligand–target binding interactions, solubility, and ADME (absorption, distribution, metabolism, and elimination) properties. This is exemplified by the (F$_3$C-X) groups (X = S, O), which contribute to the pest's overall pharmacological activity by enhancing the insect central nervous system (CNS) penetration. Moreover, the influence of fluorine on modulation of biological properties of an a.i. has been described in detail by shifts of biological activity arising from the introduction of fluorine atoms and/or fluorine-containing substituents into biological a.i.s such as the commercialized γ-aminobutyric acid (GABA)-gated chloride channel blockers (phenylpyrazoles), sodium channel modulators (pyrethroids), or inhibitors of chitin biosynthesis, type 0 (benzoyl ureas) [4c].

With the latest generation of commercialized fluorine-containing agrochemicals, two new modes of action (MoAs) could be identified for fungicides/insecticides, three novel classes for insecticides/acaricides (new Insecticide Resistance Action Committee [IRAC] sub-groups) for known MoAs, and two unknown MoAs for nematicides.

A survey of the new a.i.s (the total number was 45 commercial products) used as modern crop protection pesticides, provisionally approved by the International Organization of Standardization (ISO) during the past 10 years (2010–2020, http://www.alanwood.net/pesticides/) has shown that beside the only non-halogenated nematicide imicyafos (2010, Agro Kanesho), insecticide lepimectin (2010, Sankyo) and afidopyropen [8] (2018, Meiji Seika/BASF), herbicide tolpyralate (2017, Ishihara Sangyo Kaisha), the fungicides mandestrobin (2017, Syngenta), isofetamid [9] (2018, Ishihara Sangyo Kaisha), and picarbutrazox (2018, Syngenta), all other launched products (38 products: ~84%) are halogen substituted (10 herbicides, 15 fungicides, 13 insecticides/acaricides, and 3 nematicides). From these, 31 commercial products are fluorinated, and around 45% of them contain a trifluoromethyl group, approximately 19% a difluoromethyl, approximately 3% a trifluoromethoxy group, and 3% a difluoromethoxy group as a substituent on the phenyl or heterocyclic moieties. This enormous increase in fluorine-substituted products (Br ≪ Cl ≪ F) demonstrates the importance of halogenation, in particular fluorination, of modern agrochemicals. In this context, outstanding progress had been made in technical preparation on an industrial scale for bulk quantities of fluorine-substituted building blocks used for these latest pesticides such as 3-(trifluoromethyl)- and 3-(difluoromethyl)-1-methyl-1H-pyrazole-4-carboxylic acids for fungicides, 3-bromo-1-(3-chloro-2-pyridinyl)-1H-pyrazole-5-carboxylic acid, 2,2-difluoro-ethylamine, or trifluoromethyl-substituted pyridine moieties for insecticides or nematicides as well as substituted 4-heptafluoro-iso-propylanilines such as 1-methoxy-hexafluoro-iso-propyl for acaricides.

What is the rationale behind the strong increase in using fluorine atoms and/or fluorinated motifs in the design of modern pesticides?

A critical analysis has been carried out with the latest generation of fluorine-containing pesticides launched on the global crop protection market in the past 10 years and a few representative pesticides have been selected to illustrate in detail the contribution that the fluorine atom or special fluorine-containing substituents have made to contemporary agricultural chemistry.

11.3 Fluorinated Modern Agrochemicals

In the present chapter, a few examples are selected to illustrate how fluorine substitution is used successfully in contemporary agrochemistry. The strong influence of fluorine atoms and/or fluorinated motifs can lead to biological superiority of fluorinated a.i.s over their non-fluorinated analogs. Various selected commercial products testify to the successful utilization of fluorine in the design of agrochemicals, in particular herbicides, fungicides, insecticides, acaricides, and nematicides.

Generally, the corresponding biochemical targets or MoAs for most commercialized products have been described. These are

- the acetohydroxyacid synthase/acetolactate synthase (AHAS/ALS), the protoporphyrinogen oxidase (PPO), the cellulose biosynthesis (CBS), the very long-chain fatty acid elongase (VLCFAE), auxin, and the 4-hydroxyphenylpyruvate dioxygenase (4-HPPD) for herbicides;
- the succinate dehydrogenase (SDH, complex II), the respiratory chain (the so-called Qo-site of complex III), the polyketide synthase (PKS), and the oxysterol-binding protein (OSBP) for fungicides;
- the nicotinic acetylcholine receptor (nAChR), the ryanodine receptor (RyR), and the γ-aminobutyric acid receptor (GABA-R) for insecticides;
- the mitochondrial complex II electron transport for acaricides;
- the nematicidal SDH (complex II) for nematicides.

11.3.1 Herbicides Containing Fluorine

11.3.1.1 Acetohydroxyacid Synthase/Acetolactate Synthase Inhibitors
The first step in the biosynthetic pathway of the branched-chain amino acids is catalyzed by the enzyme acetohydroxyacid synthase (AHAS), which is also referred to as ALS [10].

The sulfonanilide pyrimisulfan (Figure 11.2) [11], containing a difluoromethylsulfonyl group, has broad spectrum, preemergence herbicidal activity against the grass *Exserohilum oryzicola*, sedge, and broadleaf weeds, with selectivity on transplanted and water-seeded rice. Its low application rate (50–75 g a.i./ha) has provided outstanding activity on major weeds of Japanese paddy fields.

11.3.1.2 Protoporphyrinogen Oxidase Inhibitors
The herbicidal enzyme protoporphyrinogen IX oxidase (Protox), which catalyzes the oxidation of protoporphyrinogen IX to protoporphyrin IX by molecular oxygen, is one of the most important targets for herbicide discovery [12]. In plants, protoporphyrin IX is the substrate for the biosynthesis of chlorophyll, which is a key pigment for photosynthesis. Inhibition of the Protox enzyme results in an accumulation of

Pyrimisulfan

Figure 11.2 Structure of the AHS/ALS inhibitor pyrimisulfan.

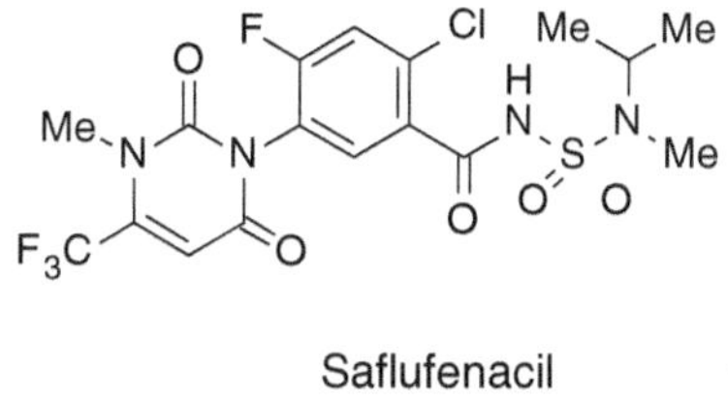

Figure 11.3 Structure of the PPO inhibitor saflufenacil.

Saflufenacil

the enzyme product protoporphyrin IX, but not the substrate, via a complex process that has not been entirely elucidated. The result of this peroxidation process is the loss of membrane integrity and leakage, pigment breakdown, and necrosis of the leaf, which leads to the death of the plant.

Saflufenacil (Figure 11.3) [13] was commercialized in 2009 as Protox herbicide for preplant burndown and selective preemergence dicotolydenous weed control in multiple crops, including corn, wheat, rice, and soybean.

The fast, cytochrome P_{450}-based *N*-dealkylation of the sulfamoyl carboxamide side chain contributes to the unusual broad crop selectivity of saflufenacil, especially in grass crops such as corn, wheat, and rice. In corn, it was found that root-absorbed saflufenacil did not translocate well to the plant shoot and that both this factor and metabolism contributed to crop selectivity [14].

11.3.1.3 Cellulose Biosynthesis Inhibitors

Cellulose biosynthesis in higher plants is essential to cell growth and division as well as tissue formation and differentiation. Any inhibitors targeting cellulose biosynthesis inhibitor (CBI) lead to severe effects on plant growth and development and they are of considerable interest as herbicides [15].

Using the so-called "toolbox" of synthetic methods a broad range of indanyl aminotriazines were synthesized and tested for their herbicidal activity as CBI in plants. By optimization of the alkylazine class it was found that a free amino group and a haloalkyl group provided the most appropriate substituents [16]. The presence of a halogen atom (Br, Cl < **F**) in combination with a 2,4-diamino-1,3,5-triazine bearing an alkyl or cycloalkyl group in the 6-position is essential for good herbicidal activity.

The (1*R*,2*S*)-configuration of the 2,3-dihydro-2,6-dimethyl-1*H*-inden-1-amine (indanyl) ring system is decisive for herbicidal activity and for a shift in the MoA from inhibition of photosystem II to inhibition of CBI. This results in the introduction of an additional chirality center by fluorine in the 1-position of the ethyl side chain. After synthesis of all eight diastereoisomers it was found that combination of the (1*R*,2*S*)-indanyl fragment with a 6-(*R*)-1-fluoro-ethyl-substituted 2,4-diamino-1,3,5-triazine resulted in the most potent compound, which is the major diastereoisomer of indaziflam (Figure 11.4).

Fluorescence microscopy confirmed the inhibition of CBI following indaziflam treatment, as well as aberrant root and cell morphology. Indaziflam increases the density of large cellulose synthase particles at the plasma membrane and reduces particle velocity, resulting in the inhibition of polymerization [17]. As the strongest innovative alkylazine herbicide, indaziflam shows a broad control of both grasses and broadleaf weeds (application rates of ~50–75 g/ha) that are resistant to other

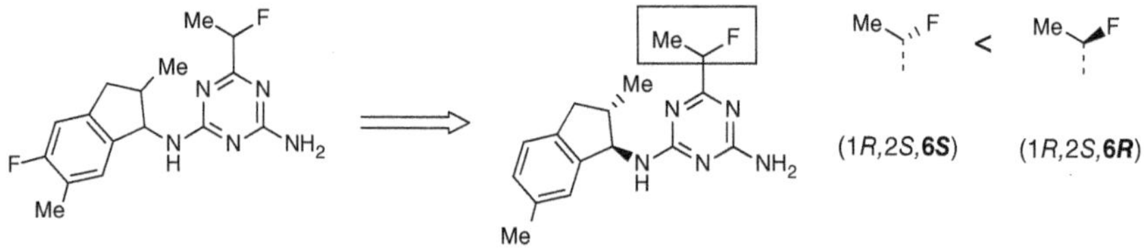

Figure 11.4 Optimization of the 5-fluoro-6-methylindanylaminotriazine resulted in the diastereomeric indaziflam.

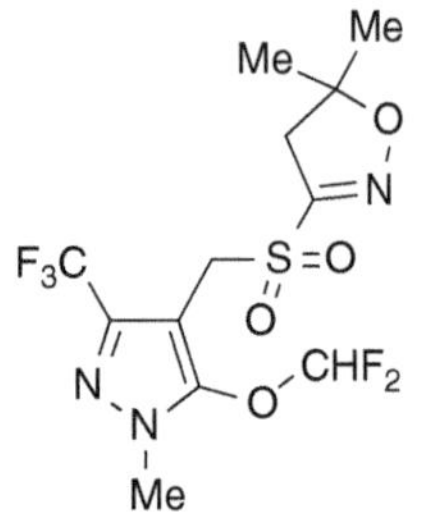

Pyroxasulfone

Figure 11.5 Structure of the VLCFAE inhibitor pyroxasulfone.

herbicidal MoAs. Its increased activity on monocotyledonous weeds provides a new alternative strategy for the long-term control of multiple invasive winter annual grasses [17]. So far, there is no evidence for any cross-resistance.

11.3.1.4 Very Long-Chain Fatty Acid Synthesis Inhibitors

Very long-chain fatty acid is inhibited by pyroxasulfone [18], which leads, for example, to an increase of fatty acid precursors in cultured rice cells.

Pyroxasulfone (Figure 11.5) contains a 5-(difluoromethoxy)-1-methyl-3-(trifluoromethyl)-1*H*-pyrazole moiety [19], enhancing the physicochemical parameters such as water solubility (3.48 mg/l at 20 °C). It can be applied in the crops corn, soybean, cotton, and wheat, and has annual grass activity at lower application rates than the more water-soluble (*S*)-metolachlor (480 mg/l at 25 °C) but also provides good control of several annual broadleaf weeds. In comparison to the soil degradation of (*S*)-metolachlor (DT$_{50}$-value = 39–63 days) the fluorinated pyroxasulfone has a slightly higher dissipation half-life DT$_{50}$-value in soil, which varied from 47 to 134 days [20].

11.3.1.5 Auxin Herbicides

Auxin mimic herbicides (or synthetic auxins) were the first selective a.i.s developed for weed control [21]. Synthetic auxin herbicides are generally known for their ability to selectively control annual and perennial broadleaf weeds in grass crops.

Discovery of aminopyralid in 1998 led to the identification of the 6-arylpicolinate (6-AP) herbicide class and its very active 6-AP member with the substitution pattern R = 2′-F, 4′-Cl (Figure 11.6).

The addition of a methoxy group at the 3′-position of the phenyl ring provided a 6-AP that exhibited a high level of herbicidal control of key dicotolydenous weeds.

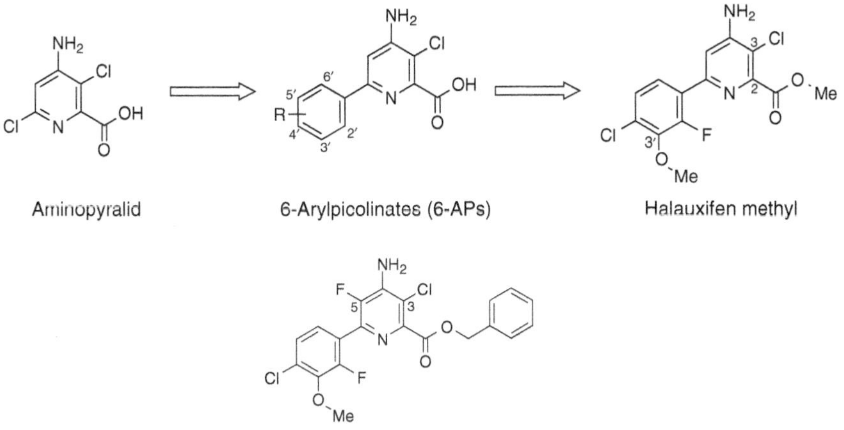

Aminopyralid   6-Arylpicolinates (6-APs)   Halauxifen methyl

Florpyrauxifen benzyl

Figure 11.6 Starting with aminopyralid the 6-arylpicolates were identified, which resulted in halauxifen methyl and florpyrauxifen benzyl.

A rapid enzymatic degradation of the phenyl 3′-OMe and pyridine 2-COOMe-group in aerobic soils leads to a 6-AP with phenyl 3′-OH and pyridine 2-COOH groups.

Enzymatic cleavage of the pyridine 2-COOMe group *in planta* of the proherbicide [22] halauxifen methyl [23] results in a pyridine type auxin having physiological and phenotypic effects similar to those induced by the natural plant hormone indole-3-acetic acid (IAA) [24].

The pyridine 2-COOH group leads to a significantly lower soil half-life range of 10–30 days and is vital for phloem-trapping as well as for binding interactions that result in a broad-spectrum herbicidal effect. A chlorine atom in the 3-position of the pyridine moiety of halauxifen methyl is important for its potent herbicidal activity; other halogen atoms such as fluorine or bromine in the 3-position are less active and the des-halogen analog is substantially less active (3-H < 3-F ~ 3-Br < **3-Cl**). Further evaluation in global field trials identified utility for dicotolydenous weed control in the wheat and barley markets.

The 5-fluorine analog of florpyrauxifen benzyl (benzyl instead of methyl, see halauxifen methyl, Figure 11.6) [25] has potent and broad-spectrum herbicidal activity with excellent selectivity in rice. The replacement of the 5-fluorine atom by hydrogen or other halogen atoms at the pyridine ring leads to a decrease of herbicidal activity (5-Br < 5-Cl < 5-H < **5-F**) against dicotolydenous weeds at an application rate of 17.5 g/ha (Figure 11.6).

11.3.1.6 Hydroxyphenylpyruvate Dioxygenase Inhibitors

The enzyme hydroxyphenylpyruvate dioxygenase (HPPD) catalyzes an early step in the tyrosine degradation pathway that is widely distributed in nature. The treatment of weeds with HPPD inhibitors causes a significant accumulation of tyrosine [26].

The HPPD inhibitor bicyclopyrone (Figure 11.7) [27] belongs to the general class of nicotinoyl cyclohexane dione herbicides. It has been launched as a selective

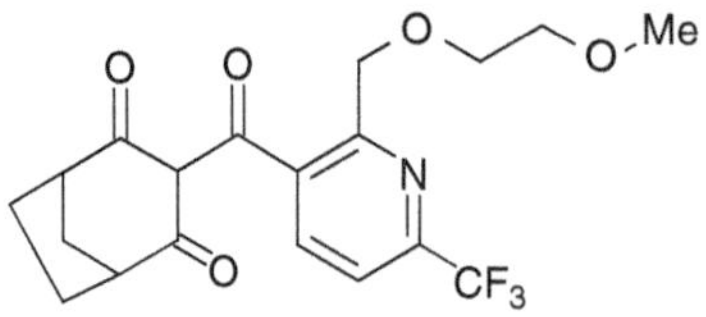

Figure 11.7 Structure of the HPPD inhibitor bicyclopyrone.

Bicyclopyrone

herbicide in corn and sugarcane for low rate control of broad leaves and grass weeds either as part of a mixture or as a single a.i.

The improved activity of bicyclopyrone compared to other herbicides such as mesotrione arises in part from the bicyclo[3.2.1]octane-2,4-dione moiety, which fits very well in the binding site of the HPPD enzyme [28] and protects the C—H bonds adjacent to both ketones from too rapid a metabolism by the cytochrome P_{450} enzymes in grass weeds.

11.3.1.7 Selected Fluorine-Containing Herbicide Development Candidates

A new PPO inhibitor, the benzoxazinone triazinone herbicide trifludimoxazin (provisionally approved by ISO) [29], is under development (see Section 3.1.2). It contains a novel *gem*-difluorinated oxazinone ring system (Figure 11.8) and is useful for controlling PPO-herbicide-resistant problem weeds such as pigweed (*Amaranthus palmeri*) and has also been shown to control burndown of grass weed such as ryegrasses (*Lolium perenne*) by foliar application prior to crop planting (e.g. maize, soybeans, small-grain cereals).

The new herbicidal class aryl pyrrolidinone anilides target dihydroorotate dehydrogenase (DHODH), a key enzyme in *de novo* pyrimidine biosynthesis, such as the development product from FMC tetflupyrolimet (provisionally approved by ISO) with (3*S*,4*S*)-configuration (Figure 11.8) [30]. The herbicide is selective for grass control in rice. The target site was discovered using a combination of forward genetic screens and metabolomics approaches and confirmed by determining intrinsic affinities of specific analogs using biochemical methods [31]. The presence of the *ortho*-fluorine atom and *meta*-trifluoromethyl group at both phenyl rings and the *N*-methyl group of the lactam moiety are required for good herbicidal activity.

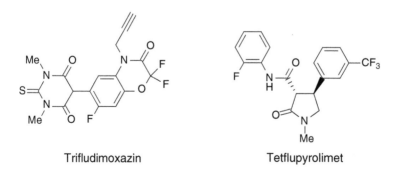

Trifludimoxazin   Tetflupyrolimet

Figure 11.8 Herbicide development candidates trifludimoxazin and tetflupyrolimet.

11.3.2 Fungicides Containing Fluorine

11.3.2.1 Fungicidal Succinate Dehydrogenase Inhibitors

Today, SDH inhibitors are a nearly four-decade-old innovation, which began in the 1970s with small narrow spectrum seed treatment agents and which started over in the 1990s with their derivatization to broad-spectrum fungicides and resulted in boscalid.

After establishment of the first fluorinated pyrazole-4-carboxamides such as bixafen and penflufen in the market, the latest six succinate dehydrogenase inhibitors (SDHIs) fungicides of this structural type possess a high level of efficacy against numerous important cereal crop pathogens such as fluxapyroxad [32], isopyrazam [33], sedaxane [34], penthiopyrad [35], benzovindiflupyr [36], or inpyrfluxam [37] have been introduced based on an economical industrial synthesis of 3-difluoromethyl and 3-trifluoromethyl-pyrazol-4-yl-carboxylic acids as intermediates (Table 11.1). They contain mostly chiral hydrophobic moieties at the "amide linker,", which are usually phenyl and thienyl groups (penthiopyrad) that are substituted in the *ortho*-position by 3,4,5-trifluorophenyl or by a larger branched or bicyclic residue. It seems that the angle defined by the amide linker and the *ortho*-substituted "hydrophobic moiety" in the phenyl or thienyl rings is important for the fungicidal spectrum and potency of SDH-inhibiting fungicides [38].

Much progress has been made in producing bulk quantities of fluorinated intermediates such as difluoroacetylfluoride and the respective chloride, difluoroacetic acid esters, and amides thereof, as well as in overcoming the regioselectivity problem of the 3-difluoromethyl-substituted 1-methyl-1*H*-pyrazole-4-carboxylic acid synthesis by ring closure reaction [39].

Except the last launched (*R*)-enantiomer inpyrfluxam, none of the other commercialized SDH-inhibiting pyrazole-4-carboxamide fungicides are produced in an enantiomerically enriched form.

Fluxapyroxad is a broad-spectrum carboxamide fungicide that can be applied in various mixtures on a wide range of crops such as cereals, fruit, vegetables, and seed treatment with a broad pathogen spectrum.

Isopyrazam, a mixture of *syn-* and *anti*-isomers, is a carboxamide developed to control black sigatoka (*Mycosphaerella fijiensis*), a leaf-spot disease in banana production.

Sedaxane has been developed for seed treatment and controls seed- and soil-borne fungi such as *Monographella nivalis*, *Typhula incarnate*, and *Rhizoctonia* spp. in a broad range of crops. Its high intrinsic activity based on the physicochemical properties (logP_{OW}-value = 3.3 at 25 °C; water solubility = 14 mg/l) is influenced by the 3-(difluoromethyl)-1-methyl-1*H*-pyrazol-4-yl-carboxamide moiety.

Penthiopyrad, which has mainly contact and only locally systemic efficacy, is active against foliar and key soil-borne pathogens such as white mold (*Sclerotinia sclerotiorum*), brown patch (*Rhizoctonia solani*), and southern blight (*Sclerotium rolfsii*), and is also active against other diseases.

In comparison with the foliar applicable isopyrazam, the benzonorbornene carboxamide benzovindiflupyr contains a dichloromethylene [=CCl$_2$]-fragment in the

Table 11.1 From boscalid and bixafen to fluorinated 3-(difluoromethyl or trifluoromethyl)-1H-pyrazole-4-yl-carboxylic acid structures of SDHI fungicides.

Common name (trade name)	Year of launch	R^1	Linker	Hydrophobic rest (R^2, R^3)
Fluxapyroxad	2010	CHF_2		
Isopyrazam[a)]	2010	CHF_2		
Sedaxane	2011	CHF_2		
Penthiopyrad	2012	CF_3		
Benzovindiflupyr	2014	CHF_2		
Inpyrfluxam	2019	CHF_2		

a) Isopyrazam in a mixture of 97% *syn*-isomer and 3% *anti*-isomer.

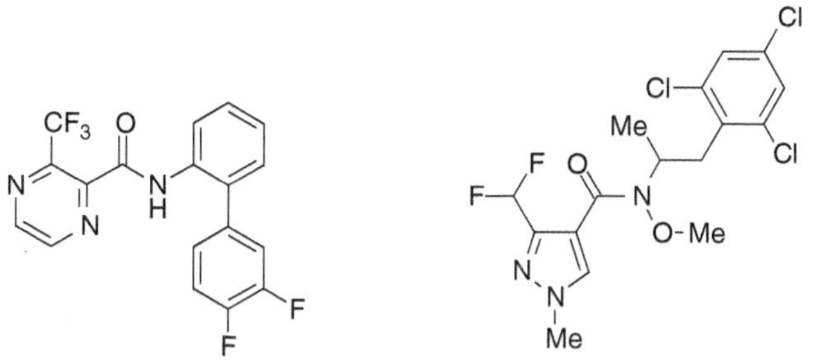

Pyraziflumid   Pydiflumetofen

Figure 11.9 Structures of the SDHI fungicides pyraziflumid and pydiflumetofen.

9-position. Benzovindiflupyr is a broad-spectrum fungicide, which can be applied as a mixture with azoxystrobin that controls rust pathogen diseases, including the speckled leaf blotch fungus (*Zymoseptoria tritici*), and has excellent activity against soybean rust (*Phakopsora pachyrhizi*).

As (*R*)-enantiomer inpyrfluxam is a broad-spectrum fungicide, especially highly effective against *Basidiomycetes*, such as rust diseases and *Rhizoctonia*. It demonstrates favorable biological properties against Asian soybean rust.

Pyraziflumid (Figure 11.9) [40] is another carboxamide with a biphenylaniline moiety making it closely related to the other SDHI fungicides such as boscalid, bixafen, and fluxapyroxad.

Therefore, its 2-(3,4-difluorophenyl)aniline derivative can be easily prepared using the Suzuki–Miyaura methodology leading to boscalid's biphenylamine or the Negishi coupling delivering the amine moiety of fluxapyroxad. It was found that pyraziflumid exhibited strong activity against *S. sclerotiorum in vitro* and had a good protective and curative effect on leaves of rapeseed. Pyraziflumid has a potential application for managing *Sclerotinia* stem rot (SSR) caused by *S. sclerotiorum* [41]. With its unique 2-trifluoromethylpyrazine-3-carboxylic acid, pyraziflumid is, after boscalid, the second SDHI fungicide that has a six-membered heterocyclic moiety as the carboxylic acid portion.

Pydiflumetofen (Figure 11.9) [42], which contains the racemic [-MeHC-CH$_2$-] fragment, an additional *N*-methoxy group, and 3-difluoromethyl-1-methyl-1*H*-pyrazole as well as 2,4,6-trifluorophenyl moieties, allows excellent control of powdery mildew and leaf spot disease. In Argentina, a mixture of pydiflumetofen and difenoconazole was introduced for the control of diseases such as powdery mildew on peppers and gray leaf spot on tomatoes. This is very effective in case of diseases that are hard to control, such as *Fusarium* and other fungi that cause serious damage to the crops.

11.3.2.2 Complex III Inhibitors

Tebufloquin (Figure 11.10) [43] belongs to the chemical class of quinolines, and is classified as unknown according to the FRAC Code List although it is reported to be an inhibitor of the respiratory chain on the level of complex III with an unknown binding site [44].

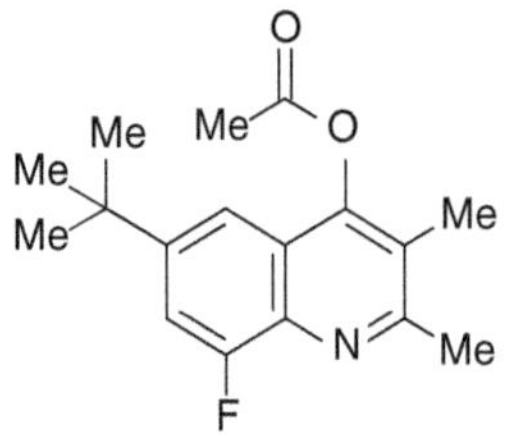

Figure 11.10 Structure of quinoline fungicide tebufloquin.

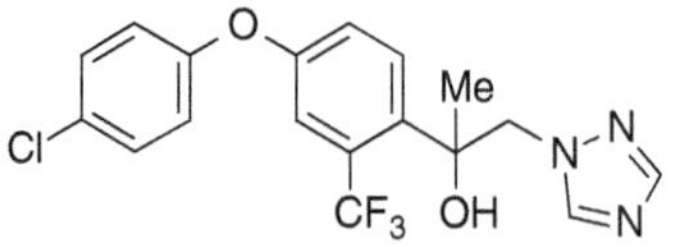

Tebufloquin

Figure 11.11 Structure of the sterolbiosynthesis inhibitor mefentrifluconazole.

Mefentrifluconazole

Therefore, it has no cross-resistance with other existing fungicides. Tebufloquin is a rather unipolar compound that demonstrates surprisingly high water solubility (0.0202 g/l) with respect to its lipophilicity ($logP_{ow}$ at 23 °C = 4.02).

Tebufloquin has been developed mainly for the control of rice blast (*Magnaporthe grisea*). The product can be used to control this disease under both protective and curative conditions with application rates of 600–800 g a.i./ha (in case of dust formulation).

11.3.2.3 Sterolbiosynthesis (Sterol-C$_{14}$-Demethylase) Inhibitors

The main MoA of 1,2,4-triazole fungicides is the inhibition of the cytochrome-P_{450}-dependent C$_{14}$-demethylation of the intermediate C$_{24}$-methylene-dihydrolanosterol in the sterol biosynthesis pathway of fungi. Today, the C$_{14}$-demethylase in sterol biosynthesis is the common target for more than 30 fungicides used in modern crop protection [45].

The racemic mefentrifluconazole (Figure 11.11) is a novel triazole fungicide inhibiting the fungal cytochrome P_{450} sterol-C$_{14}$-demethylase (CYP51A1) with high intrinsic activity and good mobility, which shows a rapid uptake and translocation via xylem enabling control of the fungal stages that have already colonized in deeper tissue layers [46].

Mefentrifluconazole provides excellent efficacy against a wide spectrum of foliar diseases including key diseases such as septoria leaf blotch (*Z. tritici*), brown rust wheat (*Puccinia triticina*), apple scab (*Venturia inaequalis*), powdery mildew (*Uncinula necator*) in grapes, and various diseases in corn, soybean, rice, turf, and vegetable.

11.3.2.4 Polyketide Synthase Inhibitors

The initial enzyme in the 1,8-dihydronaphthalene (DHN) biosynthetic pathway – a specific PKS – is an interesting target for melanin biosynthesis inhibitors (MBIs). Although PKSs with diverse functions are identified in many different organisms,

2,2,2-Trifluoroethyl[[(benzoyl)
amino]methyl]alkyl]carbamates

Tolprocarb

Figure 11.12 From [2,2,2-trifluoroethyl[[(benzoyl)amino]methyl]alkyl]carbamates to tolprocarb.

Table 11.2 Activities of tolprocarb and its derivatives during structure optimization.

R¹	R²	PKSI-A IC_{50} (μM)	MBI-A minimum inhibitory concentration (MIC) (ppm)
iso-Bu	Me	4.86×10^{-1}	100
iso-Pr	SO₂Me	4.16×10^{-1}	100
H	Me	2.01×10^{-1}	5
iso-Pr	CF₃	9.82×10^{-2}	2.5
n-Pr	Me	6.93×10^{-2}	1
iso-Pr	Cl	3.23×10^{-2}	1
iso-Pr	**Me (Tolprocarb)**	$\mathbf{3.05 \times 10^{-2}}$	**0.2**

Tolprocarb is the commercial product. All other derivatives are derivates obtained during its optimization.
Source: Data from Schindler et al. [47].

very few inhibitors of PKSs have been identified, and ever fewer inhibitors of the PKS involved in melanin biosynthesis [47].

A series of racemic [2,2,2-trifluoroethyl [[(benzoyl)amino]methyl]alkyl] carbamates were synthesized and evaluated for fungus *Pyricularia oryzae* PKS polyketid synthase inhibitory activity (PKSI-A) and melanin biosynthesis inhibitory activity (MBI-A) to verify their target sites and investigate the structure–activity relationship (SAR) (Figure 11.12) [48].

A significant correlation between PKSI-A and MBI-A was found, confirming that the inhibition of PKS activity leads directly to the inhibition of melanin biosynthesis in *P. oryzae*. This was exemplified by the ranking of activities regarding the substituents R¹ and R², respectively. The hydrogen at R¹ reduced both PKSI-A and MBI-A (R¹ = *iso*-Bu ≪ H < *n*-Pr < **iso-Pr**) (Table 11.2).

The enantiomers are clearly different – the (*S*)-enantiomer exhibited one of the highest levels of PKSI-A, whereas the (*R*)-enantiomer was inactive. Hydrophobic substituents at R² exhibited higher PKSI-A than the hydrophilic methyl sulfonyl group (R² = SO₂Me ≪ CF₃ < Cl < **Me**) (Table 11.2).

As (*S*)-enantiomer tolprocarb inhibits DHN melanin biosynthesis by inhibiting the PKS activity in the rice blast (*M. grisea*) fungus *P. oryzae*, tolprocarb is effective for rice blast control not only by foliar application but also by absorption via the roots. The MoA seems to be distinct from that of conventional melanin biosynthesis inhibitors (cMBIs).

11.3.2.5 Oxysterol-Binding Protein Inhibitors

Oxathiapiprolin [49] is the first member of a new class of piperidinyl thiazole isoxazoline fungicides with exceptional activity against diseases caused by oomycetes pathogens, resulting in excellent preventative, curative, and residual efficacy against key diseases of grapes, potatoes, and vegetables [50]. The discovery of the racemic oxathiapiprolin began with the high-throughput screening (HTS) of a purchase bis-amide compound library built around a piperidine-thiazole-carbonyl core [51]. The lead structure with >90% control of preventive and curative tomato late blight (*Phytophthora infestans*) and curative grape downy mildew (*Plasmopara viticola*) was observed at an application rate of 10 ppm (Figure 11.13).

After the first major optimization by incorporation of the 5-methyl-3-(trifluoromethyl)-1*H*-pyrazole moiety (pyrazole substitution pattern: 3-Me/5-Me ~ 3-Br/5-Br ~ 3-Cl/5-Cl < 3-CF$_3$/5-CF$_3$ ~ 3-CF$_3$/5-Cl ~ 3-CF$_3$/5-Et ~ **3-CF$_3$/5-Me**), a 20-fold increase in activity with >90% control of preventive and curative *P. infestans* and curative *P. viticola* was observed at an application rate of 10 ppm. As a mixture of 5-methyl-3-(trifluoromethyl)-1*H*-pyrazole, which has been selected as important for its xylem mobility, and a 2,6-difluoro-phenyl moiety, the oxathiapiprolin demonstrates a 10-fold increase in activity with >90% control of preventive and curative *P. infestans* and curative *P. viticola* at an application rate of 0.02 ppm. Furthermore, it is highly active against plant diseases caused by *Oomycete* pathogens.

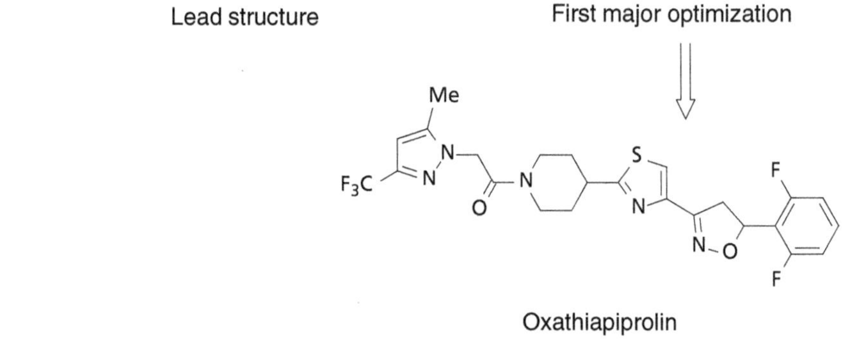

Lead structure   First major optimization

Oxathiapiprolin

Figure 11.13 Based on HTS and the right optimization strategy, the fungicide oxathiapiprolin has been identified.

An OSBP has been identified as the target for oxathiapiprolin, which is a novel MoA for fungicides. While established products need 100 g a.i. or more per hectare to efficiently control *P. infestans* and *P. viticola* in grapes, potatoes as well as vegetables, oxathiapiprolin achieves the same effect with lower use rates (12–30 g a.i./ha) by a factor of 10.

The systemic oxathiapiprolin has excellent (1) preventative activity by inhibition of zoospore release and blocking the zoospore as well as sporangia germination; (2) curative efficacy by blocking the mycelial growth within the host plant; and (3) post-infection activity by blocking mycelial growth and inhibition of further lesion expansion. In addition, as an anti-sporulant, oxathiapiprolin inhibits spore production and viability.

11.3.2.6 Selected Fluorine-Containing Fungicide Development Candidates

Currently, two further broad-spectrum SDHI fungicides are under development. The racemic fluindapyr (provisionally approved by ISO; Figure 11.14) containing a 7-fluoro-1,1,3-trimethyl-indan-4-yl moiety is structurally similar to inpyrfluxam (1,1,3-trimethyl-indan-4-yl moiety, (*R*)-enantiomer; Table 11.1). It is useful for preventive and curative control of fungal diseases, especially on cereals, soybean, rice, tree nuts, rape, maize, and tobacco, combined with very low to zero phytotoxicity (PTX) to target crops.

Isoflucypram (provisionally approved by ISO, Figure 11.14) exhibits a high level of efficacy for the control of leaf spot diseases such as speckled leaf blotch (*Septoria tritici*), rust, and other diseases on a broad range of crops. Isoflucypram originates from a mix and match approach by combining key fragments from two fungicidal active molecules. The importance of the *N*-substitution in combination with the 3-(difluoromethyl)-5-fluoro-1-methyl-pyrazole-4-carboxamide moiety was exemplified by the ranking of *N*-substituents in the biochemical complex II assay at gray mold (*Botrytis cinerea*): iso-Pr < Et, H < *cyclo*-Bu < **cyclo-Pr**.

The new neopicolinamide florylpicoxamid (Figure 11.14) [52] fungicide was inspired by the natural product UK-2A, a fungicidally active, natural metabolite isolated from *Streptomyces* sp. 517-02 [53]. The design of the (1*S*,2*S*)-diastereomer florylpicoxamid focused on retaining structural features criteria to binding at the ubiquinone Q_i target site of mitochondrial complex III of the respiratory chain

Fluindapyr Isoflucypram Florylpicoxamid

Figure 11.14 Fungicide development candidates fluindapyr, isoflucypram, and florylpicoxamid.

Saccharomyces cerevisiae. It shows control of wheat leaf blotch and other important diseases in many crops.

11.3.3 Insecticides Containing Fluorine

11.3.3.1 Nicotinic Acetylcholine Receptor Competitive Modulators

One of the insecticide molecular target sites of importance is the nAChR, which plays a central role in the mediation of fast excitatory synaptic transmission in the insect CNS. The global economic success of synthetic nAChR competitive modulators as insecticides has been broadly reviewed in various articles and book chapters over the past decade [54].

The IRAC has assigned in the Main Group 4 with the primary site of action "nAChR competitive modulators" the five chemical subgroups: 4A Neonicotinoids, 4B Nicotine, 4C Sulfoximines, 4D Butenolides, and 4E Mesoionics, which are distinct from each other.

The discovery of the insecticidal active sulfoximines combined with the launch of sulfoxaflor in 2012 has been the starting point for commercialization of fluorinated nAChR competitive modulators. Followed by butenolides (flupyradifurone, 2015) as the second class of fluorine-containing nAChR competitive modulators, the mesoionics (triflumezopyrim, 2018) complete the insecticide market as the third class.

11.3.3.1.1 *Sulfoximines*

The SAR of sulfoximines such as the racemic sulfoxaflor is directed toward the [NC-N=S(O)-]-pharmacophore. In addition, the effect of substitution (R^1) on the methylene bridge has shown that a dramatic difference in potency was observed between the unsubstituted and the methyl- or fluorine-substituted pairs ($R^1 = H \ll Me \sim F$) [55].

The substituent R^2 other than methyl did not further increase the insecticidal activity. On the other hand, the pyridine moiety of sulfoxaflor has shown that the replacement of its 6-trifluoromethyl by non-halogenated groups ($R^3 = OMe < COOMe < CN < Me$, LC_{50}-value = 2.83 ppm) resulted in a significant decrease of activity against the green-peach aphid (*Myzus persicae*). Furthermore, replacement of the 6-trifluoromethyl group by halogen atoms ($X = F < Br < I < Cl < CF_3$, LC_{50}-value = 0.11 ppm) resulted in lower activity against *M. persicae* as well (Figure 11.15).

Despite a close structural similarity to sulfoxaflor, the so-called sulfoxaflor analogs show a clear ranking according to their insecticidal potency against green-peach

| Sulfoximine insecticides | Sulfoxaflor | XF$_2$C-Sulfoxaflor analogs |

Figure 11.15 Sulfoxime insecticides such as sulfoxaflor and XF$_2$C-sulfoxaflor analogs.

Table 11.3 Activities of sulfoxaflor and XF_2C-pyridinyl analogs.

X	Green-peach aphid (*Myzus persicae*) LC_{50}-value (ppm)
CF_2CF_3	200
CF_3	45.5
Me	14.8
H	0.98
Cl	0.11
F	**0.11 (Sulfoxaflor)**

Sulfoxaflor is the commercial product. All other derivatives are derivates obtained during its optimization.
Source: Data from Loso et al. [55].

aphid (*M. persicae*) demonstrating the strong influence of X in the 6-XF_2C-pyridinyl moiety (Table 11.3). While the replacement of one fluorine in the 6-F_3C-group of sulfoxaflor by chlorine (X = Cl) resulted in a similar activity of both sulfoxaflor and closely related *N*-cyano-sulfoximine, the replacement by hydrogen and methyl already resulted in a decrease of activity by the factors 9 and ~135. Somewhat more impressive is the replacement of one fluorine in the 6-F_3C-group of sulfoxaflor by a perfluoroalkyl group such as trifluoromethyl (X = CF_3) and the long-chain heptafluoro-*n*-propyl rest (X = CF_2CF_3), which lead to a decrease of activity by the factors ~414 and 1818.

To understand the rationale behind the loss of insecticidal activity the mode of binding of sulfoxaflor and its 6-pentafluoro-analog to the acetylcholine binding protein (AChBP) from *Aplysia californica* acetylcholine binding protein (AcAChBP) has been investigated [4c]. It seems that the 6-($F_3CF_2CF_2C$)-group in 6-position of the pyridine is much larger than a F_3C-group and is unfavorable for the ligand–receptor interaction. Therefore, the result is a loss of insecticidal activity for this analog (X = $CF_2CF_3 \ll$ **F**).

11.3.3.1.2 Butenolides

The structural requirements of pyridine ring-containing butenolides such as flupyradifurone (6-Cl; $R^1 = CH_2CHF_2$, $R^2 = H$) consist of the new bioactive buteno-lide pharmacophore. According to the SAR of butenolides one or two substituents such as halogens in 5- and/or 6-position of the pyridine ring were used as promising starting points for the structure optimization steps (5-CHF_2, 6-Cl, 6-$CF_3 <$ 6-Br, 6-F < **6-Cl**) (Figure 11.16).

Substituents R^2 other than hydrogen, e.g. halogen or alkyl, did not further increase the insecticidal activity. The variation of the *N*-substituent R^1 can be more diverse; smaller alkoxy (OMe), alkyl (Me, Et), or cycloalkyl (*cyclo*-Pr) residues ensure good insecticidal activity. However, fluorine-containing substituents R^1 such as CH_2CH_2-X (X = $CH_2CF_3 < CF_3 < CH_2F <$ **F**) reflected an interesting insecticidal potency against green-peach aphid (*M. persicae*) [56].

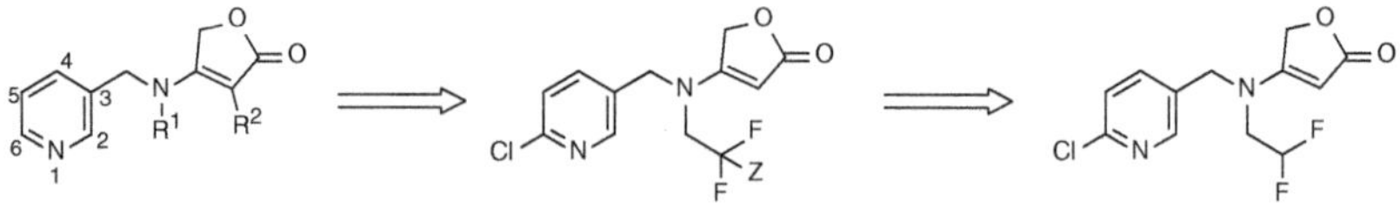

Butenolide insecticides Side chain optimization Flupyradifurone

Figure 11.16 Discovery of butenolide insecticides and side chain optimization resulted in flupyradifurone.

Further optimization of 6-chloro-pyridine-3-yl-containing butenolides having a N-2,2,2-trifluoroethyl residue (Z = F) demonstrates a very good activity against the green-peach aphid (*M. persicae*). But, replacement of hydrogen in the 2-position (Z) of the N-2,2-difluoroethyl side chain by halogen or (halogen)alkyl resulted in lower insecticidal activity against the green-peach aphid (*M. persicae*) (Z = $CF_3 \ll CHF_2 < Me \sim F < $ **H**) and mustard leaf beetle (*Phaedon cochleariae*) (Z = Me, CHF_2, $CF_3 \ll F < $ **H**), respectively. However, replacement of the N-2,2-difluoroethyl (Z = H) by a 2-chloro-2-fluoro-ethyl residue in flupyradifurone led to lower activity against both insect species (Figure 11.16).

The influence of the N-2,2-difluoroethyl side chain is remarkable and unique for the butenolide class, which could be exemplified by replacement of the N-ethyl- or N-methyl groups in the neonicotinoids, nitenpyram, clothianidin, and acetamiprid with the N-2,2-difluoroethyl group (Table 11.4).

In the case of nitenpyram, the replacement of the N-ethyl- with N-2,2-difluoroethyl group results in a 1250 times weaker insecticidal activity against cotton aphid (*Aphis gossypii*). A similar decrease of insecticidal potency was observed for the N-2,2-difluoroethyl-substituted analog of clothianidin (125 times weaker than clothianidin) and acetamiprid (10 times weaker than acetamiprid) against green-peach aphid (*M. persicae*) and southern green stink bug (*Nezary viridula*), respectively. In contrast, the replacement of the N-methyl- and N-ethyl substituents in the butenolides with the N-2,2-difluoroethyl group leads to flupyradifurone showing a five times higher insecticidal activity against cotton whitefly (*Bemisia tabaci*).

In this context, it was found that flupyradifurone shows only little or no cross-resistance in whitefly (*B. tabaci*) strains metabolically resistant to neonicotinoids such as imidacloprid. In comparison with imidacloprid as a typical neonicotinoid representative, it was found that the butenolide pharmacophore moiety in combination with the unique N-2,2-difluoroethyl residue in flupyradifurone increases its insecticidal activity and prevents the oxidative metabolization by CYP6CM1 [57]. Furthermore, it was found by creation of a homology model of sensitive aphid nAChR (*M. persicae*) that flupyradifurone can interact with tyrosine (Tyr)-residues from the α subunit in a *"H-bridge-like fashion"* to the hydroxy function of these amino acids through its N-2,2-difluoroethyl side chain [58]. This H-bridge-like interaction is not possible for the other nAChR competitive modulators investigated.

Table 11.4 Comparison of insecticidal activities of neonicotinoids with their N-2,2-difluoroethyl analogs and butenolides with flupyradifurone against different insect species.

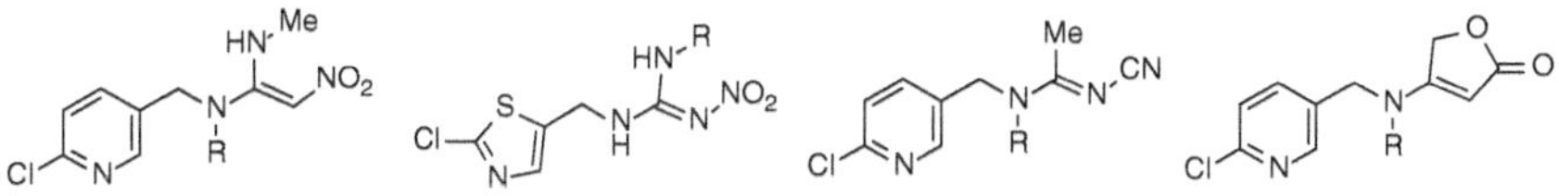

Nitenpyram (R = Et)   Clothianidin (R = Me)   Acetamiprid (R = Me)   Flupyradifurone (R = CH₂CHF₂)

IRAC class (subgroup)	Common name or IUPAC name	Tested against insect species	Insecticidal activity R = Alkyl	Insecticidal activity R = CH$_2$CHF$_2$
Neonicotinoids (4A)	Nitenpyram	*Aphis gossypii*	R = Et, 0.016 ppm a.i.	20 ppm a.i.
	Clothianidin	*Myzus persicae*	R = Me, 0.8 g a.i./ha	100 g a.i./ha
	Acetamiprid	*Nezary viridula*	R = Me, 10 g a.i./ha	100 g a.i./ha
Butenolides (4D)	4-[(Methylamino)-2 (5*H*)-furanone	*Bemisia tabaci*	R = Me, 20 ppm a.i.	4 ppm a.i.
	4-[(Ethylamino)-2 (5*H*)-furanone	*Bemisia tabaci*	R = Et, 20 ppm a.i.	4 ppm a.i.

Source: Data taken from Jeschke et al. [57].

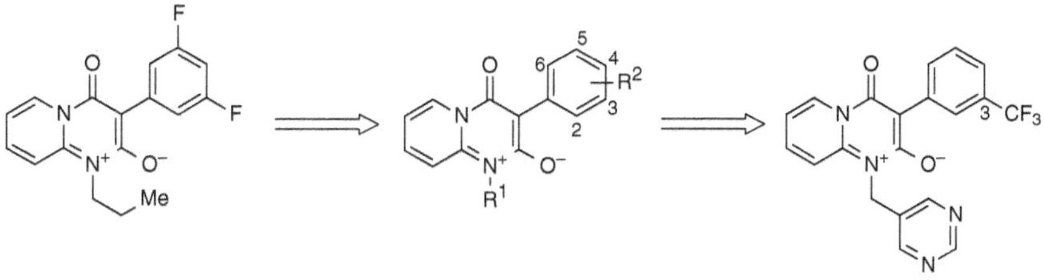

First lead structure (fungicide)   Mesoionic insecticides   Triflumezopyrim

Figure 11.17 From fungicidal active compounds to mesoionic insecticides such as triflumezopyrim.

11.3.3.1.3 *Mesoionics*

Mesoionics is a dipolar compound class that resulted from a fungicides program, in which selected candidates such as the first lead structure 3,4-dihydro-2,4-dioxo-1-(5-propyl)-3-[3,5-difluoro-phenyl]-2*H*-pyrido [1,2-a]pyrimidinium (inner salt and tautomers for the negative charge given) showed modest activity (LC$_{90}$ = 50 ppm a.i.) against the corn plant hopper (*Peregrinus maidis*) (Figure 11.17).

The first commercialized member of mesoionics triflumezopyrim [59] contains a 3-(trifluoromethyl)-phenyl moiety coupled with a mesoionic core and demonstrates good activity against susceptible and neonicotinoid-resistant hopper species in Asia such as the rice brown plant hopper (*Nilaparvata lugens*) and white-backed plant hopper (*Sogatella frucifera*).

Longer and branched *N*-alkyl residues (R^1) did not further improve the insecticidal activity. However, an increased efficacy against hopper species such as the corn plant hopper *P. maidis*, brown plant hopper (*N. lugens*), and green leafhopper (*Nephotettix cincticeps*) could be seen with replacement of *n*-propyl by fluorinated alkyl groups R^1 such as 1,1,1-trifluoro-2-propanyl or 2,2,2-trifluoroethyl residues ($R^1 = CH(CH_3)CF_3$; $R^2 = 3\text{-}OCF_3 \ll R^1 = CH_2CF_3$; $R^2 = 3\text{-}OCH_3$, LC_{90}-value = 2–10 ppm). Triflumezopyrim binds to the orthosteric site of the nAChR and this physiological action is distinct from other nAChR competitive modulators described in Section 3.3.1.

11.3.3.2 Ryanodine Receptor (RyR) Modulators

The first RyR modulators, the 1,2-benzenedicarboxamides, were reported in 1993 [60] and the anthranilic diamides in 1999 [61]. Flubendiamide (see structure in Figure 11.18) and chlorantraniliprole were launched as the first diamides with modulator activity at the RyR.

A few years ago, the modification of the lipophilic effect was the driving force for the discovery of cyantraniliprole [62]. In this case, the reduction of the log*P*-value by introduction of a cyano group (R^1) into the phenyl moiety of chlorantraniliprole by replacement of a chlorine atom (R^1) resulted in increased water solubility [61a]. In this context, an improvement of xylem systemic properties of cyantraniliprole was observed, which correlates with broad activity against a wide spectrum of lepitopteran and hemipteran pests [63].

Inspired by both successfully commercialized RyR modulators chloroantraniliprole and cyantranili-prole, the preparation of tetraniliprolc has been carried out. Tetraniliprole contains a C_1-linkered haloalkyl substituted 5-membered heterocycle at the pyrazole moiety such as 5-trifluoromethyl-2*H*-tetrazol-2-ylmethyl instead of bromine. By replacement of bromine in chlorantraniliprole ($DT_{50} = 431$ days) by the 5-trifluoromethyl-2*H*-tetrazol-2-ylmethyl moiety the soil DT_{50}-value was reduced to

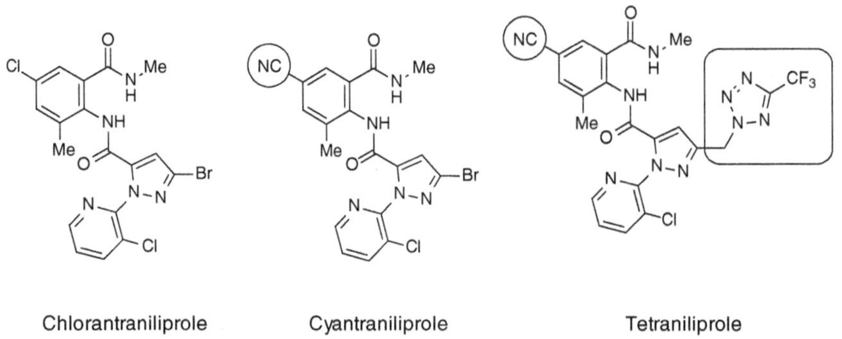

Chlorantraniliprole   Cyantraniliprole   Tetraniliprole

Figure 11.18 The anthranilic diamide insecticides chlorantraniliprole, cyantraniliprole, and tetraniliprole.

the value of 236 days, which could be further reduced significantly by introduction of a cyano group ($DT_{50} = 83$ days) (Figure 11.18).

During the structure optimization it was found that numerous further methylene-substituted 5-membered heterocycles led to high biological activity as exemplified by the [5-(trifluoromethyl)pyrazol-1-yl]methyl, [3-chloro-5-(1,1,2,2,2-pentafluoroethyl)pyrazol-1-yl] methyl, or [3,5-bis(trifluoro-methyl)-1,2,4-triazol-1-yl]methyl moieties.

Tetraniliprole demonstrates an outstanding crop protection against a broad spectrum of insect pests including lepidoptera, coleoptera, leafminer, and selected other dipera and aphids. It can be applied as foliar and soil treatment from early to late season.

11.3.3.3 GABA-Gated Cl-Channel Allosteric Modulators

The *meta*-diamides such as broflanilide [64] were discovered by structural modification of the RyR modulator flubendiamide [61b]. Change of the amide group from *ortho*- to *meta*-position resulted in a shift of the MoA (RyR modulator to GABA-Cl channel allosteric modulator, Figure 11.19). Broflanilide contains 12 "mixed" halogen atoms, i.e. 1 bromine and 11 fluorine atoms, located in the 2-fluoro-benzamide (ring A) as well 2-bromo-4-heptafluoro-iso-pyropyl-6-trifluoromethyl-phenyl (ring B) as fragments of the molecule. Perhaps this enormous halogen content has been selected probably in order to obtain lipophilicity by a high log$P$-value, essential for the GABA target.

According to the SAR of *meta*-diamides, it was found that the activity against oriental leafworm moth (*Spodoptera litura*) and diamondback moth (*Plutella xylostella*) is preferred up to $EC_{70} = 1.0$ mg a.i./l, if in ring A fluorine is in 2-position (6-F $\ll$ H $\le$ 4-F < **2-F**). On the other hand, for an activity against cotton bollworm (*Helicoverpa armigera*) up to $EC_{70} = 0.1$ mg a.i./l the combination of bromine, trifluoromethyl (2'-Br/6'-Br = 2'-CF$_3$/6'-CF$_3$ < **2'-Br/6'-CF$_3$**) together with the strong lipophilic heptafluoro-iso-propyl substituent has to be present in ring B (Figure 11.19) [65].

Broflanilide has a broad-spectrum activity (crop and non-crop pests) and can be used in different crops and in professional pest management markets, e.g. for control of lepidoptera, coleoptera, termites, ants, cockroaches, and flies. As proinsecticide [22], broflanilide is probably metabolized by CYP_{450}-mediated

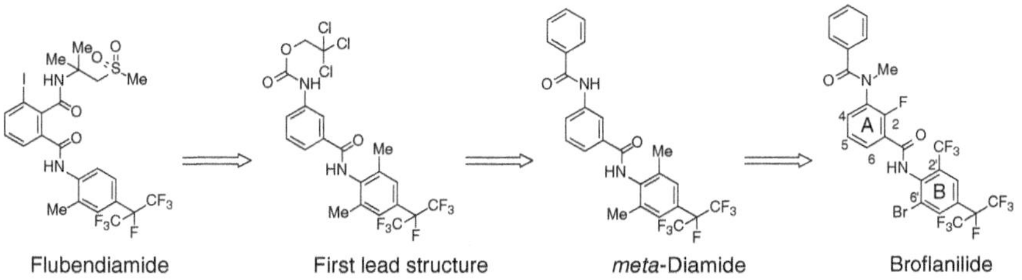

Figure 11.19 Flubendiamide (*ortho*-diamide) as lead compound for preparing *meta*-diamides such as broflanilide. Source: Katsuta et al. [65].

bioactivation to des-methyl-broflanilide (cleavage of the *N*-methyl group), which acts as a noncompetitive GABA-R antagonist on a novel site in *Drosophila* RDL GABA-Rs (resistant-to-dieldrin, RDL) [66]. Although the site of action for *meta*-diamides seems to be overlap with that of glutamate-gated chloride (GluCl) channel allosteric modulators such as avermectins and milbemycins, different MoAs have been recently demonstrated [67]. Because of the high selectivity of des-methyl-broflanilide, it has been postulated that broflanilide is effective against pests with resistance to GABA-R channel blocker-type antagonists such as fiproles.

Fluxametamid [68] is an isoxazoline insecticide (including *syn-/anti*-isomeric methoxime group) with high insecticidal activity against various Lepidoptera, Thysanoptera, and Diptera, and acaricidal activity (*Tetranychidae*). The racemic fluxametamid is based on the lead structure (R^1, R^2, R^3 = Cl, ring A) containing an iso-butyl amide group in 4-position of ring B. As a major optimization step, ring A was modified (R^1, R^3 = Cl, R^2 = H), and a 2,2,2-trifluoroethyl amide group was introduced in 4-position of ring B together with a methyl group in 5-position.

Because of the lack of a cost-efficient asymmetric technology it is probable that for the next few years, only the racemic mixture will be produced although the (*S*)-enantiomer is more active. This economic decision correlates well with the structurally related and commercial racemic isoxazoline ectoparasiticides fluralaner (Figure 11.20) [69].

From fluralaner it is known that only the (*S*)-enantiomer is active with no adverse effects caused by the (*R*)-enantiomer [70].

Fluxametamid is a ligand-gated chloride channel (LGCC) antagonist, inhibiting GABA-gated chloride channels (GABA-Cls) and GluCl channels [71]. Fluxametamid exhibited similar levels of both activities in fipronil-susceptible and fipronil-resistant arthropod pests. It was found that fluxametamid exerts distinctive antagonism of arthropod GABA-Cls by binding to a site different from those for existing antagonists. Fluxametamid has a high target-site selectivity for arthropods over mammals.

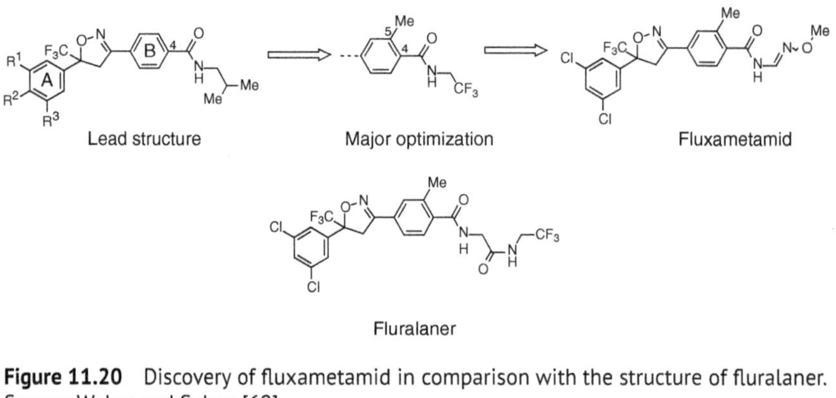

Figure 11.20 Discovery of fluxametamid in comparison with the structure of fluralaner. Source: Weber and Selzer [69].

11.3.3.4 Selected Fluorine-Containing Insecticide Development Candidates

Benzpyrimoxan (provisionally approved by ISO) [72] is an extremely promising hopper insecticide with high nymphicidal activity against rice plant hoppers such as brown plant hopper (*N. lugens*) as well as white-backed plant hopper (*Sogatella furcifera*) and a low impact on non-target organisms including pollinators and beneficial arthropods. Based on the SAR of lead structure, the activity against brown plant hopper (*N. lugens*) is preferred up to $LC_{90} = 0.3$ mg a.i./l, if R is a trifluoromethyl-group in 4-position (R = H ~ 4-F ~ 2-$CF_3 \ll 3\text{-}CF_3$ ~ 4-SCF_3 < 4-OCF_3 < **4-CF_3**) (Figure 11.21) [73].

Benzpyrimoxan shows molting inhibition different from that of known insect growth regulators (IGRs). Its field biological performance at 50–75 g a.i./ha revealed a favorable environmental profile without any resurgence and with high activity even against plant hoppers such as *N. lugens* in rice ecosystem that had developed resistance to major chemical classes of insecticide.

Oxazosulfyl (provisionally approved by ISO) is a novel sulfyl insecticide ($F_3C\text{-}SO_2$-group) processing a potent and broad-spectrum insecticidal activity for insect control in rice fields (Figure 11.13).

Isocycloseram (containing 80–100% of the (5*S*,4*R*)-isomer); provisionally approved by ISO), is a broad-spectrum insecticide and acaricide, which shows activity against lepidopteran, hemipteran, coleopteran, thysanopteran, and dipteran pest species (Figure 11.21). Similar to fluxametamid (see Section 3.3.3) isocycloseram is a GABA-gated Cl-channel allosteric modulator and acts as a noncompetitive GABA-gated chloride channel antagonist at a site different from known antagonists such as fiproles and cyclodienes.

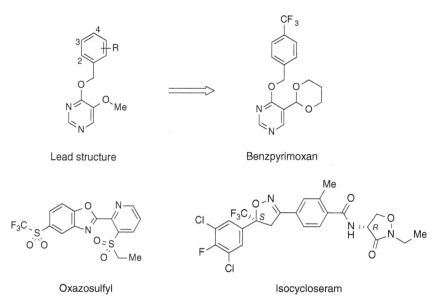

Figure 11.21 Insecticide development candidates benzpyrimoxan, oxazosulfyl, and isocycloseram.

11.3.4 Acaricides Containing Fluorine

11.3.4.1 Mitochondrial Complex II Electron Transport Inhibitors

Pyflubumide is a carboxanilide, containing a 4-(1-methoxy-hexafluoro-iso-propyl)-substituted aniline moiety [74]. This is also structurally inspired by the heptafluoro-iso-propyl residue (Z = F) from the insecticide flubendiamide (see Section 3.3.2). In this case, the 4-(1-methoxy-hexafluoro-iso-propyl) group (Z = OMe) leads to an increased lipophilicity (log$P$-value = 5.34) of the acaricide (Figure 11.22).

Both the bulky iso-butyl side chain and the N-methyl-1H-pyrazole carboxyl moiety originated from SDHI fungicides such as the lipophilic penthiopyrad (see Section 3.2.1).

During the structure optimization it was found that the replacement of a heptafluoro-iso-propyl (Z = F) residue by hexafluoro-iso-propyl (Z = H) residue is important for the acaricidal activity against the two-spotted spider mite (*Tetranychus urticae*) (Z = F ≪ OMe < **H**) (Table 11.5).

However, the heptafluoro-iso-propyl residue (Z = F) sharply decreases the acaricidal activity, because of the high lipophilicity. The hexafluoro-iso-propyl residue (Z = H) leads to the most active derivative, but the high efficacy is combined with raised toxicity. Therefore, the slightly more lipophilic 4-(1-methoxy-hexafluoro-iso-propyl) residue (Z = OCH$_3$) has been incorporated,

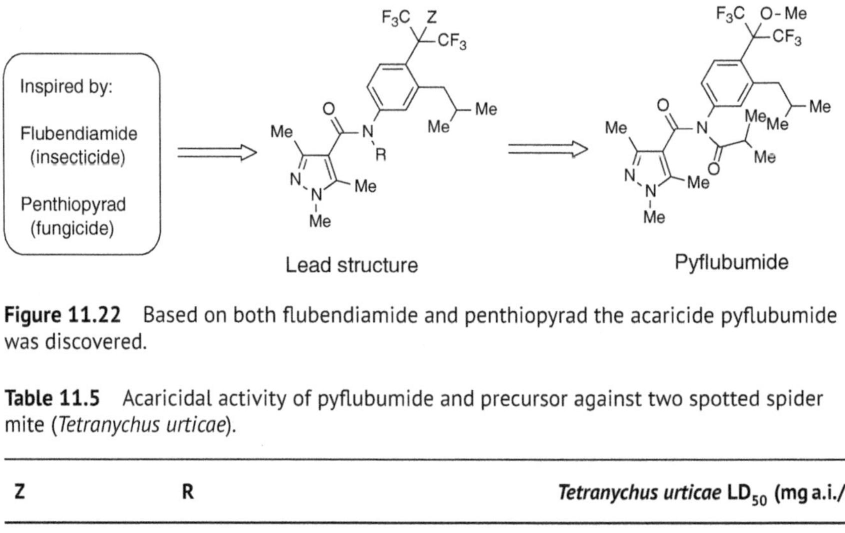

Inspired by:

Flubendiamide (insecticide)

Penthiopyrad (fungicide)

Lead structure   Pyflubumide

Figure 11.22 Based on both flubendiamide and penthiopyrad the acaricide pyflubumide was discovered.

Table 11.5 Acaricidal activity of pyflubumide and precursor against two spotted spider mite (*Tetranychus urticae*).

Z	R	*Tetranychus urticae* LD$_{50}$ (mg a.i./l)
F	H	>300
H	H	3–10
OMe	H (NH-precursor)	10–30
OMe	**CO-*iso*-Pr (Pyflubumide)**	**1–3**

Fluopyram is the commercial product. All other derivatives are derivates obtained during its optimization.
Source: Data taken from Furuya et al. [74a].

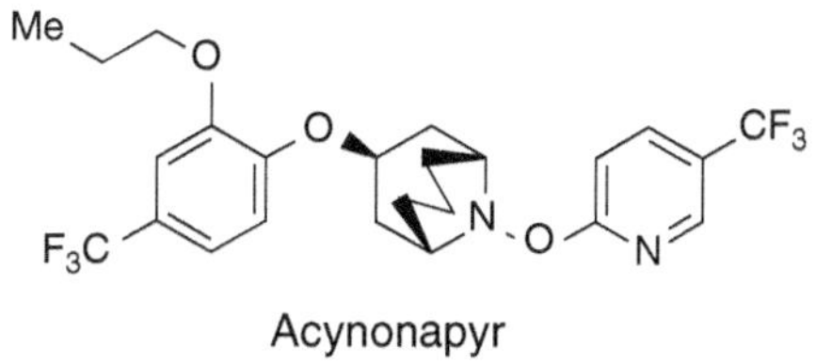

Figure 11.23 Acaricide development candidate acynonapyr.

Acynonapyr

which maintains the high acaricidal activity. Pyflubumide is a proacaricide (R = CO-*iso*-Pr) similar to the β-ketonitriles. After *N*-deacylation, the NH-metabolite (R = H) is highly inhibitory to mitochondrial respiratory chain (complex II) in possibly a different binding manner from the known β-ketonitriles [75]. Perhaps the result may have contributed to the suggestion that pyflubumide is useful for future resistance management strategies for spider mites.

11.3.4.2 Selected Fluorine-Containing Acaricide Development Candidates

In the 1990s, Zeneca described special substituted azabicyclo[3.3.1]nonanes, which have insecticidal activity against peach aphid (*M. persicae*) [76].

Two decades later, acynonapyr (provisionally approved by ISO; Figure 11.23), an acaricide with a new substituted azabicyclo [3.3.1]nonane core, demonstrates useful control of spider mites in vegetables, tea, and citrus fruits.

11.3.5 Nematicides Containing Fluorine

11.3.5.1 Nematicides with Unknown Biochemical MoA

The low toxic nematicide fluensulfone contains a [(3,4,4-trifluoro-3-buten-1-yl) sulfonyl]-fragment. This influences the motility, inhibits development, egg-laying, egg-hatching, feeding, and locomotion of plant-parasitic nematodes (PPNs) including pleiotropic actions on root-knot nematodes (RKNs) (*Meloidogyne* spp.) as well as the potato cyst nematode *Globodera pallida* and inhibits their development [77].

The SAR in fluensulfone of the fluorinated side chain is narrow and mainly focused on the 3,4,4-trifluoro-3-buten-1-yl (R = F) or 4,4-difluoro-3-buten-1-yl (R = H) residues (Figure 11.24).

The biochemical MoA of fluensulfone in nematodes is unknown so far [78]. It was found that in insects, the *gem*-difluorovinyl derivatives inhibit the β-oxidation of fatty acids in the mitochondria [79].

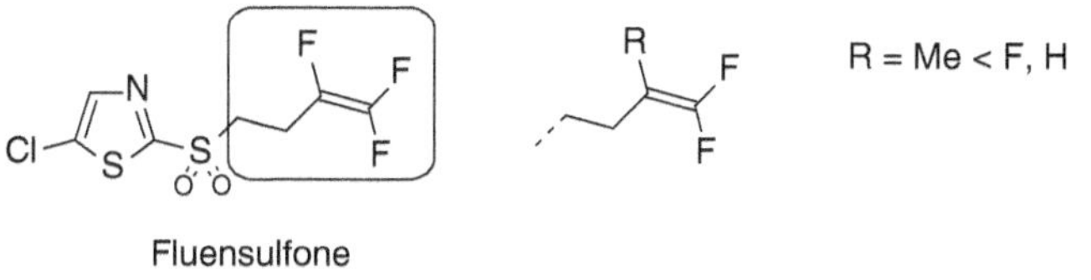

R = Me < F, H

Fluensulfone

Figure 11.24 Chemical structures of the nematicide fluensulfone including the SAR of fluoroalkenyl nematicides.

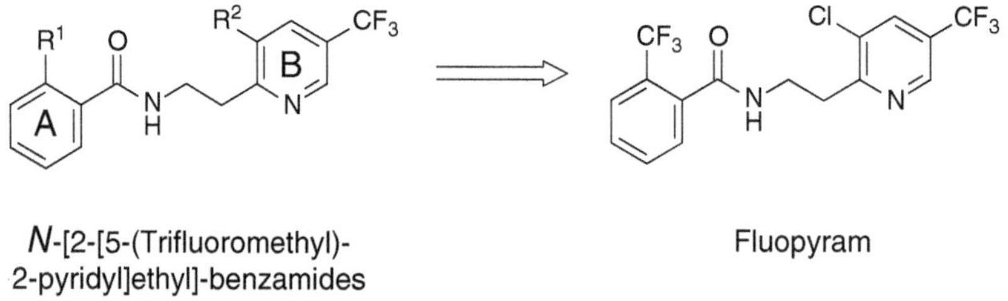

N-[2-[5-(Trifluoromethyl)-
2-pyridyl]ethyl]-benzamides   Fluopyram

Figure 11.25 From *N*-[2-[5-(trifluoromethyl)-2-propyl]ethyl]-benzamides to fluopyram.

Table 11.6 Selection of physicochemical parameters of fluopyram and derivatives.

R^1	R^2	MP (°C)$^{a)}$	Log P-value$^{b)}$
Me	Cl	107–109	2.95
CF$_3$	H	91–92	2.55
CF$_3$	CF$_3$	ND	3.48
CF$_3$	**Cl (Fluopyram)**	**106–111**	**3.15**

Data in bold type are physicochemical parameters of fluopyram. ND, not determined.
a)   Melting points taken from Ref. [81].
b)   Measurement of logP-values under acidic conditions.
Source: Modified from Suwa [81].

11.3.5.2 Nematicidal Succinate Dehydrogenase Inhibitors

The fungicidally active SDHI pyridinyl–ethyl benzamide fluopyram [80] has also been launched as nematicide. Fluopyram is highly active against PPNs and inhibits selectively the mitochondrial respiratory chain, which leads to severe depletion of cellular energy (adenosine triphosphate [ATP]) resulting in very fast immobilization of nematodes. Based on its physicochemical properties, which are clearly influenced by the chlorine atom (R^2) and both trifluoromethyl groups (R^1 in ring A and ring B) (Figure 11.25; Table 11.6), drench, in-furrow application, and soil incorporation are all possible.

Despite a similar range of the melting points, the trifluoromethyl group at the phenyl moiety (ring A) in fluopyram leads to an increase of the logP-value, which results in a stronger nematicidal activity (R^1 = Me < **CF$_3$**) against the RKN *Meloidogyne incognita*. Replacement of its chlorine atom (R^2 in ring B) at the pyridine moiety by hydrogen (R^2 = H) is correlated with a lower melting point and a lower logP-value of 2.55, whereas its replacement by the trifluoromethyl group (R^2 = CF$_3$) in ring B is correlated with a much higher logP-value of 3.48. As a consequence, a decrease of nematicidal activity (R^2 = CF$_3$ < H ≪ **Cl**) against the RKN can be observed.

11.3.5.3 Selected Fluorine-Containing Nematicide Development Candidates

Based on HTS of a compound library a lead structure with activity at a screening rate of 500 ppm against the RKN *M. incognita* was identified. The following optimization program was directed at retention of the strong nematicidal activity

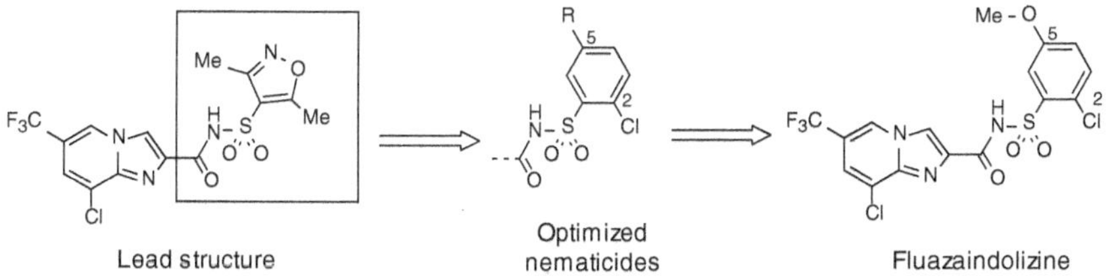

Figure 11.26 Lead structure for optimized nematicides resulted in fluazaindolizine.

and elimination of plant PTX effects [82]. For compounds containing chlorine in the 2-position of the phenyl moiety, the highest levels of activity were observed. The 2-chloro-5-substituted analogs showed good nematicidal potencies up to $EC_{50} = 2.0\,ppm$ (R = 5-OCF$_3$ $\ll$ 5-Br $\ll$ 5-Cl < 5-CF$_3$ < 5-OEt < H < **5-OMe** < 5-Me) with the single exception of the 5-trifluoromethoxy substitution ($EC_{50} = 32.6\,ppm$).

Biological characterization and field testing of the novel nematicidal sulfonamide fluazaindolizine (provisionally approved by ISO; Figure 11.26 [83]) has demonstrated excellent control of a range of PPNs, resulting in root protection and the potential for increased yield.

The nematicide is highly specific for PPNs, and studies are suggesting a novel MoA.

11.4 Summary and Prospects

The latest generation of modern crop protection compounds, including those in the development pipeline, demonstrates that the impact of fluorine-containing pesticides is still significant. The introduction of fluorine atoms or fluorinated motifs into an a.i. can lead to an improvement or a decline in biological efficacy as found, for example, in the new class of sulfoximine insecticides (sulfoxaflor) for single fluorine vs. trifluoromethyl group as pyridinyl substituents (F < **CF$_3$**). Depending on the nature of the target for a.i.s, the ligand–target binding interactions (MoA) can be influenced by the physicochemical property profile. Examples are the fluorinated propesticides halauxifen-methyl, pyflubumide, and broflanilide, which are affected by the special ADME parameter of the molecules, leading finally to biological superiority of these propesticides over their non-modified analogs [84]. With the fluorinated CBI indaziflam, a shift of the herbicidal MoA can be achieved and the butenolide insecticide flupyradifurone has the ability to overcome the detoxification enzyme CYP6CM1 in whiteflies. Furthermore, the fluorine substitution pattern can influence soil stability and/or water solubility, which was clearly demonstrated by herbicides such as pyroxasulfone, SDHI fungicides such as fluxapyroxad, sedaxane, or benzovindiflupyr, and the new nematicides fluensulfone or fluopyram. The complexity of the SAR between different a.i.s often prevents a clear prediction of whether fluorine atoms or fluorine-containing substituents will enhance biological activity. Currently, pesticides substituted with single fluorine atoms, or fluorine-containing groups such as difluoromethyl, trifluoromethyl, or difluoromethoxy are central because of their improved binding to protein-type receptors by the so-called "polar hydrophobic effect." The impact on metabolic stability of polyfluorinated isopropyl substituents on aromatic rings has been demonstrated.

Conversely, bacterial dehalogenases catalyze the cleavage of carbon–fluorine bonds, which is a key step in aerobic mineralization pathways of many fluorinated compounds, and carbon–fluorine bonds can have an extremely high stability [85].

Fluorinated molecules are therefore often resistant to degradation, representing a potential environmental challenge. In this context, fungicides and herbicides contain, in most cases, fluorine atoms, whereas nematicides and herbicides in most cases contain *"mixed"* halogen atoms, for example, chlorine and fluorine. Because of the increasing importance of SDHI fungicides in the last nine years, the number of fluorine-containing fungicides is remarkable. However, the success of fluorinated pesticides is not straightforward, and the industry is searching also for non-halogenated pesticides as exemplified by the commercialized selective herbicide tolpyralate for weed control, the SDHI fungicide isofetamid, the complex III fungicide mandestrobin or picarbutrazox with an unknown MoA, the natural product-derived insecticides lepimectine and afidopyropen. Nevertheless, there is no doubt that fluorine atoms and fluorinated motifs will remain an important tool to modulate the properties of a.i.s in modern agricultural chemistry.

References

1 Jeschke, P. (2016). *Pest Manage. Sci.* 72: 433–455.

2 Sparks, T.C. and Lorsbach, B.A. (2017). *Pest Manage. Sci.* 73: 672–677.

3 Jeschke, P. (2017). *Pest Manage. Sci.* 73: 1053–1066.

4 (a) Jeschke, P. (2010). *Pest Manage. Sci.* 66: 10–27. (b) Leroux, F., Jeschke, P., and Schlosser, M. (2005). *Chem. Rev.* 105: 827–856. (c) Jeschke, P., Gutbrod, O., and Leroux, F. (2018). The role of fluorine in the design of nicotinic acetylcholine receptor (*n*AChR) competitive modulators. In: *Fluorine in Life Sciences: Pharmaceuticals, Medicinal Diagnostics, and Agrochemicals*, 1e (eds. G. Haufe and F. Leroux), 631–651. London, Oxford: Academic Press.

5 Prakash, G.K.S. and Wang, F. (2012). *Chimica Oggi – Chemistry Today* 30: 32–38.

6 El Qacemi, M., Rendine, S., and Maienfisch, P. (2018). Recent applications of fluorine in crop protection – new discoveries originating from the unique heptafluoroisopropyl group. In: *Fluorine in Life Sciences: Pharmaceuticals, Medicinal Diagnostics, and Agrochemicals*, 1e (eds. G. Haufe and F. Leroux), 607–629. London, Oxford: Academic Press.

7 Lepri, S., Goracci, L., Valeri, A., and Cruciani, G. (2016). *Eur. J. Med. Chem.* 121: 558–670.

8 Leichter, C.A., Thompson, N., Johnson, B.R., and Scott, J.G. (2013). *Pestic. Biochem. Physiol.* 107: 169–176.

9 Zhang, Y. (2014). *Shijie Nongyao* 36: 30–32.

10 Gutteridge, S., Thompson, M.E., and Andreassi, J.L. (2019). Biochemistry of the target and resistance. In: *Modern Crop Protection Compounds*, Herbicides, 3e, vol. 1 (eds. P. Jeschke, M. Witschel, W. Krämer and U. Schirmer), 33–55. VCH-Wiley: Weinheim.

11 Yoshimura, T., Nakatani, M., Asakura, S. et al. (2011). *J. Pestic. Sci. (Tokyo)* 36: 212–220.

12 Zagar, C., Liebl, R., Theodoridis, G., and Witschel, M. (2019). Protoporphorinogen IX oxidase inhibitors. In: *Modern Crop Protection Compounds*, Herbicides, 3e, vol. 1 (eds. P. Jeschke, M. Witschel, W. Krämer and U. Schirmer), 173–211. VCH-Wiley: Weinheim.

13 Grossmann, K., Niggeweg, R., Christiansen, N. et al. (2010). *Weed Sci.* 58: 1–9.

14 Grossmann, K., Hutzler, J., Caspar, G. et al. (2011). *Weed Sci.* 59: 290–298.

15 Dietrich, H., Jones, J.C., and Laber, B. (2019). Inhibitors of cellulose biosynthesis. In: *Modern Crop Protection Compounds*, Herbicides, 3e, vol. 1 (eds. P. Jeschke, M. Witschel, W. Krämer and U. Schirmer), 387–423. VCH-Wiley: Weinheim.

16 Ahrens, H. (2015). Indaziflam: An innovative broad spectrum herbicide. In: *Discovery and Synthesis of Crop Protection Products*, ACS Symposium Series, vol. 1204 (eds. P. Maienfisch and T.M. Stevenson), 233–245. Washington, DC: American Chemical Society.

17 Sebastian, D.J., Fleming, M.B., Patterson, E.L. et al. (2017). *Pest. Manage. Sci.* 73: 2149–2162.

18 (a) Yamaji, Y., Honda, H., Kobayashi, M. et al. (2014). *J. Pestic. Sci. (Tokyo, Japan)* 39: 165–169. (b) Nakatani, M., Yamaji, Y., Honda, H., and Uchida, Y. (2016). *J. Pestic. Sci. (Tokyo, Japan)* 41: 107–112.

19 Nakatani, M., Ito, M., Yoshimura, T. et al. (2016). *J. Pestic. Sci. (Tokyo, Japan)* 41: 133–144.

20 Westra, E.P., Shaner, D.L., Westra, P.H., and Chapman, P.L. (2014). *Weed Technol.* 28: 72–81.

21 Schmitzer, P.R., Morell, M., Gast, R.E., and Weimer, M.R. (2019). New auxin mimic herbicides: 6-arylpicolinates. In: *Modern Crop Protection Compounds*, Herbicides, 3e, vol. 1 (eds. P. Jeschke, M. Witschel, W. Krämer and U. Schirmer), 343–350. VCH-Wiley: Weinheim.

22 Jeschke, P. (2016). *Pest Manage. Sci.* 72: 210–225.

23 Schmitzer, P.R., Balko, T.W., Daeuble, J.F. et al. (2015). Discovery and SAR of halauxifen methyl: a novel auxin herbicide. In: *Discovery and Synthesis of Crop Protection Products*, ACS Symposium Series, vol. 1204 (eds. P. Maienfisch and T.M. Stevenson), 247–260. Washington, DC: American Chemical Society.

24 Simon, S. and Patrasek, J. (2011). *J. Plant Sci.* 180: 454–460.

25 Epp, J.B., Alexander, A.L., Balko, T.W. et al. (2016). *Bioorg. Med. Chem.* 24: 362–371.

26 Evans, J.P. and Hawkes, T.R. (2019). Hydroxyphenylpyruvate dioxygenase (HPPD): the herbicide target. In: *Modern Crop Protection Compounds*, Herbicides, 3e, vol. 1 (eds. P. Jeschke, M. Witschel, W. Krämer and U. Schirmer), 241–252. VCH-Wiley: Weinheim.

27 Edmunds, A.J., De Mesmaeker, A., Wendeborn, S.V., Rueegg, W.T., Michel, A.M., Schaetzer, J.H., Hall, R.G., and Beaudegnies, R. (2017). Abstracts of Papers on Discovery of bicyclopyrone, *254th ACS National Meeting and Exposition*, Washington, DC, USA (20–24 August 2017). AGRO-411.

28 Brownlee, J.M., Johnson-Winters, D.H.T., Harrison, G.R., and Moran, G. (2004). *Biochemistry* 43: 6370–6377.

29 Jeanmart, S., Edmunds, A.J.F., Lamberth, C., and Pouliot, M. (2016). *Bioorg. Med. Chem.* 24: 317–334.

30 Dayan, F.E. (2019). Basel, Switzerland. *Plants* 341, 8 (9).

31 Dayan, F.E., Haesaert, G., Van Leeuwen, T. et al. (2019). *Outlooks Pest Manage.* 30: 157–163.

32 De-Paepe, I., Fritz-Piou, S., Sanyas, A. et al. (2011). *Phytoma* 649: 45–47.

33 Walter, H., Tobler, H., Gribkov, D., and Corsi, C. (2015). *Chimia* 69: 425–434.

34 Zeun, R., Scalliet, G., and Oostendorp, M. (2013). *Pest Manage. Sci.* 69: 527–534.

35 Yanase, Y., Yoshikawa, Y., Kishi, J., and Katsuta, H. (2007). The history of complex II inhibitors and the discovery of penthiopyrad. In: *Pesticide Chemistry: Crop Protection, Public Health, Environmental Safety* (eds. H. Ohkawa, H. Miyagawa and P.W. Lee), 295–303. Weinheim: Wiley-VCH.

36 Ishii, H., Zhen, F., Hu, M., and Li, X. (2016). *Pest Manage. Sci.* 72: 1844–1853.

37 Zhong, L., Zhang, F., Jiang, T. et al. (2019). *Heterocycles* 98: 637–649.

38 Sierotzki, H. and Scalliet, G.A. (2013). *Phytopathology* 103: 880–887.

39 Walter, H. (2016). Fungicidal succinate-dehydrogenase-inhibiting carboxamides. In: *Bioactive Carboxylic Compound Classes – Pharmaceuticals and Agrochemicals* (eds. C. Lamberth and J. Dinges), 405–425. Weinheim: Wiley-VCH.

40 Oda, M., Furuya, T., Morishita, Y. et al. (2017). *J. Pestic. Sci. (Tokyo, Japan)* 42: 151–157.

41 Yi-Ping, H., Xue-Wei, M., Shi-Peng, L. et al. (2018). *Pestic. Biochem. Physiol.* 45: 22–28.

42 Rajan, R., Walter, H., and Stierli, D. (2010). Preparation of pyrazolecarboxylic acid alkoxyamides as agrochemical microbiocides, WO 2010063700 A2, filed 5 December 2008 and issued 10 June 2010).

43 Hao, S., Tian, J., Xu, Y. et al. (2012). *Nongyao* 51: 410–412.

44 Hillebrand, D., Tietjen, K., and Zundel, J.-L. (2019). Fungicides with unknown mode of action. In: *Modern Crop Protection Compounds*, Fungicides, 3e, vol. 2 (eds. P. Jeschke, M. Witschel, W. Krämer and U. Schirmer), 911–932. VCH-Wiley: Weinheim.

45 Stenzel, K. and Vors, J.-P. (2019). Sterol biosynthesis inhibitors. In: *Modern Crop Protection Compounds*, 3e (eds. P. Jeschke, M. Witschel, W. Krämer and U. Schirmer), 797–844. Weinheim, Germany: VCH-Wiley.

46 Siepe, I., Strobel, D., Bryson, R., Schuster, M., Steinberg, G., Smith, J., and Kurup, S. (2019). Revysol® – fungicidal action on a microscopical level. *Poster P9.4, IUPAC International Congress*, Ghent, Belgium (19–24 May).

47 Schindler, M., Sawada, H., Tietjen, K. et al. (2019). Melanin synthesis in the cell wall. In: *Modern Crop Protection Compounds*, Fungicides, 3e, vol. 2 (eds.

P. Jeschke, M. Witschel, W. Krämer and U. Schirmer), 879–909. VCH-Wiley: Weinheim.

48 Okamoto, S., Sakurada, M., Kubo, Y. et al. (2001). *Microbiology* 147: 2623–2628.

49 (a) Cohen, Y. (2015). *PLoS One* 10 (10): e0140015. (b) Pasteris, R.J., Hanagan, M.A., Bisaha, J.J. et al. (2015). The discovery of oxathiapiprolin: A new, highly-active *Oomycete* fungicide with a novel site of action. In: *Discovery and Synthesis of Crop Protection Products*, ACS Symposium Series, vol. 1204 (eds. P. Maienfisch and T.M. Stevenson), 149–161. Washington, DC: American Chemical Society.

50 Pasteris, R.J., Hoffmann, L.E., Sweigard, J.A. et al. (2019). Oxysterol-binding protein inhibitors: Oxathiapiprolin – a new *Oomycete* fungicide that targets an oxysterol-binding protein. In: *Modern Crop Protection Compounds*, Fungicides, 3e, vol. 2 (eds. P. Jeschke, M. Witschel, W. Krämer and U. Schirmer), 979–987. Weinheim: VCH-Wiley.

51 Pasteris, R.J., Hanagan, M.A., Bisaha, J.J. et al. (2016). *Bioorg. Med. Chem.* 24: 354–361.

52 Meyer, K.G., Yao, C., Lu, Y., Bravo, K., Buchan, Z., Daeuble, J., Dekorver, K., Herrick, J., Jones, D.M., and Loy, B.A. (2019). Abstracts of Papers on Discovery of florylpicoxamid, a new picolinamide for disease control, *258th ACS National Meeting and Exposition*, San Diego, CA, USA (25–29 August 2019). AGRO-0182.

53 Shibata, K., Hanafi, M., Fujii, J. et al. (1998). *Antibiotics* 51: 1113–1116.

54 (a) Kagabu, S. (1997). *Rev. Toxicol. (Amsterdam)* 1: 75–129. (b) Jeschke, P., Nauen, R., Schindler, M., and Elbert, A. (2011). *J. Agric. Food. Chem.* 59: 2897–2908. (c) Jeschke, P., Nauen, R., and Beck, M.E. (2013). *Angew. Chem. Int. Ed.* 52: 9464–9485.

55 Loso, M.L., Benko, Z., Buysse, A. et al. (2016). *Bioorg. Med. Chem.* 24: 378–382.

56 Jeschke, P., Nauen, R., Gutbrod, O. et al. (2015). *Pestic. Biochem. Physiol.* 121: 31–38.

57 Jeschke, P., Nauen, R., Velten, R. et al. (2019). Butenolides: Flupyradifurone. In: *Modern Crop Protection Compounds*, Insecticides, 3e, vol. 3 (eds. P. Jeschke, M. Witschel, W. Krämer and U. Schirmer), 1361–1384. VCH-Wiley: Weinheim.

58 Beck, M.E., Gutbrod, O., and Matthiesen, S. (2015). *ChemPhysChem* 13: 2760–2667.

59 Cordova, D., Benner, E.A., Schroeder, M.E. et al. (2016). *Insect Biochem. Mol. Biol.* 74: 32–41.

60 Ebbinghaus-Kintscher, U., Lümmen, P., Hamaguchi, H. et al. (2019). Ryanodine receptor modulators: diamides. In: *Modern Crop Protection Compounds*, Insecticides, 3e, vol. 3 (eds. P. Jeschke, M. Witschel, W. Krämer and U. Schirmer), 1541–1548. VCH-Wiley: Weinheim.

61 (a) Lahm, G.P., Cordova, D., Barry, J.D. et al. (2019). Anthranilic diamide insecticides: Chlorantraniliprole and cyantraniliprole. In: *Modern Crop Protection Compounds*, Insecticides, 3e, vol. 3 (eds. P. Jeschke, M. Witschel, W. Krämer and U. Schirmer), 1562–1583. Weinheim: VCH-Wiley. (b) Ebbinghaus-Kintscher, U., Lümmern, P., Hamaguchi, H., and Hirooka, T. (2019). Flubendiamide. In:

Modern Crop Protection Compounds, Insecticides, 3e, vol. 3 (eds. P. Jeschke, M. Witschel, W. Krämer and U. Schirmer), 1549–1562. Weinheim: VCH-Wiley.

62 Selby, T.P., Lahm, G.P., Stevenson, T.M. et al. (2013). *Bioorg. Med. Chem. Lett.* 23: 6341–6345.

63 Shelton, K.A. and Lahm, G.P. (2015). Building a successful crop protection pipeline: molecular starting points for discovery. In: *Discovery and Synthesis of Crop Protection Products*, ACS Symposium Series, vol. 1204 (eds. P. Maienfisch and T.M. Stevenson), 15–23. Washington, DC: American Chemical Society.

64 Yoshihisa, O., Tomo, K., Fumiyo, O. et al. (2013). *Pestic. Biochem. Physiol.* 107: 285–292.

65 Katsuta, H., Nomura, M., Wakita, T. et al. (2019). *J. Pestic. Sci. (Tokyo, Japan)* 44: 120–128.

66 Nakao, T. and Banaba, S. (2016). *Bioorg. Med. Chem.* 24: 372–377.

67 Ffrench-Constant, R.H., Williamson, M.S., Emyr Davis, T.G., and Bass, C. (2016). *J. Neurogen.* 30: 163–177.

68 Asahi, M., Kobayashi, M., Kagami, T. et al. (2018). *Pestic. Biochem. Physiol.* 151: 67–72.

69 Weber, T. and Selzer, P.M. (2016). *ChemMedChem* 11: 270–276.

70 Ozoe, Y., Asahi, M., Ozoe, F. et al. (2010). *Biochem. Biophys. Res. Commun.* 391: 744–749.

71 Asahi, M., Kagami, T., Nakahira, K., Kobayashi, M., and Ozoe, Y. (2017). Mode-of-action studies of a novel ligand-gated chloride channel antagonist insecticide, fluxametamide. *254th ACS National Meeting and Exposition*, Washington, DC (20–24 August). AGRO-308308.

72 Satoh, E., Kasahara, R., Aoki, T. et al. (2017). *Int. Scholarly Sci. Res. Innov.* 11: 719–722.

73 Satoh, E. (2019). Synthesis and biological activity of a novel insecticide, benzpyrimoxan. Lecture 3.1.10. *IUPAC International Congress*, Ghent, Belgium (19–24 May).

74 (a) Furuya, T., Takashi, F., Motofumi, N. et al. (2015). Synthesis and biological activity of a novel acaricide, pyflubumide. In: *Discovery and Synthesis of Crop Protection Products*, ACS Symposium Series, vol. 1204 (eds. P. Maienfisch and T.M. Stevenson), 379–389. Washington, DC: American Chemical Society. (b) Furuya, T., Machiya, K., Fujioka, S. et al. (2017). *J. Pestic. Sci. (Tokyo, Japan)* 42: 132–136.

75 Motofumi, N., Noriaki, Y., Akiyuki, S. et al. (2015). *J. Pestic. Sci. (Tokyo, Japan)* 40: 19–24.

76 Urch, C.J., Lewis, T., Sunley, R. Raymond, L., Salmon, R., Godfrey, C.R.A., Brightwell, C.I., and Hotson, M.B. (1998). Preparation of 8-azabicyclo[3.2.1] octane, 8-azabicyclo[3.2.1]oct-6-ene, 9-azabicyclo[3.3.1]nonane, 9-aza-3-oxabicyclo[3.3.1]nonane, and 9-aza-3-thiabicyclo[3.3.1]nonane derivatives as insecticides, WO 9825924 A1, filed 6 November 1997 and issued 18 June 1998.

77 (a) Kearn, J., Ludlow, E., Dillon, J. et al. (2014). *Pestic. Biochem. Physiol.* 109: 44–57. (b) Norshie, P.M., Grove, I.G., and Back, M.A. (2016). *Pest Manage. Sci.*

72: 2001–2007. (c) Oka, Y., Shuker, S., and Tkachi, N. (2011). *Pest Manage. Sci.* 68: 268–275.

78 Maienfisch, P., Loiseleur, O., and Slaats, B. (2019). Recent nematicides. In: *Modern Crop Protection Compounds*, Insecticides, 3e, vol. 3 (eds. P. Jeschke, M. Witschel, W. Krämer and U. Schirmer), 1585–1614. VCH-Wiley: Weinheim.

79 Pitterna, T., Böger, M., and Maienfisch, P. (2004). *Chimia* 58: 108–116.

80 Lümmen, P. and Fürsch, H. (2019). Fluopyram a novel nematicide for the control of root-knot nematodes. In: *Modern Crop Protection Compounds*, Insecticides, 3e, vol. 3 (eds. P. Jeschke, M. Witschel, W. Krämer and U. Schirmer), 1630–1643. Weinheim: VCH-Wiley.

81 Suwa, A. (2008). Nematocidal agents comprising pyridylethylbenzamides and method of using the same, WO 2008/126922 A1, filed 23 October 2008 and issued 23 October 2008.

82 Lahm, G.P., Wiles, J.A., Cordova, D. et al. (2019). Fluazaindolizine: a new active ingredient for the control of plant-parasitic nematodes. In: *Modern Crop Protection Compounds*, Insecticides, 3e, vol. 3 (eds. P. Jeschke, M. Witschel, W. Krämer and U. Schirmer), 1643–1653. VCH-Wiley: Weinheim.

83 Lahm, G.P., Desaeger, J., Smith, B.K. et al. (2017). *Bioorg. Med. Chem. Lett.* 27: 1572–1575.

84 Barton, H.A., Pastoor, T.P., Baetcke, K. et al. (2006). *Crit. Rev. Toxicol.* 36: 9–35.

85 Janssen, D.B., Dinkla, I.J.T., Poelarends, G.J., and Terpstra, P. (2005). *Environ. Microbiol.* 7: 1868–1882.

12

Precision Radiochemistry for Fluorine-18 Labeling of PET Tracers

Jian Rong[1], Ahmed Haider[1,2] and Steven Liang[1]

[1]*Massachusetts General Hospital and Harvard Medical School, Department of Radiology, Division of Nuclear Medicine and Molecular Imaging, 55 Fruit Street, Boston, MA 02114, USA*
[2]*University Hospital Zurich, Department of Nuclear Medicine, Cardiovascular Gender Medicine, Rämistrasse 100, Zurich 8091, Switzerland*

12.1   Introduction

In the past several decades, positron emission tomography (PET) has been a rapidly developing nuclear imaging technique that emerged as a crucial tool in preclinical and clinical disease diagnosis and therapy monitoring [1]. PET is further used to perform pharmacokinetics and receptor occupancy studies in drug development [2–4]. Compared with medical imaging techniques that primarily provide structural and anatomical information, including magnetic resonance imaging (MRI), computed tomography (CT), and ultrasound, PET provides quantitative functional information about biological processes in target tissues and organs. To date, PET is typically used in combination with CT or MRI for anatomical orientation and tissue attenuation correction. As a noninvasive medical imaging technique, PET has a high level of sensitivity, since only 10^{-9}–10^{-6} g of radioactive PET tracer is needed for the high-quality imaging without eliciting pharmacological effects [4]. Following the administration of a PET radioligand, careful evaluation of radioactivity accumulation in the target organ allows nuclear radiologists to draw conclusions on the patient's health condition. PET imaging is accomplished by the detection of PET probes labeled with frequently used positron-emitting radionuclides including ^{11}C (carbon-11, $t_{1/2}$ = 20.4 minutes), ^{13}N (nitrogen-13, $t_{1/2}$ = 10.0 minutes), ^{15}O (oxygen-15, $t_{1/2}$ = 2.0 minutes), ^{18}F (fluorine-18, $t_{1/2}$ = 109.8 minutes), as well as several commonly used radiometals, such as ^{64}Cu (copper-64, $t_{1/2}$ = 12.7 hours), ^{68}Ga (gallium-68, $t_{1/2}$ = 68 minutes), and ^{89}Zr (zirconium-89, $t_{1/2}$ = 78.4 hours). Among these positron emitters, fluorine-18 is the most commonly used positron-emitting radionuclide, possibly owing to the widespread use of [^{18}F]FDG (2-[^{18}F]fluoro-2-deoxy-D-glucose), which is applied in clinical routine to diagnose various cancer entities, ischemic heart disease, and neurodegenerative brain disorders [1, 5, 6]. Fluorine-18 has an optimal half-life (109.8 minutes) that provides sufficient time for multistep radiosyntheses and

Organofluorine Chemistry: Synthesis, Modeling, and Applications,
First Edition. Edited by Kálmán J. Szabó and Nicklas Selander.

accumulation of ^{18}F-labeled probes in target organs. Concurrently, the half-life of 109.8 minutes limits the radiation burden for patients due to the relatively fast decay after a PET scan. Moreover, fluorine-18 has a high β^+-decay pattern (97%) and a relatively low positron energy (0.635 MeV), which confines positrons to a short travel path, thereby leading to PET images with high resolution. Notably, the reliable large-scale (several Curie) productions of [^{18}F]fluoride provide excellent molar activities. Attributed to these favorable characteristics of fluorine-18 and the rapidly growing use of PET, there will be a continuous demand for novel ^{18}F-labeled methods to incorporate fluorine-18 into highly selective PET radiopharmaceuticals for noninvasive assessment of physiological and pathological processes in the living organism.

In this chapter, we summarize major developments and breakthroughs in ^{18}F-labeling chemistry of PET radioligands in the past decade including electrophilic ^{18}F-fluorination with [^{18}F]F$_2$ and [^{18}F]F$_2$-derived reagents as well as nucleophilic ^{18}F-fluorination with [^{18}F]fluoride. Although stepwise ^{18}F-labeling methods involving the ^{18}F-labeling of a prosthetic group [7, 8] and radiofluorinations via B—, Al—, Si—, S—, and Ga—^{18}F bond formations [9–12] offer a solution for the radiolabeling of complex biomolecules, these methods are out of the scope of this book chapter.

12.2 Electrophilic ^{18}F-Fluorination with [^{18}F]F$_2$ and [^{18}F]F$_2$-Derived Reagents

[^{18}F]F$_2$ was the first electrophilic ^{18}F-fluorination reagent produced via ^{20}Ne(d,α)^{18}F or ^{18}O(p,n)^{18}F nuclear reaction, using nonradioactive [^{19}F]F$_2$ as a carrier. Importantly, the limited molar activity (the measured radioactivity per total amount in mole of compound) of the resulting [^{18}F]F$_2$ constitutes a major drawback of this approach, and typically ranges from 0.04 to 0.4 GBq/µmol (0.001–0.01 Ci/µmol). Thus, an improved preparation of [^{18}F]F$_2$ was suggested through gas-phase ^{19}F/^{18}F isotopic exchange between [^{18}F]CH$_3$F and [^{19}F]F$_2$, whereas the molar activity of [^{18}F]F$_2$ could be increased to 55 GBq/µmol (1.5 Ci/µmol) [13]. [^{18}F]F$_2$ is a highly reactive electrophilic ^{18}F-fluorination reagent that often leads to ^{18}F-labeling with poor chemo- and regioselectivity as well as low radiochemical yields (RCY, defined as the amount of activity in the product expressed as a percentage [%] of starting activity used in the considered process [14]). The frequently observed formation of isomeric by-products further complicates purification processes. In order to overcome these limitations; several electrophilic ^{18}F-fluorination reagents have been developed from [^{18}F]F$_2$, including O-^{18}F reagents ([^{18}F]FClO$_3$ [15], [^{18}F]CF$_3$OF [16], [^{18}F]CH$_3$COOF [17]), N-^{18}F reagents (N-[^{18}F]fluoropyridinium triflate [18], 1-[^{18}F]fluoro-2-pyridone [19], as well as the most recent [^{18}F]NFSI (N-fluorobenzenesulfonimide) [20] and [^{18}F]Selectfluor [21]), and [^{18}F]XeF$_2$ [22], which can be prepared through ^{19}F/^{18}F exchange between XeF$_2$ and [^{18}F]HF or via treatment of XeF$_2$ with [^{18}F]fluoride in dichloromethane at ambient temperature or in acetonitrile at elevated temperatures [23–25]. Among these electrophilic

^{18}F-fluorination reagents, *N*-^{18}F reagents such as [^{18}F]NFSI and [^{18}F]Selectfluor are more stable and easier to handle. Indeed, NFSI (*N*-fluorobenzenesulfonimide) is a commonly used mild electrophilic fluorination reagent in organofluorine chemistry. Accordingly, [^{18}F]NFSI provides the possibility for chemo-, regio-, diastereo-, and enantioselective ^{18}F-fluorinations. In 2015, Gouverneur and coworkers reported an enamine-mediated enantioselective ^{18}F-fluorination of aldehydes with [^{18}F]NFSI (Figure 12.1a) [26]. Enantioselective ^{18}F-fluorination of aldehydes was conducted with a chiral imidazolidinone and [^{18}F]NFSI. This method was used in the synthesis of (2*S*,4*S*)-4-[^{18}F]fluoroglutamic acid in 62% RCC with high enantioselectivity (>99% ee, enantiomeric excess) and high diastereoselectivity (19.1 dr) in a one-pot, three-step process including subsequent oxidation of –CHO to –COOH and deprotection. In 2017, Nodwell et al. developed a method for the selective photocatalytic ^{18}F-fluorination of nonactivated C—H bonds with [^{18}F]NFSI in branched aliphatic amino acids (Figure 12.1b) [27]. The two-step reaction involved initial hydrogen atom abstraction from the tertiary carbon in the aliphatic chain of amino acids mediated via the photoactivated decatungstate catalyst, followed by [^{18}F]fluorine atom transfer from [^{18}F]NFSI. This method provides direct access to several ^{18}F-fluorinated branched aliphatic amino acids including 4-[^{18}F]fluoroleucine (4-[^{18}F]FL), which was prepared in 23% decay-corrected RCY with no racemization. Compared with [^{18}F]NFSI, [^{18}F]Selectfluor is more reactive, yet, a remarkably selective electrophilic ^{18}F-fluorination reagent. In 2010, Gouverneur, Luthra, Solin, and coworkers developed [^{18}F]Selectfluor as a novel electrophilic ^{18}F-fluorination reagent and utilized it in the ^{18}F-fluorination of silyl enol ether and electron-rich arylstannanes in 50% and 14–18% RCCs, respectively [21]. In 2013, Gouverneur and coworkers extended the application of [^{18}F]Selectfluor to the synthesis of 6-[^{18}F]FDOPA [28]. Two Ag-mediated methods were reported – one of them involved the ^{18}F-fluorination of arylstannane precursors and subsequent deprotection (12% decay-corrected RCY) (Figure 12.1c). The second Ag-mediated method described the ^{18}F-fluorination of the respective arylboronic ester precursor and subsequent deprotection (19% decay-corrected RCY) (Figure 12.1d).

12.3 Nucleophilic Aliphatic ^{18}F-Fluorination

12.3.1 Transition Metal-Free Nucleophilic Aliphatic Substitution with [^{18}F]Fluoride

Owing to the convenient and reliable production of [^{18}F]fluoride with high molar activities, nucleophilic ^{18}F-fluorination is more attractive than electrophilic ^{18}F-fluorination with [^{18}F]F$_2$ and [^{18}F]F$_2$-derived reagents. Particularly, a high molar activity is of paramount value for biological targets with low tissue abundancy, including neurotransmitter receptors in the central nervous system (CNS). The most frequently used nucleophilic aliphatic ^{18}F-fluorination involves the displacement of an adequate leaving group with [^{18}F]fluoride (Figure 12.2a). These leaving groups can be –OTf (triflate), –OTs (tosylate), –OMs (mesylate), or halides.

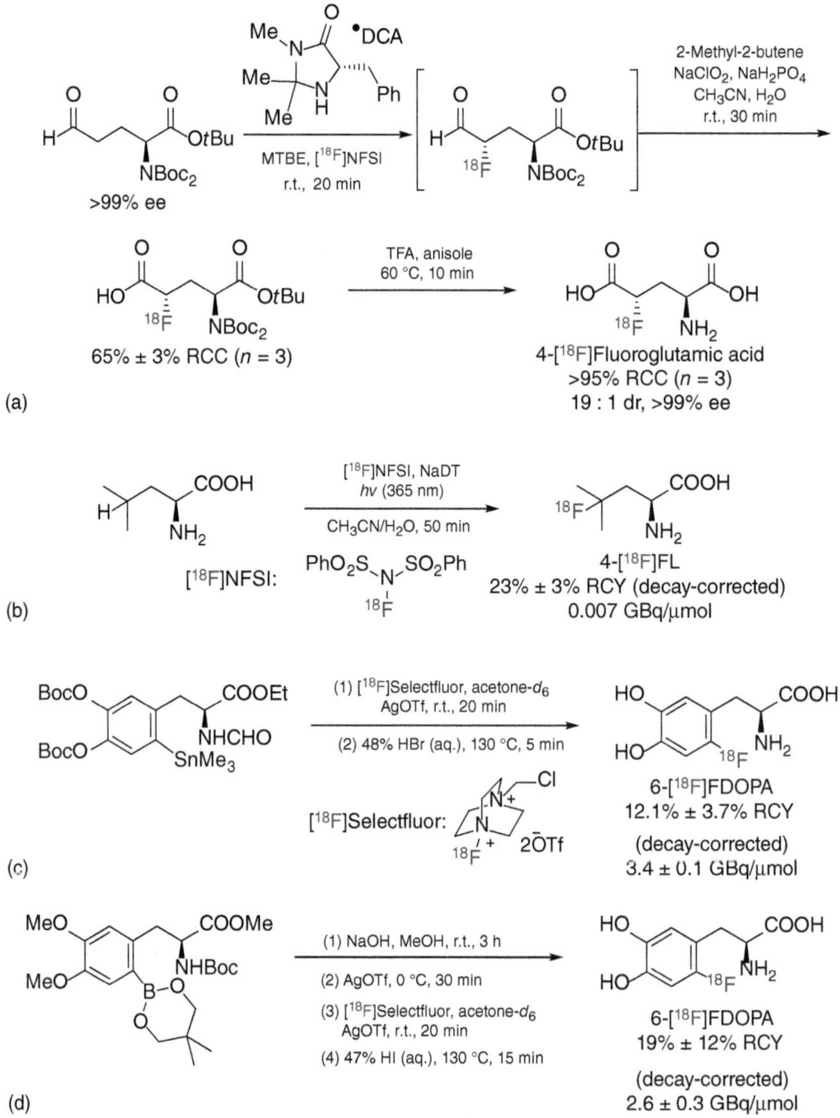

Figure 12.1 Examples of electrophilic radiofluorination via [^{18}F]NFSI or [^{18}F]Selectfluor. Source: (a) From Buckingham et al. [26]. © 2015 John Wiley & Sons. Reprinted with permission of John Wiley & Sons; (b) From Nodwell et al. [27]. © 2017 American Chemical Society. Reprinted with permission of American Chemical Society.

The leaving ability decreases in the following order: –OTf > –OTs ≈ –OMs > –I > –Br > –Cl. The most representative example for nucleophilic aliphatic substitution with [^{18}F]fluoride is [^{18}F]FDG. Initially, [^{18}F]FDG was prepared via electrophilic ^{18}F-fluorination of 3,4,6-triacetyl-D-glucal with [^{18}F]F$_2$ in low RCY and poor molar activity [32]. To date, [^{18}F]FDG is produced via treatment of commercially available tetra-O-acetyl-mannose triflate with [^{18}F]KF/K$_{222}$, followed by

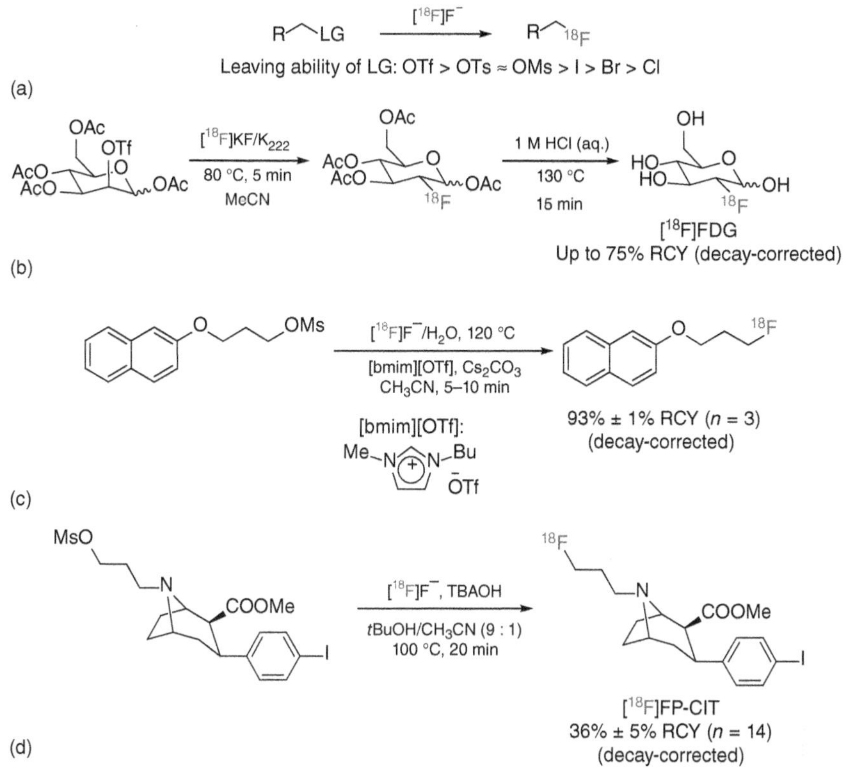

(a)

(b)

(c)

(d)

Figure 12.2 Nucleophilic aliphatic substitution reactions with [^{18}F]fluoride. Source: (b) From Hamacher et al. [29]. Reprinted with permission of Society of Nuclear Medicine and Molecular Imaging; (c) From Kim et al. [30]. Reprinted with permission of Elsevier; (d) Modified from Kim et al. [31].

deprotection, with an RCY of up to 75% (decay-corrected) (Figure 12.2b) [29]. Indeed, a plethora of other alkyl fluorides were radiofluorinated by nucleophilic aliphatic substitution with [^{18}F]fluoride. Nucleophilic aliphatic ^{18}F-fluorination is typically performed in polar aprotic solvents, such as MeCN, DMF, and DMSO to accomplish optimal ^{18}F-incorporation. Remarkably, Chi and coworkers reported on the nucleophilic ^{18}F-fluorination of aliphatic mesylates in ionic liquid (Figure 12.2c) [30]. This reaction exhibited high tolerance for water and an increased rate of radiofluorination, along with a decreased formation of undesired by-products such as alkenes, alcohols, or ethers. In 2006, the same research group further developed the nucleophilic aliphatic ^{18}F-fluorination in protic solvents by applying protic *tert*-butyl alcohol as solvent, thereby mimicking the use of ionic liquids (Figure 12.2d) [31]. Several PET tracers were prepared via this method in higher RCYs than originally described in traditional ^{18}F-fluorination approaches using polar aprotic solvents. For instance, [^{18}F]FP-CIT (*N*-(3-[^{18}F]fluoropropyl)-2β-carbomethoxy-3β-(4-iodophenyl)nortropane) was produced in 36% RCY (decay-corrected), which was substantially higher compared to its preparation in aprotic MeCN (1% RCY) [33].

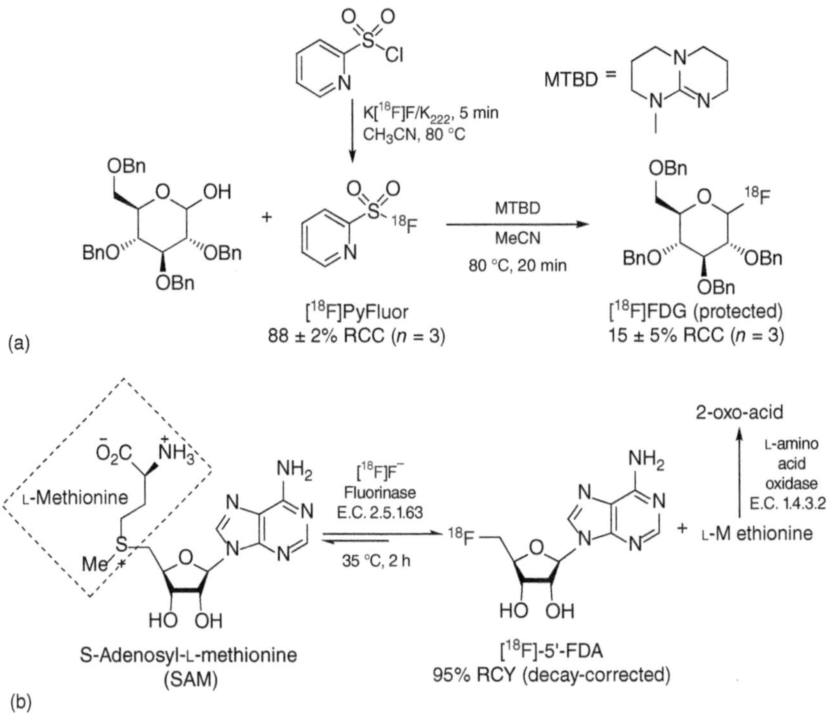

Figure 12.3 ^{18}F-Fluorination with [^{18}F]PyFluor or [^{18}F]fluoride in the presence of fluorinase. Source: (b) From Martarello et al. [35]. © 2003 John Wiley & Sons. Reprinted with permission of John Wiley & Sons.

^{18}F-Labeling of alcohols typically involves two steps, namely, the conversion of –OH (hydroxyl group) to an adequate leaving group (such as –OTs, –OMs, or –ONs) and subsequent ^{18}F-fluorination. In 2015, Doyle and coworkers reported on the deoxyfluorination reagent PyFluor (2-pyridinesulfonyl fluoride), which essentially allowed "one-pot" radiofluorination of alcohols [34]. A range of primary and secondary alcohols were smoothly deoxyfluorinated with PyFluor. In addition, the authors conducted the ^{18}F-deoxyfluorination of protected carbohydrates, thereby accomplishing the synthesis of O-Bn–[^{18}F]FDG in 15% RCC (Figure 12.3a). Complementary to these reactions, O'Hagan and coworkers developed an enzymatic ^{18}F-fluorination approach (Figure 12.3b) [35]. ^{18}F-Fluorination of SAM [(S)-adenosyl-L-methionine] was catalyzed by fluorinase (E.C. 2.5.1.63), which originated from *Streptomyces*, and yielded [^{18}F]-5'-FDA ([^{18}F]5'-fluoro-5'-deoxy-adenosine) in an RCY (decay-corrected) up to 95% [36]. In this reaction, L-amino acid oxidase (E.C. 1.4.3.2) was added to consume the by-product L-methionine, thus shifting the equilibrium toward [^{18}F]-5'-FDA. In 2014, the same research group further refined this fluorinase-catalyzed approach to afford [^{18}F]FDEA-TEG-RGD (5'-[^{18}F]fluorodeoxy-2-ethynyladenosine-labeled RGD peptides) in 12% RCY (non-decay-corrected) from the respective chloro-precursor [37].

12.3.2 Transition Metal-Mediated Aliphatic ^{18}F-Fluorination

Compared to S_N2-type aliphatic ^{18}F-fluorinations, transition metal-mediated approaches allow radiofluorination at lower temperature and provide a mild alternative to afford alkyl [^{18}F]fluorides with better functional group tolerance. In 2011, Gouverneur and coworkers described the Pd-catalyzed fluorination of allyl *p*-nitrobenzoates with Bu$_4$NF, which was successfully implemented into radiochemistry conditions using [^{18}F]Bu$_4$NF (Figure 12.4a) [38]. ^{18}F-Fluorination of allyl *p*-nitrobenzoates with [^{18}F]Bu$_4$NF in MeCN afforded cinnamyl [^{18}F]fluoride in 5–7% RCY (decay-corrected) after five minutes at ambient temperature. When cinnamyl methyl carbonate was used as precursor, the decay-corrected radiofluorination yield increased to 9–42%. In the same year, Nguyen and coworkers reported the Ir-catalyzed allylic fluorination of trichloroacetimidates [39]. This allylic fluorination of trichloroacetimidate was accomplished using [^{18}F]KF/K$_{222}$ and CSA (camphorsulfonic acid) in THF at room temperature (Figure 12.4b), affording the respective allyl fluoride in 38% RCY (decay-corrected). In 2013, Gouverneur and coworkers further reported on the Ir-catalyzed fluorination of allyl carbonates [40]. Importantly, the radiofluorination with [^{18}F]Et$_4$NF was accomplished for branched and linear ((*E*) or (*Z*)) allylic carbonates, yielding the respective products in RCYs

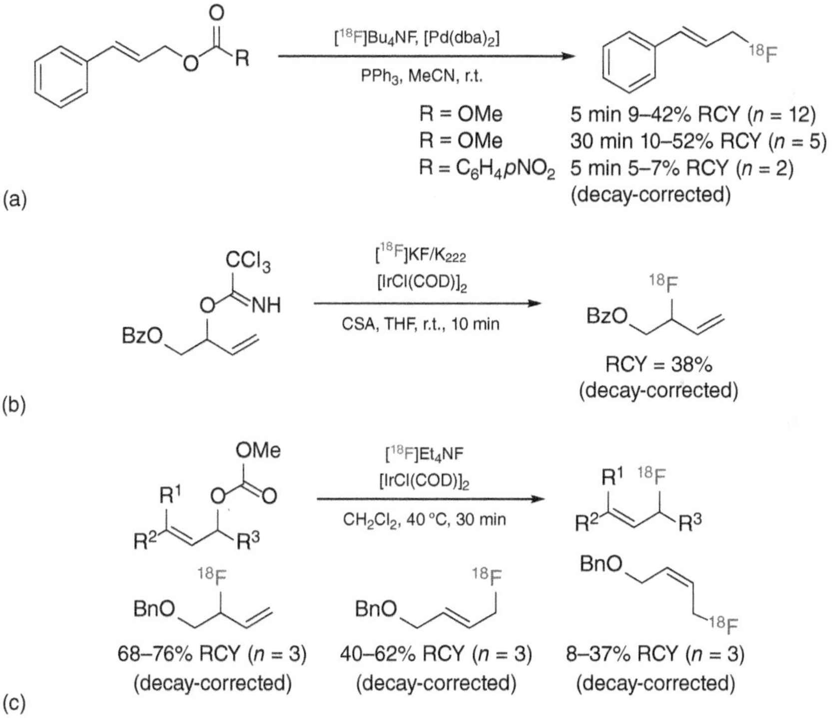

(a)

(b)

(c)

Figure 12.4 Transition metal-mediated allylic ^{18}F-fluorination. Source: (a) From Hollingworth et al. [38]. © 2011 John Wiley & Sons. Reprinted with permission of John Wiley & Sons.

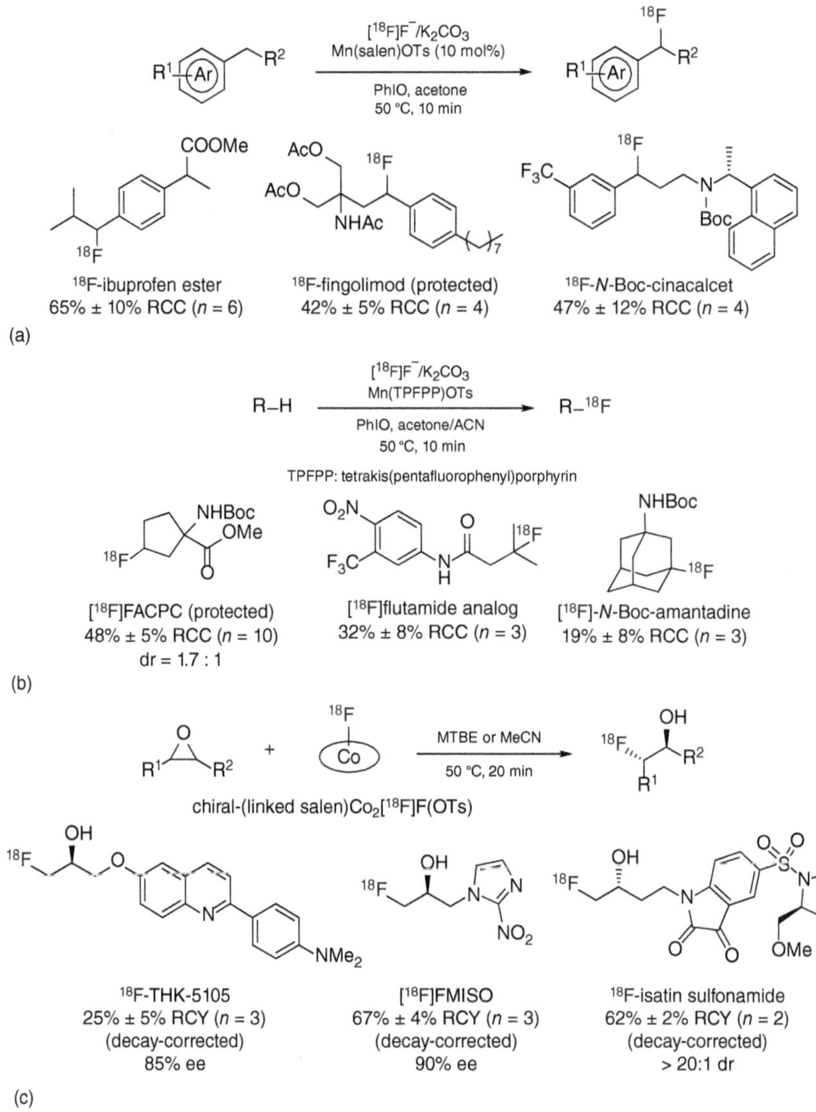

Figure 12.5 Transition metal-mediated aliphatic ^{18}F-fluorination. Source: (a) Modified from Huang et al. [41]; (b) Modified from Liu et al. [42]; (c) Modified from Graham et al. [43].

up to 76%. Moreover, position and geometry of the olefin were maintained in this reaction (Figure 12.4c).

In addition to the active allylic position, Groves, Hooker, and coworkers described a Mn-catalyzed benzylic C–H ^{18}F-fluorination (Figure 12.5a) [41]. Mn(salen)OTs was used as a catalyst that efficiently (>90%) eluted [^{18}F]fluoride ions from the ion exchange cartridge. Of note, no drying procedure was required, and a range of functional groups including halides, alkynes as well as carbonyl groups were compatible with these reaction conditions, thereby tolerating moisture and air. Indeed, several drugs were labeled via this methodology, including, but not limited to, [^{18}F]ibuprofen ester, [^{18}F]fingolimod (protected), and [^{18}F]*N*-Boc-cinacalcet (Figure 12.5a). In 2018, Groves and coworkers extended the Mn-catalyzed C–H ^{18}F-fluorination to nonactivated secondary and tertiary C—H bonds (Figure 12.5b) [42]. A more powerful catalyst, Mn(TPFPP)OTs (TPFPP, tetrakis(pentafluorophenyl)porphyrin), was employed, and a series of bioactive molecules were radiofluorinated via this method including [^{18}F]FACPC (protected), [^{18}F]flutamide analog, and [^{18}F]-*N*-Boc-amantadine in RCCs ranging from 19% to 48%. In addition to these transition metal-mediated aliphatic radiofluorinations, Doyle and coworkers developed a ring-opening reaction enabling enantioselective aliphatic ^{18}F-fluorination of epoxides with a chiral cobalt catalyst, as depicted in Figure 12.5c [43]. (*R,R*)-(Salen)Co[^{18}F]F was prepared by elution of [^{18}F]fluoride from an ion exchange cartridge with the respective tosylate precursor. Moisture and air were tolerated in this reaction, and a range of racemic epoxides were ^{18}F-labeled via epoxide ring opening in 23–68% RCYs (decay-corrected) with high enantioselectivity (>91% ee). For epoxides bearing a *Lewis* basic nitrogen or α-branching substituents, a chiral dimeric [^{18}F](linked salen)Co$_2$F(OTs) was employed to achieve high ^{18}F-labeling efficiency. Several PET tracers were prepared by this method, such as [^{18}F]THK-5105 in 25% RCY (decay-corrected) and 85% ee, [^{18}F]FMISO in 67% RCY (decay-corrected) and 90% ee, as well as ^{18}F-isatin sulfonamide in 62% RCY (decay-corrected) and high diastereoselectivity (>20 : 1 dr).

12.4 Nucleophilic Aromatic ^{18}F-Fluorination with [^{18}F]Fluoride

12.4.1 Transition Metal-Free Nucleophilic Aromatic ^{18}F-Fluorination with [^{18}F]Fluoride

Owing to the high stability of C(sp^2)—^{18}F-bonds toward *in vivo* defluorination, aryl [^{18}F]fluorides are important moieties in a wide range of PET radiotracers. Indeed, nucleophilic aromatic substitution (S$_N$Ar) with [^{18}F]fluoride is a frequently used and practical method to prepare aryl [^{18}F]fluorides. Optimal precursors for S$_N$Ar reactions with [^{18}F]fluoride typically bear a leaving group (–N$^+$Me$_3$, –S$^+$Me$_2$, –NO$_2$, –OMs, –OTs, –OTf, etc.) and an activating electron-withdrawing group such as –NO$_2$, –CN, –CF$_3$, or –CO in the *ortho* or *para* position to stabilize the Meisenheimer complex, thereby facilitating S$_N$Ar reactions (Figure 12.6a). Initially, one of the key challenges in nucleophilic aromatic radiofluorinations was the poor reactivity of deactivated electron-rich arenes. Recently, efforts have been made in the development of novel precursors for S$_N$Ar reactions with [^{18}F]fluoride to extend the substrate scope, including several novel sulfur-based precursors, triarylsulfonium salts [44–47], and diaryl sulfoxides [48]. In 2012, Ametamey and coworkers reported on the ^{18}F-fluorination of triarylsulfonium salts to afford aryl [^{18}F]fluorides [46]. Notably, this method enabled the radiofluorination of nonactivated and deactivated aryl rings with a Hammett σ_p greater than −0.170,

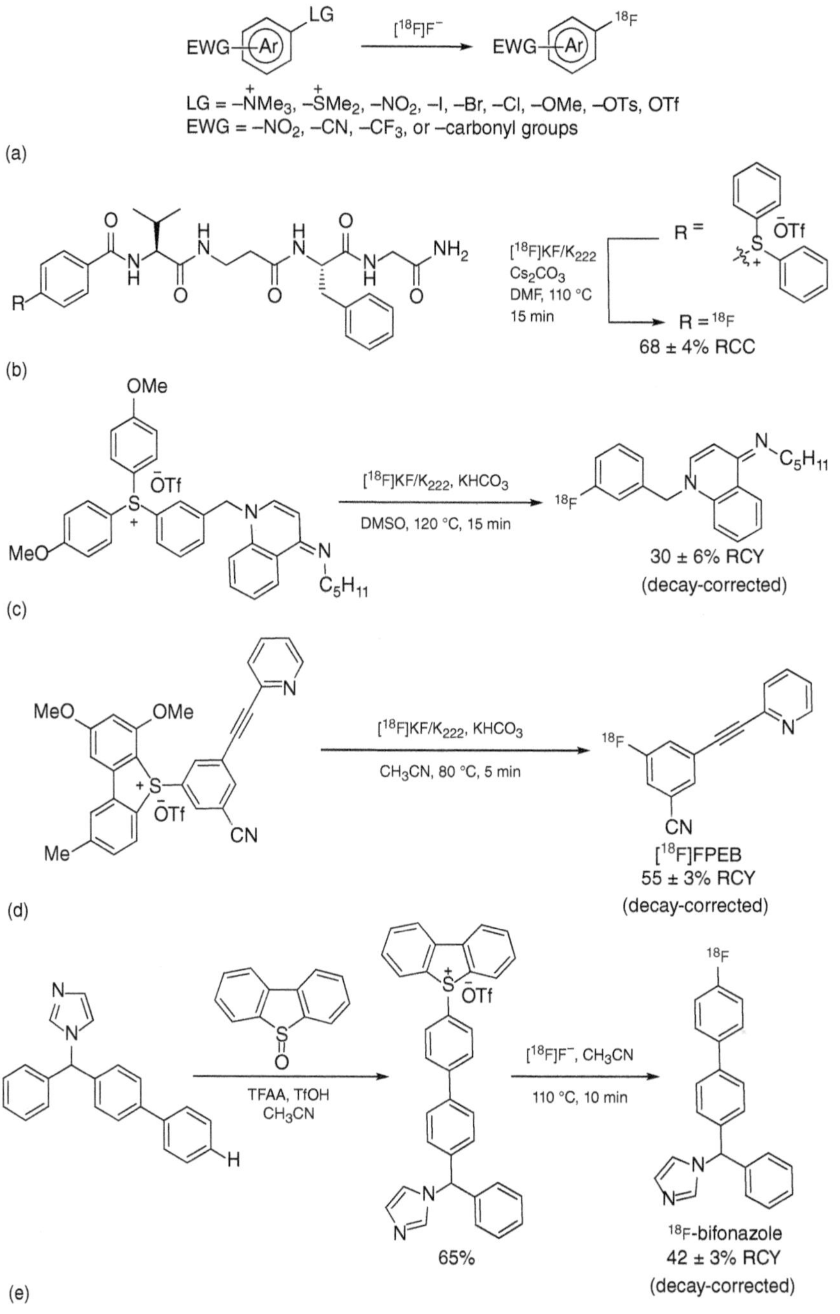

(a)

(b)

(c)

(d)

(e)

Figure 12.6 Nucleophilic aromatic ^{18}F-fluorination with [^{18}F]fluoride. Source: (c) Modified from Sander et al. [44]; (e) Modified from Xu et al. [45].

including aryl-bearing peptides, in high RCCs (Figure 12.6b). In 2015, Årstad and coworkers described the preparation of a new triarylsulfonium precursor with electron-rich spectator ligand-bearing *para*-methoxyphenyl groups (Figure 12.6c) [44]. The use of such electron-rich spectator ligands provided a broadly applicable route to prepare a broad spectrum of aryl [¹⁸F]fluorides including electron-rich substrates. In 2018, the same research group reported on a new triarylsulfonium precursor with dibenzothiophenesulfonium for ¹⁸F-fluorination (Figure 12.6d) [47]. The ¹⁸F-fluorination occurred selectively and dibenzothiophenesulfonium served as the leaving group. Several PET tracers were prepared via this method, including [¹⁸F]FPEB in 55% RCY (decay-corrected). In addition, Ritter and coworkers reported an aromatic C–H ¹⁸F-fluorination method to quickly access aryl [¹⁸F]fluorides via a two-step protocol involving arene C–H functionalization to generate aryl sulfonium salts and subsequent nucleophilic radiofluorination with [¹⁸F]fluoride (Figure 12.6e) [45]. Both electron-poor and electron-rich arenes were amenable for the site-selective C–H functionalization reaction and afforded the respective dibenzothiophenium salts with high regioselectivity and good functional group tolerance. The nucleophilic ¹⁸F-fluorination of dibenzothiophenium salts also proceeded smoothly, providing a range of ¹⁸F-labeled small molecules such as ¹⁸F-bifonazole, which was synthesized in 42% decay-corrected RCY. Of note, an electron-rich dimethoxy-substituted dibenzothiophenesulfonium served as the leaving group to achieve high selectivity of ¹⁸F-fluorination.

Besides triarylsulfonium salts, Pike and coworkers reported on the ¹⁸F-labeling of diaryl sulfoxides in 2013 (Figure 12.7a) [48]. A higher reaction temperature (150–200 °C) was required for the ¹⁸F-fluorination of diaryl sulfoxides, indicating that diaryl sulfoxides were less reactive than triarylsulfonium salts. Accordingly, this method was confined to electron-deficient aryl [¹⁸F]fluorides. Murphy and coworkers developed novel *N*-arylsydnone precursors for the ¹⁸F-fluorination of arenes bearing electron-deficient groups in 2017 (Figure 12.7b) [49]. *N*-Arylsydnones could be efficiently prepared from anilines in two steps, whereby the sydnone was inductively more electron withdrawing than the nitro functionality. Further, although sydnones were positioned away from the aryl plane, hampering full resonance with the aromatic system and facilitating radiofluorination, an activating group was still required for *N*-arylsydnones radiolabeling. Hence, the ¹⁸F-fluorination of *N*-arylsydnones was limited to electron-deficient aryl [¹⁸F]fluorides.

Owing to the broad implication of phenols in organic synthesis, direct [¹⁸F]deoxyfluorination of phenols to afford radiofluorinated arenes is a particularly appealing concept. In 2016, Ritter and coworkers described a novel concerted nucleophilic aromatic substitution mechanism, allowing the [¹⁸F]fluorination of phenols (Figure 12.8a) via the formation on an uronium complex, which is different from the classical S_NAr reaction via Meisenheimer intermediate [51]. This [¹⁸F]deoxyfluorination reaction was highly efficient in electron-deficient phenols. In 2017, the same group extended this [¹⁸F]deoxyfluorination to electron-rich phenols activated by a ruthenium complex (Figure 12.8b) [50]. Particularly, the η^6 π-activation of the CpRu(COD)Cl complex allowed the labeling of electron-rich phenols such as 4-methoxyphenol in 89% RCY (decay-corrected).

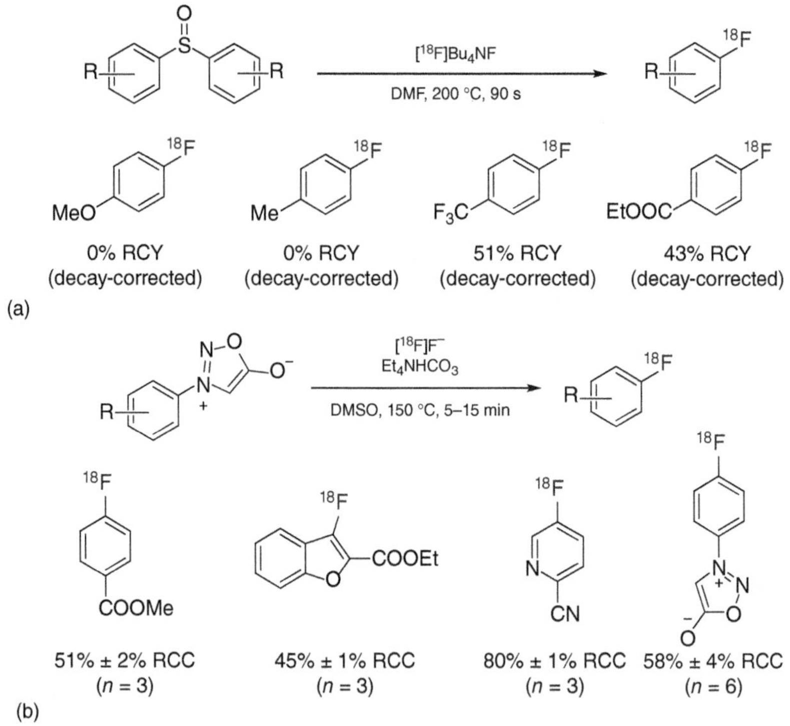

(a)

(b)

Figure 12.7 ^{18}F-Fluorination of diaryl sulfoxides (a) and *N*-arylsydnones (b). Source: (a) Modified from Chun et al. [48]; (b) Modified from Narayanam et al. [49].

^{18}F-Labeling of electron-rich arenes is typically more challenging than electron-deficient aromatics. Recently, Li, Nicewicz, and coworkers reported a direct arene C–H ^{18}F-fluorination via organic photoredox catalysis (Figure 12.9) [52]. In this reaction, arenes were labeled with [^{18}F]TBAF using TEMPO as a redox co-mediator in the presence of an acridinium-based photocatalyst and under aerobic conditions. Notably, this approach prompted the successful labeling of various deactivated aromatics bearing electron-donating substituents. PET radiopharmaceuticals prepared via this method include, but are not limited to, ^{18}F-fenoprofen (in 37% RCC over two steps) and [^{18}F]FDOPA (in 12% RCY [decay-corrected] over four steps).

Diaryliodonium salts are commonly used arylation reagents in reactions with C-, N-, O-, or S-nucleophiles and halides. In 1995, Pike and Aighirhio firstly reported on the ^{18}F-fluorination of diaryliodonium salts (Figure 12.10a) [53]. Although both electron-deficient [^{18}F]fluoroarenes and electron-rich [^{18}F]fluoroarenes were prepared via ^{18}F-fluorination of the corresponding diaryliodonium salts, unsymmetrical diaryl moieties were preferably radiofluorinated on the relatively electron-deficient aryl group. As such, [^{18}F]fluorobenzene and 1-[^{18}F]fluoro-4-methylbenzene were synthesized in 68% RCY (decay-corrected) in a ratio of 3 : 2, starting from 4-methylphenyl(phenyl)iodonium trifluoromethanesulfonate. Besides the electronic properties of the two aryl moieties, their steric

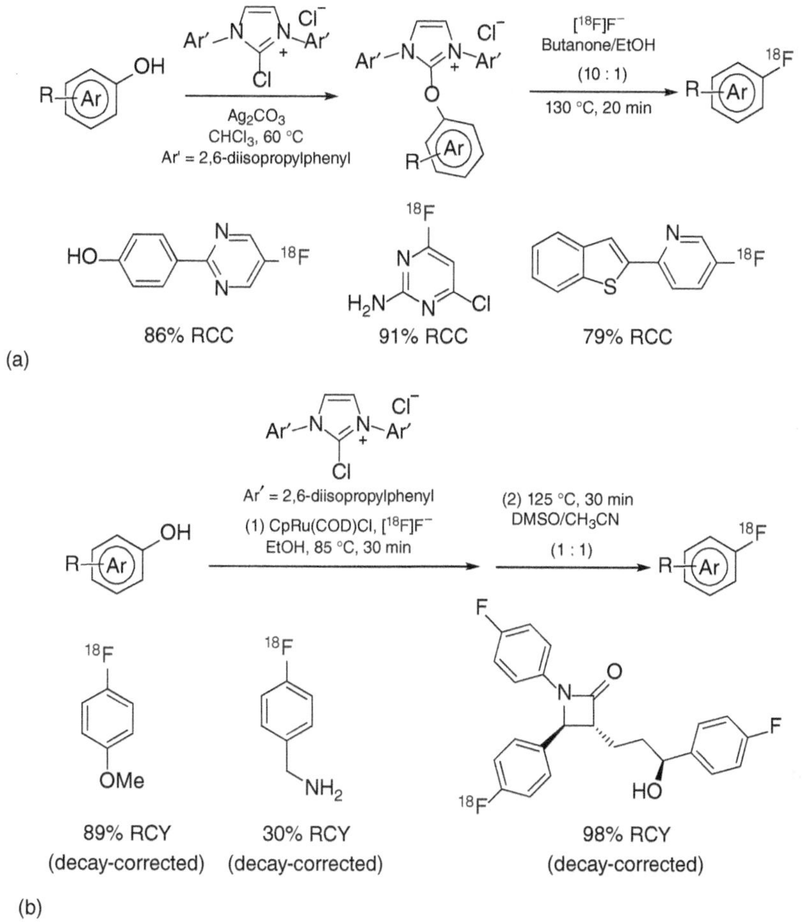

(a)

(b)

Figure 12.8 [¹⁸F]Deoxyfluorination of phenols. Source: (b) Modified from Beyzavi et al. [50].

properties significantly affected the selectivity of ¹⁸F-fluorination via diaryliodonium salts, particularly for diaryliodonium salts bearing bulky *ortho* substituents. For instance, the ¹⁸F-fluorination of 2-methylphenyl(phenyl)iodonium chloride afforded 1-[¹⁸F]fluoro-2-methylbenzene in 57% decay-corrected RCY, while the RCY (decay-corrected) of [¹⁸F]fluorobenzene was 25% (Figure 12.10b) [54]. The ability of different *ortho* substituents to promote the selective ¹⁸F-fluorination of unsymmetrical diaryliodonium salts follows the order: 2,6-di-Me > 2,4,6-tri-Me > 2-Br > 2-Me > 2-Et ≈ 2-*i*Pr > 2-H > 2-OMe. Several spectator aryl groups, including 2-thienyl [56, 57], 4-methoxyphenyl [58], 3-methoxyphenyl [58], and 4-methylphenyl [58], were used to demonstrate regioselective ¹⁸F-fluorination of unsymmetrical diaryliodonium salts. In a recent study by Chun and coworkers, the highly electron-rich 2,4,6-trimethoxyphenyl was utilized as a spectator in ¹⁸F-fluorination of aryl(2,4,6-trimethoxyphenyl)iodonium tosylates with excellent chemoselectivity (Figure 12.10c) [55]. A range of aryl [¹⁸F]fluorides

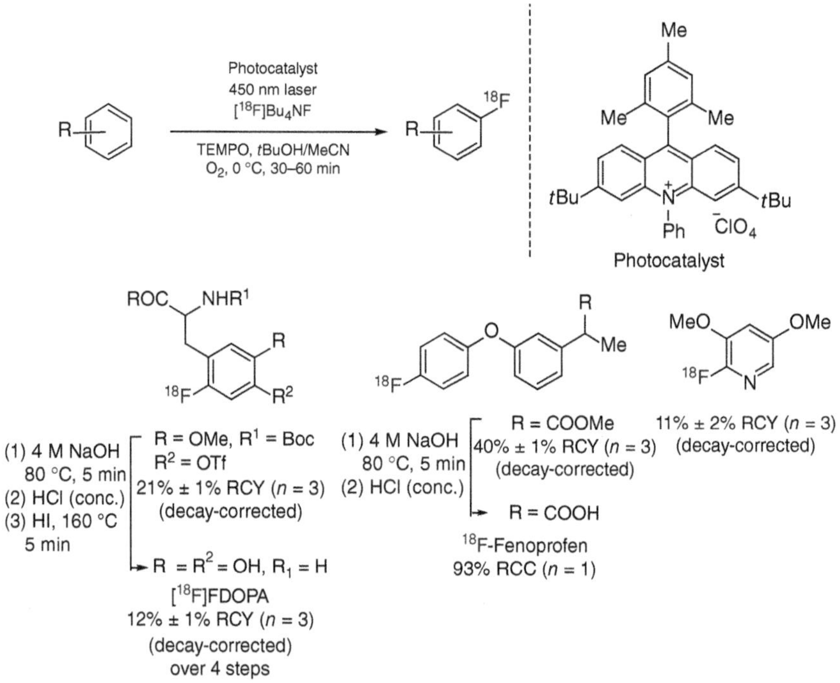

Figure 12.9 Arene C–H ^{18}F-fluorination via organic photoredox catalysis. Source: Modified from Chen et al. [52].

were prepared via this method – electron-rich 1-[^{18}F]fluoro-3-methoxybenzene, electron-deficient halogen-functionalized heteroaromatic (2-bromo-5-[^{18}F]fluoropyridine), click-labeling synthon (1-(azidomethyl)-4-[^{18}F] fluorobenzene), and [^{18}F]SFB (*N*-succinimidyl 4-[^{18}F]-fluorobenzoate) were achieved in 21%, 31%, 35%, and 20% decay-corrected RCYs, respectively. In 2016, Liang and coworkers described a novel one-pot synthesis of aryl(isoquinoline)iodonium precursors, facilitating the smooth preparation of numerous [^{18}F]isoquinolines with up to 92% RCC [59].

Notwithstanding the strenuous efforts to improve the radiofluorination of electron-rich arenes, fundamental progress has only recently been achieved when aryliodonium ylides were developed [60, 61]. In fact, aryliodonium ylides were excellent precursors with unprecedented ^{18}F-labeling efficiency in nonactivated arenes. Further, due to the lack of counterions, aryliodonium ylides were amenable to purification by normal phase liquid chromatography, facilitating large-scale precursor production. In an early report by Ermert and coworkers in 2014, ^{18}F-labeling of aryliodonium ylide with meldrum acid auxiliary generated two positional isomers via the aryne intermediate [60]. In the same year, Liang, Vasdev, and coworkers independently reported on the development of spirocyclic iodonium ylides (SCIDY) as readily available precursors, amenable to regioselective radiofluorination of nonactivated arenes (Figure 12.11a) in RCCs up to 85% [61]. Iodonium ylides with meldrum acid or barbituric acid auxiliary were often

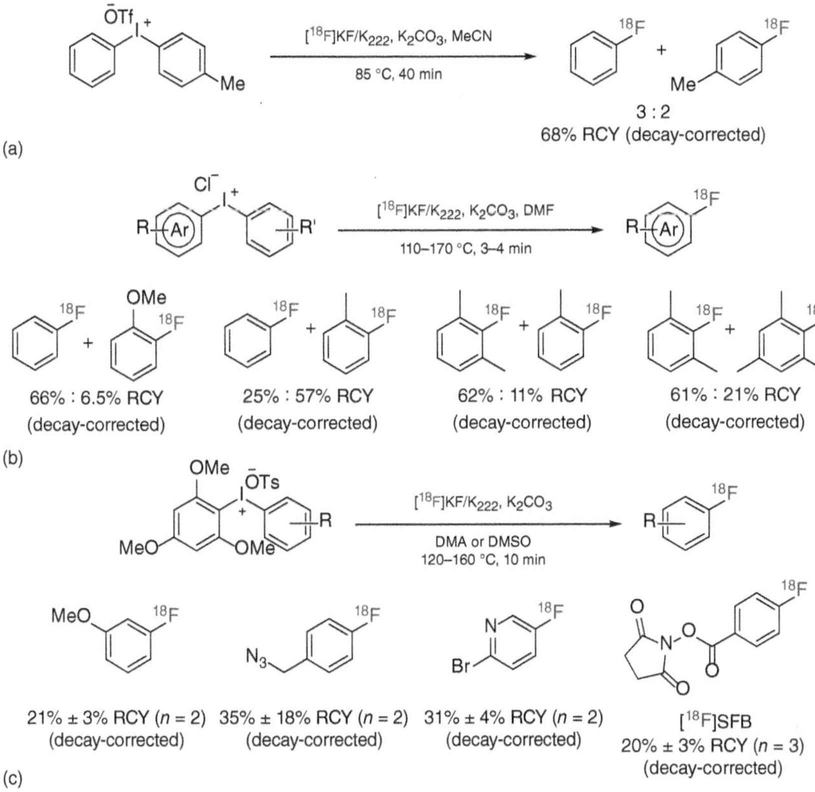

(a)

(b)

(c)

Figure 12.10 ^{18}F-Fluorination of diaryliodonium salts. Source: (a) Modified from Pike and Aighirhio [53]; (b) Modified from Chun et al. [54]; (c) Modified from Kwon et al. [55].

reported as viscous and thermally unstable oils, while SCIDY are predominantly crystalline solids with higher stability, thus facilitating precursor handling and storage. A range of aryl [^{18}F]fluorides was smoothly prepared by ^{18}F-fluorination of SCIDY precursors, including electron-rich 1-[^{18}F]fluoro-4-methoxybenzene (15% RCC), heterocyclic 3-[^{18}F]fluoropyridine (65% RCC), the click reagent 2-((2-azidoethoxy)methyl)-4-bromo-1-[^{18}F]fluorobenzene (50% non-decay-corrected RCY) [63], 5-[^{18}F]fluorouracil (11% non-decay-corrected RCY over two steps, first time prepared from nucleophilic [^{18}F]fluoride), [^{18}F]FPEB (20% non-decay-corrected RCY) [64], and ^{18}F-lorlatinib (14% decay-corrected RCY over two steps), which is an inhibitor of orphan receptor tyrosine kinase c-ros oncogene 1 that is currently being tested in clinical phase II trials for the treatment of non-small cell lung cancer [65]. Given that the thermal stability of aryliodonium ylides was highly dependent on the auxiliary group, Liang and coworkers further elucidated this concept by developing the second generation of iodonium ylides bearing a SPIAd (spiroadamantyl-1,3-dioxane-4,6-dione) auxiliary (Figure 12.11b) [62]. Indeed, SPIAd ylides outperformed all the other auxiliaries in stability tests, and a range of PET tracers were prepared via radiofluorination of SPIAd ylides, including the

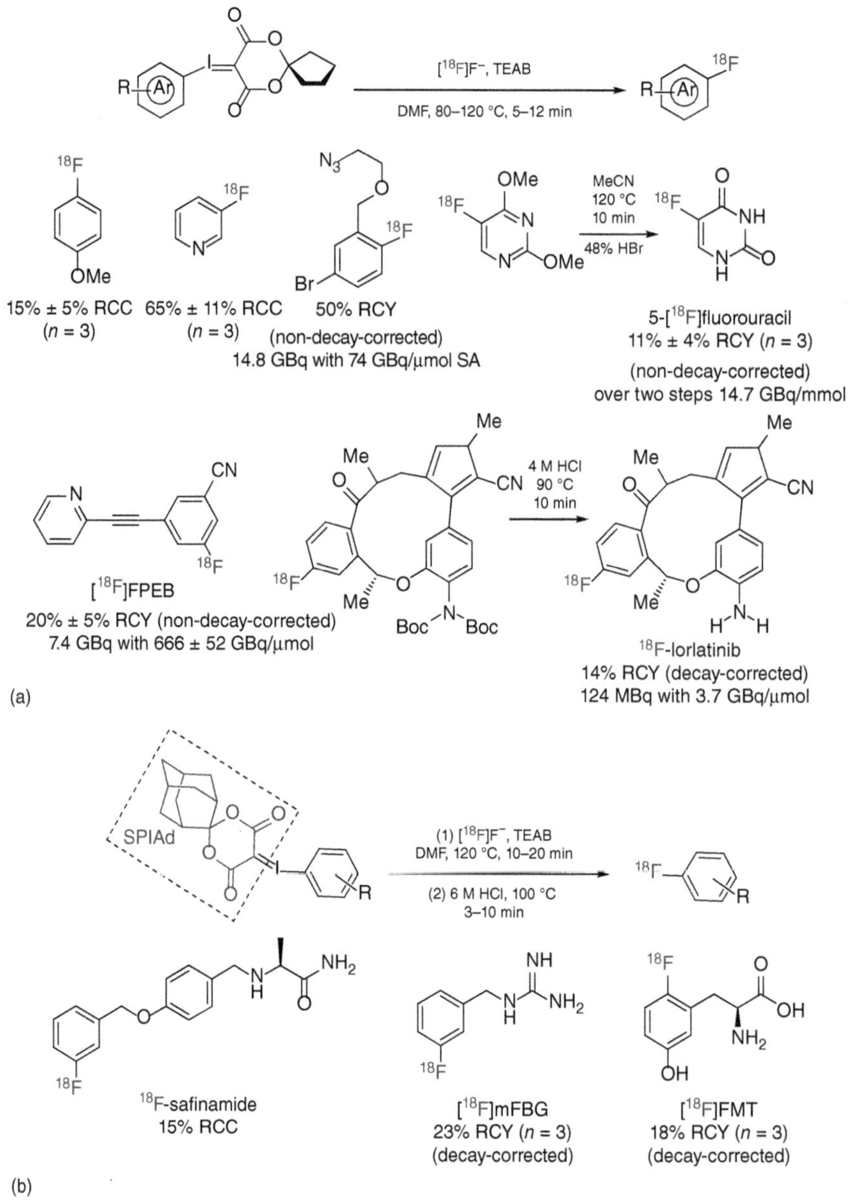

Figure 12.11 ^{18}F-Fluorination of aryliodonium ylides. Source: (b) Modified from Rotstein et al. [62].

reversible monoamine oxidase B inhibitor ^{18}F-safinamide (15% RCC over two steps), the norepinephrine transporter ligand [^{18}F]mFBG (23% decay-corrected RCY over two steps), and [^{18}F]FMT (18% decay-corrected RCY over two steps), which is a radiopharmaceutical for imaging cerebral dopamine and neuroendocrine tumors. In 2017, Riss and coworkers found that the employment of PPh$_3$ reduced

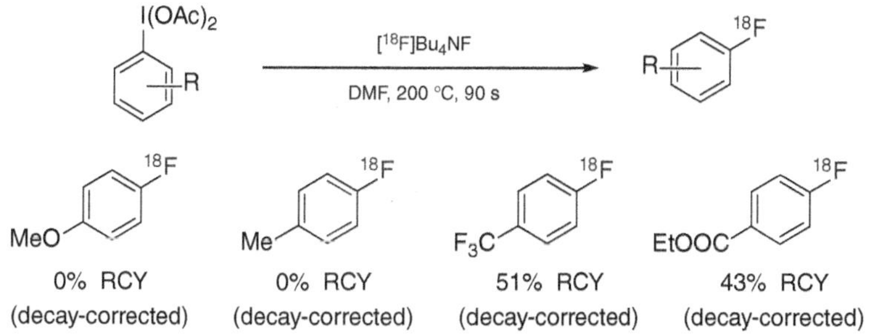

Figure 12.12 ¹⁸F-Fluorination of (diacetoxyiodo)arenes. Source: Modified from Haskali et al. [67].

the activation energy of iodonium ylide ¹⁸F-fluorinations, thereby substantially increasing reaction rates [66].

In addition to these hypervalent iodine(III) precursors-based ¹⁸F-labeling methods, Pike and coworkers reported on the ¹⁸F-fluorination of (diacetoxyiodo)arenes in 2016 (Figure 12.12) [67]. Despite the straightforward preparation of (diacetoxyiodo)arene precursors, the ¹⁸F-fluorination of (diacetoxyiodo)arenes required a high temperature (200 °C) and was limited to electron-deficient substrates.

12.4.2 Transition Metal-Mediated Aromatic ¹⁸F-Fluorination

Transition metal-mediated ¹⁸F-fluorinations are typically conducted under mild conditions, and with high functional group tolerance, thus providing an alternative approach to furnish aryl [¹⁸F]fluorides. In 2011, Ritter, Hooker, and coworkers developed an aromatic ¹⁸F-labeling method with a palladium-based electrophilic fluorination reagent as shown in Figure 12.13a [68]. The palladium complex **B**, derived from palladium complex **A** and [¹⁸F]fluoride, served as an electrophilic ¹⁸F-fluorination reagent. Particularly, the [¹⁸F]fluorine ion was transferred to palladium complex **C**, which was oxidized to a palladium(IV) aryl [¹⁸F]fluoride complex. Subsequently, aryl [¹⁸F]fluorides were achieved via C—¹⁸F bond reductive elimination. Indeed, several drugs, such as ¹⁸F-paroxetine (antidepressant, selective serotonin reuptake inhibitor) and a ¹⁸F-labeled 5-HT$_{2C}$ agonist, were prepared via this method (>1% non-decay-corrected RCY) [70]. In 2012, Ritter and coworkers developed a nickel-mediated oxidative ¹⁸F-fluorination approach (Figure 12.13b) [69]. Aryl [¹⁸F]fluorides were prepared from the nickel aryl complex **D** with aqueous [¹⁸F]fluoride in the presence of a hypervalent iodine oxidant at room temperature in less than one minute. A range of aryl [¹⁸F]fluorides, as well as alkenyl [¹⁸F]fluorides, were prepared via this method (13–38% decay-corrected RCYs). Further, they utilized the latter approach for the production of ¹⁸F-MDL100907 (3% non-decay-corrected RCY), a serotonin 2a receptor (5HT$_{2a}$) PET radioligand [60, 71], and 5-[¹⁸F]Fluorouracil (0.92 ± 0.18% non-decay-corrected RCY) with molar activities >10 Ci/μmol [72].

The radiofluorination of diaryliodonium salts under transition metal-free conditions has proved to be useful for the preparation of aryl [¹⁸F]fluorides; however, *ortho*

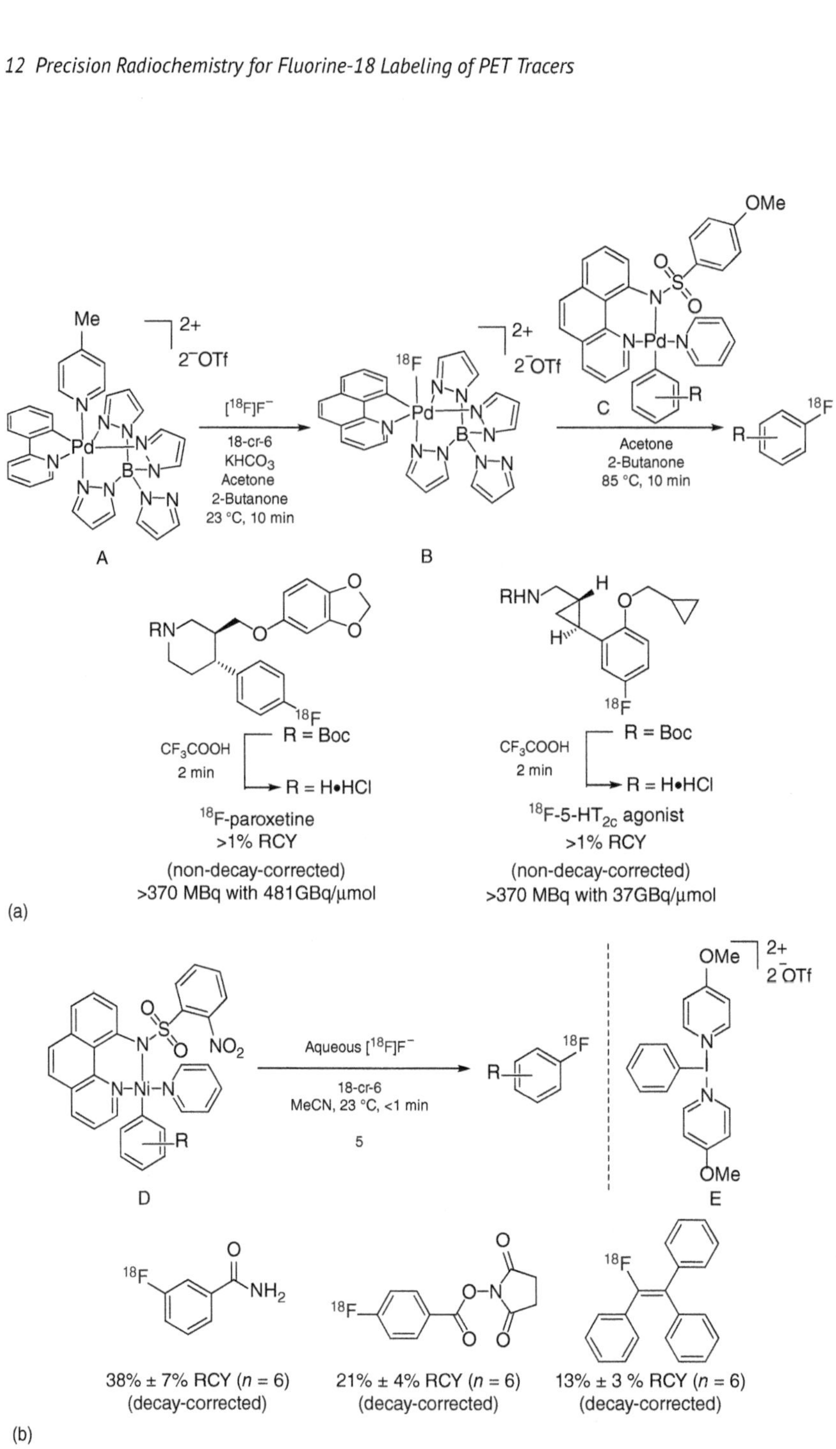

(a)

(b)

Figure 12.13 Pd- and Ni-mediated aromatic ^{18}F-fluorination. Source: (a) Modified from Lee et al. [68]; (b) Modified from Lee et al. [69].

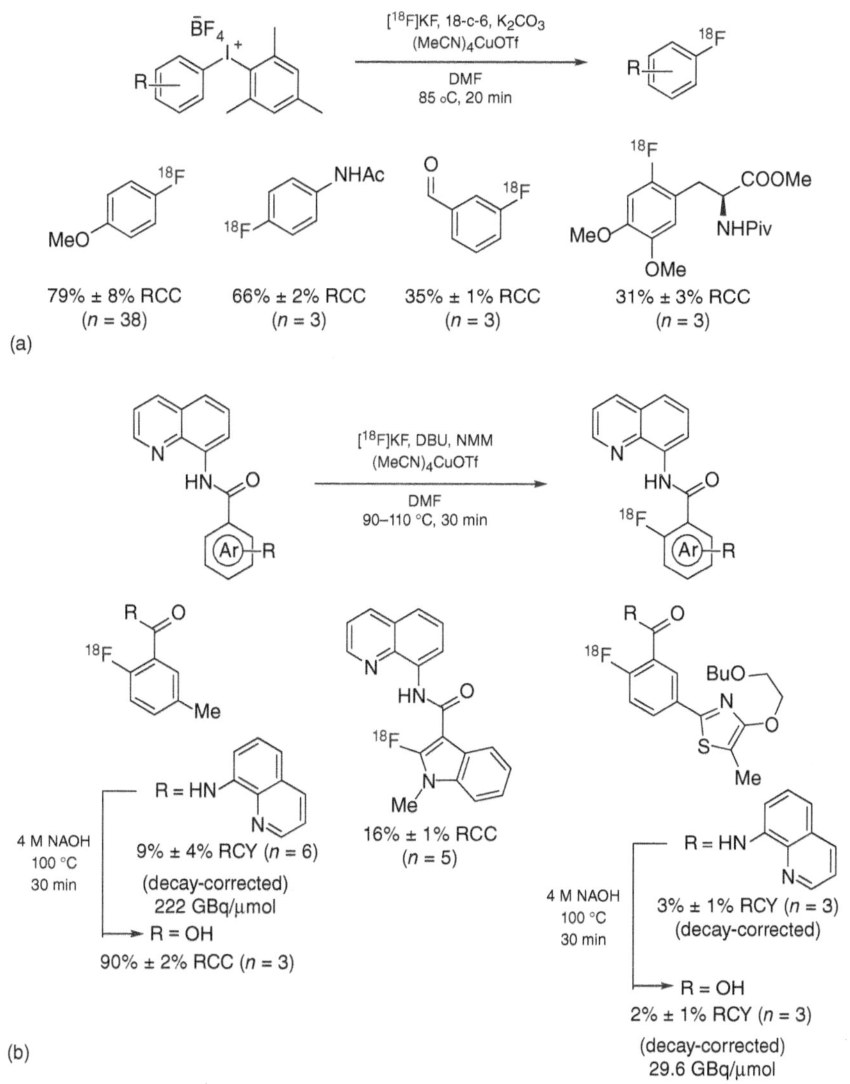

Figure 12.14 Cu-mediated aromatic ^{18}F-fluorination. Source: (a) Modified from Ichiishi et al. [73]; (b) Modified from Lee et al. [74].

substituents typically promote radiofluorination on the *ortho*-substituted arene. In contrast, Sanford, Scott, and coworkers demonstrated that copper-catalyzed radiofluorinations from mesityl(aryl)-iodonium salts resulted in opposite regioselectivity, thus favoring the introduction of fluorine-18 into arenes lacking *ortho* substituents (Figure 12.14a) [73].

Electron-rich, -neutral, and -deficient aryl moieties were amenable to radiofluorination under these reaction conditions. As a proof of concept, *N,O*-protected [^{18}F]FDOPA was prepared in 31% RCC.

Sanford, Scott, and coworkers reported on the Cu-mediated C–H ^{18}F-fluorination of electron-rich (hetero)arenes, a one-pot two-step process that involved *in situ* C–H functionalization to prepare diaryliodonium salts and subsequent ^{18}F-labeling step [75]. Recently, the same authors extended their approach by describing a Cu-mediated aminoquinoline-directed ^{18}F-fluorination of aromatic C—H bonds with K[^{18}F]F (Figure 12.14b) [74]. Particularly, the latter radiofluorination included two steps, namely, a Cu-mediated aminoquinoline-directed *ortho*-C(sp^2)–H activation, followed by ^{18}F-fluorination with K[^{18}F]F. A range of electron-neutral, -deficient, and -rich substrates were prepared employing these reaction conditions. Several 2-[^{18}F]fluoro-substituted aromatic carboxylic acids were synthesized via ^{18}F-fluorination and subsequent hydrolysis of the 8-aminoquinoline benzamides, including ^{18}F-AC261066 (a RARβ2 agonist) that was obtained in 2% decay-corrected RCY with a molar activity of 29.6 GBq/μmol (0.8 Ci/μmol).

Aryl boronic esters, aryl boronic acids, and aryl stannanes are widely used for transition metal-mediated cross-coupling reactions, and they have proven useful as precursors for ^{18}F-fluorination. In 2013, Sanford and coworkers accomplished the Cu-mediated fluorination of aryltrifluoroborates [76]. Excess fluoride (4 equiv) was used and aryl boronic esters were found to be less efficient than aryltrifluoroborates. In 2014, Gouverneur and coworkers described a novel copper-mediated ^{18}F-fluorination of aryl boronic esters (arylBPin) (Figure 12.15a) [77]. O$_2$ from the air was essential for this reaction and both electron-rich and electron-deficient arylBPins and alkenylBpins were compatible. Unprotected –OH and –NH$_2$ functionalities led to low yields due to the competitive C–O and C–N coupling side reactions. A range of PET tracers were prepared via ^{18}F-fluorination of their corresponding arylBPin precursors, including the inhibitor of the epidermal growth factor receptor—tyrosine kinase (EGFR-TK) designated ^{18}F-gefitinib (22% RCC), the tropomyosin receptor kinase (Trk) inhibitor ^{18}F-IPMICF10 (71% RCC) and ^{18}F-DAA1106 (59% decay-corrected RCY), which is currently used for the detection of striatal dopaminergic neurodegeneration [80, 81]. In 2015, Sanford, Scott, and coworkers reported on the copper-mediated ^{18}F-fluorination of aryl boronic acids (Figure 12.15b) [78]. The examples included [^{18}F]FPEB, which was obtained in 8% RCC with a molar activity of 27.8 GBq/μmol (0.75 Ci/μmol). Further, Neumaier and coworkers observed that the efficiency of copper-mediated ^{18}F-fluorination of mesityl(aryl)iodonium salts (Figure 12.14a) and aryl boronic esters (Figure 12.15a) was further improved by using a "low base" protocol to avoid decomposition of copper catalysts under the basic conditions [82]. In transition metal-mediated cross-coupling reactions, aryl stannanes exhibit a high propensity toward efficient transmetalation, thus rendering them useful precursors for transition metal-catalyzed ^{18}F-fluorination. In 2016, Murphy and coworkers accomplished the copper-mediated oxidative fluorination of aryl stannanes [83]. In the same year, Sanford, Scott, and coworkers independently reported on the successful copper-mediated ^{18}F-fluorination of aryl stannanes (Figure 12.15c) [79]. A range of electron-rich and electron-deficient aryl [^{18}F]fluorides, as well as vinyl [^{18}F]fluorides, including the serotonin receptor PET radioligand [^{18}F]MPPF

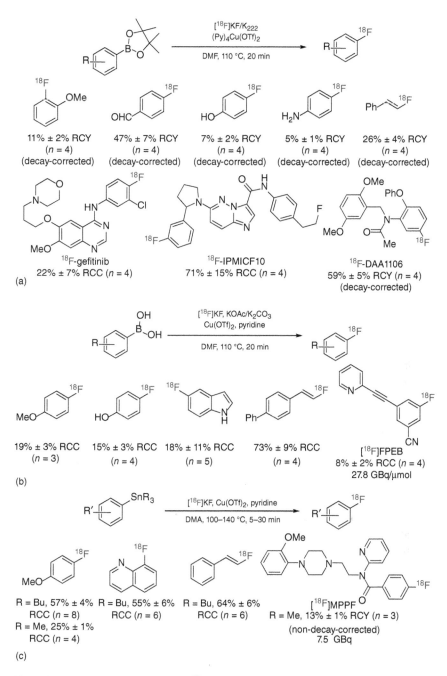

Figure 12.15 Cu-mediated aromatic ^{18}F-fluorination of aryl boronic acids, esters and aryl stannanes. Source: (a) Modified from Tredwell et al. [77]; (b) Modified from Mossine et al. [78]; (c) Modified from Makaravage et al. [79].

(13% non-decay-corrected RCY), were prepared via ^{18}F-fluorination of stannane precursors.

12.5 ^{18}F-Labeling of Multifluoromethyl Motifs with [^{18}F]Fluoride

There is a continuous development of pharmaceuticals with multifluoromethyl motifs, such as –CF$_3$, –CF$_2$H, and –SCF$_3$, resulting in an increased demand for new methods to ^{18}F-label these multifluoromethyl motifs for drug discovery and tracer development purposes. Initially, direct methods for the incorporation of fluorine-18 into multifluoromethyl motifs were based on ^{19}F/^{18}F isotope exchange (e.g. from R-CF$_3$ to R-CF$_2$^{18}F) or halide/^{18}F exchange reactions (e.g. from R-CF$_2$Br to R-CF$_2$^{18}F). Conditions for these exchange reactions were often harsh, with limited tolerance towards highly functionalized substrates. Recently, silver (I) salts were found to be an effective additive to improve halide/^{18}F exchange reactions under milder conditions. In a recent study, Gouverneur and coworkers reported the synthesis of [^{18}F]5-(trifluoromethyl)dibenzothiophenium trifluoromethanesulfonate ([^{18}F]Umemoto reagent) via Ag-mediated Br/^{18}F exchange and the subsequent oxidative cyclization process (Figure 12.16a) [84]. The ^{18}F-labeled Umemoto reagent was applied in direct radiofluorinations of unmodified peptides at the cysteine residue by the transfer of a [^{18}F]CF$_3$ group. A range of peptides were selectively labeled at their cysteine residues using the ^{18}F-labeled Umemoto reagent, including cRGDfC([^{18}F]CF$_3$) (19% non-decay-corrected RCY). Another strategy to prepare [^{18}F]CF$_3$S-labeled molecules is via the transfer of the whole –SCF$_3$[^{18}F] group. In 2015, Liang, Xiao, and coworkers accomplished the [^{18}F]trifluoromethylthiolation of alkyl electrophiles with difluorocarbene (generated *in situ*), sulfur, and [^{18}F]fluoride under mild conditions (Figure 12.16b) [85]. Difluorocarbene was generated *in situ* via decarboxylation of PDFA (difluoromethylene phosphobetaine, Ph$_3$P$^+$CF$_2$COO$^-$) under neutral conditions, and without the addition of a base or additive. The generation of [^{18}F]trifluoromethylthio anion ([^{18}F]CF$_3$S$^-$) was envisioned by reacting trifluoromethyl anion ([^{18}F]CF$_3$$^-$) with elemental sulfur (S$_8$). A range of alkyl bromides, alkyl iodides, as well as allyl bromides were efficiently [^{18}F]CF$_3$S-labeled at 70 °C in DMF. In 2017, Liang, Xiao, and coworkers further extended their work by describing the [^{18}F]trifluoromethylthiolation of α-bromo carbonyl compounds (Figure 12.16c) [86]. Mechanistically, the authors found that the difluorocarbene underwent sulfuration with elemental sulfur (S$_8$) in a first step to generate thiocarbonyl fluoride (S=CF$_2$), instead of being trapped by [^{18}F]fluoride to afford the trifluoromethyl anion ([^{18}F]CF$_3$$^-$). Subsequently, the reaction of thiocarbonyl fluoride (S=CF$_2$) with [^{18}F]fluoride gave the [^{18}F]trifluoromethylthio anion ([^{18}F]CF$_3$S$^-$). Further, the authors observed that the addition of copper (I) led to an apparent yield improvement. The unprecedented and straightforward preparation of the [^{18}F]trifluoromethylthio anion ([^{18}F]CF$_3$S$^-$) from difluorocarbene, sulfur, and [^{18}F]fluoride enabled the [^{18}F]trifluoromethylthiolation of a broad range of alkyl electrophiles including aliphatic substrates bearing electron-donating or

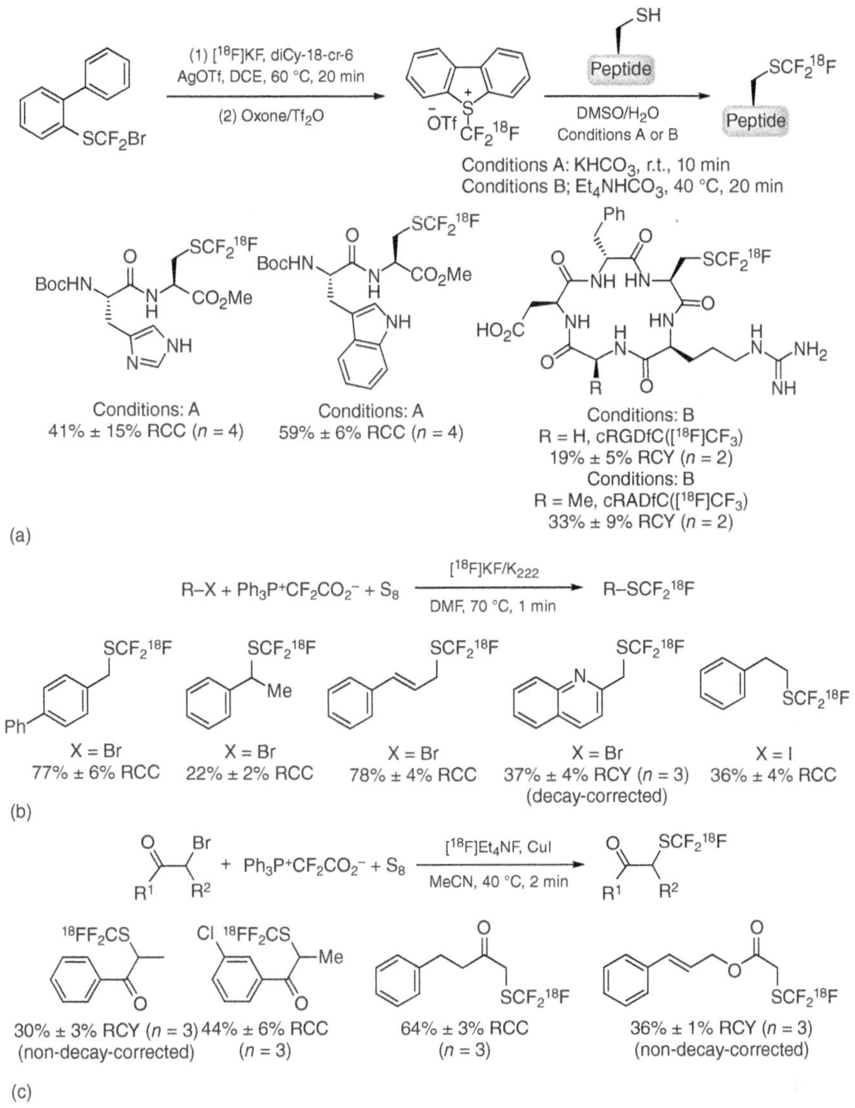

Figure 12.16 Synthesis of [^{18}F]CF$_3$S-labeled molecules. Source: (a) Modified from Verhoog et al. [84]; (b) Modified from Zheng et al. [85]; (c) Modified from Zheng et al. [86].

electron-withdrawing substituents in RCCs up to 83%. Further, this method was viable for the conversion of α-bromo carbonyl compounds to α-[^{18}F]SCF$_3$ carbonyl derivatives.

Trifluoromethyl (CF$_3$) moieties are integral elements of numerous pharmaceuticals. Radiofluorination of these moieties occurred predominantly via halide/^{18}F exchange. An alternative strategy to obtain [^{18}F]CF$_3$ bearing PET radioligands is via [^{18}F]trifluoromethylation. In 2013, Gouverneur and coworkers reported the copper-mediated [^{18}F]trifluoromethylation of (hetero)aryl iodides

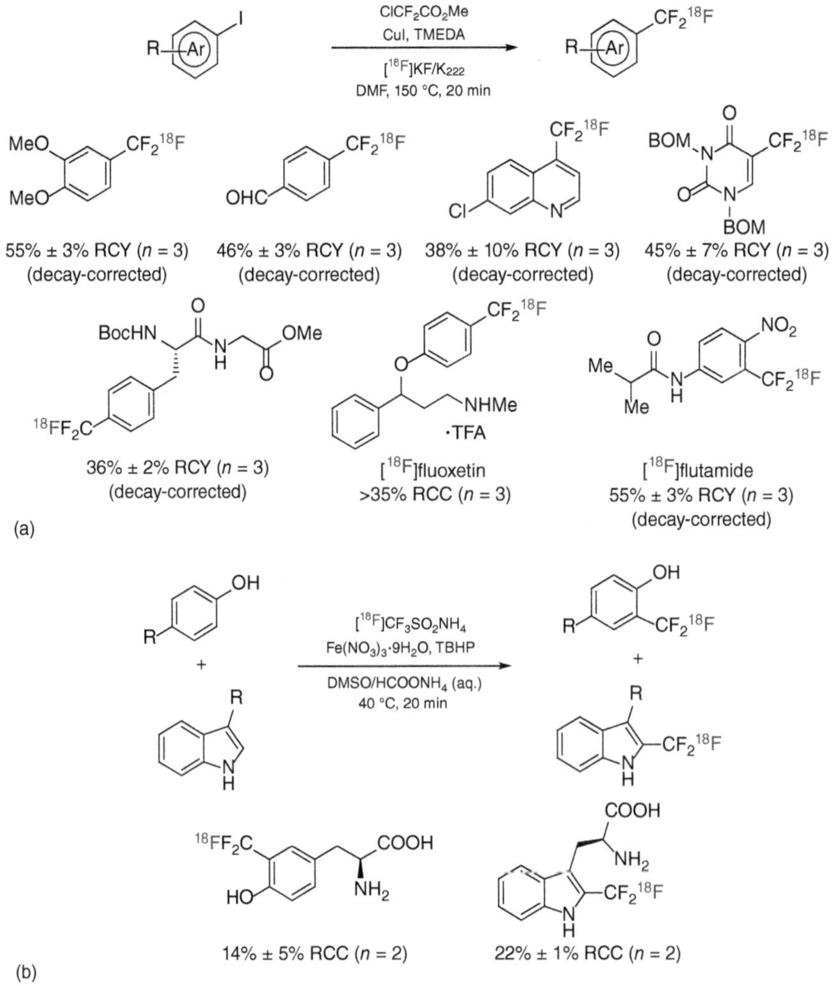

Figure 12.17 ^{18}F-Trifluoromethylation of aryl iodides (a), L-tryptophan, and L-tyrosine (b). Source: (a) Modified from Huiban et al. [87]; (b) Modified from Kee et al. [88].

with chlorodifuoroacetate and [^{18}F]fluoride (Figure 12.17a) [87]. [^{18}F]CuCF$_3$ was generated *in situ* from chlorodifluoroacetate, CuI, [^{18}F]fluoride, and TMEDA (tetramethylethylenediamine). [^{18}F]Trifluoromethylation of diverse aryl iodides and heteroaryl iodides was achieved, including *N*-Boc-protected [^{18}F]fluoxetine (35% RCC over two steps) and [^{18}F]fluoxetine (55% RCC, also known as Eulexin, nonsteroidal antiandrogen [NSAA] used primarily to treat prostate cancer).

In 2014, Riss and coworkers developed another Cu-mediated ^{18}F-trifluoromethylation approach for aryl iodides, whereby [^{18}F]CuCF$_3$ was generated from CHF$_2$I [89]. Other Cu-mediated ^{18}F-trifluoromethylation reactions, enabling the generation of [^{18}F]CuCF$_3$ from [^{18}F]HCF$_3$, were also developed [90, 91]. In addition, ^{18}F-trifluoromethylation via a radical pathway using [^{18}F]CF$_3$ radicals was considered a useful approach to afford ^{18}F-arenes. In 2020, Gouverneur and coworkers developed a novel radical ^{18}F-trifluoromethylation reagent, ^{18}F]CF$_3$SO$_2$NH$_4$ (obtained from ^{18}F]fluoride in one step), and its application in C–H ^{18}F-trifluoromethylation of tryptophan or tyrosine residues in unmodified peptides (Figure 12.17b) [88]. [18]F]CF$_3$SO$_2$NH$_4$ was prepared from a combination of ^{18}F]fluoride (^{18}F]KF/K$_{222}$), difluorocarbene (PDFA), and SO$_2$ (*N*-methyl-morpholine·SO$_2$) in DMF (110 °C, 20 minutes) with 17% decay-corrected RCY ($n = 6$). Importantly, ^{18}F]CF$_3$SO$_2^-$ was purified by a weak anion exchange cartridge (WAX) to remove organic by-products and unreacted ^{18}F]fluoride. [18]F]CF$_3$SO$_2$NH$_4$ was afforded via elution with a solution of NH$_4$OH in EtOH and further purification by reverse-phase HPLC in high radiochemical purity (>99%). Of note, ^{18}F]CF$_3$SO$_2$NH$_4$ was successfully employed in the highly chemoselective C–H ^{18}F-trifluoromethylation of peptides containing L-tryptophan and/or L-tyrosine residues with Fe(III) salts and TBHP (*tert*-butyl hydroperoxide). In fact, Gouverneur and coworkers employed the novel methodology to successfully label a series of unmodified peptides, demonstrating the utility of this approach to facilitate ^{18}F]peptide development for nuclear medicine applications.

12.6 Summary and Conclusions

In the past several decades, significant progress has been made in the development of labeling methods via direct ^{18}F-fluorination to construct aliphatic and aromatic C—[18]F bonds, and these methods provide an important toolbox for ^{18}F-labeling of PET radiopharmaceuticals.

Nonetheless, the discovery of novel radiofluorinated PET probes remains challenging due to stringent constraints that are imparted by the challenges of fluorine radiochemistry, particularly with regard to nonactivated arenes. Among the plethora of approaches for the radiofluorination of arenes, the hypervalent iodine (III) methods and copper-mediated strategies using boron or tin precursors are currently the most advanced for human use.

Given the pressing and continuous demand for novel PET radiopharmaceuticals in precision medicine, an ever-growing toolbox of radiofluorination methods is of paramount value to bridge the gap between the unmet clinical needs and the ongoing progress in modern fluorine-18 chemistry. The development of efficient and practical ^{18}F-labeling methods will be a long-standing goal for radiochemists to nourish the field of nuclear medicine and provide high-quality personalized treatment for the patient.

References

1 Gambhir, S.S. (2002). *Nat. Rev. Cancer* 2: 683–693.
2 Phelps, M.E. (2000). *Proc. Natl. Acad. Sci. U.S.A.* 97: 9226–9233.
3 Willmann, J.K., van Bruggen, N., Dinkelborg, L.M., and Gambhir, S.S. (2008). *Nat. Rev. Drug Discovery* 7: 591–607.

4 Ametamey, S.M., Honer, M., and Schubiger, P.A. (2008). *Chem. Rev.* 108: 1501–1516.

5 Ghosh, N., Rimoldi, O.E., Beanlands, R.S., and Camici, P.G. (2010). *Eur. Heart J.* 31: 2984–2995.

6 Villemagne, V.L., Dore, V., Burnham, S.C. et al. (2018). *Nat. Rev. Neurol.* 14: 225–236.

7 Ming-Rong, Z. and Kazutoshi, S. (2007). *Curr. Top. Med. Chem.* 7: 1817–1828.

8 van der Born, D., Pees, A., Poot, A.J. et al. (2017). *Chem. Soc. Rev.* 46: 4709–4773.

9 Chansaenpak, K., Vabre, B., and Gabbaï, F.P. (2016). *Chem. Soc. Rev.* 45: 954–971.

10 Krishnan, H.S., Ma, L., Vasdev, N., and Liang, S.H. (2017). *Chem. Eur. J.* 23: 15553–15577.

11 Al-Momani, E., Israel, I., Buck, A.K., and Samnick, S. (2015). *Appl. Radiat. Isot.* 104: 136–142.

12 Gómez-Vallejo, V., Lekuona, A., Baz, Z. et al. (2016). *Chem. Commun.* 52: 11931–11934.

13 Bergman, J. and Solin, O. (1997). *Nucl. Med. Biol.* 24: 677–683.

14 Coenen, H.H., Gee, A.D., Adam, M. et al. (2017). *Nucl. Med. Biol.* 55: v–xi.

15 Ehrenkaufer, R.E. and MacGregor, R.R. (1983). *Int. J. Appl. Radiat. Isot.* 34: 613–615.

16 Neirinckx, R.D., Lambrecht, R.M., and Wolf, A.P. (1978). *Int. J. Appl. Radiat. Isot.* 29: 323–327.

17 Chirakal, R., Firnau, G., Couse, J., and Garnett, E.S. (1984). *Int. J. Appl. Radiat. Isot.* 35: 651–653.

18 Oberdorfer, F., Hofmann, E., and Maier-Borst, W. (1988). *J. Labelled Compd. Radiopharm.* 25: 999–1005.

19 Oberdorfer, F., Hofmann, E., and Maier-Borst, W. (1988). *Int. J. Radiat. Appl. Instrum. Part A* 39: 685–688.

20 Teare, H., Robins, E.G., Årstad, E. et al. (2007). *Chem. Commun.* 23: 2330–2332.

21 Teare, H., Robins, E.G., Kirjavainen, A. et al. (2010). *Angew. Chem. Int. Ed.* 49: 6821–6824.

22 Sood, S., Firnau, G., and Garnett, E.S. (1983). *Int. J. Appl. Radiat. Isot.* 34: 743–745.

23 Firnau, G.S.G., Chirakal, R., and Garnett, E.S. (1981). *J. Chem. Soc., Chem. Commun.* 4: 198–199.

24 Vasdev, N., Pointner, B.E., Chirakal, R., and Schrobilgen, G.J. (2002). *J. Am. Chem. Soc.* 124: 12863–12868.

25 Lu, S. and Pike, V.W. (2010). *J. Fluorine Chem.* 131: 1032–1038.

26 Buckingham, F., Kirjavainen, A.K., Forsback, S. et al. (2015). *Angew. Chem. Int. Ed.* 54: 13366–13369.

27 Nodwell, M.B., Yang, H., Čolović, M. et al. (2017). *J. Am. Chem. Soc.* 139: 3595–3598.

28 Stenhagen, I.S., Kirjavainen, A.K., Forsback, S.J. et al. (2013). *Chem. Commun.* 49: 1386–1388.

29 Hamacher, K., Coenen, H.H., and Stöcklin, G. (1986). *J. Nucl. Med.* 27: 235–238.

30 Kim, D.W., Choe, Y.S., and Chi, D.Y. (2003). *Nucl. Med. Biol.* 30: 345–350.

31 Kim, D.W., Ahn, D.-S., Oh, Y.-H. et al. (2006). *J. Am. Chem. Soc.* 128: 16394–16397.

32 Ido, T., Wan, C.N., Casella, V. et al. (1978). *J. Labelled Compd. Radiopharm.* 14: 175–183.

33 Chaly, T., Dhawan, V., Kazumata, K. et al. (1996). *Nucl. Med. Biol.* 23: 999–1004.

34 Nielsen, M.K., Ugaz, C.R., Li, W., and Doyle, A.G. (2015). *J. Am. Chem. Soc.* 137: 9571–9574.

35 Martarello, L., Schaffrath, C., Deng, H. et al. (2003). *J. Labelled Compd. Radiopharm.* 46: 1181–1189.

36 Deng, H., Cobb, S.L., Gee, A.D. et al. (2006). *Chem. Commun.* 6: 652–654.

37 Thompson, S., Zhang, Q., Onega, M. et al. (2014). *Angew. Chem. Int. Ed.* 53: 8913–8918.

38 Hollingworth, C., Hazari, A., Hopkinson, M.N. et al. (2011). *Angew. Chem. Int. Ed.* 50: 2613–2617.

39 Topczewski, J.J., Tewson, T.J., and Nguyen, H.M. (2011). *J. Am. Chem. Soc.* 133: 19318–19321.

40 Benedetto, E., Tredwell, M., Hollingworth, C. et al. (2013). *Chem. Sci.* 4: 89–96.

41 Huang, X., Liu, W., Ren, H. et al. (2014). *J. Am. Chem. Soc.* 136: 6842–6845.

42 Liu, W., Huang, X., Placzek, M.S. et al. (2018). *Chem. Sci.* 9: 1168–1172.

43 Graham, T.J.A., Lambert, R.F., Ploessl, K. et al. (2014). *J. Am. Chem. Soc.* 136: 5291–5294.

44 Sander, K., Gendron, T., Yiannaki, E. et al. (2015). *Sci. Rep.* 5: 9941–9945.

45 Xu, P., Zhao, D., Berger, F. et al. (2020). *Angew. Chem. Int. Ed.* 59: 1956–1960.

46 Mu, L., Fischer, C.R., Holland, J.P. et al. (2012). *Eur. J. Org. Chem.* 2012: 889–892.

47 Gendron, T., Sander, K., Cybulska, K. et al. (2018). *J. Am. Chem. Soc.* 140: 11125–11132.

48 Chun, J.H., Morse, C.L., Chin, F.T., and Pike, V.W. (2013). *Chem. Commun.* 49: 2151–2153.

49 Narayanam, M.K., Ma, G., Champagne, P.A. et al. (2017). *Angew. Chem. Int. Ed.* 56: 13006–13010.

50 Beyzavi, M.H., Mandal, D., Strebl, M.G. et al. (2017). *ACS Cent. Sci.* 3: 944–948.

51 Neumann, C.N., Hooker, J.M., and Ritter, T. (2016). *Nature* 534: 369–373.

52 Chen, W., Huang, Z., Tay, N.E.S. et al. (2019). *Science* 364: 1170–1174.

53 Pike, V.W. and Aighirhio, F.I.R. (1995). *J. Chem. Soc., Chem. Commun.* 21: 2215–2216.

54 Chun, J.H., Lu, S., Lee, Y.S., and Pike, V.W. (2010). *J. Org. Chem.* 75: 3332–3338.

55 Kwon, Y.-D., Son, J., and Chun, J.-H. (2019). *J. Org. Chem.* 84: 3678–3686.

56 Ross, T.L., Ermert, J., Hocke, C., and Coenen, H.H. (2007). *J. Am. Chem. Soc.* 129: 8018–8025.

57 Carroll, M.A., Jones, C., and Tang, S.-L. (2007). *J. Labelled Compd. Radiopharm.* 50: 450–451.

58 Moon, B.S., Kil, H.S., Park, J.H. et al. (2011). *Org. Biomol. Chem.* 9: 8346–8355.

59 Yuan, Z., Cheng, R., Chen, P. et al. (2016). *Angew. Chem. Int. Ed.* 55: 11882–11886.

60 Cardinale, J., Ermert, J., Humpert, S., and Coenen, H.H. (2014). *RSC Adv.* 4: 17293–17299.

61 Rotstein, B.H., Stephenson, N.A., Vasdev, N., and Liang, S.H. (2014). *Nat. Commun.* 5: 4365–4371.

62 Rotstein, B.H., Wang, L., Liu, R.Y. et al. (2016). *Chem. Sci.* 7: 4407–4417.

63 Wang, L., Jacobson, O., Avdic, D. et al. (2015). *Angew. Chem. Int. Ed.* 54: 12777–12781.

64 Stephenson, N.A., Holland, J.P., Kassenbrock, A. et al. (2015). *J. Nucl. Med.* 56: 489–492.

65 Collier, T.L., Normandin, M.D., Stephenson, N.A. et al. (2017). *Nat. Commun.* 8: 15761.

66 Jakobsson, J.E., Grønnevik, G., and Riss, P.J. (2017). *Chem. Commun.* 53: 12906–12909.

67 Haskali, M.B., Telu, S., Lee, Y.S. et al. (2016). *J. Org. Chem.* 81: 297–302.

68 Lee, E., Kamlet, A.S., Powers, D.C. et al. (2011). *Science* 334: 639–642.

69 Lee, E., Hooker, J.M., and Ritter, T. (2012). *J. Am. Chem. Soc.* 134: 17456–17458.

70 Kamlet, A.S., Neumann, C.N., Lee, E. et al. (2013). *PLoS One* 8: e59187.

71 Ren, H., Wey, H.Y., Strebl, M. et al. (2014). *ACS Chem. Neurosci.* 5: 611–615.

72 Hoover, A.J., Lazari, M., Ren, H. et al. (2016). *Organometallics* 35: 1008–1014.

73 Ichiishi, N., Brooks, A.F., Topczewski, J.J. et al. (2014). *Org. Lett.* 16: 3224–3227.

74 Lee, S.J., Makaravage, K.J., Brooks, A.F. et al. (2019). *Angew. Chem. Int. Ed.* 58: 3119–3122.

75 McCammant, M.S., Thompson, S., Brooks, A.F. et al. (2017). *Org. Lett.* 19: 3939–3942.

76 Ye, Y., Schimler, S.D., Hanley, P.S., and Sanford, M.S. (2013). *J. Am. Chem. Soc.* 135: 16292–16295.

77 Tredwell, M., Preshlock, S.M., Taylor, N.J. et al. (2014). *Angew. Chem. Int. Ed.* 53: 7751–7755.

78 Mossine, A.V., Brooks, A.F., Makaravage, K.J. et al. (2015). *Org. Lett.* 17: 5780–5783.

79 Makaravage, K.J., Brooks, A.F., Mossine, A.V. et al. (2016). *Org. Lett.* 18: 5440–5443.

80 Preshlock, S., Calderwood, S., Verhoog, S. et al. (2016). *Chem. Commun.* 52: 8361–8364.

81 Taylor, N.J., Emer, E., Preshlock, S. et al. (2017). *J. Am. Chem. Soc.* 139: 8267–8276.

82 Zlatopolskiy, B.D., Zischler, J., Krapf, P. et al. (2015). *Chem. Eur. J.* 21: 5972–5979.

83 Gamache, R.F., Waldmann, C., and Murphy, J.M. (2016). *Org. Lett.* 18: 4522–4525.

84 Verhoog, S., Kee, C.W., Wang, Y. et al. (2018). *J. Am. Chem. Soc.* 140: 1572–1575.

85 Zheng, J., Wang, L., Lin, J.-H. et al. (2015). *Angew. Chem. Int. Ed.* 54: 13236–13240.

86 Zheng, J., Cheng, R., Lin, J.H. et al. (2017). *Angew. Chem. Int. Ed.* 56: 3196–3200.

87 Huiban, M., Tredwell, M., Mizuta, S. et al. (2013). *Nat. Chem.* 5: 941–944.

88 Kee, C.W., Tack, O., Guibbal, F. et al. (2020). *J. Am. Chem. Soc.* 142: 1180–1185.

89 Rühl, T., Rafique, W., Lien, V.T., and Riss, P.J. (2014). *Chem. Commun.* 50: 6056–6059.

90 van der Born, D., Sewing, C., Herscheid, J.D.M. et al. (2014). *Angew. Chem. Int. Ed.* 53: 11046–11050.

91 Ivashkin, P., Lemonnier, G., Cousin, J. et al. (2014). *Chem. Eur. J.* 20: 9514–9518.

Index

Organofluorine Chemistry: Synthesis, Modeling, and Applications,
First Edition. Edited by Kálmán J. Szabó and Nicklas Selander.
© 2021 WILEY-VCH GmbH. Published 2021 by WILEY-VCH GmbH.